Heredity Produced

Transformations: Studies in the History of Science and Technology

Jed Z. Buchwald, general editor

Jesuit Science and the Republic of Letters
Mordechai Feingold, editor

Down from the Mountain: The Birth of Naval Architecture in the Scientific Revolution, 1600–1800
Larrie D. Ferreiro

Wireless: From Marconi's Black-Box to the Audion
Sungook Hong

The Path Not Taken: French Industrialization in the Age of Revolution, 1750–1830
Jeff Horn

Harmonious Triads: Physicists, Musicians, and Instrument Makers in Nineteenth-Century Germany
Myles W. Jackson

Spectrum of Belief: Joseph von Fraunhofer and the Craft of Precision Optics
Myles W. Jackson

Affinity, That Elusive Dream: A Genealogy of the Chemical Revolution
Mi Gyung Kim

American Hegemony and the Postwar Reconstruction of Science in Europe
John Krige

Conserving the Enlightenment: French Military Engineering from Vauban to the Revolution
Janis Langins

Picturing Machines 1400–1700
Wolfgang Lefèvre, editor

Secrets of Nature: Astrology and Alchemy in Early Modern Europe
William R. Newman and Anthony Grafton, editors

Heredity Produced: At the Crossroads of Biology, Politics, and Culture, 1500–1870
Staffan Müller-Wille and Hans-Jörg Rheinberger, editors

Historia: Empiricism and Erudition in Early Modern Europe
Gianna Pomata and Nancy Siraisi, editors

Nationalizing Science: Adolphe Wurtz and the Battle for French Chemistry
Alan J. Rocke

Heredity Produced

At the Crossroads of Biology, Politics, and Culture, 1500–1870

edited by Staffan Müller-Wille and Hans-Jörg Rheinberger

The MIT Press
Cambridge, Massachusetts
London, England

MIT Press books may be purchased at special quantity discounts for business or sales promotional use. For information, please e-mail special_sales@mitpress.mit.edu or write to Special Sales Department, The MIT Press, 55 Hayward Street, Cambridge, MA 02142.

This book was set in Stone serif and Stone sans by SNP Best-set Typesetter Ltd., Hong Kong. Printed and bound in the United States of America.

Library of Congress Cataloging-in-Publication Data
Heredity produced : at the crossroads of biology, politics, and culture, 1500-1870 / Staffan Müller-Wille and Hans-Jorg Rheinberger, editors.
p. cm.—(Transformations)
Includes bibliographical references and index.
ISBN-13: 978-0-262-13476-7 (hardcover : alk. paper)
1. Heredity—History. I. Müller-Wille, Staffan, 1964– II. Rheinberger, Hans-Jörg.
QH431.H434 2007
576.5—dc22

2006023090

10 9 8 7 6 5 4 3 2 1

Contents

Preface ix

Introduction 1

1 Heredity—The Formation of an Epistemic Space 3
Staffan Müller-Wille and Hans-Jörg Rheinberger

I Heredity in the Legal Context 35

2 From Clan to Kindred: Kinship and the Circulation of Property in Premodern and Modern Europe 37
David Warren Sabean

3 Resemblance, Paternity, and Imagination in Early Modern Courts 61
Silvia De Renzi

4 Continuity and Death: Literature and the Law of Succession in the Nineteenth Century 85
Ulrike Vedder

II Heredity and Medicine 103

5 The Medical Origins of Heredity 105
Carlos López-Beltrán

6 Erasmus Darwin and the "Noble" Disease (Gout): Conceptualizing Heredity and Disease in Enlightenment England 133
Philip K. Wilson

7 Degeneration and "Alienism" in Early Nineteenth-Century France 155
Laure Cartron

III Natural History, Breeding, and Hybridization 175

8 Figures of Inheritance, 1650–1850 177
Staffan Müller-Wille

9 Duchesne's Strawberries: Between Growers' Practices and Academic Knowledge 205
Marc J. Ratcliff

10 The Sheep Breeders' View of Heredity Before and After 1800 229
Roger J. Wood

IV Theories of Generation and Evolution 251

11 Speculation and Experiment in Enlightenment Life Sciences 253
Mary Terrall

12 Kant on Heredity and Adaptation 277
Peter McLaughlin

13 The Delayed Linkage of Heredity with the Cell Theory 293
François Duchesneau

14 On the Shoulders of Generations: The New Epistemology of Heredity in the Nineteenth Century 315
Ohad S. Parnes

V Anthropology 347

15 *Las Castas*: Interracial Crossing and Social Structure, 1770–1835 349
Renato G. Mazzolini

16 Acquired Character: The Hereditary Material of the "Self-Made Man" 375
Paul White

17 "Victor, l'enfant de la forêt": Experiments on Heredity in Savage Children 399
Nicolas Pethes

18 Sui generis: Heredity and Heritage of Genius at the Turn of the Eighteenth Century 419
Stefan Willer

Epilogue 441

19 The Heredity of Poetics 443
Helmut Müller-Sievers

List of Contributors 467
Index 473

Francisco José de Goya y Lucientes, "Asta su abuelo (Up to his grandfather)" (Los Caprichos, n° 39), Aquatint, 1799. Courtesy Biblioteca Nacional de España, Madrid.

Preface

The essays assembled in this book reflect both the problems discussed and the results obtained during the first phase of a long-term, collaborative research project carried out at the Max-Planck-Institute for the History of Science (Berlin) since 2001 under the heading of "A Cultural History of Heredity." The project aims at studying the juridical, medical, cultural, technical, and scientific practices and procedures in which knowledge of heredity became materially entrenched in different ways and by which it unfolded its often unprecedented effects over a period of several centuries. In its *longue durée* and transdisciplinary character, such a project is vitally dependent on the collaboration of experts from a broad range of disciplines, covering cultural history in its various subdomains of science, technology, medicine, economy, law, anthropology, and the arts.

Two workshops devoted to the management and reflection of hereditary phenomena from the late seventeenth to the middle of the nineteenth century were conducted to bring together such experts, and the present book took shape on the basis of deliberations conducted during these workshops.[1] We decided to assemble a selection of workshop contributions and invited other contributions to systematically cover issues that came to be foregrounded in our discussions. In particular, they concern aspects of marriage regulation, property transmission, and kinship models in the legal context; the transmission of diseases as conceptualized in medicine; the roles played by natural history, breeding, and hybridization in narrowing down the recurrence of characters; the impact of systems of generation and theories of evolution; and the way the incipient discourse on humans—anthropology—relied on and shaped the perception of transgenerational phenomena.

Aside from the participants in the two workshops, the project has profited from discussions with individual scholars. In particular we would like to thank Raphael Falk (Hebrew University, Jerusalem), Jean Gayon (University Paris 1-Sorbonne), Chris Hann (Max-Planck-Institute for Social Anthropology, Halle), Jonathan Harwood

(Manchester University), Manfred Laubichler (Arizona State University), Yoshio Nukaga (University of Tokyo), Maaike van der Lugt and Charles Miramon (École des Hautes Études des Sciences Sociales, Paris), as well as Sigrid Weigel (Centre for Literary Studies, Berlin). Finally, we would like to express our appreciation to Antje Radeck and Robert Meunier for preparation of the manuscript, as well as to Katrina Dean for language editing. The project "A Cultural History of Heredity" was generously supported by the government of the Principality of Liechtenstein (Karl-Schaedler-Fund).

Note

1. For documentation of these workshops and up-to-date information on the project, see http://www.mpiwg-berlin.mpg.de/en/HEREDITY/index.html.

Heredity Produced

Introduction

1 Heredity—The Formation of an Epistemic Space

Staffan Müller-Wille and Hans-Jörg Rheinberger

1.1 Generation and Heredity

Today, *heredity* is defined as the transmission of characters and dispositions in the process of organic reproduction. It appears to be of such paramount importance for the makeup of individual organisms, and to be sustained by such immediate and eye-catching evidence, that it is hard to imagine that it was not always of central concern to the life sciences. Yet, until the mid-eighteenth century, not only the concept of heredity but also that of organic reproduction itself was absent from speculations about living beings. "Until that time," as François Jacob remarked in his book *La logique du vivant* (1970), "living beings did not reproduce; they were engendered. . . . The generation of every plant and every animal was, to some degree, a unique, isolated event, independent of any other creation, rather like the production of a work of art by man."[1] There is a claim implicit in this remark that has been confirmed by a number of historians of biology. Until the end of the eighteenth century—according to some even until the advent of advanced cytological observations and Mendelism around 1900—hereditary transmission was not separated from the contingencies of conception, pregnancy, embryonic development, parturition, and lactation.[2] Similarity between progenitors and their descendants arose simply because of the similarity in the constellation of causes involved in each act of generation.

It is in this sense, for example, that William Harvey (1578–1657), in his *Anatomical Exercises on the Generation of Animals* (1651), maintained that the "work of the father and mother is to be discerned both in the body and mental character of the offspring,"[3] while he simultaneously expressed the view that "it is well said that the sun and moon engender man; because, with the advent and secession of the sun, come spring and autumn, seasons which mostly correspond with the generation and decay of animated beings."[4] The organic processes of transmission, development, nutrition, and adaptation were not distinguished in early modern theories of generation, nor were, in consequence, inherited, connate, and acquired properties of organisms. Nature and nurture, or heredity and environment, were not yet seen as oppositions.

Tales of monstrous births due to astral influences and maternal imagination convey this early modern perspective. They have often been retold to witness the period's interest in phenomena deviating from nature's normal course, due to divine intervention or exceptional constellations of natural causes.[5] We would like to add an example showing it was possible to maintain this perspective in the naturalistic framework of the later eighteenth century as well. Laurence Sterne's (1713–1768) novel *The Life and Opinions of Tristram Shandy, Gentleman* (1760) begins with the episode of its unhappy hero's conception: Tristram Shandy's father had the habit of fulfilling his marital obligations on the eve of the first Sunday each month, and always after having wound up the big clock in the hallway. On one of these occasions, while Tristram was just about to be conceived, his wife interfered with the usual routine by exclaiming, "Pray, my dear, have you not forgot to wind up the clock." The bewilderment and distraction produced in her husband by that insensitive question was the unfortunate start for the miserable life of Tristram Shandy. As the latter retrospectively reasons at the beginning of the novel,

> I wish either my father or my mother, or indeed both of them, as they were in duty both equally bound to it, had minded what they were about when they begot me; had they duly consider'd how much depended upon what they were then doing;—that not only the production of a rational Being was concern'd in it, but that possibly the happy formation and temperature of his body, perhaps his genius and the very cast of his mind;—and, for aught they knew to the contrary, even the fortunes of his whole house might take their turn from the humours and dispositions which were then uppermost.[6]

The concept of organic production highlighted in this example leaves no room for anything substantial to be said beyond the singular event of procreation, understood as an individual, separate act. Accordingly, metaphors of alchemy and art, including the mechanical arts, pervaded the discourse of generation in the seventeenth and eighteenth centuries.[7] René Descartes (1596–1650), for example, described the initial formation of the fetus as a process in which the male semen "is fermented and concocted by maternal heat, its parts entering a subtler mixture."[8]

In contrast to metaphors from the arts and crafts, another metaphor, that of heredity, only began to gain currency with respect to organic reproduction in the first half of the nineteenth century.[9] One is entitled to speak of a metaphorical application of the term *heredity* here, because it was introduced from the legal sphere, where it was used synonymously with *inheritance* or *succession*, corresponding to the Latin *hereditas*. In this sense, the term *heredity* referred to the regulations concerning the passing on of properties and positions, or to the quality of thus being heritable, a sense that has since become archaic.[10]

To be sure, phenomena that would now count as manifestations of heredity—such as the passing on of physical or mental peculiarities along familial lines, both paternal and maternal, or the recurrence of distinctive traits that skip a generation—had by no means gone unnoticed prior to the end of the eighteenth century. Nevertheless, these phenomena were not accounted for by metaphors of legal inheritance, the sole exception being the limited but notable case of hereditary diseases. Hereditary diseases (*morbi haereditarii*) had been noticed as such since antiquity, being defined as diseases that occur only in certain families, in analogy to the possessions passed on within a particular family.[11] Yet, even here, it seems that the use of the term *hereditary* did not imply the transmission of abstract properties, but rather the persistence of concrete causes of generation. Thus Jean Fernel (1497?–1558) maintained in his *Medicina* (1554) that a son is "as well inheritor of his [father's] infirmities as of his land."[12] A similar perspective is also evident in discussions of similarities between grandchildren and their grandparents in the Aristotelian and Galenic traditions. The problem was not to explain how properties were transmitted, but rather to explain how the same causal agents that once had been involved in the generation of ancestors apparently could remain active in the generation of their remote descendants.[13]

This is a curious historical fact, because inheritance seems such a handy and suggestive metaphor to address phenomena of character transmission. In its original legal context, inheritance refers to the distribution of possessions over time, more specifically, to the distribution of possessions along a path outlined by a set of rules and taxonomies specifying the conditions under which those possessions, on the death of a certain individual, may be passed on to other individuals. These taxonomies usually distinguish kin groups with respect to the deceased person, and the rules specify which of these groups are entitled to receive which portions of the inheritance, if any. Thus inheritance regulations may state that property is always passed on to the first-born son, instituting the so-called majorat, or that it is to be distributed in equal parts to all children to the exclusion of, for instance, nephews and nieces.

The analogies with organic heredity seem evident here, because kinship and descent resonate as much with biological as with social relationships. However, a metaphorical transposition of the concept of heredity from the legal sphere to that of organic reproduction would presuppose in the first place that organic reproduction is conceived as a process in which properties are transmitted from ancestors to descendants. And it is this specific perspective that was missing in the premodern discourse of generation. Observations of, and reflections on, the momentous generative act like those in the work of Aristotle, Harvey, or Descartes were clearly not designed to forge conceptions of organic heredity. Genealogies had to be constructed, distinctions drawn, and connections made among individual generative acts to produce the kind of structure that could meaningfully be addressed by the metaphor of heredity.

The appearance of metaphors of heredity in the realm of the living by the end of the eighteenth century therefore indicates more than a mere change in wording. It points to an antecedent *longue durée* development in the course of which the reproduction of organic beings, in contrast to their mere production, became recognizable as a domain governed by laws of its own. And this again means that practices had to develop by which the physical relations between organisms, instituted by individual generative acts, were ascertained, either actively by experiment or passively by keeping records. Conceptualizing organic reproduction as "heredity," one might say, presupposes that the generation of organisms is regarded as a domain regulated by structures or forces extending beyond the momentous act of generating an individual being.[14]

In an essay titled *A Theory of Heredity*, published in 1876 and arguably one of the founding documents of modern hereditary thought,[15] Francis Galton (1822–1911) repeatedly expressed the idea that heredity could only be explained by means of presupposing an underlying and enduring structure. He named this structure a "stirp" (derived from *stirpes*, Latin for "root") and identified it with the "sum-total of the germs, gemmules, or whatever they may be called, which are to be found . . . in the newly fertilised ovum."[16] In illustrating the character of this structure Galton took recourse to some awkward-sounding analogies, comparing it, among other things, to a "post office":

> Ova and their contents are, to biologists looking at them through their microscopes, much what mail—bags and the heaps of letters poured out of them are to those who gaze through the glass windows of a post office. Such persons may draw various valuable conclusions as to the postal communications generally, but they cannot read a single word of what the letters contain. All that we may learn concerning the constitutents of the stirp must be through inference, and not by direct observation; we are therefore forced to theorise.[17]

Galton explicitly stressed that his comparisons were not meant to be "idle metaphors, but strict analogies . . . worthy of being pursued, as they give a much needed clearness to views on heredity."[18] A closer look at the comparison just quoted reveals what was at stake with this call for concrete models of heredity. The comparison with a post office works on two levels. On the first level, heredity is described as a complex arrangement of various elements. The mailbag stands for the cytological space of the fertilized ovum, a space "not exceeding the size of the head of a pin,"[19] but providing the only physiological link between ancestors and their descendants.

The letters stand for the elements of the "stirp" that fill this cytological space, yet at the same time are part of a larger system of "postal communications" between ancestors and descendants. In short, heredity is defined by a structure, the "stirp," which comprises both genealogical and cytological relationships since the former must be regarded as being somehow represented in the latter. Galton expressed this idea in the following terms: "Everything that reache[s a descendant] from his ancestor must have been packed in his own stirp."

On a second level, Galton's comparison with a post office singles out heredity as a subject of scientific research. The post office, with its windows, stands for the various tools and technologies that scientists employ to make sense of hereditary phenomena. Microscopes alone, as Galton expressly admits, were certainly not enough. The complex configuration invoked in describing heredity as a system of postal communications observed through the windows of a post office suggests that it was an equally complex configuration of technologies that gave contours to heredity. To mention only the most obvious example of these technologies, addressing heredity in its genealogical dimension required reliable means of recording genealogical data. The comparison of this configuration of technologies with a post office, moreover, shows it fulfilled two functions. First, it contained heredity within its own phenomenal space, separating it from other biological phenomena such as development or nutrition. Second, it made heredity observable by establishing relationships among the elements of heredity rather than observing these elements directly. The "letters" may remain unreadable to the observer, but his position allows him to say something about the "postal communications generally." In contrast to other subjects of biological research, which can be addressed as "epistemic things" within the narrow confines of an experimental setting,[20] heredity, as characterized by Galton, can be called an "epistemic space"—a domain of research to be mapped out by taxonomies and regularities, rather than an individual object of research to be identified by determining its properties and functions.[21]

The overarching theme of this book is to explore the various historical developments and cultural realms that contributed to the formation of the epistemic space of heredity. François Jacob portrayed the passage from generation to heredity, in a Foucauldian fashion, as a succession of epistemes separated by sharp epistemological breaks.[22] In contrast, this book will emphasize long-term incremental developments that went on in a variety of often unrelated cultural contexts. In this introduction, we will try to draw an overall picture of this *longue durée* development, by summarizing the findings of the individual contributions that follow. In the first section we delineate the chronological and thematic scope of the book. In the second section we sharpen the contours of the historiographic problem implied by accounting for the development of a concept that did not yet have a fixed and definite general meaning. Finally, we outline the major historical developments responsible for the formation of the epistemic space of heredity.

1.2 The Distribution of Early Modern Heredity

There are a number of valuable histories of genetics written from the perspective of a history of ideas. François Jacob's *La logique du vivant* (1970; English translation 1993), Robert Olby's *Origins of Mendelism* (1966; second edition 1985), Ernst Mayr's *The Growth of Biological Thought* (1982), and Peter Bowler's *The Mendelian Revolution* (1989) have set lasting standards in this respect. There are also some elaborate, far from whiggish histories written from a disciplinary perspective. These include Hans Stubbe's *Kurze Geschichte der Genetik* (1963; English translation 1972), Leslie C. Dunn's *A Short History of Genetics* (1965), and Elof Axel Carlson's *The Gene: A Critical History* (1966). It is, however, revealing that the latter two histories begin with the twentieth century and thus confine themselves to genetics as a discipline. More recent publications on historical and conceptual aspects of genetics are the collected volume *The Concept of the Gene in Development and Evolution* (2000), edited by Peter Beurton, Raphael Falk, and Hans-Jörg Rheinberger, Evelyn Fox Keller's *The Century of the Gene* (2000), and Lenny Moss's *What Genes Can't Do* (2003). They review the history and prehistory of genetics largely from an epistemological perspective. This book takes a fresh approach to the history of heredity, drawing on a wealth of well-researched case studies, in order to embrace the cultural history of heredity. This involves tracing the emergence of the knowledge of heredity in broader social and historical contexts, from a wide synchronic and diachronic perspective.

One reason for such a book is that heredity has become entrenched as a fundamental notion of twentieth-century biology. This creates both the illusion that heredity must have been recognized as such since time immemorial, and the converse view, that there was no concept of heredity to speak of before the advent of genetics. This book attempts to counteract both of these exaggerations by bringing to the foreground the early history of hereditary thought. It forms the first part in a planned series of volumes that will cover this history from the early modern period to the present. The first volume in the series extends over the period in which heredity, formerly a concept restricted to the realm of law, began to be applied as a metaphor in matters of organic reproduction and successively became a concept of central importance to the life and human sciences. We identify the beginning of this period with the emergence of racial classifications in early sixteenth-century Spain and Portugal, and its end with the appearance of general biological theories of heredity in the second third of the nineteenth century.

This chronological bracket is by no means supposed to mark sharp breaks or to reflect a linear development. It rather reflects the thesis of this book that concepts of heredity initially developed in widely different ways and independently of each other in the context of a variety of knowledge regimes like medicine, natural history, or breeding. Their merging into one domain subject to a general theory of heredity in mid-nineteenth-century biology was a historically and culturally contingent process whose prehistory needs to be unfolded in all its entanglements. To reiterate, inheritance patterns, including some of the more complex ones, had not gone unnoticed in the early modern period, nor had they in antiquity and the Middle Ages.[23] However, the recognition of intergenerational similarities remained scattered in literature and distributed over various genres, a state of affairs that effectively inhibited the formation of a generalized notion of heredity.

The first thing to be noted about this distribution of hereditary knowledge is an asymmetry. While the biological notion of heredity, in retrospect, appears to have been a surprisingly late achievement, inheritance and its regulation formed a focus of legal and political debate much earlier. According to a well-known thesis advanced by historical anthropologist Jack Goody, it was primarily the Church, with its dependence on donations from individuals, that began to interfere in early medieval times with traditional strategies of inheritance—adoption, cousin marriage, concubinage, marriage with affines (relatives by marriage), or the remarriage of divorced persons—to bring about an alienation of family holdings.[24] The Gregorian reforms of the eleventh century in particular achieved this aim by instituting celibacy among priests and by prohibiting marriage between kin up to the seventh degree. In this process, kinship became defined according to the much more inclusive Germanic system, which calculated kinship degree among relatives by counting the number of generations back to the first common ancestor, rather than the number of generative acts that lay between them, as in the Roman system (figure 1.1). Moreover, kinship was even further extended by including both cognates, or relatives by blood, and affines, or relatives by marriage, under such calculations of kinship degrees. The emerging territorial state—competing with the Church but also adopting similar bureaucratic and property-holding forms—readily stepped in to add its force to these regulations.[25] By the end of the seventeenth century any alliance between families through marriage was thus effectively subject to dispensations from the Church or state authorities.[26]

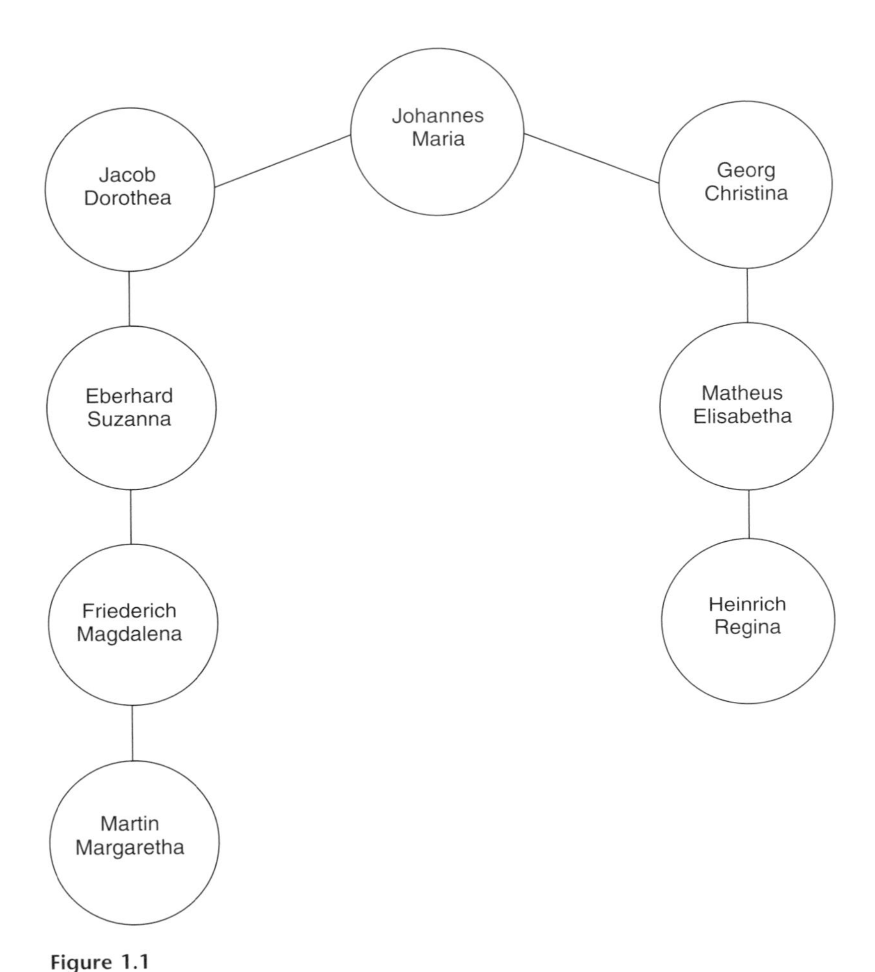

Figure 1.1
Diagram illustrating the Roman and Germanic system of determining degrees of consanguinity. According to the Germanic system Friederich and Regina would be relatives in the third degree, because they are separated from their first common ancestor (Johannes) by three generations; according to the Roman system, in the sixth degree, because six "generations" (in the sense of generative events) lie between them. From David Warren Sabean, *Kinship in Neckarhausen 1700–1870* (Cambridge University Press, 1998). Courtesy Cambridge University Press.

The result of these developments was, on the one hand, a veritable outburst of genealogical activities. One important innovation was the introduction of the *ipse* or *ego*—representing an individual rather than the sibling group or *truncus* of earlier times—as the departing point for kinship calculations in genealogical tables and diagrams.[27] On the other hand, noble families in particular tried to counteract the dispersion of family property by preferring the agnatic or male descendent line equivalent to the medieval *lignage*. This emphasis on the male line came at the expense of female relatives as well as collateral kin like cousins in the transmission of property, office, or title. This may account for the fact that, as far as we could see, the aristocratic concept of "noble blood" played a surprisingly small role in the emergent discourse of heredity, although genealogical data on the occurrence of such features as the "Hapsburg lip" or hemophilia in noble families was available. In fact, the point of emphasizing the noble lineage, often with a privilege accorded to the first-born son, was to exclude the majority of "blood" relatives, including later-born sons, collaterals, daughters, and matrilineal kin from inheritance to avoid the dispersal of family property.[28]

Ironically, these strategies led to a further emphasis on the individual at the expense of "natural" rights of kin groups, at least conceptually.[29] It is on this basis, also, that primogeniture came under increasing attack in the eighteenth century for its injustice, culminating in one of the French Revolution's central claims, namely, that no generation should force its regulations on future generations.[30] The *Code civil*, consequently, contained inheritance regulations based on equal rights among individual heirs and full divisibility of property according to a mathematical formula based on an analysis of kinship degrees.[31] Though primogeniture was partly reinstated during the nineteenth century, it had come under lasting pressure. Instead, the early nineteenth century saw a resurfacing of marriage to close kin, preferably marriages between cousins, to hold together family property in transmission and to establish close familial networks of mutual trust and loyalty in order to cope with capitalist conditions of wealth and property accumulation.[32] The Darwin and Wedgwood families provide formidable examples of this strategy, including the concurrent fear of degeneration.[33]

Without claiming expert knowledge of these extremely complex and regionally diverse developments in inheritance regulations, we have tried to give an overview of some of their major trends to derive an important point. Outside the life sciences proper, there existed a well-developed, lively, and controversial discourse on inheritance, which incorporated a wealth of definitions, taxonomies, and calculation procedures.[34] And yet, as we have already stated, it remains a fact that these rich semantics were not taken up and deployed by natural historians or philosophers to analyze biological phenomena. Even the naturalistic justifications that sometimes found their way into the legal and political discourse—the justification of the inclusion of affines under incest regulations by the "unity of flesh and blood" supposedly instituted by marriage,[35] or the denunciation of primogeniture as "monstrous" by French revolutionaries[36]—were not, as far as we can see, immediately reflected in medical or philosophical accounts of generation.[37]

It is nevertheless possible to identify a trend for heredity *avant la lettre*. There were several, highly specific, and altogether separate domains of knowledge, which became increasingly structured during the early modern period by the de facto recognition of the hereditary transmission of differential characters. These include (1) the classification of diseases as hereditary in nosography, which became a focus of medical theorizing during the eighteenth century;[38] (2) natural history, botany in particular, where, from the late seventeenth century on, the definition of taxonomic characters increasingly relied on their constancy;[39] (3) animal and plant breeding, where professional breeders began to "mold" organisms for specific, distinctive features from about the mid-eighteenth century on;[40] (4) the recruitment of hybridization experiments and family histories to probe the role of the sexes and the validity of preformation in late seventeenth- and eighteenth-century theories of generation;[41] and (5) anthropology, which began to study physical difference, notably skin color, and its physiological and historical origins to account for human diversity.[42]

While the distribution of hereditary knowledge over separate domains once again shows that hereditary phenomena in the realm of the life sciences had not really gone unnoticed before the end of the eighteenth century, it also evinces what did not exist: a general concept of heredity binding these domains together. Such a general concept of heredity emerged only slowly in the second third of the nineteenth century, culminating in the works of Prosper Lucas (1805–1885), Charles Darwin (1809–1882), and Galton. Carlos López-Beltrán has illustrated this transition by pointing to a linguistic shift. While the use of the adjective *hereditary* can be dated back to antiquity in the context of nosography (*maladies héréditaires*), a transition to a nominal use (*hérédité*) took place only from the 1830s onward, first among French physiologists and physicians, then in other European scientific circles. This shift indicates a reification of the concept, or, in López-Beltrán's words, the establishment of a "structured set of meanings that outlined and unified an emerging biological conceptual space."[43] It also implies a concomitant shift, namely the erosion of a set of very ancient distinctions with respect to similarities between parents and offspring, which the modern notion of heredity systematically cuts across. Distinctions had been made between specific versus individual, paternal versus maternal, ancestral versus parental, normal versus pathological similarities, and even between similarities pertaining to the left and the right halves of the body.[44] Such distinctions gave way to a generalized notion of heredity that focused on elementary traits or dispositions independent of the particular life forms they were part of, whether pathological or normal, maternal or paternal, individual or specific.

This leaves us with three parallel developments during the period this book surveys, which we will try to relate and explain in the remainder of this introduction. First, there existed a relatively early obsession with questions of heredity in the legal and political sphere. Second, traditional distinctions of intergenerational similarities were increasingly eroded. Third, heredity as a biological concept emerged at a relatively late stage in the process. Relating these developments to each other involves a historiographic problem, namely, to explain the evolution of a knowledge regime while it still lacks a coherent linguistic, conceptual, or institutional framework. We will address this problem in the next section.

1.3 Heredity as a Knowledge Regime

The perspective this book takes is neither one of a conventional history of ideas nor one of social history. Instead it explores the emergence of specific practices, the shaping of standards, taxonomies, and arguments, the evolution of architectures of hereditary knowledge, and the conjunctions of these elements in a variety of social arenas. "Heredity," according to this perspective, is more than and quite different from what came to be regarded and respected as the scientific discipline of "genetics." The book is less about the history, let alone prehistory, of a scientific concept, theory, or discipline, than about the history of a broader knowledge regime, in which a naturalistic concept of heredity gradually took shape and developed to the extent that it today affects all domains of society.[45]

Centering on the emergence of a knowledge regime, rather than a concept, theory, or discipline, has two main consequences. First, heredity need not be treated as a notion with a definite meaning to begin with. We have become acquainted with a conception of heredity that expresses the permanence of form over generations, not necessarily leading, but nevertheless lending itself, to naturalistic justifications of political authority and cultural conservatism.[46] Instead of acknowledging or criticizing such reifications, heredity is treated throughout this book as a knowledge regime that always exhibited internal conceptual dynamics in terms of changing classifications and shifting perceptions of causalities. It makes a difference, for example, whether inheritance of individual traits on the one hand and specific forms on the other are held separate, or if they are conflated; whether heredity is considered "soft," and thus to some extent reversible, or "hard," and thus irreversible. We will also see how these structures varied over time and only eventually acquired a disciplinary drive that was fueled by a core concept of heredity.

Second, and in consequence, the conceptual dynamics of knowledge of heredity are explored with regard to the concept's dependence on practices and institutions. Acclimatization experiments in the eighteenth century, for instance, were used to distinguish constant, species-specific characters from characters varying individually with external conditions like climate and soil. However, this very practice also resulted in setting the stage for identifying characters that were peculiar to certain individuals within a species and yet remained constant even under varying external conditions.[47] This is an instance of a feature that can be observed repeatedly in the history of efforts to make sense of hereditary phenomena. The experimental realization of a generally accepted dichotomy with regard to generational persistence leads to the discovery of another puzzle. In the very process of its experimental confirmation and acceptance, the dichotomy is challenged by the delineation of phenomena that are in need of alternative descriptions and explanations. The taxonomies and causalities subsumed under heredity, if put into practice and institutionalized, furnish the very conditions for contrary evidence and the resolution of the ensuing conceptual challenges.

An illustration of these points is provided by the social system of *castas* that was introduced in Spanish and Portuguese America in the sixteenth century. This classification scheme resulted from attempts to find a measure by which legal and social status could be allocated to the various sections of colonial society. It sharply contrasted with the dominant discourse on generation as outlined in section 1 above. It was primarily based on a classification of people according to skin color, to a lesser degree also on hair form and eye color. Children resulting from mixed marriages were positioned in this scheme in analogy to the simple mechanism of color mixing, implying "blending" as the causal relation connecting traits of parents with traits of their offspring. As Georges-Louis Leclerc Comte de Buffon (1707–1788) put it with respect to this system: "D'un mulâtre et de la mégresse vient le quarteron, qui a trois quarts de noir et un quart de blanc."[48] During the eighteenth century, the system of castas found expression in a rich, pictorial genre with pictures devised as sets arranged in serial or tabular form (see figs 15.1–15.5). Each of the pictures shows a mixed couple and their child, and each bears an inscription that states the components entering the mixture—that is, each parent's casta, and the result, the child's casta.[49]

Despite its rigid appearance, the castas system remained in constant flux throughout the early modern period, as witnessed by a rich proliferation of castas terms. And yet it was not despite, but because it was so rigidly based on an abstract classification according to color, and on blending as an equally abstract mechanism, that the castas scheme could cope with this proliferation. The distinction according to colors—white, black, and brown—analytically defined the positions for all sorts of intermediate and more complicated cases. And the mechanism of blending offered a unified explanation for their coming about, insofar as it could be regarded as operating independently of particular circumstances. To determine the casta of a person, it therefore sufficed to know the castas of his or her parents. Due to its analytic and quasi-mechanical character, the castas system could absorb a wealth of new phenomena while remaining stable in its basic outlines. This system accorded legal status to the members of a society in flux, reflected in the depiction of the castas in their legally sanctioned costumes and occupations, "hooked," as it were, onto the underlying mechanism. It therefore provides one of the first instances for the metaphorical transfer of legal conceptions of inheritance to the realm of biological reproduction and vice versa.[50]

The castas system furnishes a palpable example of a knowledge regime of heredity. It clearly occupies a position at the intersection of the epistemic, political, and cultural realm and, in fact, organizes that very intersection. There are two minimal conditions that allow us to speak about the castas system as knowledge of heredity. First, the transmission of physical properties from generation to generation forms a limited and autonomous discourse that finds expression in a certain genre, in this case a visual one. Second, the system exhibits a certain taxonomic structure and a set of propositions about the causal relationships connecting the entities posited in this taxonomy.[51] The idiosyncratic and highly localized origin of the castas system shows, moreover, that no clear line can be drawn from where it becomes legitimate to speak of particular instances of hereditary thought.[52] It might be legitimate, for instance, to regard discussions about the apparent transmission of individual characters from grandparent to grandchild, scattered throughout texts on generation since antiquity, as a nascent knowledge regime of heredity. However, one may equally well deny that these occurrences constitute a full-fledged knowledge regime because they were only weakly developed, in terms of both contextual delimitation and conceptual structure.[53]

The attention paid to heredity as a knowledge regime is the reason that the essays in this book follow each other neither in a historical nor in a thematic order. Instead a structure was chosen conforming to the cultural domains—law, medicine, natural history, breeding, biology, anthropology—in which different phenomena came to be conceptualized in terms of heredity. Aspects of these domains became subsequently modified, partially conjoined, and finally integrated to form the subject of a general concept of heredity. Throughout the period considered here, however, this concept remained largely implicit, in constant flux, and contested.[54] It may be indicative of this conceptual flux that until the middle of the nineteenth century several, but only scattered and very tentative, attempts were made at developing registration forms and representational schemes through which variables of heredity could be visualized and handled in terms of data collection, display, and processing. We have mentioned one example already, the depiction of castas. Another example, from the late eighteenth century, is the attempt at a genealogical order combining information on descent and hybridization to account for the emergence of strawberry varieties.[55] Medical notions of heredity in early nineteenth-century France were backed by statistical tables, which in turn became tools for the management of populations in the broader context of a hygienic regime of collective health improvement.[56] Finally, breeders started to keep logbooks that allowed them to track the selective production of progeny both backward to parental and forward to filial generations.[57]

Thus a variety of symbolic means for measuring the barely charted grounds of hereditary phenomena emerged at the end of the period considered. The consolidation of such inscription and registration procedures is a topic that will be pursued in detail in the coming volumes. It is certainly not too early, however, to suggest that the development of these inscription technologies was a necessary condition for the full development of the epistemic space that Francis Galton addressed in his *A Theory of Heredity*. It remains to be seen why they were taken up in the first place and transferred to the domain of scientific inquiries into generation, a domain that had been able to do without them for two thousand years.

1.4 The Genesis of Heredity as a Biological Concept

The central historiographic problem of the present book can be put as follows. How is it that the phenomenon of hereditary transmission, which, from a contemporary perspective, seems so important and so tangible in its effects, was subjected to systematic conceptualization so late? The answer that emerges in the following chapters may be surprising. The concept of heredity did not result from a growing attention to mere similarities between parents and offspring, from an obsession of the scientific mind with regularity at the expense of contingency and complexity. The example of the Latin American castas system indicates that the contexts that gave rise to a knowledge of heredity were much more specific. It indicates that the knowledge regime of heredity started to unfold where people, objects, and relationships among them were set into motion.

To put our thesis differently, the problem that heredity came to address was not the constancy of species, but the fluctuating patterns and processes that structure life at the subspecific level. This shift of attention was the result of a mobilization of early modern life in a variety of social arenas. Mobilizing plants and animals, for instance, was a precondition for being able to distinguish between inherited and environmentally induced traits in organisms. Only when organisms were actually removed from their natural and traditional agricultural habitats could environmental differences manifest themselves in trait differences, and only then could heritable traits manifest their steadiness against a background of environmental change. A number of examples could be cited: breeding new varieties for specific marketable characteristics; efforts to acclimatize plants and animals in new habitats; the exchange of specimens among botanical and zoological gardens; experiments in fertilization and hybridization of geographically separated plants and animals; the dislocation of Europeans and Africans that accompanied colonialism; and the appearance of new social strata in the context of industrialization and urbanization. All of these processes interlocked to relax and sever traditional cultural and natural ties to provide the material substrate for the emerging discourse of heredity.

It is true that the principle of "like engenders like" had been around since the earliest times of Greek poetry and philosophy, as an expression for what ought to happen as a rule.[58] This "law," however, remained unanalyzed. It lacked the kind of inner structure that might have provoked the productive application of a metaphor, which in its original context of legal regulations of property transmission possessed such complex semantics. It was in the anthropological writings of Immanuel Kant that conceptions of heredity first began to acquire a specific biological meaning. The reason for this proliferation of meanings was not the similarity that offspring exhibit with regard to their parents in general. It was rather a narrowly circumscribed, highly specific phenomenon, namely, the existence of distinct races in the human species distinguished by traits that blended in hybrids but were invariably transmitted to offspring even under changed, environmental conditions.[59]

Such a phenomenon undercut the distinction of specific forms and individual peculiarities. Characterizing classes at a subspecific level, racial characters belonged to the individual peculiarities that interfered with the universality of species. Yet, these peculiarities were being infallibly reproduced generation by generation and seemed to be subject to the same kind of regularity that governed species. To account for this, Kant brought together natural law and contingent, family history in his concept of *Vererbung*. The dispositions or *Anlagen* for hereditary traits were included from the very beginning in the original organization of ancestors, thus, in a sense, being preformed and not acquired. But once they had been expressed as actual traits in reaction to a change in environment and were, in a sense, acquired epigenetically, they were permanently and irrevocably transmitted.

The way Kant set up the problem and the way he advanced a solution can be regarded as prototypical for the emergence of heredity. The problem was not the constancy of species forms but the patterns of variety that structure life at a subspecific level. As long as such patterns coincided with locally circumscribed environments, they were readily explained by the permanence of ties between living beings and their "natural places." In these cases, it is the place that "inherits" its inhabitants and impresses its character on them. Only when these ties were dissolved in favor of a variety of relationships between forms, places, and modes of transmission did a need arise for a complex metaphor like heredity to be applied in order to account for the proliferating phenomena of change and stability.

The motivation to apply and explore the concept of heredity in the face of a mobilization of social and natural ties, effected through transplantations and hybrid unions, can be observed in all the cultural subfields discussed in the contributions to this book. From what we learn of part I about discussions of inheritance regulations in legal and political arenas, it seems to be the increasing importance of mobile over landed property that brought heredity to the fore as a hotly debated legal problem in the first place.[60] In consequence, patriarchal authority, primogeniture regulations, the status of the illegitimate, the values of real estate, and the power of money to abstract inheritance from land became foci of intense conflict.[61] One way such conflicts tended to be resolved—of particular relevance to our book's topic, the genesis of heredity—was by turning to nature. Internal resemblances in "temperament," manifested by resemblances in habits or diseases, began to be considered a reliable source of genealogical evidence.[62]

"Temperament" or "diathesis," both referring to constitutional dispositions toward specific diseases, formed the core concept of discussions among physicians about hereditary diseases in the eighteenth and nineteenth centuries.[63] It was among physicians, French doctors and physiologists in particular, that the noun heredity (*hérédité*) was first widely adopted. Hereditary diseases had received growing attention throughout the eighteenth century, culminating in two prize competitions run by the Royal Society for Medicine (Paris) around the 1790s. A set of distinctions were fleshed out and clarified in the process, turning the noun *heredity* into "the carrier of a structured set of meanings that outlined and unified an emerging biological concept."[64] Congenital, connate, and acquired diseases, observational criteria such as homochrony referring to timing in the outbreak of hereditary diseases, causal concepts like latency, and various conceptions of "soft" versus "hard" heritability, were among these distinctions.

The reasons for these developments are complex, but two stand out. First, hereditary diseases were brought to bear on the political arena. As Gianna Pomata observed, "a recurrent feature of the medical discourse on heredity in the eighteenth and early nineteenth century is the critique of the aristocratic family." Interest in the eighteenth century was mostly focused on "noble" maladies, like gout, believed to be "softly" inherited by overconsumption.[65] This focus shifted in the nineteenth century, after the aristocratic family model had come under lasting pressure, to "degenerative" diseases like phthisis (tuberculosis) and madness. These were ascribed to the rapidly growing class of landless and poor migrating to urban centers, and believed to be subject to "hard" heredity.[66] Now, "these classes, rather than the aristocracy, were perceived as a threat to the social order."[67]

The second set of reasons for changes in medical views of hereditary diseases reside in the new social role physicians acquired at the turn of the nineteenth century. The new scientific profile that they sought made it advisable to define thoroughly where the art of medicine met its limits in incurable, constitutional, and thus heritable diseases.[68] New responsibilities of physicians for public hygiene led to the definition of dangers such as heritable diseases, which lay hidden in the populace and which, consequently, only the expert could address. Finally, hospitalization and medical statistics made possible new forms of representation—medical topographies and chronicles—through which not only individual cases or family histories, but a representative population became visible. The genealogical and epidemiological data thus accumulated were certainly one of the most important prerequisites to disentangling the more complex patterns of familial diseases, and to telling them apart from diseases caused by local influences or differences in lifestyle.[69]

The role that the hospital played for the constitution of the discourse of heredity in medicine can be compared with that of botanical gardens and menageries in natural history. We have already mentioned that one of the necessary conditions for separating environmental from hereditary factors was that organisms had to actually be removed from their native environments. This was what botanical gardens and menageries effectively did by accumulating living specimens from all over the world under a regime of more or less controlled conditions.[70] The exchange of specimens among these institutions increased the possibility of detecting even more complex and opaque hereditary patterns, like atavisms, character segregation, or mutations. It is therefore no wonder that gardens and menageries, although instituted originally for the descriptive purposes of natural history, also formed the first loci for approaching heredity experimentally, a research tradition that was to lead up to Mendel's famous experiments.[71]

Similar experiments in transplantation and hybridization were carried out by breeders, which became increasingly organized in professional societies in the eighteenth century and exchanged their breeds over wide geographic areas.[72] However, the intricate relationship between naturalists and breeders also demonstrates how strong institutional and social obstacles to a unified view of heredity remained until well into the nineteenth century.[73] In the Linnaean tradition horticultural varieties were not regarded as a proper subject for botanists, and it was only from the 1830s that naturalists were ready to adopt for their experiments the genealogical recording techniques that breeders had developed.[74] Their attitude of attending to individual traits in populations in order to "mold" their creatures for marketable traits was especially counterintuitive to naturalists. The latter would remain interested in the origin, permanence, and possible transformation of species rather than in individual variations throughout the nineteenth century.[75]

Eventually it was the erosion of institutional barriers, such as those between naturalists and breeders or medical practitioners and natural scientists, that played an important role in paving the way for the discourse of heredity. In the long run, such erosions allowed for a perspective from which both the specific and the individual, both the normal and the pathological—in short, both regular and deviant forms—would appear to be determined by the same set of natural laws governing organic reproduction.[76] The "life of the species" (*Leben der Gattung*), to borrow a term from the German naturalist Carl Friedrich Kielmeyer (1765–1844),[77] that is, the vital processes connecting a multiplicity of beings, became the subject of a new science that would receive the name *biology* around 1800.

Both Michel Foucault and François Jacob have identified "organization" as the key concept constitutive of the "new" science of biology.[78] We do not question the importance of this conceptual innovation. But it seems pertinent for a full understanding of what constituted biology to recognize the interindividual and intraspecific dimensions that most of its concepts gained around 1800.[79] Instead of referring, as in earlier times, to individual bodies, organic functions like generation, growth, development, nutrition, and sensation were increasingly perceived as reproductive functions physically constituting the unity of species.[80] As Buffon put it programmatically in his influential *Discourse on the Manner of Studying and Expounding Natural History* (1749): "The history [of the species] ought to treat only relations, which the things of nature have among themselves and with us. The history of an animal ought to be not only the history of the individual, but that of the entire species."[81] Reflecting this change in perspective is the concept of generation, which in the early nineteenth century acquired its subsidiary meaning of a set of individuals born about the same period.[82]

The focus on organization that emerged around 1800 is only seemingly a paradoxical concomitant of the growing attention paid to the "life of the species." As a matter of fact, reproduction was at the very heart of Kant's influential concept of a "natural purpose" (*Naturzweck*), by which he tried to determine the causality specific to organized beings.[83] With regard to reproduction proper, conceived as the production of an organized being through another organized being, this point of view was to lead to a growing focus on the internal organization of the germ. It is, after all, the germ that provides the physical link in reproduction. Two complementary frames of how to conceive of the germ's potency to bring forth beings of its own kind were formulated in the eighteenth century, and persisted side by side throughout the nineteenth century. The first one, paradigmatically formulated in Johann Friedrich Blumenbach's (1742–1850) concept of "vital force" (*Bildungstrieb*), conceived of heredity as a force acting at a distance, analogous to Newtonian gravitation. The second, paradigmatically formulated in Buffon's concept of "organic molecules" (*molécules organiques*), saw heredity as residing in organized matter that was transmitted from one generation to the other.[84] The latter paradigm is also exemplified in the growing prominence of solidist models of the causation of hereditary diseases in early nineteenth-century medicine, which sought the cause of disease in structural lesions rather than an imbalance of humors.[85]

Thus we see heredity coming to be inserted at the intersection of the life of the species and the life of the individual in the context of a science of the living that began to take form around 1800. The institutions mentioned earlier—botanical gardens, museums, breeders' societies, hospitals, and medical administrations—were instrumental in gaining this biological perspective. But they were certainly not sufficient to forge a concept of heredity. Additional conjunctions occurred between two domains that may be roughly circumscribed as descriptive approaches in natural history and medical statistics on the one hand and experimental approaches in physiology and pathology on the other. It is, we believe, because of the lack of such conjunctions that one field in which we might have expected an early emergence of hereditary theories—cytology—actually shows a surprisingly late engagement with such theories.[86]

With respect to the long-standing separation of natural history and physiology, a distinction made by the French physiologist Claude Bernard (1813–1878) is particularly revealing. In his *Leçons sur les phénomènes de la vie commune aux animaux et aux végétaux* (1878), Bernard distinguished between a "chemical" and a "morphological" or "organizing synthesis" in organisms. The former consisted in vital functions effected by physicochemical processes, and was thus accessible to experimentation.

The latter consisted in the concatenation and mutual subordination of such vital functions in the reproduction of organic forms, and eluded, at least in the eyes of Bernard, experimental intervention.[87] It is equally revealing that one of the earliest protagonists of a full-fledged theory of heredity, Charles Darwin, entertained a lifelong interest in colonial organisms, which allowed him to apply analogies from entities above the level of the individual organism like species and populations to entities below that level, including buds, cells, and gemmules.[88]

This leads us to another arena in which the discourse of heredity took shape. Enlightenment discussions about the nature and organization of living matter frequently brought together many of the disparate elements of the knowledge regime of heredity described in the preceding paragraphs. Philosophy, emancipating itself from theology, law, and medicine during the eighteenth century, certainly lacked control through experiment and the rigor of later theorizing in biology. But it was precisely because of this that philosophy could easily transgress boundaries between knowledge domains, between the practical knowledge of breeders and physiological speculation. Philosophy invested its subjects with concerns that went far beyond those of particular professions and were of political and theological significance. The debate about preformation in the eighteenth century, for instance, touched on questions like: whether matter was wholly passive or by itself endowed with active forces; whether the formation of individuals was subject to superordinate powers or whether individuals were autonomous and equal; and whether or not the sexes had equal parts in the generation of offspring.[89] Challenges to patriarchal order with its privileges and particularities passed on in lineal, male descent provide much of the background. Formulated by the Enlightenment, and epitomized in the French Revolution, such questions underlay the increasing tendency in the early nineteenth century to anchor the attribution of social and cultural status in a universal natural order defined by the distribution of hereditary dispositions.[90]

It does not, therefore, come as a surprise that anthropology in particular became one of the "hot spots" of debates about heredity. Clearly, this was a field that could not be directly accessed by experiment, the only substitute, though with its own irresolvable aporias, being the observation of "savage children."[91] The exponentially growing number of ethnographic reports from regions outside Europe, however, had opened a "veritable 'laboratory of human nature'" facilitating a "'natural history' of man" at the end of the eighteenth century.[92] A curious focus of early physical anthropology was its preoccupation with skin color, an external trait par excellence. In the early modern period, black skin color had become identified with Africa and with lineal descent from one of the sons of Noah, Ham.[93] By the end of the seventeenth century the focus shifted from mere assertions of such differences to the study of their anatomical and historical origin.[94]

With respect to heredity it was the system of castas, in particular, that provided one of the earliest models for its conceptualization and "a vast field of 'pre-Mendelian' investigation."[95] It evolved into a universal scheme of racial classification, as first put forward by Carolus Linnaeus (1707–1778) in his *Systema naturae* (1735), and then supported by various theories on the origin of differences within the human species. Race, like the concept of caste, its correlate in cultural history, became a biopolitical notion that resorted to physiological reproduction in order to explain the uneven reproduction of wealth, power, and opportunities. The same strategy can be observed in the conflict between nature and nurture that began to revolve around cases of "savage children" in the early nineteenth century, in the Victorian ideology of the "self-made man," and, finally, in the discussions on the origin and (self-)reproduction of "genius" that became prominent around 1800, long before Galton published his *Hereditary Genius* (1869).[96] In all of these disparate arenas of literary as well as scientific culture, the specific and the individual were conflated in elementary dispositions. While manifesting themselves in select individuals, hereditary dispositions were omnipresent within the species, and accounted for what appeared as an oxymoron: the reproduction of difference.[97]

1.5 Conclusion

The various domains of the knowledge regime of heredity described in the previous section as giving rise to a discourse of heredity, were not brought together according to the model of "influence." Rather, conjunctions between these domains came about by means of a domino effect that mobilization in one domain had on another. The growth of a class depending on mobile property evoked a culture of leisure collecting and breeding. The importation of plants for collection purposes inspired attempts at their acclimatization for economic purposes. The breeder, with his successes in establishing marketable strains of plants and animals, provided the role model for the "self-made man." These connections point less to a "culture" or "episteme" of heredity that suddenly emerged around 1800, than to a piecemeal relaxation of social and natural ties in several cultural subfields. It is only by their subsequent conjunction that a discernible field of phenomena was outlined that eventually, in the mid-nineteenth century, came to be addressed by theories of heredity.

Although this state of affairs makes it difficult, even impossible, to draw a unitary picture of a prehistory of heredity, it is possible, in hindsight, to characterize the result.

As a point of departure we choose a quote that Galton almost certainly had in mind when he praised his cousin's approach in *A Theory of Heredity*. Darwin says, in *Variation of Animals and Plants under Domestication* (1868),

> It is probable that hardly a change of any kind affects either parent, without some mark being left on the germ. But on the doctrine of reversion [i.e., the reappearance of heritable traits beginning with the second generation] . . . the germ becomes a far more marvellous object, for, besides the visible changes which it undergoes, we must believe that it is crowded with invisible characters, proper to both sexes, to both the right and left side of the body, and to a long line of male and female ancestors separated by hundreds or even thousands of generations from the present time: and these characters, like those written on paper with invisible ink, lie ready to be evolved whenever the organization is disturbed by certain known or unknown conditions.[98]

Two aspects of Darwin's theory of heredity, evident from this passage, are remarkable when compared with Harvey's view of organic reproduction with which we started. First, the extent to which Darwin already endorsed a view of heredity that abstracted from personal relations between parents and offspring is obvious. While conceding the possibility of inheritance of acquired properties, Darwin makes it clear that the true carriers of the properties to be inherited are not the parents themselves, but submicroscopic entities that circulate, from generation to generation, among individuals within one and the same species.[99] Second, and more fundamentally, the passage quoted evinces a peculiar inversion in comparison with early modern conceptions of organic production. While the latter emphasize the vertical dimension of lineal descent—in which parental organisms actually *make* their offspring—Darwin invokes an image where the horizontal dimension dominates, the dimension of a common reservoir of dispositions, passed down from the sum total of ancestors, redistributed among individuals of one generation, and competing now, in the present, for their realization.[100] We take these two aspects as the fundamental hallmarks of modern hereditary thought.

It might not be too farfetched to see analogies here with two important aspects of capitalist economy: alienation and circulation. The analogies with the new kinship and inheritance system, which we described in the second section as evolving with the onset of modernity, are striking. We would like to emphasize something else, however. Darwin invoked the constellation of virtually thousands of generations represented in the microscopic space of the fertilized egg to describe "the most wonderful object in nature," the germ. This description points to the complexity of the problem of heredity. One can take both Darwin's and Galton's attempts to address heredity as a "space" quite seriously and speak of an "epistemic space" of heredity that came into being in the mid-nineteenth century. In contrast to other subjects of biological research, which can be characterized as "epistemic things" in the sense of being circumscribed entities confined within individual experimental settings, heredity depended on a vast configuration of distributed technologies and institutions connected by a system of exchange: botanical gardens, hospitals, chemical and physiological laboratories, genealogical and statistical archives. Capitalism and bourgeois culture certainly facilitated the various conjunctions that made this configuration possible. Heredity as an epistemic space, however, was not just a construction to justify underlying socioeconomic developments. As we probably are only beginning to realize today, in times where genetic screening, testing, and patenting pervade all sectors of social and economic life, and with the synthetic powers of genomics on the horizon, the epistemic space that heredity came to constitute has reconfigured life in its entirety.

Notes

1. Jacob [1970] 1993, 19–20. For the origin of the (biological) concept of reproduction in Buffon's work, see McLaughlin 2001, 173–179.

2. Russell 1986, 40; Allen 1986; Churchill 1987; Maienschein 1987; Bowler 1989, 6; López-Beltrán 2004b, 19–20.

3. Harvey [1651] 1847, 363.

4. Harvey [1651] 1847, 367; on Harvey's theory of generation see Gregory 2001 and Müller-Wille, chapter 8, this volume.

5. Pinto-Correia 1997, chapters 3 and 4; Park and Daston 1981; Lugt 2004; on maternal imagination, see De Renzi, chapter 3, this volume.

6. Sterne 1760, 1.

7. Jacob [1970] 1993, 24–25.

8. Descartes [1701] 1996, 507, our translation.

9. Rey 1989, 7; López-Beltrán 2004a.

10. The *Oxford English Dictionary* (online edition) lists 1540 as the earliest occurrence of *heredity* in the legal sense, marking it as "obsolete," while the earliest references for *heredity* in the modern, biological sense date from 1863 (Herbert Spencer's *Principles of Biology*) and 1869 (Francis Galton's *Hereditary Genius*). *Black's Law Dictionary* (7th ed., 1999) defines *heredity* as "archaic" for "hereditary succession; an inheritance." The American *Corpus Juris Secundum* (vol. 39A) defines heredity only as "a universal law of organic life."

11. Lugt, forthcoming.

12. Lopéz-Beltrán, chapter 5, this volume.

13. See Lesky 1950, 148–155; Müller-Wille, chapter 8, this volume.

14. Gayon 1995.

15. Olby 1985, 55–63; Gayon 1998, 105–106.

16. Galton 1876, 330.

17. Galton 1876, 331.

18. Galton 1876, 336.

19. Galton 1876, 330.

20. Rheinberger 1997.

21. On the importance of mapping strategies in twentieth-century genetics, see Rheinberger and Gaudillière 2004.

22. See the introduction to Jacob [1970] 1993.

23. For overviews see Lesky 1950; Stubbe 1965, chapters 1–3; Lugt and Miramon, forthcoming.

24. Goody 1983, 123.

25. Goody 1983, 136–144.

26. Sabean 1998, chapter 3.

27. Goody 1983, 142–146; on the cultural and political functions of early modern genealogy, see Heck and Jahn 2000.

28. Goody 1983, 120–123.

29. This does not mean that members of the extended family were in practice excluded from family property. The construction of *lignages* must rather be seen as a way families organized themselves around property; on this complex, and regionally diverse, issue, see Sabean, chapter 2, this volume.

30. Vedder, chapter 4, this volume.

31. *Napoleons Gesetzbuch* 1808, 312–323, 346–367.

32. Sabean, chapter 2, this volume.

33. Browne 2002, chapter 8; White, chapter 16, this volume.

34. Al-Khwarizmi's (780–850) algebra treatise was largely dedicated to the solution of inheritance problems; see O'Connor and Robertson 1999.

35. Sabean 1998, 70–71.

36. Vedder, chapter 4, this volume.

37. The only social arena where legal and medical notions of inheritance interacted was that of forensic medicine; see De Renzi, chapter 3, this volume.

38. Olby 1993; López-Beltrán, chapter 5, this volume.

39. Glass [1959] 1967; Larson 1994, chapter 3; Müller-Wille, chapter 8, this volume.

40. Roberts 1929; Zirkle 1935; Russell 1986; Ratcliff, chapter 9, this volume; Wood, chapter 10, this volume.

41. Glass, Temkin, and Strauss [1959] 1967; Roger [1963] 1993, 81–91; Rey 1989; Terrall, chapter 11, this volume.

42. Gould 1981; Hannaford 1996; Braude 1997; Mazzolini, chapter 15, this volume.

43. López-Beltrán, chapter 5, this volume.

44. See Lesky 1950; Stubbe 1965, chapters 1–3; Rey 1989.

45. With *knowledge regime* we are adopting a term that Dominique Pestre introduced to avoid the problems that the narrower, yet more historically variable connotations of *science* create for any attempt at writing a *longue durée* history of science; see Pestre 2003, 31–37.

46. See, for example, Jordanova 1995, 375.

47. Müller-Wille, chapter 8, this volume.

48. Buffon [1749–1767] 1971, 352. Cf. Mazzolini, chapter 15, this volume.

49. Katzew 2004, chapter 3; Mazzolini, chapter 15, this volume.

50. Katzew 2004, chapter 2; Mazzolini, chapter 15, this volume.

51. We are adopting these minimal conditions from Mary Hesse's network approach to theories (Hesse 1974, chapters 1 and 2), without, however, buying into the holistic assumptions that underlie this approach.

52. The castas system, though intricate, was certainly a product of folklore rather than science, and was not meant, at its origin, to serve as a universal racial theory; see Canizares Esguerra 1999.

53. See Terrall, chapter 11, this volume.

54. On concepts in flux, see Elkana 1970.

55. Ratcliff, chapter 9, this volume.

56. Cartron, chapter 7, this volume.

57. Wood, chapter 10, this volume; White, chapter 16, this volume.

58. Stubbe 1965, 10–12.

59. McLaughlin, chapter 12, this volume.

60. Sabean, chapter 2, this volume.

61. Vedder, chapter 4, this volume.

62. De Renzi, chapter 3, this volume.

63. Olby 1993.

64. López-Beltrán, chapter 5, this volume.

65. See Wilson, chapter 6, this volume. On gout as the "patrician malady," see Porter and Rousseau 1998.

66. Cartron, chapter 7, this volume.

67. Pomata 2003, 151.

68. Waller 2002; Cartron, chapter 7, this volume.

69. La Berge 1992; Williams 1994; Cartron, chapter 7, this volume.

70. Müller-Wille 2003.

71. Olby 1985, chapters 1 and 2; Larson 1994, chapter 3.

72. Russell 1986; Wood and Orel 1996; Wood, chapter 10, this volume.

73. Ratcliff, chapter 9, this volume.

74. Wood, chapter 10, this volume.

75. Olby 1979; Bowler 1989, chapter 5.

76. Canguilhem [1966] 1991, 29–46.

77. Kielmeyer [1793] 1993, 5.

78. Foucault 1966, 238–245; Jacob [1970] 1993, chapter 2.

79. See Coleman 1971 for a portrait of nineteenth-century biology that misses this dimension.

80. See Ritterbush 1964, chapter 5; Roger [1963] 1993, 567–582; Jacob [1970] 1993, 88–92; Lenoir 1982, chapter 1; Larson 1994; Spary 2000, chapter 3.

81. Quoted from Lyon and Sloan 1981, 111 (Premier Discours [1749], 30).

82. Weigel et al. 2005; Parnes, chapter 14, this volume.

83. Jacob [1970] 1993, 88–89; McLaughlin 1990, 44–51.

84. On Blumenbach's concept of *Bildungstrieb* see Lenoir 1980 and McLaughlin 1982; on Buffon's concept of *moules intérieures* see Roger [1963] 1993, 542–558, and Ibrahim 1987.

85. López-Beltrán, chapter 5, this volume.

86. See Terrall, chapter 11, this volume; Duchesneau, chapter 13, this volume.

87. Rheinberger 1994.

88. Hodge 1985.

89. Roger [1963] 1993, epilogue; Roe 1981; Terrall, chapter 11, this volume.

90. This is the sense in which Michel Foucault diagnosed a passage from a "principle of alliance" to a "dispositif of sexuality" around 1800 (Foucault 1976; see also Foucault 1991).

91. Pethes, chapter 17, this volume.

92. Sloan 1995, 113–114.

93. Braude 1997.

94. Mazzolini 1994.

95. Mazzolini, chapter 15, this volume.

96. Pethes, chapter 17, this volume; White, chapter 16, this volume; Willer, chapter 18, this volume.

97. Müller-Sievers, chapter 19, this volume.

98. Darwin [1868] 1988, 30–31.

99. This view was later canonized in Johannsen 1911; see Rheinberger and Müller-Wille 2004.

100. See Parnes, chapter 14, this volume.

References

Allen, Garland E. 1986. T. H. Morgan and the split between embryology and genetics, 1910–1926. In T. Horder, I. A. Witkowski, and C. C. Wylie, eds, *A History of Embryology*. Cambridge: Cambridge University Press.

Beurton, Peter, Raphael Falk, and Hans-Jörg Rheinberger, eds. 2000. *The Concept of the Gene in Development and Evolution: Historical and Epistemological Perspectives*. Cambridge: Cambridge University Press.

Bowler, Peter J. 1989. *The Mendelian Revolution: The Emergence of Hereditarian Concepts in Modern Science and Society*. Baltimore: Johns Hopkins University Press.

Braude, Benjamin. 1997. Sons of Noah. *The William and Mary Quarterly, 3rd Ser.* 54:103–142.

Browne, Janet. 2002. *Charles Darwin: The Power of Place: Volume II of a Biography*. New York: Knopf.

Buffon, George Louis Leclerc Comte de. [1749–1767] 1971. *De l'homme*. Ed. Michèle Duchet. Paris: François Maspero.

Canguilhem, Georges. [1966] 1991. *The Normal and the Pathological*. Trans. C. R. Fawcett. New York: Zone Books.

Canizares Esguerra, Jorge. 1999. New world, new stars: Patriotic astrology and the invention of Indian and Creole bodies in Colonial Spanish America, 1600–1650. *American Historical Review* 104:33–68.

Carlson, Elof Axel. 1966. *The Gene: A Critical History.* Philadelphia: Saunders.

Churchill, Frederick B. 1987. From heredity theory to "Vererbung": The transmission problem, 1850–1915. *Isis* 78:336–364.

Coleman, William. 1971. *Biology in the Nineteenth Century: Problems of Form, Function, and Transformation.* Wiley History of Science Series. New York: Wiley.

Darwin, Charles. [1868] 1988. *Variation of Animals and Plants under Domestication, Volume II.* Vol. 20. The Works of Charles Darwin. Ed. P. H. Barrett and P. B. Freeman. New York: New York University Press.

Descartes, René. [1701] 1996. Primæ cogitationes circa generationem animalium et nonulla de saporibus. In C. Adam and P. Tannery, eds, *Oeuvres de Descartes,* vol. 11. Paris: Vrin.

Dunn, Leslie C. 1965. *A Short History of Genetics: The Development of Some of the Main Lines of Thought, 1864–1939.* New York: McGraw-Hill.

Elkana, Yehuda. 1970. Helmholtz' "Kraft": An illustration of concepts in flux. *Historical Studies of the Physical Sciences* 2:263–298.

Foucault, Michel. 1966. *Les mots et les choses: Une archéologie des sciences humaines.* Paris: Gallimard.

———. 1976. *Histoire de la sexualité, Vol. 1: La volonté de savoir.* Paris: Gallimard.

———. 1991. Faire vivre et laisser mourir: La naissance du racisme. *Les Temps Modernes* 535:37–61.

Galton, Francis. 1876. A theory of heredity. *Journal of the Anthropological Institute* 5:329–348.

Gayon, Jean. 1995. Entre force et structure: genèse du concept naturaliste de l'hérédité. In J. Gayon and J.-J. Wunenburger, eds, *Le paradigme de la filiation.* Paris: L'Harmattan.

———. 1998. *Darwinism's Struggle for Survival: Heredity and the Hypothesis of Natural Selection.* Cambridge: Cambridge University Press.

Glass, Bentley. [1959] 1967. Heredity and variation in the eighteenth century concept of the species. In B. Glass, O. Temkin, and W. L. Strauss, eds, *Forerunners of Darwin, 1745–1859,* 144–172. Baltimore: Johns Hopkins Press.

Glass, Bentley, Owsei Temkin, and William L. Strauss, eds. [1959] 1967. *Forerunners of Darwin, 1745–1859.* Baltimore: Johns Hopkins Press.

Goody, Jack. 1983. *The Development of the Family and Marriage in Europe.* Cambridge: Cambridge University Press.

Gould, Steven J. 1981. *The Mismeasure of Man.* New York: Norton.

Gregory, Andrew. 2001. Harvey, Aristotle and the Weather Cycle. *Studies in the History and Philosophy of the Biological and Biomedical Sciences* 32:153–168.

Hannaford, Ivan. 1996. *Race: The History of an Idea in the West*. Baltimore: Johns Hopkins University Press.

Harvey, William. [1651] 1847. Anatomical Exercises on the Generation of Animals. In *The Works of William Harvey*. Translated from the Latin with a Life of the Author by Robert Willis. London: Sydenham Society.

Heck, Kilian, and Bernhard Jahn, eds. 2000. *Genealogie als Denkform in Mittelalter und Früher Neuzeit*. Tübingen: Niemeyer.

Hesse, Mary. 1974. *The Structure of Scientific Inference*. London: Macmillan Press.

Hodge, Jonathan S. 1985. Darwin as a lifelong generation theorist. In D. Kohn, ed., *The Darwinian Heritage*. Princeton, NJ: Princeton University Press.

Ibrahim, Annie. 1987. La notion de moule intérieur dans les théories de la génération au XVIIIe siècle. *Archives de Philosophie* 50:555–580.

Jacob, François. [1970] 1993. *The Logic of Life: A History of Heredity*. Trans. B. E. Spillmann. Princeton, NJ: Princeton University Press.

Johannsen, Wilhelm. 1911. The genotype conception of heredity. *American Naturalist* 45:129–159.

Jordanova, Ludmilla. 1995. Interrogating the concept of reproduction in the eighteenth century. In F. D. Ginsburg and R. Rapp, eds, *Conceiving the New World Order*. Berkeley: University of California Press.

Katzew, Ilona. 2004. *Casta Painting: Images of Race in Eighteenth-Century Mexico*. New Haven, CT: Yale University Press.

Keller, Evelyn Fox. 2000. *The Century of the Gene*. Cambridge, MA: Harvard University Press.

Kielmeyer, Carl Friedrich. [1793] 1993. *Ueber die Verhältnisse der organischen Kräfte unter einander in der Reihe der verschiedenen Organisationen*. Ed. K. T. Kanz. Marburg: Basilisken-Presse.

La Berge, Ann F. 1992. *Mission and Method: The Early XIXth Century French Public Health Movement*. Cambridge: Cambridge University Press.

Larson, James L. 1994. *Interpreting Nature: The Science of Living Form from Linnaeus to Kant*. Baltimore: Johns Hopkins University Press.

Lenoir, Timothy. 1980. Kant, Blumenbach, and vital materialism in German biology. *Isis* 71: 77–108.

———. 1982. *The Strategy of Life: Teleology and Mechanics in Nineteenth-Century Germany*. Chicago: University of Chicago Press.

Lesky, Erna. 1950. *Die Zeugungs- und Vererbungslehren der Antike und ihr Nachwirken*. Abhandlungen der Geistes- und Sozialwissenschaftlichen Klasse der Akademie der Wissenschaften und der Literatur in Mainz, Jahrgang 1950, Nr. 19. Wiesbaden: Franz Steiner.

López-Beltrán, Carlos. 2004a. In the cradle of heredity: French physicians and l'hérédité naturelle in the early nineteenth century. *Journal of the History of Biology* 37:39–72.

———. 2004b. *El sesgo hereditario: Ámbitos históricos del concepto de herencia biológica*. México: Universidad Nacional Autónoma de México.

Lugt, Maaike van der. 2004. *Le ver, le démon et la vierge: les théories médiévales de la génération extraordinaire*. Paris: Les Belles Lettres.

———. Forthcoming. Les maladies héréditaires dans la médecine savante médiévale. In M. van der Lugt and C. de Miramon, eds, *L'hérédité à la fin du Moyen Age*. Florence: Sismel, Micrologus Library.

Lugt, Maaike van der, and Charles de Miramon, eds. Forthcoming. *L'hérédité à la fin du Moyen Age*. Florence: Sismel, Micrologus Library.

Lyon, John, and Phillip R. Sloan. 1981. *From Natural History to the History of Nature: Readings from Buffon and His Critics*. Notre Dame: University of Notre Dame Press.

Maienschein, Jane. 1987. Heredity/development in the United States, circa 1900. *History and Philosophy of the Life Sciences* 9:79–93.

Mayr, Ernst. 1982. *The Growth of Biological Thought: Diversity, Evolution, and Inheritance*. Cambridge, MA: Belknap Press.

Mazzolini, Renato G. 1994. Il colore della pelle e l'origine dell'antropologia fisica (1492–1848). In R. Zorzi, ed., *L'epopea delle scoperte*. Florence: Olschki.

McLaughlin, Peter. 1982. Blumenbach und der Bildungstrieb: Zum Verhältnis von epigenetischer Embryologie und typologischem Artbegriff. *Medizinhistorisches Journal* 17:357–372.

———. 1990. *Kant's Critique of Teleology in Biological Explanation: Antinomy and Teleology*. Lewiston, ME: Edwin Mellen Press.

———. 2001. *What Functions Explain: Functional Explanation and Self-Reproducing Systems*. Cambridge: Cambridge University Press.

Moss, Lenny. 2003. *What Genes Can't Do*. Cambridge, MA: The MIT Press.

Müller-Wille, Staffan. 2003. Nature as a marketplace: The political economy of Linnaean botany. *History of Political Economy* 35 (supplement: N. De Marchi and M. Schabas, eds, *Oeconomies in the Age of Newton*): 154–172.

———. 2005. Konstellation, Serie, Formation: Genealogische Denkfiguren bei Harvey, Linnaeus und Darwin. In S. Weigel et al., eds, *Generation: Zur Genealogie des Konzepts–Konzepte der Generation*. Munich: Fink.

Napoleons Gesetzbuch Code Napoléon: *Einzig officielle Ausgabe für das Königreich Westphalen.* 1808. Straßburg: F. G. Levrault.

O'Connor, John J., and Edmund F. Robertson. 1999. Abu Ja'far Muhammad ibn Musa Al-Khwarizmi. *The MacTutor History of Mathematics Archive.* www-groups.dcs.st-and.ac.uk/~history/Mathematicians/Al-Khwarizmi.html.

Olby, Robert C. 1979. Mendel no Mendelian? *History of Science* 17:53–72.

———. 1985. *Origins of Mendelism.* 2nd ed. Chicago: University of Chicago Press.

———. 1993. Constitutional and hereditary disorders. In W. F. Bynum and R. Porter, eds, *Companion Encyclopedia of the History of Medicine.* London: Routledge.

Park, Katharine, and Lorraine Daston. 1981. Unnatural conceptions: The study of monsters in sixteenth- and seventeenth-century France and England. *Past and Present* 92:20–54.

Pestre, Dominique. 2003. *Science, argent et politique: Un essai d'interprétation.* Paris: INRA.

Pinto-Correia, Clara. 1997. *The Ovary of Eve: Egg and Sperm and Preformation.* Chicago: University of Chicago Press.

Pomata, Gianna. 2003. Comments on Session III: Heredity and Medicine. In *Conference: A Cultural History of Heredity II: 18th and 19th Centuries*, Prepoint 247, 145–152. Berlin: Max-Planck-Institute for the History of Science.

Porter, Roy, and Georges Sebastian Rousseau. 1998. *Gout: The Patrician Malady.* New Haven, CT: Yale University Press.

Rey, Roselyne. 1989. Génération et hérédité au 18e siècle. In C. Bénichou, ed., *L'ordre des caractères: Aspects de l'hérédité dans l'histoire des sciences de l'homme*, 7–41. Paris: Sciences en situation.

Rheinberger, Hans-Jörg. 1994. Morphologie bei Claude Bernard. *Aufsätze und Reden der Senckenbergischen Naturforschenden Gesellschaft* 41:137–150.

———. 1997. *Toward a History of Epistemic Things: Synthesizing Proteins in the Test Tube.* Stanford, CA: Stanford University Press.

Rheinberger, Hans-Jörg, and Jean-Paul Gaudillière, eds. 2004. *Classical Genetic Research and Its Legacy—The Mapping Cultures of Twentieth-Century Genetics.* London: Routledge.

Rheinberger, Hans-Jörg, and Staffan Müller-Wille. 2004. Gene. *Stanford Encyclopedia of Philosophy.* Ed. E. N. Zalta. Winter 2004 ed. http://plato.stanford.edu/entries/gene/.

Ritterbush, Philip C. 1964. *Overtures to Biology: The Speculations of Eighteenth Century Naturalists.* New Haven: Yale University Press.

Roberts, Herbert F. 1929. *Plant Hybridization before Mendel*, Princeton, NJ: Princeton University Press.

Roe, Shirley. 1981. *Matter, Life, and Generation: Eighteenth-Century Embryology and the Haller-Wolff Debate*. Cambridge: Cambridge University Press.

Roger, Jaques. [1963] 1993. *Les sciences de la vie dans la pensée française du XVIIIe siècle: La génération des animaux de Descartes à l'Encyclopédie*. Paris: Albin Michel.

Russell, Nicholas. 1986. *Like Engend'ring Like: Heredity and Animal Breeding in Early Modern England*. Cambridge: Cambridge University Press.

Sabean, David Warren. 1998. *Kinship in Neckarhausen, 1700–1870*. Cambridge: Cambridge University Press.

Sloan, Phillip R. 1995. The gaze of natural history. In C. Fox, R. Porter, and R. Wokler, eds, *Inventing Human Science: Eighteenth-Century Domains*. Berkeley: University of California Press.

Spary, Emma C. 2000. *Utopia's Garden: French Natural History from Old Regime to Revolution*. Chicago: University of Chicago Press.

Sterne, Laurence. 1760. *The Life and Opinions of Tristram Shandy, Gentleman*. www.gifu-u.ac.jp/~masaru/TS/contents.html.

Stubbe, Hans. 1965. *Kurze Geschichte der Genetik bis zur Wiederentdeckung der Vererbungsregeln Gregor Mendels*. H. Stubbe, ed., Genetik: Grundlagen, Ergebnisse und Probleme in Einzeldarstellungen, Vol. 1. Jena: VEB Fischer.

Waller, John C. 2002. "The illusion of an explanation": The concept of hereditary disease, 1770–1870. *Journal of the History of Medicine and Allied Sciences* 57:410–448.

Weigel, Sigrid, Ohad Parnes, Ulrike Vedder, and Stefan Willer, eds. 2005. *Generation: Zur Genealogie des Konzepts—Konzepte von Genealogie*. Munich: Wilhelm Fink.

Williams, Elizabeth A. 1994. *The Physical and the Moral: Anthropology, Physiology, and Philosophical Medicine in France, 1750–1850*. Cambridge: Cambridge University Press.

Wood, Roger, and Vitezslav Orel. 1996. *Genetic Prehistory in Selective Breeding: A Prelude to Mendel*. Oxford: Oxford University Press.

Zirkle, Conway. 1935. *The Beginnings of Plant Hybridization*. Philadelphia: University of Pennsylvania Press.

I Heredity in the Legal Context

2 From Clan to Kindred: Kinship and the Circulation of Property in Premodern and Modern Europe

David Warren Sabean

This chapter deals with changes in inheritance and kinship dynamics in Europe over the long term from the early modern period into the nineteenth century. I am going to argue that a shift took place during the second half of the eighteenth century away from a system that conceptualized relations vertically to one that conceptualized them horizontally, from kinship organized through descent and consanguinity to kinship organized around alliance and affinity.

While my intent is to look at broad historical shifts in the articulation of inheritance practices with the dynamics of kinship, there are great complexities and regional differences, of course, that have to be taken into consideration. And there are quite different issues to be dealt with when one takes class into account, since the ways families reproduced themselves in the town or the countryside, on the noble estate or the peasant farm, among officeholders or courtiers, entail considerable differences. A great deal of discussion about early modern inheritance remains at the level of legal norms, and for the purposes of comparison with notions of heredity, the analysis of doctrines, codes, laws, legal decisions, and innovations in the instruments of property holding is of considerable use. Nonetheless, some of the most important new research shows that practice cannot be deduced from legal norms or regional customs and that law can be a very flexible instrument. Understanding the logic of familial relations and shifts in family organization leads us to rethink the doctrinal history of law.[1] Just how norms and practices of inheritance interface with other cultural ideas is an open question, one that will for the most part have to be set aside here, but that remains as a desideratum for future research.

2.1 Restricting Inheritance, 1600–1750

There is considerable evidence that from the late Middle Ages through the seventeenth century at all levels of society a process took place to restrict the number of heirs to property or the circle of people who had access to familial property. This was almost always associated with an accent on agnatic relationships, such that generations of family members became associated with male lineages, patrilines, or clanlike structures defined through links from fathers to sons. This generalization flies in the face of all those regions that practiced equality of inheritance, a point to which I will come back. Yet it seems clear that a systematic institutionalization of primogeniture or inheritance restricted to one child, either the eldest, the eldest son, or a child chosen by the parents, developed in fits and starts from region to region and from class to class, sometimes earlier and sometimes later, but for the most part during the period from the fifteenth to the late seventeenth century. The famous case of peasant families in the Pyrenees or the Massif Central in France is a case in point. There, it has now been established, the practice of *unicité* (inheritance by one child only) was not a product of Roman law or an age-old custom, but a system of practices that developed its final form in the fifteenth and sixteenth centuries.[2] As an example of a region that contrasts with these areas of small holdings and intense civic life, it was in the last decades of the fifteenth and first decades of the sixteenth century that the substantial farms of the south German Upper Swabia region established an inheritance system favoring the youngest son.[3] In the fourteenth and early fifteenth centuries, farms had splintered through a process of partible inheritance into relatively small pieces, a process reversed in all the small territories of the region by the end of the fifteenth century. But first, before settling on a unique son to favor in the inheritance process, some of them made a prior move. From my early days in the archives, I remember the most impressive charter from the monastery of Weingarten that I have ever seen, loaded with a large array of seals, according to which women had never had the right to inherit any farms or parts of farms—the weightiness of the document obviously belying the facts of the situation.[4]

In other words, one of the chief means of restricting inheritance during this period was first to exclude women, a move repeated in different social classes and in different European regions during the period. This does not mean that women were always excluded. For example, in the Gévaudan region, discussed by Lamaison and Claverie, what they call "patrimonial lines" developed among farmholders, whereby no particular heir was at the center of practices of succession but rather the patrimony itself, which sometimes, even in the presence of a male heir, fell to a daughter.[5] What is important to stress, however, even in this case, is the exclusion of all the cadets (i.e., younger children) from succession. Eileen Spring in 1993 surveyed the legal issues and practices associated with entail and the strict settlement in England among the aristocratic and gentry classes.[6] A crucial point to understand is that in the Middle Ages, despite common law favoring primogeniture, families often provided well for younger children, including daughters. But even more significant, with the rules concerning dower and inheritance by females in the absence of a male heir, considerable amounts of property would have fallen into the hands of women: Spring suggests a figure of 40 percent.[7] Yet, from the late Middle Ages onward, the history of property law and familial practice was in the direction of strict primogeniture, patrilineality, and patriarchal rule, with the process taking its final form at the beginning of the eighteenth century.

According to Spring, the "extraordinary decline" in female succession led to familial relations being organized around succession from generation to generation of large estates that only marginally were charged with providing for younger sons, daughters, and widows.[8] Two features of the system should be emphasized. The first is the device of the "strict settlement," whose final form emerged around 1650. Essentially, it was a contract set up between the oldest son and his father on the son's *marriage*,[9] and spelling out the charges in advance for which the estate would be liable for all the children yet to be born. Thus sentiment and testamentary freedom were to be precluded from the beginning (see Vedder, chapter 4, this volume). The second feature derives from the history of dower, jointure, and portion. Essentially dower rights—which amounted to a third of the husband's estate in the Middle Ages—were done away with in favor of a practice whereby the family of the bride provided a portion, to which the groom answered with a jointure, a sum to be drawn on in the case of his earlier death. From the sixteenth to the seventeenth century, the ratio of portion to jointure rose from 5:1 to 10:1.[10] All during the marriage, the husband held the wife's portion and received the income from it. The upshot of this system was to throw the entire costs of maintaining a wife and settling a widow back onto her own family. All of this simply reinforced primogeniture and patrilineal ideas. While changes to this system were fought vigorously by landed families throughout the nineteenth century, the restrictive, patrilineal inheritance practices were inimical to the new capitalist forces and were challenged by nineteenth-century reformers seeking to establish rules of equal division of property among the heirs.[11]

There are parallels among other classes and in different European regions. A similar story can be told about the Saxon princes during the early modern period. As in England, it was actually the contract at marriage that was the crucial instrument in deciding the property rights of women. Ute Essegern has studied fifteen marriages within the ruling house of electoral Saxony during the seventeenth century.[12] In each case, along with a marriage settlement that spelled out in detail exactly how the portion (*Ehegeld*) and jointure (*Morgengabe*) were to be accounted for, the princess provided a *Verzichtserklärung*, a document giving up all her rights to inheritance and succession in her home territory. Essegern suggests that the amount of the portion provided by her family was related to the significance of the rights she was giving up. In any event, the exclusion of women at the moment when they married out was a device to support the concentration of property and regalian rights among male heirs, and just as with rural Upper Swabia and aristocratic England, the restriction of female rights in the interest of patrilineal principles preceded the move to primogeniture or single-son inheritance. There is another strong parallel with the English example. What the family of the bride provided as a portion far outweighed what the groom provided as jointure, although in Saxony the wife continued to be entitled to a dower (*Leibgeding*). This dower simply corresponded in value to the interest on the wife's portion, originally provided by her family. Taking the data from all fifteen marriages, it is clear that as among the English aristocratic families, the wife's family essentially absorbed all the costs of the wife's establishment and dowerage.[13]

Bernard Derouet has been carrying out a series of detailed comparative studies of inheritance practices, family organization, and communal practices in rural France. His work on the southern French regions of impartible inheritance emphasizes a number of features already familiar to us from English and German aristocratic practices. He has also shown that areas of equal male inheritance (with the exclusion of women) like the Bourbonnais are to be seen as having more in common with impartible than the partible systems. After all the differences and distinctions are accounted for, these areas of restricted inheritance have in common the exclusion of the endowed child from rights to inheritance or a share in the economy of the familial enterprise, and once again, the crucial document was the marriage contract. Both the systems of *héritage integrale* and equal male inheritance took their final forms at the beginning of the early modern period, in some places concluding the process of transition only in the seventeenth century. The effect was to create strongly patrilineal forms of succession, to prevent the partitioning of land, and to organize social relations around a property that persisted over time, giving cohesion to social memory and form to the circulation of goods and people in the neighborhood and region.[14]

While there are significant parallels in the different examples offered here, there are also significant differences. In the mountainous areas of France, "place" was at the center of the system. It was the farm enterprise that determined the dynamics of family, rather than the particular connection through kinship. In fact, Derouet claims, the principle of the French impartible inheritance systems is "residence," which he contrasts with "filiation" in the systems of equal inheritance. The "house" is the key player, with ascribed obligations and exchanges with certain other houses carried along through time, irrespective of the particular kinship relationships and alliances of the moment. The house gives the name to its members and offers the key structural support for both its members and its interactions with other similar units.[15] Contrast this with new constructions of blood that appear among English landowners in the sixteenth century, where the bilateral confluence of different lines becomes obscured by the emphasis on the male line, such that aristocratic and gentry men conspire to disinherit their daughters in favor of their collateral male kin. Symbolically, at least, this comes down to the name that has to be passed on. By the eighteenth century, the terms of the strict settlement required that the male successor change his name in order to assume the inheritance.[16]

These considerations suggest different ways of conceiving of inheritance, the constitution and configuration of line and lines, the symbolics of blood, and the ends and strategic concerns of property devolution. One can distinguish three different disciplinary ways of conceiving of intergenerational connection, devolution of substance, and kinship relation: the theological, the medical, and the legal. To take the theological model first: In the early modern period, throughout Catholic and Protestant Europe (with England as an exception as far as law but not practice is concerned), there were widespread marital prohibitions, forbidding marriage within quite an extensive range—second or third cousins.[17] One had to marry outside, with someone who was "un-familiar," someone outside the group descended from great- or great-great-grandparents. This is a negative way of describing those to which one had recognized ties of obligation. This theological representation developed a method of calculation that essentially spoke of shared substance, which diminished with distance but which gave no priority to relatives from the maternal or the paternal side. The medical discourse of the time offered both support and contradiction to this theological reasoning.[18] Galenists thought that in the process of generation, both males and females produce seed. *Sperma* was substantially blood, so intercourse was conceived of as a mingling of blood. This position fits quite well into the theological assumptions by suggesting equality of inherited substance. In contrast to this, the Aristotelian manner of representing generation fits much more clearly into a patrilineal model. It was constructed in terms of dissimilarity of substance, for the man contributed form through his sperm—understood as active, as idea, as thought—and the woman produced matter, blood, a passive substance on which the male sperm acted. These medical models offered different ways of thinking through connections in the different regimes of inheritance, and in at least one theological argument, these different ways seem to have played an important role. Galenists marshaled arguments against marriage with the wife's sister and brother's wife as essentially the same relationship, while Aristotelians thought that they were quite dissimilar and were far more ready to give up the prohibition of the wife's sister than they were that of the brother's wife. They objected to the *confusio seminis* of two brothers in one receptacle. Finally, the legal codes, instruments, and customs could work along models similar to the theological or the medical ones.[19] What needs to be underlined, however, is that law could be a very flexible instrument for the purposes of familial policies—the same legal norm could be used in quite different contexts to quite different ends. On the other hand, in Eileen Spring's account of the English law of property, for example, attempts to solve the problem of primogeniture while caring for the younger children led to constant innovations in legal instruments and judicial opinions.[20]

This flexibility becomes particularily evident when we turn to partible systems of inheritance. It is apparent that dividing property among all the children regardless of sex, models the flow of property in a parallel manner to the way theologians modeled the flow of blood or substance. While a unilineal way of reckoning can think of relationships passed along through one sex, systems of hereditary equality ought not, in principle, to be able to isolate such a group. Kinship terminology here introduces notions of an "ego-focused" kinship group, a kindred or *parentèle*—that is, the set of relatives proceeding out from a particular individual, each kindred being unique, with no two people having exactly the same one.[21] This contrasts with a model of kin descended from a single ancestor who are conscious of themselves as a group and who can be designated as a "clan." Many of the legal definitions and instruments for the administration of partible inheritance gained clarity and system during the late Middle Ages and the early modern period. In northern France—for example, where broad regions followed practices of equal inheritance—the customs were collected and published in response to royal decree during the fifteenth and sixteenth century.[22] To give a German example, a sixteenth-century Württemberg commission compiled all the customs of every village of the realm and ostensibly then created a unified law of inheritance based on an abstraction from the differences or synthesis of their commonalities.[23] In both cases, however, we know now that the very act of collection repressed many details, created uniformity where none had existed before, and gave much opportunity for intervention on the part of lawyers. The homogenization of practices as cultures of devolution developed took place both in partible and impartible inheritance regions in the transition to the early modern state apparatus.

There was one widespread instrument of family property policy to be found throughout partible inheritance areas in Europe that emphasized the collective right of descendants regarding the ownership of lineage property, namely, the *retrait lignager*.[24] The systematic development of this institution, like many of the things discussed so far, began in the late Middle Ages and spread throughout the early modern period. Essentially the right of retrait offered collateral kin the right to preemt the purchase of a strip of land or a building that was being offered for sale by a relative but—and this is important—at the market price. There was no hint in the practice of the claim to a particular family price. The right simply gave all those who descended from the original owner and who might have inherited the particular plot the right of access to ownership. Although the custom is to be understood as providing mutual familial rights to property along lines of descent, it did not restrict individual rights to ownership. In Württemberg during the course of the eighteenth century, a completely rational market for land developed, with any strip offered for sale at three successive village auctions. It was only after the top bid had been accepted that a relative could intervene and purchase it at the stated price. And, of course, it was only land that had descended through inheritance that was subject to such purchase. Anything bought on the market and resold had no kinship rights attached to it. To deepen the consideration of systems of equality, it is useful to make a clear distinction between inheritance and succession. While inheritance might be subject to equality, succession might not. For example, in a powerful or well-to-do peasant kindred, only one member might sit on the local council or court or serve among the magistrates. And there might well be more informal functions or positions subject to succession by only one member of the family. Furthermore, in many situations siblings, cousins, and siblings-in-law sold their bits of inheritance to someone in their group, who might be the designated successor to the father, the one chosen to serve among the magistrates, or the one to concentrate on farming, while others sought careers in other occupations. Regions where equality of inheritance emerged seem most often to have offered multiple occupations and a lively land market, which gave the opportunity both to create stability of a line (many families with many generations occupying the same positions as magistrates or village headmen) and the founding of new branches. Mobility in such a situation went together with well-integrated kindreds.[25] So even in regions of equality, where the flow of property came down through males and females, there were institutions that developed from the late Middle Ages onward to recognize lineal thinking, to crystallize out groups of agnatically related kin, and to make succession in the male line possible.

A comparative study of all of the ways that families concentrated succession in Europe remains to be carried out. But in surveying the different examples offered so far, it ought to be clear that one of the breaks between the Middle Ages and the early modern period is the rise of a series of mechanisms to ensure the regulated descent of property and to define more and more carefully how the rights of inheritance and succession were to be sorted out. The trajectory for many regions and strata was to restrict the number of siblings who had access to the main family goods (see De Renzi, chapter 3, this volume). There are various ways that this could be carried out, although even without clear rules, many groups developed familial decision-making practices that put the patrimony at the center of their strategies.[26] Essentially family policy was oriented toward allowing just enough males to inherit as the economic conditions and the property mass allowed. The general trend among aristocratic families in Europe was to concentrate property in such a way that daughters had ever fewer claims on the substance of the estate and to keep the bulk of the property under the governance of one male heir.[27] In some ways, the practice was most rigorous in Spain, where the form of entail known as fidei commissum goes back essentially to the beginning of the sixteenth century and made its way to Austria around 1600 and in the course of the seventeenth century to Hungary.[28]

Heinz Reif and Christoph Duhamelle have studied the nobilities that controlled the eccelesiastical territories in northwest Germany and in the Rhineland.[29] Here, in the course of the seventeenth and eighteenth centuries, the access to ecclesiastical preferment and office in the territories became restricted to people of noble birth. And over time, what constituted noble birth became itself ever more restricted. At the height of the system, any individual claiming noble status had to demonstrate sixteen quarterings—that is, all of his great-great-grandparents had to have been nobles. A feature of all of southern Europe was the development of urban patrician classes, which restricted access to one son of a family and insisted on noble status. In Milan, for example, a Congregation of Orders was established in the sixteenth century to monitor admission to the patriciate.[30] By the early eighteenth century, it was necessary to show a hundred years of continuous family residence as well as noble origins.

To sum up the trend: official and ecclesiastical families organized a great deal of their social exchanges around goods that they controlled. Increasingly from the late middle ages onward, urban oligarchies or merchant companies developed, which controlled the access to power, position, or resources largely through descent and succession. That did not mean that there was no mobility, but most mobility was a matter of the slow multigenerational rise of families who carefully used their resources to further the family as a whole. The chief point here is that descent, succession, and inheritance determined life chances to a large extent and that families were organized around what one can conceptualize as "stable" properties. In such a situation, the word *clan* comes to mind, and sometimes the language of the time—*Sippe*, *race*, clan—suggests in one way or another social relations of a descent group. Clan emphasizes descent from a common ancestor but does not preclude hierarchalization. In the English case, for example, younger sons and cousins were waiting in the wings to assume the family property, and Eileen Spring suggests the notion of "blood" as the key conceptual metaphor for the family. In the German ecclesiastical territories, the successors to noble properties and their clerical brothers, uncles, nephews, and cousins coordinated their familial resources under the leadership of the estate holders. Gérard Delille, in his exhaustive study of kinship, office, and property in southern Italy, has shown how cadet branches acted as clients to the older branches.[31] Urban patriciates and merchant monopolists throughout Europe created tightly organized political and financial relationships based on principles of "downward devolution." In peasant areas, such as the Gévaudan, the property of cadet branches circulated back to the main branch after several generations, while the main house kept the junior branches in clientage. As for the issue of women and the main estate, among English and German aristocracies, considerable amounts of property were hived off in dowries to maintain the out-marrying daughters/sisters. The rise of primogeniture and other forms of agnatic descent should not be interpreted in an individualistic sense, yet it is clear that more research is needed to understand how the new dynamics of succession choreographed individual family members in an elaborate dance around the family estate.

2.2 Alliance and Affinity, 1750–1870

Many of the support structures for social reproduction described in the previous section were taken away or no longer continued to play the same role beginning in the decades around 1800, although the origins of the shift in many areas go back to earlier decades in the eighteenth century. Again, this was a long-drawn-out process that affected different classes and different regions at different rates. To risk overgeneralization, I would characterize this shift as one from vertical to horizontal relationships, from "clan" to "kindred," from status to contract, from families organized around stable properties to families organized around capital, from downward devolution to horizontal exchange. It is not, of course, that people no longer thought in terms of descent. In fact bourgeois genealogical investigation only burgeoned during the nineteenth century, and one can find a continuing language of line or lineage in the literature of the period (see Vedder, chapter 4, this volume). Still, the weight of interaction shifted to a broad interaction of kin related by blood but especially by marriage. Affinity was indeed "elective," with elements of choice, opportunity, familiarity, and desire. Knowledge of horizontally linked kin grew considerably during the period. And the new marriage system set up linked families (patrilines), which interacted socially in newly intense ways.

My suggestion is that this also has a great deal to do with property. The new states coming out of the French Revolution or the Congress of Vienna reorganization of the German map did away with property in office. No one any longer had a right to office by descent or inheritance. Nevertheless the bureaucratic class reproduced itself, this time by furthering the ambitions of its young men through a network of interrelated families, promoting and placing them by arranging access to education, contacts with powerful individuals, and economic support during the long climb into secure positions. In a similar way, the new conditions of a capitalist economy changed the way families related to property. The issue was no longer to keep an estate intact but to grasp new opportunities offered by financial markets, trade, and commodity production. The new entrepreneur can only be understood in terms of the accumulation of capital through a network of kin. One can see, for example, with the mining and manufacturing firms on the Rhine, that each generation backed several young men, reorganizing their families around those who proved successful. Families created far-flung networks, exchanging children for education, as service personnel, and eventually as managers and owners of firms. If *clan*, based on unilineal descent, is the term that comes to mind for the earlier period, then *kindred*, comprising bilateral and allied kin, seems to work best for the nineteenth century. It is in this context that the new conditions of wealth and property thrust the weight of family dynamics toward alliance and affinity. It was then that certain forms of critique against nepotism and corruption and against primogentiture and entails became politically effective (see Vedder, chapter 4, this volume).[32]

This reorganization of kinship and the relationship of the family to property is complex and many-sided. Here I want to take up only two aspects in order to illustrate some of the issues of the change from an early modern to a modern system of alliance. One is the rise in endogamy from the middle decades of the eighteenth century and the other is a curious shift in the representation of incest during the same period. Examining these two related phenomena will allow us to return at the end to questions of property management and family dynamics.

Again, the best way to proceed is with examples, in this case one from a south German village (Neckarhausen) and the other from rural Naples.[33] In these two examples, from quite different cultures and environments, the same kind of kinship reorganization took place in the later part of the eighteenth century, characterized in the first place by the rise of endogamous marriages, repeated alliances between the same families (thought of in terms of patrilines), perhaps best described in a shorthand way as "cousin" marriages. Taking the beginning of the eighteenth century as the starting point for analysis of Neckarhausen, I have found that villagers indeed often married people linked to them through kinship but only through networks of affines. This form, found all over Europe, has been given the name of *rechaining* by Zonabend and has been brilliantly analyzed by Segalen, Derouet, and Delille, among others.[34] What never occurred in Neckarhausen in this earlier period was marriage among consanguines (or blood relatives). There were no first-, second-, or third-cousin marriages, no marriages to close affines (wife's sister, brother's wife, brother's wife's sister, sister's husband's sister), and none to blood relatives of a deceased spouse. Thus, alliances struck in one generation could neither be replicated in the same generation nor in the next one, two, or three. Marriage typically connected families of differential wealth by connecting brothers-in-law to each other as patrons and clients. This marriage system can be labeled *exogamous* in two senses: it coupled partners who were not related to each other by blood and it made systematic alliances between families over time impossible, continually breaking up class solidarity.

This structure contrasts markedly with that at the beginning of the nineteenth century. The population had doubled (to 740), with the class of artisanal producers and farm laborers growing considerably. Many villagers became involved in wage-dependent labor outside Neckarhausen, and the village was undergoing agricultural intensification and capitalization. With the new mobility came the possibility of marrying outside the village in a much larger "field." It was easier to avoid relatives and kin in a larger village and a context of mobility, yet by this time a tight, endogamous marriage system—in both senses of the word—had been constructed. Instead of wealthy villagers marrying poor as before, newlyweds came to match their properties more or less exactly.

We can take marriages from the 1820s as an example—25 percent were with kin. By the 1860s 50 percent would be with kin. The difference with the early eighteenth century had to do with the rise of consanguineal kin as marriage partners. Already in the 1740s and 1750s, second cousins married each other, a practice well in place by the 1780s. By 1800, many people sought out first cousins. Marriages between second cousins reproduce alliances struck by grandparents, while those between first cousins repeat exchanges made in the parents' generation. Also new were marriages with the deceased wife's sister, other close affines, and sibling exchange. By the 1820s only one family in ten failed to renegotiate an already-established alliance.

Exchanges took place within a pattern that emphasized descent traced through men. The alliances were between "patrilines," but there was no particularly preferred marriage: members of two patrilines made several marriages over a short space of time, which called for another spurt of exchanges a generation or two down the line. The point was to strengthen and redouble social ties that would atrophy if not recast. The system was open and flexible and allowed for reproducing relationships already constructed and for the construction of new ones. Such a tight endogamous pattern of alliances is "modern," not archaic, certainly in the sense of being developed during a period of capitalized agriculture and wage labor. And it was closely tied to the formation of class relations in the village. Class differentiation went together with kin integration. Endogamous marriage alliances reproduced from generation to generation took place in a context of population rise, capitalization and intensification of agriculture, class differentiation, regional mobility, integration into wider markets, and decimation and increasingly rapid turnover of landed property.

There are several reasons for these changes, among which was the rise of a vigorous market in land, which worked as a distribution mechanism to integrate a developing set of relations among consanguineal kin.[35] Given the intestate rules, which narrowed rights to lineal claimants, inheritance was a poor means for integrating kin in a situation of extreme fragmentation of land and growing stratification. The shift from an inheritance-driven system to one balanced more by looser exchanges of marriage partners and land among allied kin gave a familial function to the market. In the situation of an ever-growing market, networks between landholders were necessary to control and channel the access to resources. The development of cousin marriages among the landholders coincided closely with the opening up of the market. Allied kin developed coordinating linkages among themselves to control the distribution of resources. Around 1700, only about 10 percent of property transfers were outside the immediate family, and cousins played no role in the market. By the 1820s as much land changed hands over the market as through inheritance. Cousins were major players. Around 1800 artisans and construction workers in their turn began to adopt the new system of alliances. At a time when landholders began to marry first cousins, they began to marry second cousins. Here there was a difference: they tended to marry cousins in other villages, developing widespread geographic networks adapted to the conditions of mobile labor.

With the outline of the history of kinship in Neckarhausen as our guide, I want to show that what happened there was not unusual in a European context. I have chosen for comparison a study of rural Naples by Gérard Delille, who has examined closely villages in the Valley of the Irno, a region of small and medium peasant cultivators producing wine, citrus, and other fruit.[36] In the area under consideration the shift from an exogamous to an endogamous marriage alliance system followed a similar course to that of Neckarhausen. Delille describes two essentially different systems of marriage alliance for this region, one for the late fifteenth to the end of the seventeenth centuries and another developing progressively in the eighteenth century, emerging full blown in the nineteenth. As in Neckarhausen (and Germany as a whole) until the eighteenth century, the earlier system was characterized by negative rules—one was not allowed to marry within a set, wide range of kin. But here the same phenomenon of "rechaining" that has been demonstrated for many areas in France and in Neckarhausen could be found. While developing linked houses through marriages along affinal chains could create solidarities in one generation, with the exogamy rules of the Church, repeated alliances down the generations were precluded.

In the Kingdom of Naples, the rise of consanguineal marriages went together with the destruction of lineages at the same time as the same kinds of marriages ended the older form of clientage and brought about the restructuring of village politics in Neckarhausen. Delille shows that the system of lineages that had expanded and solidified during the sixteenth century and into the first half of the seventeenth century began to break down during the period of economic and demographic crisis between 1680 and 1730. By 1798, the old system was simply no longer there, having been replaced by isolated families and groups of no more than two or three hearths. Along with this change went ever-increasing rates of consanguinity and the selection of ever-closer consanguines as marriage partners. One even finds examples of uncles marrying nieces. And the restrictions regarding close affines disappeared as well—many men married sisters of their deceased wives.

Delille puts the change in kinship dynamics into the same context as has been found for Neckarhausen. For one thing, the market in land expanded considerably, and the price per unit of land rose to unheard of heights. As far as the possessing classes were concerned, the new endogamy was designed to prevent dispersion of property, but the poorer classes abandoned the old forms of reciprocity as well. Delille says that the phenomenon can be interpreted as the sign of a more and more profound fracture separating the different social classes. Propertied groups married in a more and more restricted circle of kin in order not to disperse and divide wealth.

What the material from the Kingdom of Naples and Neckarhausen shows is that the new class/kin endogamy was designed to provide multiple forms of exchange and the broad coordination of a class in its effort to manage credit, land markets, officeholding, and corruption, all of which could only have been done by real but flexible structures and a well-coordinated system of reciprocities.[37] We have, then, two contrasting systems that succeeded one after the other—one built around clientage and vertical integration of groups ("clans") and one built around class and horizontal integration, perhaps no longer of "groups" but of flexibly coordinated strata (kindreds).

These two similar narratives connecting two widely separated places suggest questions to ask about long-term trends in Europe as a whole. The rise of endogamy is less well documented than its fall around 1900. Most of the relevant studies, often carried out by biologists and geneticists, begin their analyses at some point in the late nineteenth or early twentieth century.[38] All of the studies concur in the description of a high point in consanguineal marriages reached between 1880 and 1920, with a regular and sometimes abrupt decline to a point in the 1950s when such marriages became insignificant almost everywhere. A detailed study of four Pyreneen villages based on a complete family reconstitution showed that until 1790 there were few consanguineal marriages. After that, rates rose to a peak between 1890 and 1914.[39] Kühn's study of two Eifel villages found a similar rise.[40] Alström studied first-cousin marriages in Sweden as a whole between 1750 and 1844 based on dispensation records (dispensations were first allowed in 1680). Before 1750 rates were under 0.2 percent. They rose to 1 percent in 1800 and reached 1.5 percent in 1844 when the requirement to get a dispensation was abrogated.[41] Jean-Marie Gouesse has examined the rates of endogamy for Catholic Europe during the early modern period through to the twentieth century on the basis of papal and episcopal records of dispensations.[42] Until the end of the seventeenth century there were few. The rates of dispensation only rose rapidly in the eighteenth century. Taking three dates 100 years apart (1583–84, 1683–84, 1783–84), the ratios were 1 : 11 : 55. But the trend continued. Between the 1760s and 1860s, for example, the rates for France increased elevenfold. Gouesse's summary of the trend seems right—in Europe endogamy rose toward the end of the eighteenth century, reaching a high point between the mid-nineteenth century and World War I, and fell rapidly from the 1920s on.

Taking all the data together, it appears that for Catholic and Protestant Germany, Catholic Italy, Spain, France, and Belgium, and Protestant Sweden and Norway, the overall trend in the rise of endogamous marriage was similar. There was no statistically significant endogamy before the eighteenth century anywhere in continental Europe. The 1740s was the crucial decade in Neckarhausen for the change. In Sweden, France, Neckarhausen, and the four villages in the Pyrenees studied by Bourgoin and Khang, cousin marriage and other forms of close kin alliances were well in place by 1800 and only rose in numbers to reach a peak sometimes earlier and sometimes later but most often during the 1880s. Whatever relationship one uses to track the rise (uncle/niece, brother-/sister-in-law, first cousins, affines), the overall trend appears to have been the same throughout wide areas of Europe. However, different areas, different occupational groups, and different classes created forms of alliance quite different from each other. While some may have relied on reiterated first-cousin exchanges, others made use of more extended consanguines, and still others integrated kindreds through highly flexible forms of affinal alliance.[43] Published rates of consanguinity for urban areas follow the same trend but are consistently lower than for rural areas. The most famous such study was carried out by George Darwin (married to a cousin and offspring of a cousin marriage) on London and English towns in 1875. He established rates of around 4 percent for first-cousin marriages for the middle classes, peerage, and landed gentry, much higher than for the urban classes as a whole—between 1.5 and 2 percent.[44] All of these forms began to be utilized in the eighteenth century and became crucially important for social organization in the nineteenth century—at different rates but everywhere.

2.3 Sentiment and Incest

In restructuring the alliance system, new mechanisms had to be put into place to channel familial energies and regulate socially sanctioned marital choices, which brings us to the topic of the representation of incest. With the new alliance system, families became the focal point for developing sentiment, managing cultural style, and directing erotic desires. Socialization into the aesthetics of choice was all the more important, given the fundamental problem of managing the flow of capital in the system of alliance. It would seem that the period from 1740 to 1840 was one where brothers and sisters schooled themselves in sentiment and developed for each other a language of passionate affection and love.[45] Attachment for a future spouse grew out of feelings and moral style developed among siblings or sets of cousins who grew up together.[46] The incredible outpouring of correspondence among pairs of siblings during the period offers us insight into the practices of the new intimacy.[47] So too do the scads of novels, epic poems, plays, and theological treatises concerned with sorting out legitimate and illegitimate feelings brothers and sisters shared.[48]

Quite a few commentators around 1800 tried to work out the differences in feelings toward the sister and the wife. Some put the issue in Kantian terms, suggesting that with one's wife there was always an objective moment that instrumentalized the relationship. The theologian Carl Ludwig Nitzsch thought that that element was sex.[49] For him, the sexual drive was completely selfish, but he further thought that sexual desire developed only after a benevolent disposition was formed within the family—setting up proper objects of desire. And he makes two points. First, marriage in order to reduce its instrumental core ought to consider the alliance of two families, implying ties binding people from within the same cultural milieu. Second, whatever comes out of a marriage, the tenderness between spouses never attains the level of intensity characteristic of siblings. Love between a brother and a sister is the model of purity, of selflessness, of a relationship as end in itself.

A crucial part of the new discourse about sentiment involves a consideration of marriage as something that takes place among people on the same cultural, class, and stylistic plane, a union of true equals and true intimates. If exogamy rules of the seventeenth century enforced marriage with the "stranger," the new dynamics involved a search for the familiar, an attachment to a mirroring self. The developing brother-sister imaginary in the context of the shift in kin relations from vertical to horizontal underscored a system of marriage exchange that stressed homogamy—the search for the same rather than the other.[50] The intense structuring of new social milieus through reiterated social and cultural exchanges of allied families made cousins objects of desire—often cousins raised in the same household. Work by Christopher Johnson on a large French bourgeois family network pushes the analysis in this same direction.[51] In the extended family he studied, the rise of close, erotically charged brother-sister ties provided a new central focal point for familial dynamics, and the language of cousinship became conflated with that of siblingship. One sister (whose letters of longing for her brother bordered on the incestuous, according to Johnson) wrote to her brother about his impending marriage to their cousin: "habituated from your childhood to your chérie as a sister and she loving you as a brother, you have developed an affection that can only end with life itself." Later in their marriage, the cousin/wife addressed her husband in her letters as "my love, my friend, my spouse, my brother."

The issue of "same" and "other" is a very complex problematic for the period. At the same time that cultural milieus structured around the dynamics of interconnected families developed—those milieus within which marital choice was shaped and desire given focus—another discourse began to model male and female in terms of otherhood, although both were often understood in correlation with each other or seen as two parts of a necessary unity.[52] Hegel's contrasting of sibling and marital relations can stand for many other texts from the period. I see them as symptomatic of a social and cultural situation where like was seeking out like, where ever more active familial life provided cultural sites for the formation of desire, and where schooling in emotion and sentiment connected the problematic of sister and wife together. This is precisely what occurs in the middle of Hegel's *Phenomenology*. It is widely known, I think, that the relationship between Hegel and his younger sister was very intense. Just after his marriage at forty, she had a nervous breakdown and was in an asylum for more than a year. Soon after Hegel died, she wrote a letter to his widow about Hegel's childhood and personality development, then committed suicide. It seems to me that this is foreshadowed in Hegel's discussion of wife and sister in the chapter on the "ethical world." He suggests that the emotional tie for a woman was to marriage itself, but hardly to the particular husband in question—at least in an "ethical" household. Or perhaps feeling is not really the issue, for her relationships "are not based on a reference to this particular husband, this particular child, but to *a* husband, to children *in general,*—not to feeling, but to the universal." Everything is different, however, with respect to the brother, because there is no sexual desire that disturbs the recognition of self: "The moment of individual selfhood, recognizing and being recognized, can here assert its right, because it is bound up with the balance and equilibrium resulting from their being of the same blood, and from their being related in a way that involves no mutual desire. The loss of a brother is thus irreparable to the sister, and her duty towards him is the highest."[53] In this passage, Hegel is essentially making the same point as theologian Nitzsch, putting the stress on the horizontal and sentimental relationships, rather than on the vertical and hierarchical ones. He has to be seen in the context of redirecting the flow of property and learning to manage the complexities of a reconfigured system of alliance.

Perhaps a systematic study of inheritance in the nineteenth century should deal with such issues as the French civil code, which established equal inheritance as a norm (resisted in complex ways in parts of France); the attack on entail by bourgeois reformers in England and the disentailing laws after 1820 in Spain; the marriage of heir and heiress in the Pyrenees or Gévaudan, which violated the old rules of exogamy and the lineal stability of houses; or the abolition of the sale of offices in France (see Wilson, chapter 6, this volume). It seems to me that however practices, institutions, and laws might be at the center of analysis, looking at the way property circulated both between generations and within generations presents a larger context in which to examine the issues (see Vedder, chapter 4, this volume). And a further issue to be pursued is changes in the nature of property and wealth. At the core of the problem of both kinship and property in the cultures of early modern and nineteenth-century Europe are concepts of lineage, race, clan, tradition, affinity, alliance, kindred, and many other terms that capture ways of organizing both thinking and practices on vertical and horizontal grids. It was not just landholders who developed strategies of lineage in the early modern period but officeholders, merchants, pastors, university professors, priests, and a host of others who organized themselves around different forms of property, which they attempted to maintain through familially controlled access. Modernization by creating new forms of wealth and rationalized state management did not undercut family and kin attempts to control the flow of resources, but helped foster new strategies of alliance and a new ideology of affinity stressing horizontal rather than vertical ties. This process put more weight in social reproduction on grasping new opportunities than on devolution of fixed estates.

Notes

1. Just this point is made in Spring 1993, 3–4.

2. Derouet 1995, 675–676, 678, 685–686; Derouet 1993, 467; Derouet 1998, 228–229.

3. This is discussed in Sabean 1972.

4. I did not note down the reference at the time, but the document is found in the source collection (*Bestand*) of the monastery housed in the state archives in Stuttgart.

5. Claverie and Lamaison 1982. See the detailed analysis of their material in Sabean 1998, 407–416.

6. This is the argument throughout Spring 1993.

7. Spring 1993, 93.

8. Spring 1993, 144.

9. Spring 1993, chapter 5.

10. Spring 1993, 50–52.

11. Spring 1993, 66–67.

12. Essegern 2003, 116–134.

13. Essegern 2003: I have made the calculations on the basis of the figures she provides on pp. 123–125.

14. Derouet 2001, 350–359; Derouet 1997b, 89–90; Derouet 1998; Derouet and Goy 1998; Derouet 1993; Derouet 1989, 174–177, 191–196; Derouet 1995, 685; Derouet 2003, especially 32.

15. Derouet 1989, 176, 191–192; Derouet 2001, 352–353; Derouet and Goy 1998, 119–120.

16. Spring 1993, 95–97, 149.

17. On the issues of marriage prohibitions and the reckoning of kinship, see Sabean 1998, 63–89.

18. For the issues dealing with notions of blood, see Laqueur 1990, 35–43.

19. See De Renzi, chapter 3, this volume. For suggestive ideas on law, see Derouet 1997a.

20. Spring 1993, 29, 67–69, 142–143, 179.

21. See the remarks on "kindred" and "ego-focused" groups in Sabean 1998, 21, 43, 45, 85, 88, 417–419, 421.

22. Derouet 1997a, 370–372.

23. Hess 1968.

24. Derouet 2001, 338–344; Sabean 1990.

25. On these issues, see Derouet 2001, 348–349, 359–363; Derouet 2002, 305–317; Derouet 1997b, 76–78, 80–83, 85–87.

26. Hurwich 1993.

27. Habakkuk 1953, 3.

28. Carr 1953; Schenk 1953; Macartney 1953.

29. Reif 1979; Duhamelle 1998.

30. Roberts 1953.

31. Delille 2003.

32. The argument here is dealt with at greater length in Sabean 1998, chapters 2, 22–23.

33. The complex argument outlined here is found in Sabean 1998. Delille's work is summarized and analyzed in Sabean 1998, 399–407.

34. Zonabend 1981, 311–318; Jolas, Verdier, and Zonabend 1970, 17–22; Segalen 1991.

35. The land market in Neckarhausen is treated at greater length in Sabean 1990. Compare the analysis in Derouet 2001.

36. Delille 1985.

37. It was in the new interrelationships of family and economy that issues of heredity, inheritable traits, and creative energies were worked out. See the suggestive ideas along these lines in White, chapter 16, this volume.

38. A complete bibliography is found in Sabean 1998. For a few select studies, see Orel 1932; Saugstad 1977; Calderon 1983; Abelson 1979; Sutter and Goux 1962; Serra and Soini 1959.

39. Bourgoin and Khang 1978.

40. Kühn 1937.

41. Alström 1958.

42. Gouesse 1986.

43. Segalen 1991.

44. Darwin 1875.

45. Among many other writings on the subject, see Prokop 1991, especially vol. 1, 52–53, 78–85.

46. Sabean 1998, 449–508.

47. A good example from among many is offered by Reiff 1979.

48. Titzman 1991.

49. Nitzsch 1880.

50. This is discussed at length in Sabean 1998, chapters 21–23.

51. Johnson 2002.

52. Hausen 1976.

53. Hegel [1807] 1949, 476–477.

References

Abelson, Andrew. 1979. Population structure in the Western Pyrennees II: Migration, the frequency of consanguineous marriage and inbreeding, 1877 to 1915. *Social Biology* 26:55–71.

Alström, C. H. 1958. First cousin marriages in Sweden 1750–1844 and a study of the population movement in some Swedish sub-populations from the genetic-statistical view point. *Acta Genetica* 8:295–369.

Bourgoin, Jacqueline, and Vu Thien Khang. 1978. Quelques aspects de l'histoire génétique de quatre villages pyrénées depuis 1740. *Population* 33:633–659.

Calderon, R. 1983. Inbreeding, migration and age at marriage in rural Toledo, Spain. *Journal of Biosocial Science* 15:47–57.

Carr, Raymond. 1953. Spain. In Albert Goodwin, ed., *The European Nobility in the Eighteenth Century: Studies of the Nobilities of the Major European States in the Pre-Reform Era*, 43–59. London: Adam and Charles Black.

Claverie, Elisabeth, and Pierre Lamaison. 1982. *L'impossible marriage: Violence et parenté en Gévaudan, XVIIe, XVIIIe et XIXe siècle*. Paris: Hachette.

Darwin, George H. 1875. Marriages between first cousins in England and their effects. *Journal of the Statistical Society* 38:153–184.

Delille, Gérard. 1985. *Famille et proprieté dans le royaume de Naples (XVe–XIXe siècle)*. Bibliothèque des écoles françaises d'Athènes et de Rome 259. Démographie et société 18. Rome: Écoles françaises de Rome/Paris: École des hautes études en sciences sociales.

———. 2003. *Le maire et le prieur: Pouvoir central et pouvoir local en Méditerranée occidentale (XVe–XVIIIe siècle)*. Bibliothèque des écoles françaises d'Athènes et de Rome 259, 2. Civilisations et Sociétés 112. Rome: Écoles françaises de Rome/Paris: École des hautes études en sciences sociales.

Derouet, Bernard. 1989. Pratiques successorales et rapport a la terre: Les sociétés paysannes d'ancien régime. *Annales ESC* 44/1:173–206.

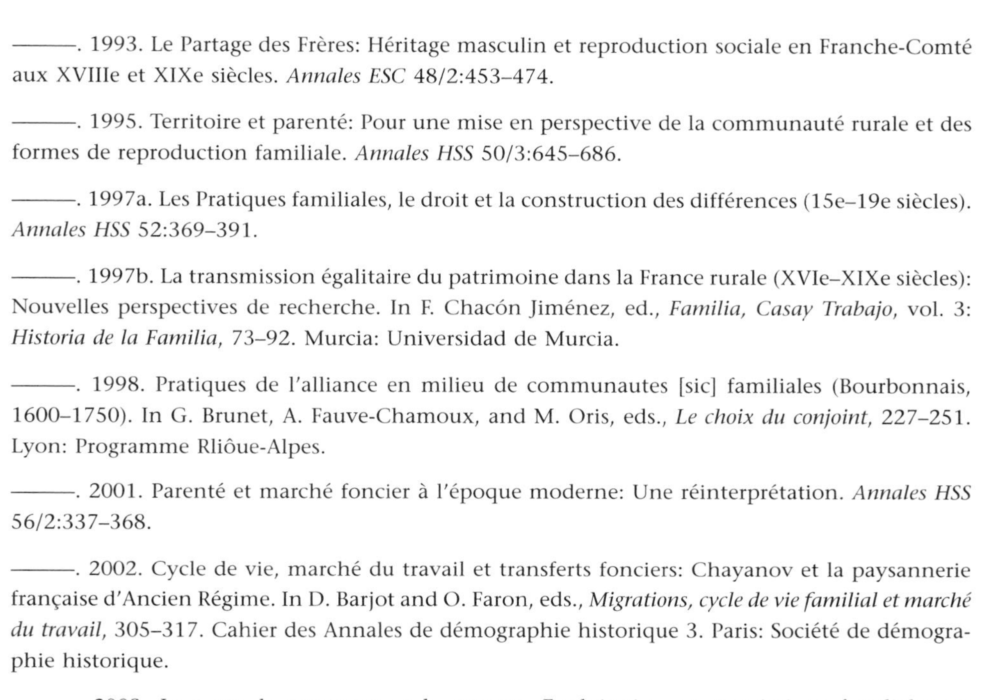

———. 1993. Le Partage des Frères: Héritage masculin et reproduction sociale en Franche-Comté aux XVIIIe et XIXe siècles. *Annales ESC* 48/2:453–474.

———. 1995. Territoire et parenté: Pour une mise en perspective de la communauté rurale et des formes de reproduction familiale. *Annales HSS* 50/3:645–686.

———. 1997a. Les Pratiques familiales, le droit et la construction des différences (15e–19e siècles). *Annales HSS* 52:369–391.

———. 1997b. La transmission égalitaire du patrimoine dans la France rurale (XVIe–XIXe siècles): Nouvelles perspectives de recherche. In F. Chacón Jiménez, ed., *Familia, Casay Trabajo,* vol. 3: *Historia de la Familia,* 73–92. Murcia: Universidad de Murcia.

———. 1998. Pratiques de l'alliance en milieu de communautes [sic] familiales (Bourbonnais, 1600–1750). In G. Brunet, A. Fauve-Chamoux, and M. Oris, eds., *Le choix du conjoint,* 227–251. Lyon: Programme Rliôue-Alpes.

———. 2001. Parenté et marché foncier à l'époque moderne: Une réinterprétation. *Annales HSS* 56/2:337–368.

———. 2002. Cycle de vie, marché du travail et transferts fonciers: Chayanov et la paysannerie française d'Ancien Régime. In D. Barjot and O. Faron, eds., *Migrations, cycle de vie familial et marché du travail,* 305–317. Cahier des Annales de démographie historique 3. Paris: Société de démographie historique.

———. 2003. La terre, la personne et la contrat: Exploitation et associations familiales en Bourbonnais (XVIIe–XVIIIe siècles). *Revue d'histoire moderne et contemporaine* 50/2:27–51.

Derouet, Bernard, and Joseph Goy. 1998. Transmettre la terre: Les inflexions d'une problématique de la différence. *Mélanges de l'école française de Rome. Italie et Méditerranée (MEFRIM)* 110:117–153.

Duhamelle, Christophe. 1998. *L'héritage collectif: La noblesse d'Église rhénane: 17e–18e siècles.* Recherches d'histoire et de sciences sociales 82. Paris: Editions de l'Ecole des Hautes Etudes en Sciences Sociales.

Essegern, Ute. 2003. Kursächsische Eheverträge in der ersten Hälfte des 17. Jahrhunderts. In Martina Schattkowsky, ed., *Witwenschaft in der frühen Neuzeit: Fürstliche und adlige Witwen zwischen Fremd- und Selbstbestimmung,* 116–134. Leipzig: Leipziger Universitätsverlag.

Goodwin, Albert, ed. 1953. *The European Nobility in the Eighteenth Century: Studies of the Nobilities of the Major European States in the Pre-Reform Era.* London: Adam and Charles Black.

Gouesse, Jean-Marie. 1986. Mariages de proches parents (XVIe–XXe siècle): Esquisse d'une conjuncture. In *Le Modèle familiale européen: Normes, déviances contrôle du pouvoir,* 31–61. Actes des séminaires organisés par l'école française de Rome 90. Rome: École francaise de Rome Palais Farnèse.

Habakkuk, H. J. 1953. England. In Albert Goodwin, ed., *The European Nobility in the Eighteenth Century: Studies of the Nobilities of the Major European States in the Pre-Reform Era*, 1–21. London: Adam and Charles Black.

Hausen, Karin. 1976. Die Polarisierung des "Geschlechtscharaktere"—Eine Spiegelung der Dissoziation von Erwerbs- und Familienleben. In Werner Conze, ed., *Sozialgeschichte der Familie in der Neuzeit Europas*, 363–393. Stuttgart: Klett.

Hegel, Georg Wilhelm Friedrich [1807]. 1949. *The Phenomenology of Mind*. Translated, with an introduction and notes, by J. B. Baillie. 2nd ed. London: Allen & Unwin.

Hess, Rolf-Dieter. 1968. *Familien- und Erbrecht im württembergischen Landrecht von 1555 unter besonderer Berücksichtigung des älteren württembergischen Rechts*. Stuttgart: Kohlhammer.

Hurwich, Judith J. 1993. Inheritance practices in early modern Germany. *Journal of Interdisciplinary History* 23:699–718.

Johnson, Christopher. 2002. The sibling archipelago: Brother-sister love and class formation in nineteenth-century France. *L'HOMME: Zeitschrift für Feministische Geschichtswissenschaft* 13/1:50–67.

Jolas, Tina, Yvonne Verdier, and Françoise Zonabend. 1970. "Parler famille." *L'homme* 10: 5–26.

Kühn, Arthur. 1937. Inzucht und Auslese in zwei Eifeldörfern. *Archiv für Rassen- und Gesellschaftsbiologie* 31:482–505.

Laqueur, Thomas. 1990. *Making Sex: Body and Gender from the Greeks to Freud*, 35–43. Cambridge, MA: Harvard University Press.

Macartney, C. A. 1953. Hungary. In Albert Goodwin, ed., *The European Nobility in the Eighteenth Century: Studies of the Nobilities of the Major European States in the Pre-Reform Era*, 126–139. London: Adam and Charles Black.

Nitzsch, Carl Ludwig. 1800. *Neuer Versuch über die Ungültigkeit des mosaischen Gesetzes und den Rechtsgrund der Eheverbote in einem Gutachten über die Ehe mit des Bruders Witwe*. Wittenberg und Zerbst: Zimmermann.

Orel, Herbert. 1932. Die Verwandtenehen in der Erzdiözese Wien. *Archiv für Rassen- und Gesellschaftsbiologie* 26:249–278.

Prokop, Ulrike. 1991. *Die Illusion vom grossen Paar*. 2 vols. Frankfurt: Fischer-Taschenbuchverlag.

Reiff, Heinz. 1979. *Westfälischer Adel 1770–1860: Vom Herrschaftsstand zur regionalen Elite*. Göttingen: Vandenhoeck und Ruprecht.

Roberts, J. M. 1953. Lombardy. In Albert Goodwin, ed., *The European Nobility in the Eighteenth Century: Studies of the Nobilities of the Major European States in the Pre-Reform Era*, 60–82. London: Adam and Charles Black.

Sabean, David Warren. 1972. *Landbesitz und Gesellschaft am Vorabend des Bauernkrieges*. Stuttgart: Fischer.

———. 1990. *Property, Production and Family in Neckarhausen 1700–1870* Cambridge: Cambridge University Press.

———. 1998. *Kinship in Neckarhausen 1700–1870*. Cambridge: Cambridge University Press.

Saugstad, Letten Fegersten. 1977. Inbreeding in Norway. *Annals of Human Genetics* 40:481–491.

Schenk, H. G. 1953. Austria. In Albert Goodwin, ed., *The European Nobility in the Eighteenth Century: Studies of the Nobilities of the Major European States in the Pre-Reform Era*, 102–125. London: Adam and Charles Black.

Segalen, Martine. 1991. *Fifteen Generations of Bretons: Kinship and Society in Lower Brittany 1720–1980*. Trans. J. S. Underwood. Cambridge: Cambridge University Press.

Serra, Angelo, and Antonio Soini. 1959. La consanguinité d'une population: Rappel de notions et de résultats—Application à trois provinces d'Italie du Nord. *Population* 14:47–72.

Spring, Eileen. 1993. *Law, Land, & Family: Aristocratic Inheritance in England, 1300–1800*. Chapel Hill: University of North Carolina Press.

Sutter, Jean, and Jean-Michel Goux. 1962. Évolution de la Consanguinité en France de 1926 à 1958 avec des données récentes détaillés. *Population* 17:683–702.

Titzman, Michael. 1991. Literarische Strukturen und kulturelles Wissen: Das Beispiel inzestuöser Situationen in der Erzählliteratur der Goethezeit und ihre Funktionen im Denksystem der Epoche. In Jörg Schönert, ed., with Konstantin Imm and Joachim Linder, *Erzählte Kriminalität: Zur Typologie und Funktion von narrativen Darstellungen in Strafrechtspflege, Publizistik und Literatur zwischen 1770 und 1920*, 229–281. Studien und Texte zur Sozialgeschichte der Literatur, vol. 27. Tübingen: Niemeyer.

Zonabend, Françoise. 1981. Le très proche et le pas trop loin: Réflexions sur l'organisation du champ matrimonial des sociétés structures de parenté complexes. *Ethnologie française* 11:311–318.

3 Resemblance, Paternity, and Imagination in Early Modern Courts

Silvia De Renzi

3.1 "The Hereditary," Resemblance, and Family Ties

In 1634 a case was brought in Rome before the Tribunal of the Sacra Rota, the highest ecclesiastical court in Europe, which also functioned as the court of appeals in civil cases. Marco Antonio Vittorio was challenging the verdict of another tribunal in a paternity case in which he was indirectly involved. Some time earlier, a girl, Margarita, had claimed to be the daughter of Marco Antonio's late brother Giovanni Battista and demanded that Marco Antonio should provide her with maintenance and later with a dowry. To make their final decision as to whether Marco Antonio should comply with Margarita's request, the judges of the Rota had to establish, among other things, whether she was Giovanni Battista's daughter. To do so they assessed a range of proofs, including the girl's resemblance to the deceased. But the value of resemblance as evidence was one of the main points contested by the parties. While the girl's lawyers used it to make their case, Marco Antonio's lawyers claimed that resemblance did not carry much weight in paternity disputes.

In the continental legal system judges could ask the advice (*consilium*) of expert witnesses on specific matters. Early modern physicians might be asked to give advice on resemblance in cases of adultery and paternity, as the presence of the topic in one of the first medicolegal treatises shows.[1] The judges of the Rota turned to Paolo Zacchia, who was the celebrated author of several volumes of *Quaestiones Medico-Legales*; these would become the most influential work on legal medicine for the next century. In his consilium, Zacchia argued that resemblance does carry weight: his view was decisive and, confirming the sentence of the first court, the Rota ruled in favor of Margarita.[2]

In this chapter I intend to use the case of Margarita as a starting point to explore the legal and natural philosophical debates that made it possible for resemblance to be both invoked and rejected as evidence in seventeenth-century tribunals. Since antiquity, natural philosophers and physicians had discussed the resemblance of children to parents in terms of physical features, moral disposition, and pattern of diseases within the dominant framework of Aristotelian-Galenic natural philosophy.[3] Different views about how male and female semen and blood determined the resemblance of the child to either parent were expressed and are an example of what Carlos López-Beltrán has described as a concern for "the hereditary"; resemblance and its causes preceded what only much later started to be thought of in terms of transmission of traits.[4] But in the early modern period, resemblance was also a key concept for understanding nature, and as I will show, family similarity was understood as part of a broader category of phenomena of resemblance that could not always be explained by the action of semen or blood.

In this context, much energy was spent explaining how maternal imagination could interfere with the process of generation and result in the birth of offspring who did not look like their parents and were therefore considered deformed children or monsters. It has been argued that behind the interest in the generation of monsters lay a broader concern with the deep uncertainty surrounding paternity. While monsters were regarded as the result of the usurpation of the proper formative role of fathers by maternal imagination, resemblance to the father "imposes a temporary, if superficial order, while masking a fundamental, primordial disorder. And what resemblance conceals, the monster unmasks."[5]

If generation was a delicate process, so too were the social concerns connected with it, paternity and the transmission of status and property. As in the rest of Europe, in seventeenth-century Italy, male primogeniture, with the exclusion from inheritance of younger sons as well as daughters, was becoming the dominant pattern of inheritance among aristocratic families and one of the main bases for their social identity. This was a departure from the tradition of Roman law that guaranteed equal rights to the sons and did not exclude daughters from inheritance. Moreover, the implementation of the new rules was patchy and did not eradicate the common tendency of family members to challenge and question inheritance arrangements, which were typically put at risk by the appearance of illegitimate children. Even when the latter were simply claiming the maintenance allowed to them by canon law, the negotiations surrounding their filiation (*filiatio*) might easily end up in court.[6]

Resemblance and maternal imagination were among the resources on which legal experts drew. Exploring the use of these concepts in the courtroom gives us new insights into how the physical and social aspects of generation were intertwined. It also allows us to investigate physicians' contributions to the debate on "the hereditary" from a perspective that has been little explored, that of their practice as expert witnesses. Medical knowledge was certainly an important component in the early modern approach to hereditary phenomena. For example, physicians might be asked by their noble patients for advice on marriages in case of obvious deformities or diseases.[7] And yet, as López-Beltrán has maintained, these phenomena did not occupy center stage, and even tended to be "explained away," until the end of the eighteenth century due to the dominance of humoral theory in medicine. The strong emphasis in humoral medicine on the fundamental influence of the environment—in the form of the six nonnaturals—meant that nothing was so fixed in the inherited temperament that it could not be altered by the nonnaturals. By contrast, the nineteenth-century notion of heredity implied that a set of inherited traits not only shaped the makeup of offspring, but also determined their life, diseases, and death. For this view to emerge, major assumptions about temperament and the boundaries between body and environment would have to change.[8]

However, Paolo Zacchia's consilium on the case of Margarita and his *Quaestiones Medico-Legales* show that as early as the seventeenth century, the notion of temperament could rather be used to argue for the persistence of similar features in parents and children. As I demonstrate in this chapter, urged by legal controversies and by an extreme use of the theory of imagination, Zacchia made temperament the main resource with which to understand resemblance, question the role of maternal imagination, and turn nature into a stable and reliable source of evidence in paternity disputes. In his hands, temperament could ultimately ensure the fulfillment of paternal obligations. If overall consensus about resemblance as evidence of filiation was not reached, Zacchia certainly did not "explain away," but rather addressed head on, the complicated knot of practical and theoretical issues surrounding "the hereditary."

3.2 Illegitimacy and Evidence of Paternity in the Legal Tradition

The legal rights of illegitimate children were one of the issues on which there was divergence between, on the one hand, the Roman and canon law that constituted the *Ius Commune*, the foundation of legal teaching in universities, and, on the other hand, the statutes that had been compiled in medieval cities, the *Ius Proprium*.[9] The Ius Commune distinguished between children *iusti* or *legitimi*, children born in wedlock; *naturales*, children born to parents who were living together but were not married; and *spurii*, children born either from an occasional union, or from adultery. Although both naturales and spurii were illegitimate, the former could in principle inherit a fraction of their father's possessions, while the latter were completely excluded from paternal inheritance.[10] In contrast to this, the statutes of some Italian cities established that even spurii had a right to inherit, under certain conditions. Furthermore, canon law had introduced an important correction to the Roman law by stating that all children, including illegitimate ones, were entitled to maintenance.[11]

Seventeenth-century Rome was characterized by a significant sex imbalance with a larger population of men—mostly clerics supposedly embracing celibacy—than women. As a consequence, the population of prostitutes was large.[12] Furthermore, as the center of Catholicism, Rome attracted thousands of pilgrims from all over the world. This led to occasional encounters resulting in an increase in the population of mixed descent.[13] Judging by the size of the area where foundlings were cared for in the Hospital of Santo Spirito, illegitimate children were in abundance, and of course many of them were not abandoned.[14] Offspring of concubines could live with their parents, and in cases of adulterous relations between masters and servants, it was probably common for the resulting children to live in the father's house. With the provision of the Council of Trent for compulsory registration of all births and baptisms, it was in theory easy to keep track of these irregular situations. However, only very gradually did record keeping become consistent and monitored, and the format of the record more uniform.[15] In the course of the seventeenth century, records of baptism would increasingly be used in paternity cases.[16] Traditionally, proof of filiation had been found elsewhere.

An appropriate point of entrance to the intricate issue of proof of filiation as discussed in Counterreformation Italy is *De Nothis, Spuriisque filiis liber*, published in 1550 by Gabriele Paleotti, at the time professor of law in Bologna, but on his way to becoming a judge of the Rota in Rome.[17] While arguing in favor of illegitimate children and their right to alimentary support, Paleotti reviewed centuries-old discussions about the status of evidence of filiatio. With regard to the father, filiatio could never be established with genuine proof, but only with some probability and by conjectures. Jurists were used to having to deal with conjectures and presumptions rather than full proof, and a system had been developed by which judges would add up conjectures of different weight.[18] Following Roman law, Paleotti stated that for children born to a married woman, the strongest presumption was that they were the offspring of the husband, even if there was evidence that the woman had been unfaithful. A second kind of conjecture was the father's behavior toward the child, including economic provision for him or her. The third conjecture was based on how the supposed father called the child; if he addressed him or her as his son or daughter, this was evidence, though not very strong, that the child was the offspring of the man in question.

Finally, Paleotti indicated widespread reputation of the identity of the child's father as providing strong evidence, though he conceded that jurists had not reached consensus on this.[19] Paleotti's conjectures and presumptions were all social, and he did not even mention resemblance.[20]

Although lack of reference to resemblance was by no means unusual among jurists at the time, a different tradition coexisted, in which resemblance was indeed mentioned and regarded as a possible indication of paternity. It was, however, quite controversial. In his comments on the Roman law about children as heirs, the illustrious jurist Bartolo da Sassoferrato (1313–1357) dealt with the hypothetical case of a widow who marries again only a few days after the death of her first husband and delivers a baby nine months later. Who should be identified as the baby's father?[21] Bartolo reported the view of the jurist Giacomó Belvisi, who argued that in assessing filiatio one should start from the resemblance of the child, and failing this, other clues should be taken into account.[22] Bartolo, however, dismissed Belvisi's opinion on the grounds that a woman's thought (*cogitatio*) at the time of conception has the power to make the child similar to someone other than the actual father. He therefore argued that other physical clues should be preferred, mainly signs of conception and duration of pregnancy.[23]

Bartolo's comment on the unreliability of resemblance as evidence remained very influential in subsequent legal literature and practice. In one of the cases that make up the five *centuria* published with his *De Arbitratiis iudicum quaestionibus* (1576), Giacomo Menochio (1532–1607), one of the most authoritative Italian jurists of the time, reviewed the evidences of filiation.[24] To the four conjectures that we have seen reported by Paleotti—marriage, the father's behavior toward the child, how the alleged father calls the child, and reputation—Menochio added a fifth conjecture, "resemblance or likeness." After assessing a number of authorities who, following Belvisio, seemed to have accepted resemblance as evidence, Menochio claimed to be more persuaded by Bartolo and his followers, who denied its value on account of the role of the mother's imagination.[25]

In addition to Bartolo's objection, resemblance was problematic on other grounds. Menochio interspersed his review with some typical examples of the vagaries of resemblance, including the puzzling cases of black children born to white mothers and resembling their black grandparents, as well as the bizarre cases of people who strongly resembled each other although they had no family connections and lived far apart. His sources for the latter category of examples were Pliny and the writer Valerius Maximus, who had reported anecdotes of uncanny resemblance between people who were not relatives. Each furthermore reported how such resemblances had been slyly exploited, especially by people of a lower status who had used resemblance to impose themselves on illustrious families.[26]

This theme is made more explicit in the discussion of resemblance as a conjecture for filiatio carried out by Giuseppe Mascardo (?–1588), another sixteenth-century Italian jurist famous for his work on legal evidence.[27] In his account, the stories of deception included the cases of Queen Semiramis, whose resemblance to the king's son allowed her to rule the Assyrian empire for many years; and that of a Renaissance jester who was so similar to Sigismondo Malatesta, the ruler of Rimini, that he could claim to be his son. But Mascardo also observed that among polygamous populations resemblance was regarded as the best way of identifying one's children. He acknowledged that many of the most acute naturalists held the view that resemblance mattered, and used animals' generation of like offspring to prove their point.[28]

On the one hand, then, Menochio and Mascardo expressed contemporary uneasiness about resemblance, which they saw as linked to the socially and politically sensitive issue of doubles and impostors.[29] On the other hand, reference to the polygamous people's reliance on resemblance conveyed the message that it could work as a family identifier. This would not have escaped readers at a time when the Church was imposing a stricter family discipline and made the Tridentine rituals surrounding marriage and birth compulsory.[30]

A remarkable ambiguity is detectable in the works of these jurists. Having rejected, on the authority of Bartolo, the possibility of using resemblance as evidence of filiatio, they seem then to concede that if not a proper *conjectura*, it should at least be regarded as providing grounds for "some sort" (*qualemqualem*) of suspicion of filiatio. Resemblance was especially relevant in conjunction with other circumstances, such as cohabitation of the alleged mother and father.[31] That resemblance could be the effect of a variety of causes made it fallacious and too hazardous to be used as full legal evidence, but it was not without significance. Jurists would rank it at the lowest level of their steep hierarchy of legal evidence, while regarding the usual series of "social" indications of filiatio, such as reputation and the treatment of children, as much stronger. As often happened with ambiguous legal evidence, assessment was left to the judge's discretion.

We are now in a better position to understand the irresolution of the judges of the Rota in the case of Margarita in 1634. The legal literature included both acceptance and rejection of resemblance as evidence and the parties in the case had mobilized both sides of this tradition. To resolve the issue, the highly respected judge Clemente Merlino turned to a physician's expertise.

3.3 The Causes of Resemblance

By the time Zacchia was asked to give his view on the case of Margarita, he had already discussed the issue of the likeness and unlikeness of offspring in the first book of his *Quaestiones Medico-Legales* (1621), which dealt with issues linked to pregnancy and delivery.[32] At the start of the section he acknowledged that, with only a few exceptions, jurists tended to understand resemblance as the work of imagination and therefore did not consider it proper proof.[33] By contrast, to him resemblance had a completely different cause and one that made it strong legal evidence.

Zacchia developed his argument using a very inclusive notion of resemblance. To him the resemblance of children to their parents was just one case within a wide range of phenomena of resemblance, from that between humans and animals to that between strangers. Zacchia's reference to the resemblance between humans and animals shows how physiognomy was culturally pervasive and a powerful resource for thinking about nature.[34] However, while Zacchia certainly shared the interest of his contemporaries in physiognomy, and more generally considered resemblance as a multifaceted phenomenon to be found throughout nature, he did not embrace any of the hermetic or Neoplatonic theories that made resemblance one of the fundamental principles of nature. On the contrary, his engagement with the issue was cast in a strictly Aristotelian mode of argument and terminology, evident in his concern with causation.[35]

The goal that Zacchia set himself was to find a cause that was so inclusive as to provide an explanation for all the kinds of resemblance under consideration. His first move was to distinguish between internal and external resemblance, the former pertaining to a deep-seated agreement (*convenientia*) in temperament, the latter to the resemblance of body parts. Internal resemblance between two individuals, Zacchia believed, is caused by a similarity in their whole temperament, that is, the proportion of the qualities in their body. The temperament results from the action of the innate heat and can be transmitted through the seeds of both mother and father. Although Zacchia endorsed Galen's dual theory of seed, overall he seems to have been less interested in how semen's action brings about resemblance than in establishing the deepest and most inclusive cause of resemblance, the convenientia of temperament and innate heat. Convenientia in temperament is followed by resemblance in moral habits, observed Zacchia. He further argued that the link between temperament and habit is so strong and unequivocal that where we detect a similarity of moral customs, we can assume an underlying similarity of temperament.

Internal resemblance is usually also followed by external resemblance and this applies not just to children and parents, but to people living in the same country. Where the original temperament is preserved—this happens more often outside Europe—"men are born who look like each other."[36] Following the traditional notion of resemblance, Zacchia also argued that another consequence of sharing temperament was a common pattern of diseases: parents who suffered from epilepsy or melancholy would generate children with the same diseases. However, there were complications. It was possible that a child who was affected by the same diseases as his father did not share his features, and vice versa. This was due to a disjunction between temperament of the whole body and temperament of each bodily part. A child with the same general temperament of his father would have his moral disposition and his diseases of the whole body. But the temperament of his head might be similar to that of his mother, in which case his face would look like hers and he would suffer from the same head pain. External resemblance, then, implies an underlying internal resemblance, but the relationships between the two are not straightforward. Furthermore, what really distinguishes internal and external resemblance is that the latter can occasionally be affected by imagination, while the former cannot.[37]

None of this was original. By resorting to the key Galenic concepts of temperament and innate heat, Zacchia could satisfy his need to find an all-embracing cause for the wide range of phenomena of resemblance with which he and his contemporaries were concerned. So, if the resemblance between strangers had remained an unsettling and unresolved phenomenon, Zacchia invoked the fact that agreement of the whole temperament can occur by accident as well as by nature.[38] The distinction between internal and external resemblance, the strong link between internal resemblance and temperament, and the infrequency with which according to Zacchia children not resembling their parents are born, were the building blocks of his view and had important consequences in the legal arena. However, to make sense of what Zacchia was gaining by putting temperament at center stage, it is necessary to consider his discussion of imagination, which is briefly included in the first book of the *Quaestiones*, and emerges with particular clarity in his consilium in the case of Margarita.

3.4 Nature as Evidence

At the Rota, the main argument used by Marco Antonio's lawyers to reject as evidence Margarita's resemblance to Giovanni Battista was that the girl's mother might have had intercourse with another man while thinking of Giovanni Battista. Due to maternal imagination, the girl's resemblance to Giovanni Battista was therefore not evidence of his fatherhood, because Margarita might nonetheless be the daughter of another man.[39]

In the early modern period, imagination was the topic of intense discussion, from philosophical debates on the hierarchy of human faculties and their means of action, to theological disputes in relation to witchcraft and divination.[40] These controversies provided the main framework within which more specific disputes over maternal imagination took place. Maternal imagination loomed large in learned and non-learned explanations of the likeness and unlikeness of children to parents. If Aristotelian philosophers and Galenic physicians clashed over the issue of the existence and function of female semen, the theory of female imagination cut through the debate and recognized women's active role. A philosopher such as Tommaso Campanella (1568–1639) used imagination and children's resemblance as a springboard to reject Aristotle's view of generation, while the belief that women's emotions had a substantial impact on the shape of their offspring was popular and widespread.[41]

The persistence of the theory of maternal imagination has been explained in connection with the pervasiveness of patriarchal values, in particular disquiet about the female generative force and uncertainty over paternity. If women's role could not simply be expunged, the theory of maternal imagination ensured that the female contribution would be regarded as a cause of disruption in the process of generation.[42] Yet, although certainly very popular, the theory of maternal imagination did not go undisputed. Zacchia's argument and the authors he mobilized in his discussion both in the *Quaestiones* and in the consilium allow us to appreciate why this was the case.

To Zacchia, contemporary views of imagination as a force shaping the look of offspring were flawed on a number of grounds. One problem concerned the time when imagination might act. According to some, imagination at the time of conception was enough to determine the shape of the child, hence the popular advice to hang pictures in bedrooms. On this question Zacchia was not at all convinced: at the time of conception there is nothing on which to act, because those parts that could be shaped in one way or another only emerge a few days later.[43] Zacchia was here drawing on the work of Emilio Parisano, the Paduan physician who is perhaps best remembered for his opposition to the theory of the circulation of the blood, but who also wrote extensively on generation.[44] Zacchia's conclusion was very clear: "Indeed, not only does imagination not operate at all at the time of conception, but almost never, except on extremely rare occasions, does it operate during the entire pregnancy either."[45]

A further problem that Parisano had already pointed out had to do with the means by which the imagination acts. Zacchia reminded his readers that in the classification of faculties *assimilatio* is the activity of the *facultas formatrix* that belongs to the natural faculties, while imagination is to be placed among the animal faculties. Therefore to argue that imagination can affect the appearance of the fetus reveals a fundamental misunderstanding of the hierarchy and functions of faculties. Although Zacchia does not refer to the further problem of the separation of the fetus from the mother's body that had bothered Parisano, he was clearly not happy with what he regarded as a lack of rigor in the imagination theory.[46]

Once the claim that imagination is responsible for resemblance was put to the test, too many questions remained unresolved. It could be conceded that imagination acts as an accidental, not a natural cause of resemblance, but this distinction led Zacchia to the thorny issue of the relations between different kinds of causes. His analysis is detailed, but the main point emerges clearly. An accidental cause can never prevent the action of a natural cause and given that the facultas formatrix present in the parents' seeds is a natural cause, which can account for all the phenomena of resemblance, it is absurd to think that an accidental cause—imagination—could affect the natural, formative cause.[47] Zacchia accepted that imagination could have some effect on external resemblance only, but he was adamant that it cannot affect the facultas formatrix. Furthermore, if we regarded imagination as the universal cause of resemblance, then it would be necessary to conclude that children always take after the imagined person rather than the father, and that the facultas formatrix acts erratically. But most of the time children look like their parents and are the product of the regular action of the facultas formatrix; only rarely are monsters—or children strongly dissimilar to their parents—born. Concluding his consilium in the case of Margarita, Zacchia wrote:

> And a reply to the opposite claims is clear from the things I have said, which totally reject imagination as a cause of resemblance, because assimilation is the work of the formative faculty and pertains to natural faculties only, not to the animal ones, as does the imagination. And imagination does not possess the means with which to form anything, and has nothing in common with the formative faculty. And as to the experiences that they present, if we consider them one by one it will be easily clear that they prove nothing and that even if they are accepted as true, they are effects monstrous in nature, on which no certain rule can be based.[48]

This was the crucial point of Zacchia's whole analysis. It was rooted in his firm conviction that nature acts regularly and consistently and that what happens most of the time can always be explained by natural causes. In opposing different kinds of causes, such as *causa per se* and *causa per accidens*, Zacchia was using a scholastic distinction with which his physician and jurist readers were certainly familiar.[49] But in stressing the merely accidental nature of imagination as a cause and its fundamental difference from the ordinary operations by which generation is brought about, Zacchia was hinting at a difficulty in the theory of maternal imagination that recent studies seem to have overlooked and that emerges in a reference he made to *De fascinatione libri tres* of Bishop Leonardo Vairo published in 1589.[50]

Imagination had been invoked to explain a range of phenomena associated with witchcraft—for example, the alleged power of women to cast spells (*fascinatio*) and their alleged encounters with the devil. Theologians involved in the heated discussions on witchcraft were particularly keen to draw a clear distinction between unusual and even bizarre phenomena that had natural explanations and phenomena that could be accounted for only by the devil's intervention.[51] The main aim of Vairo's book was precisely to tell different kinds of fascination apart and provide a proper theological interpretation. He took issue with the ways imagination had been mobilized. On the one hand, there were those who advocated imagination as a natural explanation of phenomena like forecasting the future, which to him was nothing but the result of either supernatural powers as in the prophets, or the devil's intervention. But on the other hand, imagination had been used to account for the resemblance of children to their parents. In this case, Vairo argued, the imagination theory was to be rejected simply on the grounds that resemblance of children has a very clear and natural cause, parents' seed, which makes it unnecessary to invoke other, ambiguous, factors.[52]

Zacchia's readers, especially the judges of an ecclesiastical tribunal such as the Rota, would have immediately appreciated the significance of his reference to the authoritative bishop Vairo. Imagination was potentially a dangerous concept in the double sense that it touched on issues of demonology and that an excessive appeal to it could muddle things that should rather be kept clearly apart. The boundaries delimiting the natural (what happens most of the time), the preternatural (what is anomalous and happens occasionally), and the supernatural (what happens through the intervention of God only) were one of the most debated issues of the day.[53] It obviously had major theological implications, because the definition of the devil's action and the assessment of miracles depended on these distinctions. But it is revealing that Vairo should use the resemblance of children to parents to argue that in the presence of perfectly adequate natural causes, it was wrong to complicate things by introducing unnecessary causes. This was also Zacchia's point, although advocating clear distinctions between the natural, the preternatural, and the supernatural could acquire opposite meanings in a theologian's and a physician's argument. Indeed, Zacchia had a very explicit position on the issue of miracles, which in seventeenth-century Rome was at the top of the agenda. He forcefully stated that, contrary to what many people think, miracles occur rarely, and most phenomena counted as miracles can in fact be explained by natural causes. Of course he was keen to stress that by this he did not mean to imply the restriction of God's power, but to avoid any undue inflation of his intervention. Zacchia's view of resemblance as the result of natural and regular causes was integral to his general attitude, which tended to expand the space occupied by the natural at the expense of both the preternatural and the supernatural. [54]

The distinction between what happens most of the time and what happens occasionally in nature had a further specific significance in the courtroom, because it overlapped with the broader issue of the relations between rules and exceptions in the administration of justice.[55] Jurists were trained to apply norms to the almost infinite variability of individual cases, but when issues related to the body or nature were concerned, the question of what should be considered normal or exceptional had to be put to other experts. Physicians' competence in establishing what happens most of the time and what happens only rarely or by accident had the effect of securing nature as evidence. If nature was a dependable witness, then physicians were certainly best suited to interrogate her.[56] In the case of paternity, by reducing the imagination to an unusual and secondary cause and by replacing it with the regularity and consistency of natural causes, resemblance acquired the status of strong evidence.

Most of the time children do resemble their parents. To bring this home to his readers, Zacchia quoted Aristotle's claim that children who do not resemble their parents are in a certain sense monsters, but in his interpretation, the word *monster* simply stood for an extremely rare event.[57] As a result, an unexpected unlikeness should be attributed to infidelity.[58] However, by the same token, resemblance of illegitimate children to their alleged father could become a compelling piece of evidence in favor of their claims, as Zacchia argued in the case of Margarita. Here, his lengthy discussion was mainly aimed at demolishing the argument of Marco Antonio's lawyers. Their interpretation of the imagination theory was not unusual: contemporary literature, both learned and popular, reported cases of children born from adulterous relationships but resembling the husband of their mother because during intercourse her fear of discovery had made her think of her husband.[59] We have seen that judges were cautious in admitting resemblance as evidence, and the opportunistic use of it by the parties probably made them even more suspicious. Zacchia's argument for a simpler and more direct cause of resemblance was meant to assist them in their task. While the theory of imagination could lead to a situation in which hardly anything could ever be established about family links, to argue for temperament and innate heat as the causes of resemblance meant securing reliable evidence in paternity disputes. If the regular work of nature was to be trusted, patriarchal order would be better protected, but so would the rights of illegitimate children.

However, Zacchia was aware that his view on resemblance faced a major problem. Making resemblance the basis on which filiation could be claimed would allow anyone who by accident looked similar to somebody else to claim to be his kin.[60] Of course this was hardly acceptable, and Zacchia's distinction between internal and external resemblance was to become the solution to the problem. External resemblance was problematic also because, as Zacchia had conceded, it could be affected by the imagination. But by no means could internal resemblance be affected by the imagination, because it was the direct result of the action of the innate heat in producing temperament. So internal resemblance could and should become the ground on which to assess filiation.[61] While conceding that it was not possible to perceive such resemblance directly with the senses, Zacchia argued that knowledge of it can be obtained through two very strong indications, resemblance of habits—for example, courage or proclivity to gambling—and resemblance in health and sickness.[62] Being the most direct expression of temperament, these two main phenomena constituting "the hereditary" were now joined together. Of the two, resemblance of diseases offered the better evidence, though physicians might be required to assess it. It was in conjunction with internal resemblance that the easier-to-assess external resemblance could also become a reliable indication.[63] Especially if compared with the kinds of evidence discussed in the legal tradition, both habits and diseases appeared to be "natural" rather than "social" proof of filiation, because they were both the result of temperament. By making them the most reliable evidence, Zacchia had transformed nature into the ultimate witness.

3.5 Conclusions

Zacchia's view on resemblance was accepted by the judges adjudicating the case of Margarita, and they ruled in favor of the girl. However, a survey of some paternity cases adjudicated by the Rota between the mid-seventeenth century and the first half of the eighteenth century show that judges felt uneasy about using resemblance as evidence. In the main, they continued to assess the four conjectures listed in the legal tradition—marriage, the father's behavior toward the child, how the alleged father calls the child, and reputation—to which they increasingly added documentary evidence such as parish registers and family books.[64] Doctors did not take up Zacchia's innovative proposal to take seriously resemblances as natural proof of filiation either. The authoritative works on legal medicine by Michael Bernard Valentini and Hermann Friderich Teichmeyer in the eighteenth century indicate that they were aware of Zacchia's notion of internal resemblance, but expressed reservations about its legal use on the ground that both the moral disposition and diseases can only be assessed once the child has grown into an adult.[65]

Such a limited impact of Zacchia's proposal is puzzling, for throughout the eighteenth century he remained the most authoritative source in medicolegal practice. A plausible cause of this reaction on the part of both jurists and physicians is the persistence of the theory of imagination and its social advantages. The endorsement of the theory of imagination by German physicians acting as expert witnesses has been explained by the continuing preoccupation with the preservation of legitimate family lines vis-à-vis adultery. Faced with a striking dissimilarity to the mother's husband, they would find in the theory of imagination a good justification for sticking to the old presumption in favor of legitimacy.[66] However, where did this conservative preference for maintaining the order of legitimacy leave the rights of illegitimate children? In the early eighteenth century, judges of the Rota were still keen to implement canon law and give illegitimate children alimentary support. Again and again they repeated that with regard to claims of maintenance, which did not encroach on the claim to inheritance of a third party, evidence of filiation might be light (*levis*), and to this end the old "social" conjectures, to which documentary evidence could now be added, proved sufficient.[67] There was no need to go any further in trying to secure paternity more firmly.

Throughout the early modern period resemblance was invoked and assessed in the courtroom, though in combination more than competition with other kinds of evidence of paternity. Clearly it was regarded as problematic. Some of the complications were related to the theory of imagination, which made resemblance the effect of unpredictable forces with little or no link to the actions of accountable individuals. Physical resemblance was also fraught with other cultural difficulties, including a pervasive uneasiness with the phenomenon of the double and impostures. In the context of the deep uncertainties surrounding paternity, the reluctance to take resemblance into account protected legitimate families, because the unlikeness of a child to the husband of the mother would not be considered evidence of her adultery. Awareness of the difficulties associated with paternity and a flexible regime of legal proof also meant that the requirements for evidence of filiation varied according to the claim made before the court. Overall, judges felt more at ease with "social" conjectures based on witnesses' testimony and documentary evidence that had always enjoyed a high status and was now increasingly available. The primacy of these conjectures is an indication of the legal and social framework within which paternity was placed and the relative absence of specific concern with "natural" evidence.

All this might be regarded as further confirmation of the lack of engagement with the phenomena of "the hereditary" that scholars have signaled for the early modern period. Well into the eighteenth century jurists and physicians seem to have regarded the link between "the hereditary," kinship, and inheritance in weak terms. Yet, I have shown that room for a different approach to resemblance already existed in the legal tradition. Furthermore, contemporary natural philosophical and medical knowledge provided Zacchia with the intellectual resources with which to overturn the prevailing explanations of resemblance, expunge imagination from the courtroom once and for all, and turn nature into strong evidence. His position emerges clearly even amidst his constant review of competing positions. Obviously, we need to regard Zacchia's approach to resemblance in his own terms. For example, his effort to interpret the resemblance of strangers as comparable to the resemblance of children to their parents shows just how different contemporary perceptions of bodies and kinship were from ours. Yet, Zacchia's aim to turn nature into a witness is beyond question. For him, we might say, hereditary phenomena, in particular hereditary diseases, which were the most direct and unadulterated expression of temperament, had the potential to become the most reliable, though still practically difficult to show, evidence of filiation. This was certainly part of Zacchia's project to make the natural the basis for medicolegal practice, and ultimately the fundamental assumption of the legal system and the foundation on which justice could be administered. Theological and legal authorities—who were themselves concerned about the boundaries of the natural—would agree with his project, though their final aims might be different.

Notes

1. Codronchi 1597. Fischer-Homberger 1983, devotes a whole section to resemblance.

2. Zacchia's *consilium* on the case of Margarita and the final *decisio* of the Rota are in Zacchia 1661, tomus posterior, 269–271, 478–480. The case is briefly discussed, among others, in Santangelo Cordani 2003, who provides a comprehensive survey of the legal debate surrounding paternity in the Rota jurisprudence and briefly discusses the issue of resemblance.

3. Roger 1997, 62–70. On Galen's theory of generation, see Boylan 1986.

4. See López-Beltrán 2002 and chapter 5, this volume.

5. Huet 1993, 34. The literature on monsters is growing; for the pre-Enlightenment period, see Céard 1977; Park and Daston 1981; Daston and Park 1998; Knoppers and Landes 2004. Without denying the early modern obsession with monsters, it was babies who resembled the "wrong" father who would more often stir first puzzlement and then legal controversy. Indeed since antiquity many of the stock examples of resemblance gone astray are linked to cases of women unduly charged with adultery.

6. See Casanova 1997; Kuehn 1997, 2002; Pomata 2002. On legal practices in seventeenth-century Rome, see Ago 1995, 1998. For a discussion of primogeniture in Europe over a longer period, see Sabean, chapter 2, and Vedder, chapter 4, this volume. For an analysis of the natural philosophical assumptions on which the construction of lineages was based, see Pomata 1994.

7. López-Beltrán 1994, 217.

8. The complex way this was debated in the eighteenth century has been explored by López-Beltrán 2002; for stimulating comments on his discussion, see Pomata 2003.

9. How to reconstruct the hierarchy between the two sets of laws in medieval and early modern legal practice is a matter of controversy among historians; see Kuehn 1997.

10. Pecorella 1968.

11. Pecorella 1968; Marongiu 1966. Scholars disagree on the extent to which local statutes actually gave broader inheritance rights to illegitimates. Pecorella 1968, 553, stresses the preoccupation of local authorities with illegitimacy and their consequent restriction of inheritance rights; more recently, Kuehn 1997, 271, has shown how illegitimates' rights were used to preserve a family's property in a direct line.

12. Sonnino 2000.

13. I owe this consideration to Staffan Müller-Wille.

14. See Schiavoni 1991.

15. Records of baptism had started well before the Council of Trent, because they were linked to policing guild membership and to municipal fiscal control. After Trent, records of baptism acquired the function of preventing improper marriages, for instance between people linked through godfathers or godmothers; see Corsini 1972.

16. On the importance given to legal documents as evidence of paternity, see Santangelo Cordani 2003.

17. On Paleotti, see Prodi 1959.

18. On presumptions and conjectures in the continental legal theory and practice, see Alessi Palazzolo 1979; Maclean 1992.

19. Paleotti 1550, ff. 28v.–35r.

20. Paleotti would later address the issue of resemblance as evidence, though in the rather different context of his discussion of the use of family portraits in legal controversies; see Paleotti 1961, 337. For a discussion of portraits and genealogy, see Mann and Syson 1998; Corradini 1998. The modest importance of physical resemblance in this context is stressed by Labrot 1990.

21. Bartolo discussed the evidence of *filiatio* in his commentary on passage D.28.2.29 of Justinian's *Digest*. On Bartolo da Sassoferrato, see Calasso 1964. On Belvisio, see Caprioli 1966.

22. Belvisio as quoted in Bartolo 1562, 266.

23. Bartolo 1562, 266.

24. Menochio 1576, ff. 118r.–122v.

25. Menochio 1576, f. 122r.

26. See Pliny's *Historia naturalis*, 7.12; Valerius Maximus, *Factorum et dictorum memorabilium libri IX*, IX, 14.

27. Mascardo 1593, vol. 2, ff. 117v.–118r.

28. Mascardo 1593, vol. 2, f. 117v.

29. For an influential discussion of the early modern notion of imposture and its relation to identity, see Davis 1983.

30. Prosperi 1996, especially 650–666.

31. Menochio 1576, f. 122v; Mascardo 1593, vol. 2, f. 118r.

32. I have used the 1661 edition, in which the discussion on resemblance is in tomus prior, 90–101.

33. Zacchia 1661, tomus prior, 98.

34. One of the most influential works on physiognomy was Giovan Battista Della Porta's *De Humana Physiognomonia* 1586, on which see Vasoli 1990; Caputo 1990; Haskell 1993, especially 60–67.

35. Foucault has made resemblance one of the key features of the Renaissance episteme; for a discussion of Foucault's failure to engage with the prevalent Aristotelian logic and philosophical framework, see Maclean 1998.

36. Zacchia 1661, tomus prior, 96.

37. Zacchia 1661, tomus prior, 95–97.

38. Zacchia 1661, tomus prior, 95.

39. Zacchia 1661, tomus posterior, 269.

40. On imagination in Renaissance philosophy, see Park 1974; Fattori and Bianchi 1988. On imagination and witchcraft, see Clark 1997; Romeo 1990.

41. On maternal imagination, see Roodenburg 1988; Rublack 1996. On Campanella, see Giglioni 1998.

42. While recognizing the force of women in generation, the theory of maternal imagination represented it as subverting a number of different hierarchies. See Huet 1993, 24; see also Shildrick 2000 and Finucci 2001.

43. Zacchia 1661, tomus prior, 98.

44. Parisano 1623; on Parisano, see Roger 1997, 101–103.

45. Zacchia 1661, tomus prior, 98.

46. Zacchia 1661, tomus posterior, 270.

47. Zacchia 1661, tomus posterior, 270–271.

48. Zacchia 1661, tomus posterior, 271; for further discussion in similar terms, see tomus prior, 478.

49. The types of causes mentioned by Zacchia are included in Goclenius's popular philosophical lexicon; see Goclenius [1613] 1964. On causes in early modern medicine, see Maclean 2002.

50. Zacchia 1661, tomus prior, 93.

51. Clark 1997, especially 151–311; see also Romeo 1990, who briefly discusses Vairo's work.

52. Vairo 1589, 63–99.

53. See Daston and Park 1998; Clark 1997.

54. Zacchia 1661, tomus prior, 259–260. On Zacchia's position on supernatural phenomena, see Lavenia 2004.

55. On the different ways norms and exceptions were discussed in medicine and the law, see Maclean 2000.

56. On the effort made by Zacchia to boost medical authority, see Colombero 1986; De Renzi 2002.

57. Zacchia 1661, tomus prior, 96.

58. Zacchia 1661, tomus prior, 98.

59. See Finucci 2001.

60. Zacchia 1661, tomus prior, 99.

61. Zacchia 1661, tomus prior, 99.

62. Zacchia 1661, tomus prior, 99.

63. Zacchia 1661, tomus prior, 99.

64. I have looked at paternity cases included in four collections of *decisiones* taken by the Rota and published between 1637 and 1759; see Buratto 1637, Celso 1668, Gutierrez 1735, and De Luca 1759.

65. Valentini 1722; Teichmeyer 1723.

66. Fischer-Homberger 1983, 264–266.

67. On the different kind of evidence required in cases of alimentary support, see Santangelo Cordani 2003, 1980–1983.

References

Ago, R. 1995. Ruoli familiari e statuto giuridico. *Quaderni storici* 30 (1): 111–133.

———. 1998. *Economia barocca: Mercato e istituzioni nella Roma del Seicento*. Rome: Donzelli.

Alessi Palazzolo, G. 1979. *Prova legale e pena: La crisi del sistema tra Evo Medio e Moderno*. Naples: Jovene.

Bartolo. 1562. *In secundum tomum Pandectarum Infortiatum Commentaria*. Basel: Froben.

Boylan, M. 1986. Galen's conception theory. *Journal of the History of Biology* 19 (1): 47–77.

Buratto, M. 1637. *Sacrae Rotae Romanae Decisiones*. Rome: Ex Typographia Rever. Camerae Apostolicae.

Calasso, F. 1964. *Bartolo da Sassoferrato*. In *Dizionario Biografico degli Italiani*. Rome, Istituto della Enciclopedia Italiana 6:640–669.

Caprioli, S. 1966. *Belvisi Giacomo*. In *Dizionario Biografico degli Italiani*. Rome, Istituto della Enciclopedia Italiana 8:89–96.

Caputo, C. 1990. Un manuale di semiotica del Cinquecento: Il *De humana Physiognomonia* di Giovan Battista Della Porta. In M. Torrini, ed., *Giovan Battista Della Porta nell'Europa del suo tempo*, 69–92. Naples: Guida.

Casanova, C. 1997. *La famiglia italiana in età moderna: Ricerche e modelli*. Rome: La Nuova Italia Scientifica.

Céard, J. 1977. *La Nature et les prodiges: L'Insolite au XVIe siècle*. Geneva: E. Droz.

Celso, A. 1668. *Decisiones Sacrae Rotae Romanae coram Amgelo Celsó*. Rome: Ex Typographia Reverendae Camerae Apostolicae.

Clark, S. 1997. *Thinking with Demons: The Idea of Witchcraft in Early Modern Europe*. Oxford: Oxford University Press.

Codronchi, G. 1597. *De vitiis vocis libri duo . . . Cui accedit consilium de raucedine, ac methodus testificandi in quibusuis casibus medicis oblatis, postquam formulae quaedam testationum proponantur.* Frankfurt: Apud haeredes Andreae Wecheli, Claudium Marnium & Ioannem Aubrium.

Colombero, C. 1986. Il medico e il giudice. *Materiali per una storia della cultura giuridica* 16:363–381.

Corradini, E. 1998. Medallic portraits of the Este: *Effigies ad vivum expressae.* In N. Mann and L. Syson, eds., *The Image of the Individual: Portraits in the Renaissance,* 22–39. London: British Museum.

Corsini, C. A. 1972. Nascite e matrimoni. In *Le fonti della demografia storica in Italia: Atti del seminario di demografia storica 1971–1972,* vol. 1, part 2, 647–699. Rome: Comitato Italiano per lo studio dei problemi della popolazione.

Daston, L., and K. Park. 1998. *Wonders and the Order of Nature 1150–1750.* New York: Zone Books.

Davis, N. Zemon. 1983. *The Return of Martine Guerre.* Cambridge, MA: Harvard University Press.

De Luca, G. B. 1759. *Mantissa Decisionum Sacrae Rotae Romanae ad Theatrum Veritatis & Justitiae Cardinalis De Luca.* Venice: Ex Typographia Balleoniana.

De Renzi, S. 2002. Witnesses of the body: Medico-legal cases in seventeenth-century Rome. *Studies in History and Philosophy of Science* 33A:219–242.

Fattori, M., and M. Bianchi, eds. 1988. *Phantasia-Imaginatio: V Colloquio Internazionale del Lessico Intellettuale Europeo.* Rome: Edizioni dell'Ateneo.

Finucci, V. 2001. Maternal imagination and monstrous birth: Tasso's *Gerusalemme Liberata.* In V. Finucci and K. Brownlee, eds., *Generation and Degeneration: Tropes of Reproduction in Literature and History from Antiquity to Early Modern Europe,* 41–77. Durham, NC: Duke University Press.

Fischer-Homberger, E. 1983. *Medizin vor Gericht: Gerichtsmedizin von der Renaissance bis zur Aufklärung.* Bern: Verlag Hans Huber.

Giglioni, G. 1998. Immaginazione, spiriti e generazione. La teoria del concepimento nella *Philosophia sensibus demonstrata* di Campanella. *Bruniana & Campanelliana. Ricerche filosofiche e materiali storico-testuali* 4 (1): 37–57.

Goclenius, R. [1613] 1964. *Lexicon Philosophicum quo tanquam clave Philosophiae fores aperiuntur.* Hildesheim: Georg Olms Verlagsbuchhandlung.

Gutierrez, J. 1735. *S. Rotae Romanae decisiones recentissimae et selectissimae.* Geneva: Sumpt. Fratrum de Tournes.

Haskell, F. 1993. *History and Its Images: Art and the Interpretation of the Past.* New Haven, CT: Yale University Press.

Huet, M.-H. 1993. *Monstrous Imagination*. Cambridge, MA: Harvard University Press.

Knoppers, L. L., and J. B. Landes, eds. 2004. *Monstrous Bodies/Political Monstrosities in Early Modern Europe*. Ithaca, NY: Cornell University Press.

Kuehn, T. 1997. A late medieval conflict of laws: Inheritance by illegitimates in *Ius Commune* and *Ius Proprium*. *Law and History Review* 15 (2): 243–273.

———. 2002. *Illegitimacy in Renaissance Florence*. Ann Arbor: University of Michigan Press.

Labrot, G. 1990. Hantise généalogique, jeux d'alliances, souci esthétique: Le portrait dans les collections de l'aristocratie napolitaine (XVIe–XVIIIe siècles). *Revue Historique* 284:281–304.

Lavenia, V. 2004. "Contes des bonnes femmes." La medicina legale in Italia, Naudé e la stregoneria. *Bruniana & Campanelliana. Ricerche filosofiche e materiali storico-testuali* 2:299–317.

López-Beltrán, C. 1994. Forging heredity: from metaphor to cause, a reification story. *Studies in History and Philosophy of Science* 25 (2): 211–235.

———. 2002. Natural things and non-natural things: The boundaries of the hereditary in the 18th century. In *A Cultural History of Heredity I: 17th and 18th Centuries*. Berlin: Max-Planck-Institut für Wissenschaftsgeschichte.

Maclean, I. 1992. *Interpretation and Meaning in the Renaissance: The Case of Law*. Cambridge: Cambridge University Press.

———. 1998. Foucault's Renaissance episteme reassessed: An Aristotelian counterblast. *Journal of the History of Ideas* 59 (1): 149–166.

———. 2000. Evidence, logic, the rule and the exception in Renaissance law and medicine. *Early Science and Medicine* 5:227–257.

———. 2002. *Logic, Signs and Nature in the Renaissance: The Case of Learned Medicine*. Cambridge: Cambridge University Press.

Mann, N., and L. Syson, eds. 1998. *The Image of the Individual: Portraits in the Renaissance*. London: British Museum.

Marongiu, A. 1966. Alimenti: Diritto intermedio. In *Enciclopedia del Diritto*, vol. 2, 21–24. Milan: Giuffrè.

Mascardo, G. 1593. *Conclusiones probationum omnium quae in utroque foro quotidie versantur*. Frankfurt am Main: Impensis haeredum Sigismundi Feyrabendij.

Menochio, G. 1576. *De Arbitrariis iudicum quaestionibus & causis libri duo*. Frankfurt am Main: Sigismundus Feyerabend.

Paleotti, G. 1550. *De Nothis, Spuriisque filiis liber*. Bologna: Apud Anselmum Giaccarellum.

———. 1961. Discorso intorno alle immagini sacre e profane. In P. Barocchi, ed., *Trattati d'arte del Cinquecento fra Manierismo e Controriforma*, vol. 2, 116–509. Bari: Laterza.

Parisano, E. 1623. *Nobilium Exercitationum libri duodecim De Subtilitate.* Venice: Apud Evangelistam Deuchinum.

Park, K. 1974. *The Imagination in Renaissance Psychology.* MA thesis, University of London.

Park, K., and L. J. Daston. 1981. Unnatural conceptions: The study of monsters in sixteenth- and seventeenth-century France and England. *Past and Present* 92:20–54.

Pecorella, C. 1968. Filiazione: Parte storica. In *Enciclopedia del diritto*, vol. 17, 449–456. Milan: Giuffrè.

Pomata, G. 1994. Legami di sangue, legami di seme: Consanguineità e agnazione nel diritto romano. *Quaderni storici* 29:299–334.

———. 2002. Family and gender. In J. A. Marino, ed., *Early Modern Italy 1550–1796*, 69–86. Oxford: Oxford University Press.

———. 2003. Comments on the papers given by Phillip Wilson, John C. Waller, and Laure Cartron. In *A Cultural History of Heredity II: 18th and 19th Centuries*. Berlin: Max-Planck-Institut für Wissenschaftsgeschichte.

Prodi, P. 1959. *Il Cardinal Gabriele Paleotti (1522–1597).* Rome: Edizioni di Storia e Letteratura.

Prosperi, A. 1996. *Tribunali della coscienza: Inquisitori, confessori, missionari.* Turin: Einaudi.

Roger, J. 1997. *The Life Sciences in Eighteenth-Century French Thought.* Ed. Keith R. Benson. Stanford, CA: Stanford University Press.

Romeo, G. 1990. *Inquisitori, esorcisti e streghe nell'Italia della Controriforma.* Florence: Sansoni.

Roodenburg, H. W. 1988. The maternal imagination: The fears of pregnant women in seventeenth-century Holland. *Journal of Social History* 21:701–716.

Rublack, U. 1996. Pregnancy, childbirth and the female body in early modern Germany. *Past & Present* 150:84–110.

Santangelo Cordani, A. 2003. L'accertamento della paternità tra dottrina e prassi all'indomani del Concilio di Trento: Uno sguardo alle *Decisiones* della Rota Romana. In A. Padoa Schioppa, G. Di Renzo Villata, and G. P. Massetto, eds., *Amicitiae pignus: Studi in ricordo di Adriano Cavanna, Tomo Terzo*, 1949–1987. Milan: Giuffrè.

Schiavoni, C. 1991. Gli infanti "esposti" del Santo Spirito in Saxia di Roma tra '500 e '800: Numero, ricevimento, allevamento e destino. In *Enfance abandonnée et société en Europe: XIVe–XXe siècle—Actes du colloque international . . . Rome, 30 et 31 janvier 1987*, 1017–1064. Rome: École Française de Rome.

Shildrick, M. 2000. Maternal imagination: Reconceiving first impressions. *Rethinking History* 4:243–260.

Sonnino, E. 2000. Le anime dei romani: Fonti religiose e demografia storica. In L. Fiorani and A. Prosperi, eds., *Roma, la città del papa: Vita civile e religiosa dal giubileo di Bonifacio VIII al giubileo di papa Wojtyla*, Storia d'ytalia. Annali 16. 327–364. Turin: Einaudi.

Teichmeyer, H. F. 1723. *Institutiones medicinae legalis vel forensis*. Jena: Sumptibus Ioh. Felicis Bileckii.

Vairo, L. 1589. *De fascino libri tres*. Venice: Apud Aldum.

Valentini, M. B. 1722. *Corpus Juris Medico-Legale*. Frankfurt am Main: Sumptibus Johannis Adami Jungii Typis Matthiae Andreae Viduae.

Vasoli, C. 1990. L' "analogia universale": La retorica come "semeiotica" nell'opera del Della Porta. In M. Torrini, ed., *Giovan Battista Della Porta nell'Europa del suo tempo*, 31–52. Naples: Guida.

Zacchia, P. 1661. *Quaestionum Medico-Legalium Tomus Prior [posterior]*. Lyons: Sumptibus Ioannis-Antonii Huguetan & Marci-Antonii Ravaud.

4 Continuity and Death: Literature and the Law of Succession in the Nineteenth Century

Ulrike Vedder

This chapter explores the lines of transmission between generations. It deals with what is passed on from one generation to another, the ways this is done, and who the agents mediating between the generations are. Because I intend to approach the discourse of inheritance and succession by considering how it is broached in literature, I will focus on literary rather than legal texts. However, in dealing with these questions, I will pay close attention to the legal system as one important cultural system of regulation, in particular the law of succession. This law seeks to determine the transfer of material possessions and the relationships underlying such transfers. Since we are dealing with definitions of family and kinship, and how these are reproduced and projected into the future, the discourses of nature and law clash time and again. The figure of the bastard, which I will consider in this chapter, is a valid case in point, since the bastard recurs as a figure of conflict along the lines of inclusion and exclusion. The concept of the bastard designates, by way of exclusion, what is proprietary and native, and what is not and therefore alien. We will see that the authority of the law—that is, its power to determine and sanction—is challenged in literature, just as the boundaries drawn by the laws of nature are unsettled when passion, guilt, and power (forces inseparable from inheritance) are brought into play.

4.1 Law, Nature, Literature

The law of succession governs the relationship between genealogy and legitimacy, kinship and property, and affiliation and transfer between generations. However, genealogical and legal relationships are structured by semiotic systems that need to be interpreted in order to become valid. This is one reason literature is a preferred arena for debating issues concerning genealogy and how it plays into property law and kinship relations. On the one hand, it is worth noting that a fictionalizing impetus takes place time and again in the law and its dispensation. Specific ways of dealing with authority and authorship, with rhetoric and fiction, with interpretation and the production of meaning, play their part in the complex relationship of law and literature. Both law and literature require and allow for the development of strategies to posit and repeal the rules of succession. Both adopt a range of procedures to typify and represent events as cases. By means of narration, case histories make reality unequivocal or ambiguous, define normal cases in relation to exceptions, and enforce, differentiate, and abolish legal and aesthetic norms.[1]

These various aspects of law and literature form a rich nexus of issues concerning the law of succession, where the problems of defining basic terms such as *kinship* and *property* become particularly apparent. So too do the intricate entanglements of property and passions that are involved in the transmission between generations. Reasons for conflict include the difficulties of determining points of rupture and devolution, of negotiating inclusion and exclusion, affiliation and alienation. These difficulties also include the fundamental doubt of paternity that turns every indication of paternity into a fiction. Or, to put it differently, they include what amounts to the potential conflict between maternal knowledge about the events of family reproduction, on the one hand, and the bequeathing of name and property along patrilineal lines of succession, on the other.[2] As we will see in one of Balzac's novels, these conflicts are conceived in terms of a gender war in such a way that the issue of inheritance cannot be resolved by a legal settlement.

The law of succession expands through time and absorbs the past, builds on tradition, and determines directives for the future of a family order. Another element of the law's impetus to fictionalize lies in the fact that it casts existing law into the future, since each final will reckons with a situation that has not (yet) occurred, namely, death. This problem is aggravated by a specific settlement of inheritance that is projected not only onto the next generation, but onto the open, unfinished future, while it remains nonetheless tied to a lineage endowed with a past that is still effective and considered great and worth preserving. In Germany and France this is the "majorat," or the entailed estate. The majorat subjects all future generations to a form of succession that privileges the "firstborn of the paternal line" by way of the authority of the estate's founder and thus preserves the indivisibility of the estate. It intends to transmit and make immortal the dignity, virtue, and glory of a family by safeguarding the family estate—real estate, in most cases—against fragmentation and loss, as much as against financial speculation and indebtedness. Thus, according to whichever economic point of view is taken, family property is transformed either into a security with maintained value or into locked or unrealizable capital.

Instituting a majorat, which became more and more common in Germany from the seventeenth century onward, thus determines that the special funds of commodities and assets are to be indivisible and inalienable. This occurs through a private declaration of volition that is also a legal transaction.[3] According to its reformulation in the General Common Law of the Prussian States (June 1, 1794), assets would go to "the next in line in the family, according to grade, . . . among several close enough, however, to the elder." As a later commentary states, this was to be determined according to "individual succession within the agnatic line, as only the former preserves wealth and only the latter the name."[4] What we recognize here is that the law produces a rhetorical community—the family to be perpetuated—and points to the rhetorical essence of supposedly natural communities. At the same time, the law of succession and the majorat in particular are supposed to naturalize legal communities. The legal institution of the majorat indicates the status of "nature" in two directions. First, the majorat refers to nature in determining succession in terms of blood relationship. Succession is only able to occur within the male line of relatives—that is, within the patrilineal filiation of the same blood. Second, the majorat naturalizes this perpetuation "for all eternity," so that property is seen to be virtually sanctioned by the law of nature.

The majorat constitutes a regulated means of consolidating a family estate, and of giving material goods a temporal meaning. In turn, such an entailed estate materializes time through the succession of a family. As a "material medium"[5] for the continuity of a family and its assets, the majorat establishes a particular form of genealogy, which is charged with a high conflict potential that makes it of interest for literature. This potential results, first, from the injustice that only the firstborn inherits. Second, it stems from the founder's claim to authority by committing succeeding generations to a specific form of tradition for all eternity. Third, conflicts might arise from the problems of defining terms such as *family, paternity*, and *parental love*, and from the concomitant difficulties of producing signs and their interpretations. Literature makes use of these uncertainties and complications. This can go as far as questioning not only the legitimacy of progenies, which means calling into question family affiliation, but, as we will see in one of Achim von Arnim's stories, it can go as far as challenging the legitimacy of the laws of nature, which amounts to calling into question affiliation to the human species.

During the French Revolution, legislation referred to the new general principles of freedom and equality, and aspired to a "morcellement," a dividing up of the large estate holdings in the hands of the landed aristocracy. Accordingly, the so-called substitutions—the indivisible and inalienable inheritance by primogeniture—were abolished. There was a conspicuous rhetoric of excess and unnaturalness in the ensuing debates: "monstre des substitutions," "monstrueuse inégalité," "monstruosité politique."[6] In these debates, furthermore, the present was set against the adherence to the past as well as against the majorat being cast into a future, for which the present kept being sacrificed. Thus, there is mention of the Ancien Régime's regulations of the law of succession "that subordinates the interests of the living people to the caprice of the dead" and where "the [present] generation . . . finds itself constantly sacrificed to the generation that is not yet."[7] These time gaps become an insoluble problem in the literary texts embracing the majorat. The revolutionary legal texts, by contrast, look for a solution that is both unequivocal and final. Thus the Convention states in Article One of the "Décret sur les substitutions" of November 14, 1792: "All substitutions are suspended and prohibited in future." However, this did not mean that the majorat had been done away with for good. There were numerous clashes in the first half of the nineteenth century, both in France and in the German states, such as Prussia, which led to the alternating abolition and the more or less clandestine reconstitution of the majorat.[8]

Arguments in favor of and against the majorat are also raised in the literary texts that I will look at in this chapter. The rhetoric of the law and of jurisdiction come into specific operation in these texts—be it in trial scenes or in drawing up settlements. Beyond such instances of legal discourse, the literary narratives of inheritance and succession of the first half of the nineteenth century deal more widely with the law's power to define and interpret what constitutes family and the linking together of generations. Conversely, these narratives are also about how legal norms are made legitimate in the face of threatening disruptions of property arrangements and the succession of generations. These disruptions arise as part and parcel of the transmission and recurrence of guilt, death, and passions inscribed in the majorat's very conception of succession.

In the following section, I will discuss the legal-literary order and how genealogies run their course in three texts that deal with the antiquated legal institution of the majorat. In the fiction written around 1800, legally sanctioned paternal authority not only enters a state of crisis, but leads straight to the downfall of families, as we will see in E. T. A. Hoffmann's *Das Majorat* (1817). In this work and others, an entirely different social principle takes the place of the law: the power of money, or respectively, capital. This process is demonstrated in a polarized manner in Achim von Arnim's *Die Majoratsherren* (1819). In this story, the power of money, however, makes the exclusiveness of the "old family" in some way contingent, a theme developed in Honoré de Balzac's novel *Le Contrat de mariage* (1835). Both narratives feature main characters that are hybridized, even creolized, and as such are presented as contaminations of the legal discourse and of the discourse of race. My treatment of these three stories of the demise of old family lineages will amount to an account of the end of genealogies.

4.2 The Genealogy of Death: Law, Writing, Fiction in the Work of E. T. A. Hoffmann

In Hoffmann's "Das Majorat" (1817), the lineage of "Freiherr Roderich" comes to an end, at the end of the text, without any descendants. All members of the family have either been murdered treacherously, have fallen in battle, or have died in accidents or from grief. The ancestral castle lies in ruins and the rich landholding has fallen into the hands of the state. The *splendor familiae et nominis*, which is mentioned repeatedly in all arguments in favor of the majorat, features as a splendid aim in this story. Ironically, family reputation has become more and more tainted precisely by the majorat and has ultimately evaporated. A failed operation, which is not explained in any detail but has obviously occurred in the field of "occult sciences,"[9] leads Freiherr Roderich to seek the cause of his misfortune in his "ancestors' guilt," because they had left the ancestral castle. Furthermore, he decrees a majorat, "in order that, for the future at least, the representatives of the family might be induced to reside at Rolandsitten."[10] Determining the future through the law of succession is thus supposed to bind together house and lineage, all the more necessary since the family "had already spread its branches into foreign territories."[11] In addition the injustice of the majorat and its consequences, shackling all these creeping branches to one place and its laws of inheritance, fatally causes the individual branches and hence the entire family pedigree to break off.

In Hoffmann's story, the majorat is marked by its almost imminent lethal effect, taking hold of all the founder's descendants. Both of the two sons and their four descendants, as well as two wives and the manor's steward, die due to the majorat, whereas not one single birth or fathering is recorded in the course of the story. The majorat thus consists in the recurrence of death. The sons die their fathers' deaths, and the first son's murderer keeps returning to the room of death—first as a sleepwalker, then as a dying man, and finally as a ghost. What this amounts to is in fact a genealogy of death and of the undead. In other words, just as the naturalization of the legal institution of the majorat leads to a rhetoric of unnaturalness and the monstrous, so too in literature does that which has been regulated in supposedly "rational" legal terms in Hoffmann go over increasingly to the "irrational," specifically to the genre of ghost and visionary stories. Later in this chapter we see how the spectral metaphor is similarly evoked in the work of Arnim.

In Hoffman's "Das Majorat," the new beginning that Wolfgang, the founder's son, hopes for turns out to be impossible. He had married abroad against his father's wishes and under the false common-law name *Born*, meaning "spring" or "well." Yet Wolfgang named his son Roderich again, after his own father. Besides names and properties, there are various other agents that mediate between the generations. Letters announcing death, letters written from the deathbed, and other written documents negotiate the past and future. Certificates of baptism, certified excerpts from church registers, the sovereign's confirmation of the endowment, tally books, and documents left behind—such as wills, collections of correspondence, or confessions—proliferate. These documents compete for interpretation, from which the factuality of the law can begin to transpire. When the legitimacy of Wolfgang's son—who appears by surprise as an issue from the secret marriage and who is initially suspected of being an illegitimate child not entitled to inheritance—is called into question, litigation proceedings are required to establish facts of law and nature. Court interpretation is required to determine how the endowment of the entailed estate and the will are connected, as much as to ascertain the conclusiveness of the various proofs of identity.

When the first-person narrator, a young "Justizmann," visits the ruins of the castle at the end of the text as the sole survivor of the story, it is not as if he has claimed victory in a historical process, as one might be tempted to gather from the analogy of the collapsed ancestral castle with the demise of the aristocracy. Rather, he is caught up in the narration of a genealogy of death, a fact made concrete by Roderich inviting the narrator to be buried alongside the noble family in the same tomb.

4.3 Mixture and Exclusion in the Work of Arnim

In Arnim's story "Die Majoratsherren" (1819), which was written almost at the same time as Hoffmann's "Das Majorat," the narrator seeks to elude such entanglements. Despite the many blendings and meddlings that traverse the text, Arnim's narrator is separated from the age of the majorat and the lineage that collapses with it. He situates himself after the French Revolution, when, in the frame of the story, he discusses the social transformation that the French Revolution entailed and that severed the glorious prerevolutionary past—the age of the majorat—from the postrevolutionary present—a monotonous and poor age. This shift is symbolized in the majorat house, which, by the end of the story and of the French Revolution, will have fallen into the hands of Vasthi, an unscrupulous old Jewish woman.

In dealing with Arnim's story, I will focus on the key role of time and temporality as well as on the mixing and separation of what is native and proprietary from what is not. Both aspects are crucial to considering the issues of inheritance and transmission by succession, since both are to be made governable by the law of succession. However, the organizing principle of the majorat fails on two levels. First, within the family, the majorat's order of succession is ignored in treacherous ways. Second, failure occurs in historical-systematic terms, since the economy displaces the law—in its function as a dominant principle that rules the organization of society.

The story of the young, indecisive majorat lord who visits his substantially older cousin—a grotesque, decrepit lieutenant, who spends his life waiting for a spouse and an inheritance—only lasts for four days. During this interval of days the past history that had occurred thirty years before is revealed. At that time, the old majorat lord, who was without a male heir, had given away his only child, a newborn daughter, who was not entitled to inherit, to a Jewish "horse-coper" (horse dealer),[12] who called the child Esther and took her in as his foster daughter. Concurrently, the old lord secretly adopted the illegitimate son of a lady-in-waiting as his own to cheat the rightfully entitled nephew, the already-mentioned cousin of the young majorat lord, out of the majorat. This cousin is portrayed as being as immutable as the majorat house that has been unoccupied for thirty years but where the clocks are still wound up—and as lifeless. The lieutentant, for example, still goes out on the same walks in his worn-out uniform coat and clings to unchanging snuff and visiting habits. Neither his preoccupation with heraldry nor his collection of coats of arms is able to infuse his life with any of the old glory or hope for the future. These interests are contaminated by economic necessity, which requires him to copy the collection, stick the coats of arms on paper with care, and sell them to earn a living.

The protection that the majorat sets against any form of change thus also amounts to a constraint of not being able to change anything. Both older protagonists are subjected to this constraint. The cousin is stuck with the old lady-in-waiting (also the young majorat lord's mother), for whose love he has been waiting in vain for thirty years, just as he has for the entailed estate. Similarly, the two younger protagonists—the young majorat lord and the horse-coper's foster daughter, Esther—are both permeated by the ghosts of the past. Each fails to live in the present, which does not belong to him or her. This practice of living in the past and for the future ultimately leads to death, even in cases where the attenuation of life trajectories at first appears to have been vindicated. The old cousin does indeed go on to take possession of the entailed estate and marry the lady-in-waiting, moving into the majorat house with her and a horde of animals—in pointed contrast to children. However, life there, which he has desired so much for thirty years, is hell. The new majorat lord is humiliated by his spouse and her horde of animals until one day he passes away unnoticed and ingloriously.

This tale draws attention to the issue of the present in genealogical and "law-of-succession" narratives. The generations that the majorat chains together are conceived as a continuum of the past and the future—without assigning more to the present than maintaining the majorat in due form. In Hoffmann, the present consists of administering the majorat and its documents and tally books. In Arnim, the loss of a present that is fixed by the past brings forth "a hollow space in time,"[13] which can be identified in topographic terms as the unoccupied majorat house with its clocks ticking away emptily.

Vasthi, the old Jewish woman who is Esther's foster mother, makes use of the hollow space of this empty present. In the wake of the Revolution, after the elimination of the majorat, the release of the Jews, and "under foreign rule,"[14] Vasthi emerges as a war profiteer to take over "the Entailed house and turn it into an ammonia factory . . . through the favour of the new government."[15] To emphasize the point, the last sentence of the story reads: "Credit came to take the place of feudal right."[16] In Arnim's story, the economy comes to occupy the place of the law. Monetary transaction with its stereotypical Jewish connotation takes the place of the feudally marked real estate holding. Inalienable real estate—charged with family history—is displaced by its speculative, abstracting monetary value. This makes affiliation with the lineage contingent, a contingency that the text produces as a blurred distinction between what is proprietary and what is alien.

Thus, the young majorat lord, who as a visionary is hardly capable of distinguishing between reality and vision, is a bastard in terms of matrimonial law and the law of succession. Esther is not only the foster daughter of a Jewish "horse-coper (*Rosstäuscher*)"—who earns a living if not necessarily from *täuschen* (deceit) then from *tauschen* (exchange)—but she is also an image of her Christian mother. In the act of dying, she is cast as a mythical, hybrid creature consisting of a Jewish angel of death and a Christian shining light. The two illegitimate children, the young majorat lord and Esther, reveal their exchanged and deceptive origins in a vision: "You are I and I am you."[17] The self can no longer be identified clearly, neither within family boundaries—given the multiplication of forms of fatherhood and motherhood as natural, adoptive, stepparents and foster parents—nor within the human species. In the end, the cousin, caricatured as a turkey, is obliged to wait on the lady-in-waiting's cats and dogs at the majorat banquet table.

Given such narrative blendings of boundaries, one particular boundary that is brought into focus time and again in the text gains significance and explosive force. This is the boundary between the town and the Jewish ghetto, which can only be crossed at the expense of death. The boundary between Christians and Jews, which first is introduced in geographic terms, is then sharpened and naturalized in a racist manner as that between the human and the nonhuman. This naturalized racial distinction is drawn when the Jewish woman, Vasthi, is strangling Esther, her Christian foster daughter, and again when Vasthi appears as a monstrous sharpened silhouette: "like the cut-out card faces . . . held before a light. . . . She looked like no human creature but rather like a vulture which, long warmed by God's merciful sun, pounces with all that harboured fire upon a dove."[18]

The crisis of family authority and its inheritance is transferred to anti-Jewish denunciation to effect a shift from the decrepitness of feudal-patrilineal transmission, caricatured in the grotesque drawings of the elderly cousin and his lady-in-waiting, to a takeover by "foreigners" and their financial resources. As a universal means of exchange, money abstracts any tangible value—such as, for example, that of entailed real estate, from its family-historical significance. Because it is exchangeable in general terms, money devalues the specifically individual and proprietary. In Lacan's terms, it becomes "the most annihilating signifier,"[19] standing for something it then replaces. This process becomes even more "unnatural" in Arnim's story as family disorder, destruction, and annihilation radiate from the mother's position in the system (Esther's mother, the lady-in-waiting, Vasthi). Honoré de Balzac's novel *Le Contrat de mariage*, which I will discuss next, offers another account of the way the discourse of law (and lawfulness) and the discourse of nature (and race) compete with each other through money, which is in turn bound up with gender.

4.4 Classification and Gender War in the Work of Balzac

In Balzac's novel (first published separately as *Fleur des pois* in 1835, then republished in 1842 in the *Comédie humaine* series as one of the "Scènes de la vie privée"), a contract between spouses is concluded in which the establishment of a majorat features as a key clause that will ultimately ruin the marriage. What is highly significant about this contract is that it is concluded between two "hybrids," of which one, as we are told, is assigned to the social order—a "social mongrel"—and the other to the natural order in which "the Creole nature is a thing apart." One of the parties to the contract is Paul de Manerville, a rich heir from landed-gentry stock, gentle and naive, whom the text describes as "un métis social,"[20] a "social mongrel." Paul owes this designation to his inability to exercise any male, aristocratic power. Though he is forced to follow the customs of the aristocratic habitus, he pursues the entirely unaristocratic and bourgeois, sentimental ideal of love, desiring wedlock with a woman who stands by him, shares his thoughts and secrets, and forms a unity with him in order to lead a married life shaped by intimacy, love, and naturalness: "A heart to which I can confide my business troubles and tell my secrets."[21]

Provincial society nicknames Paul "Fleur des pois," "pea blossom" (which was what the novel was first called), which is an early eighteenth-century designation for an "homme à la mode." At the beginning of the nineteenth century, when the novel is set, however, this name is no longer "à la mode," but an anachronism that the novel uses as a marker of a political and social conflict. To be named a "pea blossom" is considered an honor by "the royalists," whereas it evokes mockery among the "liberal circles"[22]—not least, perhaps, because the pea is a hermaphrodite. The pea flower is known to have acquired a key position in systematics at the end of the nineteenth century, among other things in Mendel's experiments and creation of "plant hybrids" (recorded in his *Versuche über Pflanzenhybride*, first published in 1866). Mendel chose the *fleur des pois*, because it allows one to reconstruct and calculate—and thus produce by experiment—the purity and hence proportions of a mixture in an optimal manner. This association is, it must be admitted, anachronistic, but it draws attention to Balzac's arrangement of the novel as an experiment, specifically as an experiment in the domain of nature, as well as in the social and legal spheres. Balzac's experiment concerns the unity of the species and the difference between the genders, insofar as two hybrids are supposed to unite. In *Le Contrat de mariage*, the "métis social" chooses a stunningly beautiful Creole, whose inheritance from her deceased father, a businessman, has long been squandered and who is thus in search of a marriage that will bring her profit in economic and social terms. She and her mother are not only Creoles, but female, and these two features of "nature" make up their dangerous character: "The Creole nature is a thing apart . . . an attractive nature withal, but dangerous."[23]

These dangerous hybrid formations transcend the modes of the natural and the social, since their boundaries and transformations are constantly subject to discursive negotiations that Balzac explores in different ways. Negotiations occur in the area of the law within the novel itself, and in the sciences in the programmatic preface to the entire *Comédie humaine*. In the "Avant-Propos" (written in 1842), Balzac ponders his dream, a "chimère,"[24] as he calls it, of the comparability of the animal, "l'Animalité," and the human, "l'Humanité." As is well known, Balzac also refers to Georges Cuvier's dispute in the Académie with Étienne Geoffroy Saint-Hilaire (1830) about the relationship between unity and variety.[25] Balzac follows Geoffroy Saint-Hilaire's position of a *unité de composition* when he states that there is only one animal, one model for all organized beings.[26] Tracing back all milieu-determined species to a fundamental, unified model encourages Balzac to undertake his program of comprehensively cataloging society, of classifying the social, in the *Comédie humaine*, on the analogy of the organizing principles of Georges-Louis Leclerc Comte de Buffon's *Histoire naturelle* (1749–1788). He then goes on to cite the "drames," "hasards," and "confusion" that stand in the way of such organizing principles. Contrary to the writing of literature, which he at first conceives as the application of a theory, Balzac's novels in the *Comédie humaine* are precisely about the inscrutabilities of classification, about how coincidences and passions intervene in the drawing up of boundaries and distinctions between the various genera, and about the inconsistency and lack of uniformity of the *composition*. This becomes apparent in *Le Contrat de mariage*, not only by featuring hybrid characters, but by negotiating a contract between the two family lawyers that reaches more and more into the arena of warfare. In other words, and to quote Balzac's: "I have painted every misfortune of a wife: it is time to show the pain of the husband."[27]

And this is where we find the calculating—and in this sense "unnatural"—Creole, who will emerge victorious from the battle waged over the contract between spouses, in which the newly inserted majorat is supposed to act as a barrier preventing the flow of money from the male to the female side. The Manerville's long-standing family lawyer had after all managed to bring the heir's estate through the French Revolution and the wars unscathed. But neither the lawyer nor the amorous husband is equipped for the gender war that the young wife and her mother wage, which entails the husband's complete and utter financial ruin, and ends in his departure for Calcutta, while his mother-in-law takes possession of his rich commodities and assets.

Classifying human beings into social species, on the analogy of classifying animals into zoological species, which Balzac plans in the preface to his cycle of novels, comes up immediately against the difficulty of gender difference, which compounds the problem of classification.[28] Whereas Buffon had managed to describe the females of a particular species in a few lines in his *Histoire naturelle*, Balzac makes it quite clear that one annotation on the female "variant" will not do in cataloging the social species. Balzac's basic hypothesis, in which he follows Geoffroy Saint-Hilaire's, is that there is a unity of the genus, whose wealth of variations is developed by each particular milieu and that can, given the analogies between these variations, ultimately be put down to the notion of unity again. This hypothesis can no longer be sustained once the gender issue is taken into account, because Balzac states that in society a woman is not merely the female version of the male, as can be seen in the fact that two human beings living in one household may be totally unlike.[29]

The programmatic text of the "Avant-Propos" does not go into more detail about this discrepancy, which explodes the analogy of *humanité* and *animalité* and thus, too, the notion of the unity of the human species as a basis of a social world that can be cataloged. By contrast, the novel acts out this area of conflict, by inserting a legal construct—in the shape of a contract between spouses—in the place of a mediating natural entity. Within this contract, the majorat appears as the unifying central principle, with whose assistance all differences between the sexes and families are supposed to be resolved by way of the law of succession. However, the majorat actually causes the conflict to erupt into a passionately waged war and to ultimately put both conflict and marriage beyond reconciliation. It is the Creole's young lawyer—a modern profiteer of speculative transactions, who later goes on to marry "a rich mulatto"[30]—who manages to pull off a surprise coup by casually inserting a clause that the old lawyer fails to take seriously because he presumes the scenario to which it refers is entirely unrealistic. This clause sketches the case of an absence of descendants, in which case the majorat is to be transferred into the spouses' community of property. As it happens, the young wife brings about this very case and makes this fiction a reality by evading the generation of descendants. The descendants no longer die of the majorat, but they are not even brought into life.

The clause that is slipped in ex post facto, inserted into a comprehensive majorat contract like a foreign body, thus becomes the decisive moment in the spouses' sexual and genealogical life. Therefore it is a single word only, spoken on the morning after the wedding night, by which, as the young bride tells her mother, her "revanche" has started. We are not told which word it was, but the fact that the young husband's ruin goes ahead according to plan is clear enough evidence that it was one of negation, such as *no*, perhaps even *never*, expressing the bride's refusal of "natural reproduction" and denial of conjugal rights. As Balzac scholarship has shown, the old lawyer would have been able to prevent the husband's downfall by legal means, since the subsequently inserted contractual clause was in fact ineffective.[31] Balzac, however, demonstrates that it is the institution of the majorat itself that is ineffective; the fact that it is drawn up at all leads to destruction.

4.5 Conclusion

In the works of Balzac, Hoffmann, and Arnim, giving up the majorat means giving up the conception of direct and linear transmission between the generations. This voluntary and involuntary surrender is manifest in four areas. First, loss of property and title leads to a loss of position. Second, being cut off from the family name and thus losing affiliation to what is proprietary leads to a loss of identity. Third, the dissolution of resemblance and essence mark a breach in the lines of transmission. Finally, forcing some potential successors out of time and disallowing their futures robs the whole family of "perpetual continuity." The three literary texts, each in its own way, highlight three ways in which genealogical thinking was troubled, or even thwarted. First, inheritance was undone by the unpredictable passions and entanglements that are always transmitted alongside the estate and that the three texts that I have considered set in motion. Second, shifting political economies provide a background to the increasing abstraction of the property that is to be entailed. This alienation has a tangible shape as real estate and gold nuggets in Hoffmann, is expressed through the transition from a feudal system to credit in Arnim, and appears as a speculative capital investment in the shape of bonds and shares in Balzac. Third, succession is troubled by the fact that all three texts present figures of the hybrid, the bastard, and the undead. Each of these characters defies categorization and thus thwarts the majorat and its underlying principles. The hybrid hinders determination of affiliation. The bastard undermines determination of the perpetuation of the proprietary. The undead challenges the rational temporal consistency of one generation succeeding another.[32]

Read as texts on heredity, the two stories and novel respectively show the crisis of the "genealogical imperative"[33] in an epoch marked by political, economic, cultural, scientific, and juridical upheavals. Moreover, with their concentration on the legal institution of the majorat as an extreme means of preserving a continuity—even if that leads to death—they reflect the conflicts between the discourses of nature and law in different ways. Hoffmann was a lawyer and focused on the fatal changes that the protagonists undergo when they become heirs of the majorat. Arnim as a "Majoratsherr" himself (his grandmother founded a majorat) documented the tensions between the compelling immobility of the inheritance and the financial constraints that reinforce his disapproval of the Ancien Régime as well as his anti-Judais.[34] Balzac, with his programmatic interest in cataloging society in a scientific way, turns his tale into a rather ironic narration of a young and yet decrepit heir whose familial continuity in its biological and juridical dimensions is lost in the end thanks to the majorat.

These narrations of natural, social, and economic death, or of an at-best undead living, through the majorat—though it promises not only continuity but even control over eternity by controlling succession—may also throw light on the relation between specific laws of succession and inheritance on the one hand and heredity in the broader sense as a knowledge regime on the other. To understand this relationship, we might take into account the shifts in the nature of inheritance and kinship dynamics that David Sabean analyzes in chapter 2 of this book. Viewed from this perspective, the specific juridical case of the majorat can be seen as a burning mirror that makes visible the aporia of a continuous transmission of property following intergenerational rules. These rules are fixed once and for all to keep alive an inalienable property for all time—a process in which death plays an unavoidable role. This aporia of a movement of transmission on behalf of death is unsolvable by the abolition of the specific legal institution, yet marks all conceptualizations, laws, rules, and discourses of hereditary transmission in a more or less implicit way. Through exploration of this transition, literature gains its narrative potential.

Notes

The translation of this chapter was provided by Mark Kyburz.

1. See Freeman and Lewis 1999; Posner 1988; Weisberg 1992.

2. See Eigen 2000. For the problem of resemblance as a symptom of the questionable character of all paternities, see De Renzi, chapter 3, this volume; Müller-Sievers 2005.

3. Eckert 1992, 24.

4. Gierke 1909, 104.

5. Mangold 1989, 228.

6. See Eckert 1992, 180–185.

7. Eckert 1992, 202.

8. See Dietze 1926; Sybel 1870; Eckert 1992.

9. Hoffmann 1826, 5–6.

10. Hoffmann 1826, 5–6.

11. Hoffmann 1826, 6.

12. Arnim 1990, 11.

13. Oesterle 1988, 29.

14. Arnim 1990, 41.

15. Arnim 1990, 41–42.

16. Arnim 1990, 42.

17. Arnim 1990, 24.

18. Arnim 1990, 37.

19. Lacan 1991, 37.

20. Balzac 1976, vol. 3, 530.

21. Balzac 1897, 17.

22. Balzac 1897, 23.

23. Balzac 1897, 145.

24. Balzac 1976, vol. 1, 7.

25. See the annotations of the editor in Balzac 1976, vol. 1.

26. "Il n'y a qu'un animal. . . . un seul et même patron pour tous les êtres organisés" (Balzac 1976, vol. 1, 8).

27. "J'ai peint toutes les infortunes des femmes: il est temps de montrer aussi la douleur des maris" (Balzac: Lettres à l'Etrangère, vol. 1, 275; quoted in Perrod 1968, 221).

28. "La description des Espèces Sociales était donc au moins double de celle des Espèces Animales, à ne considérer que les deux sexes" (Balzac 1976, vol. 1, 9).

29. "Dans la Société la femme ne se trouve pas toujours être la femelle du mâle. Il peut y avoir deux êtres parfaitement dissemblables dans un ménage" (Balzac 1976, vol. 1, 8–9).

30. Balzac 1897, 176.

31. See Perrod 1968.

32. On the figure of the bastard as a symptom of the conflict between the discourse of law and the discourse of nature around 1800, see Vedder 2002.

33. Tobin 1978.

34. See Riedl 1994.

References

Arnim, Achim von. [1819] 1990. *Gentry by Entailment*. London: Atlas Press.

Arnim, Achim von. [1819] 1991. Die Majoratsherren. In Gisela Henckmann, ed., *Erzählungen*, 211–251. Stuttgart: Reclam.

Balzac, Honoré de. [1835] 1976. Le Contrat de mariage. In Pierre-Georges Castex, ed., *La Comédie humaine*, vol. 3, 527–653. Paris: Gallimard.

Balzac, Honoré de. [1842] 1976. Avant-Propos. In Pierre-Georges Castex, ed., *La Comédie humaine*, vol. 1, 7–20. Paris: Gallimard.

Balzac, Honoré de. 1897. *The Marriage Contract*. London: Caxton.

Dietze, C. von. 1926. Fideikommisse. In Ludwig Elster, ed., *Handwörterbuch der Staatswissenschaften*, vol. 3, 993–1006. Jena: Fischer.

Eckert, Jörn. 1992. *Der Kampf um die Familienfideikommisse in Deutschland: Studien zum Absterben eines Rechtsinstituts*. Frankfurt am Main: Lang.

Eigen, Sara P. 2000. A mother's love, a father's line: Law, medicine and the 18th-century fictions of patrilineal genealogy. In Kilian Heck and Bernhard Jahn, eds., *Genealogie als Denkform in Mittelalter und Früher Neuzeit*, 87–107. Tübingen: Niemeyer.

Freeman, Michael, and Andrew D. E. Lewis, eds. 1999. *Law and Literature*. Oxford: Oxford University Press.

Gierke, Otto. 1909. Fideikommisse (Geschichte und Recht der Fideikommisse). In Johannes Conrad, ed., *Handwörterbuch der Staatswissenschaften*, vol. 4, 104–114. Jena: Fischer.

Hoffmann, E. T. A. [1817] 1985. Das Majorat. In Wulf Segebrecht and Hartmut Steinecke, eds., *Sämtliche Werke,* Vol. 3: *Nachtstücke und andere Werke 1816–1820*, 199–284. Frankfurt am Main: D. Klassiker-Verlag.

Hoffmann, E. T. A. 1826. Rolandsitten; or The dead of entail. In Robert P. Gillies, ed., *German Stories: Selected from the Works of Hoffmann, de la Motte-Fouqué, Pichler, Kruse, and Others*, vol. 2, 1–175. Edinburgh/London: William Blackwood/T. Cadell.

Lacan, Jacques. 1991. Das Seminar über E.A. Poes "Der entwendete Brief." In Norbert Haas, ed., *Schriften*, vol. 1, 7–60. Weinheim/Berlin: Quadriga.

Mangold, Hartmut. 1989. *Gerechtigkeit durch Poesie: Rechtliche Konfliktsituationen und ihre literarische Gestaltung bei E.T.A. Hoffmann*. Wiesbaden: Deutscher Universitäts-Verlag.

Müller-Sievers, Helmut. 2005. Ahnen ahnen: Formen der Generationenerkennung in der Literatur um 1800. In Sigrid Weigel, Ohad Parnes, Ulrike Vedder, and Stefan Willer, eds., *Generation: Zur Genealogie des Konzepts—Konzepte von Genealogie*, 157–169. Munich: Fink.

Oesterle, Günter. 1988. "Illegitime Kreuzungen": Zur Ikonität und Temporalität des Grotesken in Achim von Arnims "Die Majoratsherren." *Etudes Germaniques* 169:25–51.

Perrod, Pierre-Antoine. 1968. Balzac et les "Majorats": De la brochure sur Le Droit d'Ainesse au Contrat de mariage. *L'Année balzacienne* 9:211–240.

Posner, Richard. 1988. *Law and Literature: A Misunderstood Relation*. Cambridge, MA: Harvard University Press.

Riedl, Peter Philipp. 1994. ". . . das ist ein ewig Schachern und Zänken . . .": Achim von Arnims Haltung zu den Juden in den "Majorats-Herren" und anderen Schriften. *Aurora* 54:72–105.

Sybel, Heinrich von. 1870. *Geschichte der Revolutionszeit von 1789–1800*. Vol. 4. Düsseldorf: Buddeus.

Tobin, Patricia D. 1978. *Time and the Novel: The Genealogical Imperative*. Princeton, NJ: Princeton University Press.

Vedder, Ulrike. 2002. Zwillinge und Bastarde: Reproduktion, Erbe und Literatur um 1800. In Ulrike Bergermann, Claudia Breger, and Tanja Nusser, eds., *Techniken der Reproduktion: Medien—Leben—Diskurse*, 167–180. Königstein/Taunus: Helmer.

Weisberg, Richard. 1992. *Poethics and Other Strategies of Law and Literature*. New York: Columbia University Press.

II Heredity and Medicine

5 The Medical Origins of Heredity

Carlos López-Beltrán

Such as the temperature of the father is, such is the son's, and look at what disease the father had when he beget him, his son will have after him . . . as is as well inheritor of his infirmities as of his land.

—Jean Fernel 1554; quoted in Burton 1621

Either no causation governs this matter (of hereditary descent) or else the most minute constitutional peculiarities of children are caused by similar elements in their parents. Then let parents learn and remember that their prospective children will be the very images of themselves, reflected in all their shades of feeling and phases of character . . . inheriting similar tastes, swayed in similar passions, governed by kindred sentiments, debased by the same vices, ennobled by like virtues, adorned by kindred charms and graces, and endowed with similar moral powers and intellectual capabilities.

—Fowler 1848

The concept of biological heredity began with that of *human* heredity. As Jean Fernel's (1497–1558) famous assertion reveals, the first expression of this notion was a metaphorical one built on the analogy between transmission of wealth and other legal goods and the common observation of physical similarity between parents and offspring ("Like begets like")—in this case, the undesirable similarity of disease. Human heredity was thus first articulated with the observation that genealogical relationships between persons imply more than social bonds: there are also physical rapports manifested by looks and by health. The inference that something could be passed on from parents to offspring in the act of generation that accounts for the perceived similarities is natural enough for us. However, for those living under different physiological and theological frames, in which similarities could be accounted for in different ways or dismissed as accidental or irrelevant, this was a new idea.[1] Similarly, the metaphor that was found suitable for such transmission was far from natural when it was first deployed. As land, craft, wealth, and title were inherited, so too could body and mind, or some of their features, be inherited. This idea took a long time to become acclimatized in modern biological thought. It was adopted in several different fields whose paths coincided toward the third decade of the nineteenth century, when our modern concept of heredity finally acquired its basic structure.

The idea of a human biological heredity was negotiated among a range of social actors between the early modern period and the mid-nineteenth century, and its status shifted in consequence from a simple suggestive metaphor to a full-fledged natural law. The ontological drift was from the superficial transmission of general features, such as land or temperament, to the deep-rooted persistence of peculiar powers and dispositions, linked in an essentialist way to family and race. The two ends of this *longue durée* negotiation are neatly captured by the quotes by Fernel and the American phrenologist Orson Squire Fowler (1809–1887) above.[2] The quotes also define the main territory in which such negotiations took place. The biological fact of inheritance emerged in relation to medical questions about hereditary transmission of disease. The epochal shift from a metaphorical use to a causal, explanatory notion, as I have also noted elsewhere, is reflected by a linguistic shift—that is, the adoption of the noun *heredity* in several European languages after the second decade of the nineteenth century. The ubiquitous noun *heredity* that we use today only acquired its modern conceptual frame as a structured, causal, mechanistic presence in the biological realm once the reification of the metaphor was ubiquitous.[3] My aim in this chapter is to highlight how an important set of developments that helped materialize biological heredity took place within the context of European medical preoccupations with hereditary transmission of disease. Physiological theory developed to sort out if a proclivity, or a disposition to acquiring a given ailment, could in fact be passed on from ancestors to descendants.

From antiquity until the eighteenth century, the adjective *hereditary* was loosely employed when a given trait was found to characterize a family or another genealogical group. I have proposed we should maintain the adjectival terminology, using the formula "the hereditary" to refer to the set of phenomena naturalists and physicians could associate with temperamental similarity. These include resemblance to the father and mother, general temperamental dispositions, peculiar gaits, and physical marks like moles or large heads.[4] The reference to the hereditary nature of a trait occurred, however, with much more frequency and consistency when anomalies, moral or physical, were the subject. "Hereditary gout" or "hereditary depravities" were more common formulas than their positive counterparts. In fact, as Maaike van der Lugt and I have discussed, it is quite likely that the use of the hereditary metaphor filtered into European languages through translations of Arabic medical treatises, mainly Avicenna, when translating their notion of *haereditarii morbi* (hereditary diseases).[5]

I will maintain in this chapter that some of the most important steps that moved the notion of hereditary transmission of physical features from the early modern accidental perspective to the nineteenth-century essentialist approach are found among European medical men. The early modern medical notion of hereditary diseases was a main hinge by which the metaphorical flow of structured meaning passed from the legal to the biological sphere. The *longue durée* efforts of physiological clarification of this notion became the motive to keep the transmission metaphor alive for many decades, until it became developed into the powerful explanatory tool of nineteenth-century naturalists, physiologists, and, on another path, of psychiatrists, social reformers, and novelists of decline.[6]

To show this I will focus on a series of essays written in response to two prize competetions issued by the Parisian Société Royale de Médecine in 1788 and 1790, a period coinciding with the start of the revolutionary period in France and the redefinition of the role of the medical profession entailed by the revolution. I first describe the background to this competition in seventeenth- and eighteenth-century medical theories of inheritance. I then try to show that an understanding of the much more structured, and theoretically sophisticated, notion of hereditary transmission we find in the answers to the prize competitions provides a better historical understanding of the apparently sudden appearance of a biological concept of heredity in the first decades of the nineteenth century. The conceptual space in which naturalists framed their preoccupations concerning hereditary transmission of physical features, especially after the first three decades of the nineteenth century, was a space carved out by medical men in the first instance. Heredity, I repeat, started out as *human* heredity, in particular, as human pathological heredity.

5.1 Early Modern Physiologies of Hereditary Transmission

The characterization of a given disease as hereditary poses an immediate puzzle to the physician. The metaphor of inheritance implies the existence of a fact of transmission. A given failure in the health of a person is being seen as attributable to a particular presence in his body of a trait (a mark, a flaw) that was transmitted by at least one of his parents to him through physiological routes linked directly to the act of generation. This inference is not necessarily supported by every framework of physiological explanation, but the metaphor points strongly toward this notion. The existence of family patterns in the occurrence of certain diseases, which had been known since antiquity, was only one indication of the possibility that a bequeathal of diseases did in fact occur. Under a hippocratic-galenic, humoral view of disease this transmission could be portayed in different fashions. Most hereditary diseases were linked to chronic, basic, temperamental ailments rooted in in fundamental imbalances acquired by the individual during his "first formation." At that moment (under the prevailing dual seminal view of generation) the seminal contributions of both parents became fused and stabilized in the new individual's temperament. Of course many circumstantial influences, including climatic and emotional influences and those of the imagination,[7] could alter the outcome—the child's temperament. The high variability in the physical and moral attributes of children from the same parents could testify to this reality. Nevertheless the tendency, observed in the process of like engendering like, of reproducing the parents' temperamental characteristics, was a given regularity that physicians took for granted. This tendency included the reproduction of some of the diseases that ailed one's parents at the same period of life in which they had been afflicted.

Previous observations of these phenomena motivated Michel de Montaigne (1533–1592), when struck by the same disease his father had had at the same age, to pose the question of transmission:

> What monster is it that this teare or drop of seed whereof we are ingendred brings with it, and in it the impressions, not only of the corporal forme, but even of the very thoughts and inclinations of our fathers? Where doth this droppe of water contain or lodge this infinite number of formes? And how bear they these resemblances of so rash and unruly a progresse, that the childe's childe shall be answerable to his grandfather, and the nephew to his uncle?[8]

This basic question produces several physiological and causal puzzles. If physical and moral resemblance between parents and offspring is a normal outcome of processes of reproduction, is communication of hereditary disease a side effect, so to speak, effected by the same route and causes? Are such diseases the product of specific morbid humoral materials that can be eradicated from the family line, or are they just the product of holistic, general temperamental weakness that befalls some of the children in a constitutionally prone family?

A hippocratic-galenic explanation of hereditary transmission of disease, particularly exploited by Paracelsians,[9] was to attribute the familial pattern of some diseases to the communication of peculiar, noxious, poisonous physical elements like taints, bad humors, and salts. These elements found their way through the "pangenetic" route, and could jump, via the semen, from generation to generation, participating in the original matter from which the new organism would be formed each time. Such noxious materials could have specific evil effects, and thus be responsible for the outbreak of a particular kind of disease, or they could possess a general capacity to damage any part of the system they happened to land on, triggering different diseases according to the moment and place they become active. As can be seen in Montaigne's puzzlement, a particularly mysterious phenomena in these considerations was the latency shown by the causal factors involved. Both resemblances and hereditary diseases could remain hidden for many years during an individual's life, and show themselves with uncanny timing at exactly the same age every time around. They could also skip one or several generations, only to reappear in distant relatives.

There is a problem in the physiological equating of normal resemblances and pathological ones. However, the hypothesis in the latter case that what is hereditarily transmitted is not the disease itself but a constitutional disposition to it, clearly made by physicians since the seventeenth century, eventually promoted the unification of pathological and nonpathological hereditary transmission. All these considerations made the sorting out of hereditary physiological influences particularly difficult, especially in a period when physicians were trying to eradicate any appeal to magical or mysterious influences from their scientific outlook. In the context of the crisis that hippocratic-galenic physiology was suffering in the seventeenth and eighteenth centuries in European schools, the questions around hereditary transmission of disease acquired increasing relevance.

At some point in the eighteenth century physicians embarked on a conceptual quest to make sense of the idea of a hereditary disease, and restrict its boundaries as clearly as possible. European physicians felt the need, long before other naturalists, to focus on genealogical patterns of trait transmission. A main objective for them was to have criteria to distinguish diseases that could properly be called hereditary from those that, while sometimes adopting similar patterns of occurrence, were not transmitted in the act of generation and thus had to be set apart. As a consequence, observation of the characteristics and development of candidate diseases became increasingly crucial to the discussion. The timing and duration of their attacks were seen as important telling factors concerning the origin and ultimate cause of the disease. A disease acquired at the moment of conception by parental influence—the main criteria for its hereditariness for most authors[10]—would typically be a constitutional, chronic disease. It would manifest itself at the same age, more or less, in the offspring as it had on the affected parent. The more analytic authors tried to discuss and eliminate the other possible nonhereditary, causal influences.

In the process of clarifying what the adjective *hereditary* meant to them, medical men began to give structure and causal meaning to what had started out as a metaphorical, purely descriptive term. As I have detailed elsewhere, this process seems to have gathered momentum after 1600, when medical physiologists were trying to adjust their traditional hippocratic-galenic views of disease to the less dogmatic and more empirically oriented scientific environment of the post-Renaissance.[11] As a basis for the discussions of hereditary disease, the collection of relevant ancient and modern evidence, registered in the form of more or less trustworthy cases in the literature, increased enormously during these two centuries (1600–1800). The accumulation of case histories eventually became in itself a very powerful tool against the skeptics, who from different quarters tended to dismiss the claims of specific hereditary transmission of physical ailments as delusions or physiological impossibilities. During the eighteenth century several important medical authors were still highly skeptical about the real possibility of there being any causal communication of anomalous features through the seminal fluids or by other means between the temperament or constitution of parents and that of the children. In his youth, in 1748, the French military surgeon and later lifetime secretary to the Académie royale de Chirurgie—Antoine Louis (1723–1792)—articulated these doubts very lucidly in his influential *Dissertation . . . sur la question . . . comment se fait la transmission des maladies héréditaires?*[12]

In consequence, physicians interested in hereditary transmission embarked on the production of theoretical distinctions based on the accumulated evidence, distinctions that began to give the subject a more sophisticated profile. Their need to provide physiologically possible causal routes to disease made them realize that a simple external pattern, familial, group, regional, or other, was not enough for claiming a hereditary cause for a disease. In other words, their need to establish clear criteria for distinguishing the hereditary diseases, which were mostly identified with constitutional ones, from those "acquired" after conception, both congenital and postnatal, forced them to focus on the peculiarities of original source, timing of appearance, and timing of recurrences of the disease. By the end of the eighteenth century, a series of quite sophisticated distinctions, and definitions of what is meant for a disease and a physical feature in general to be hereditary, were circulating among medical men, mainly but not exclusively in France.

The roots of the criteria developed by eighteenth-century physicians in their attempts to define the hereditary can be traced back to the writings of Hippocrates and Aristotle, but these criteria became increasingly stringent and logical with respect to the exclusion of the nonhereditary during this time. The more subtle and careful thinkers among medics wanted to claim as properly hereditary only certain constitutional characters and diseases. Coming from either of the parents, these characters and diseases encompassed those that established themselves as part of the organic constitution before the first solidifications of the seminal humors in the womb gave rise to the new individual in all its complexity. For other authors, sources of parental humoral influences besides the seminal ones had to be considered within the realm of the hereditary. For example, maternal influences via the placenta were considered hereditary, and for the more inclusive or least discriminatory writers on the subject, also the influence of lactation on the newborn's constitution. Nonhumoral sources of influence, like the emotions and imagination of the mother, were discussed as well.[13]

5.2 *Les maladies héréditaires* at the Threshold of the Revolution

In the years 1788 and 1790, in a momentous period in French history, the Parisian Société Royale de Médecine called for two successive essay competitions in which the issue to be addressed was the reality, evidence, and theoretical support for a belief in the existence of hereditary diseases. The tone in which the questions were set made it clear that the Society wanted some real theoretical progress and did not want the competitors to rely on unsupported ancient tales and assumptions around the phenomena of hereditary transmission of disease. The questions set by the Society in the public session of February 27, 1787, with a prize of "600 livres," were: "Determine 1) if there are truly hereditary diseases," and "2) if it is in the power of medicine to prevent their development or to heal them after they break out."[14] The permanent secretary of the Société Royale, Felix Vicq D'Azyr (1748–1794), took a keen personal interest in the organization and the outcome of these competitions. Judging from the emphasis that the Société Royale placed on the essay prizes, it seems very likely that the hereditary transmission of disease had become an issue that surpassed its mere theoretical importance. French physicians associated with the Philosophes, like Pierre Jean George Cabanis (1757–1808), Phillipe Pinel (1745–1826), and Vicq D'Azyr, were struggling to transform the profession into a more socially relevant one.[15] Hereditary transmission of disease, which up to that moment had been of concern only in relation to family medicine, in some cases associated with royal families,[16] was being considered as a public hygiene problem. The skeptical challenge that the surgeon Antoine Louis had raised in a similar context of a prize competition issued by the academy of Dijon in 1748 was another main motivation for the Société Royale. Having posed with such clarity the challenges facing any real physiological attribution of a hereditary transmission of disease, and Louis himself having acquired such a notorious position within the ranks of the rival Académie royale *de Chirurgie*, physicians of the Société Royale had to produce a plausible rationale for the inheritance of disease. This would preferably be framed in terms of an acceptable solidist account, which both a surgeon like Louis and an anatomist like Vicq D'Azyr would prefer.

The focus that the Société Royale forced on the hereditary transmission of disease in this critical moment produced, I believe, an enlightening collection of works reflecting the theoretical difficulties facing the notion of hereditary transmission in general. The collection of essays sent to the competition, mostly unpublished manuscripts, remain a valuable source for elucidating the hereditary thinking of the period. The high standards set by the jury were rewarded. Finding that the participants of 1788 had not produced acceptable critical analysis of the question posed, the jury in the first contest, Vicq D'Azyr among them, decided not to grant the prize, but to call for a rerun with the same questions but raising the prize money to 800 livres in 1790.[17] In summing up, the judges in the first contest detailed their discontent:

> Most of the participants made the supposition rather than proving the existence of hereditary diseases; thay have not defined enough their nature. We need to know if some morbific vices are really and individually transmitted from parents to children, or if the diseases that are called hereditary, are not rather a consequence of the conformation of the organs, which in parents and children must, due to their structure, be subject to the same affections. It is mainly towards the existence and nature of these diseases that their research must be directed to.[18]

This wording was motivated by the weakness found in many of the essays that adopted a humoralist view. As I have mentioned, earlier skepticism about the physiological reality of hereditary transmission of disease voiced by Antoine Louis and other influencial physicians had particularly exploited such weaknesses. They had argued that diseases that seem hereditary are only due to common defects of conformation of the solid parts, and there is no clear way the individual conformation of the parents' bodies can have a regular, direct causal effect on the conformation of the embryo. Transmission of disease through humoral influences, even if sometimes derived from a parent, cannot be said to have any special route or character that would justify their separation into a different category. As a result, those physicians argued, there is no real basis for a physiological interpretation of the old analogy of inheritance in relation to disease.[19]

Another set of judges confronted a new and more carefully argued set of pieces in 1790. According to them, at least four dissertations were of very high quality. These were the dissertations written by Pierre-Joseph Amoreux (1741–1824), Alexis Pujol (1739–1804), Jean-François Pagès (1760–1803), and Joseph-Claude Rougemont (1756–1818).[20] Although not agreeing in all aspects, these works gave strong and authoritative accounts of the reasons why a physician of the late eighteenth century could and should defend the principle of hereditary transmission of disease or, as some of them insisted, a disposition to disease, regardless of what some skeptics argued. Yet the Société Royale was still not fully satisfied. Despite clarification of the subject in the aforementioned essays, the judges stated that the question of inheritance was not yet solved. There remained a need of "nouveaux éclaircissemens" that medical men would need to apply themselves to with renewed zeal. "Isolated observations," they added, "considered separately cannot have but a very limited degree of usefulness. It will not be but on their gathering and their comparison that we will be able to give them their value."[21] This invitation to surpass the casuistic method that had plagued the subject for too long, and to gather the evidence in a more cumulative fashion, was a clear recognition that the consensus among wider sectors of the medical community was changing, away from their old, case-based strategy. In several competition essays, authors had acknowledged that Antoine Louis had a point when he questioned the use of tales and anecdotes to justify belief in hereditary disease, and was right when he wrote that "the principles which form the true theory of medicine can only be acquired by fastidious researches and long and difficult labor."[22]

Though not completely satisfied, the Société Royale commissioners decided to grant the prize to the Caribbean-born physician Joseph-Claude Rougemont, whose contribution, according to the judges, treated all aspects of the question, and made an "exact and strict analysis of all the writings and all the facts related to the problem posed." For instance, he carefully distinguished hereditary diseases from those that the child can contract within the mother's womb or during birth. Some shortcomings in method, they added, were balanced by the clarity he brought to the whole subject.[23] Unfortunately the manuscript of this essay is missing from the archives, and Rougemont's text was never published in French. Because he was Professor of Anatomy and Surgery at Bonn University at the time, his work on hereditary disease was published in Frankfurt (1794), in a German translation, and was a classic on the subject in early nineteenth-century Germany.[24]

An "accesit" prize was given to Amoreux, the son of a well-known physician, who taught most of his life at Montpellier. While the young Amoreux's great historical erudition was highlighted, he failed, according to the judges, on the "prophilactique & curatif" aspects. Both Pujol and Pagès received honorary mentions. Pujol, who lived and worked at Castres, and won several prizes with his essays in Société Royale competitions, had extended his first version considerably for the 1790 rerun of the competition to explain inheritance of diseases. Though considered "trés Bon" and "érudite" by the judges, he refused to eliminate the theological bits, which according to him, were the reason he did not win. Pagès's original manuscript is also missing from the archives, but it was published in the *Encyclopédie Méthodique* in 1798. Thus only a few of the essays written for the competitions found their way into print. Perhaps the most widely read was Pagès's piece. Together with Rougemont's winning essay, this essay was among the more modern and better argued, and it was the solidist position that received a backing in both essays. Amoreux and Pujol, in contrast, had tried to integrate both humoralist and solidist positions in their contributions.

Several of the contenders in the second round began their pieces expressing a certain degree of outrage at the 1788 judges' decision to demand further proof of the existence of hereditary transmission of disease, claiming that the overwhelming number of indubitable cases gathered in the literature—and witnessed during their practice—should be enough to persuade any reasonable person. Puzzled by his first essay's failure to convince the judges, Amoreux, for example, had reworked it considerably for the 1790 contest, adding more "evidence from authority" as well as more facts collected from a wide range of sources. Amoreux, Ladevère, and Pujol, among others, tried to justify their inductive procedures, and felt—as physicians who regularly generalized from accumulations of cases—betrayed by the Société Royale's siding with Louis in this issue. They believed that carefully selected cases, accompanied by sound physiological and pathological knowledge, did provide a good basis for making general statements about hereditary transmission. The adequate selection of medical authorities, and especially the experience that many years of practice—sometimes with different members of the same family—gave to the best observers among medical men, clearly provided them with a sound basis for belief. On the other hand, a sheer accumulation of cases alone was not enough because it never would compensate for the combination of a well-trained eye with a well-read mind that good, old-style physicians had.

As for theoretical progress, some important analytic features were already in place. When reading the 1788 entries, one finds in several authors an insistence on the need for a clearcut distinction between "congenital" and "connate" (*conné*) causation in order to sort out the hereditary from the nonhereditary.[25] This feature, as mentioned by the judges, was present not only in the winning essay but in several other 1790 essays. The judges praised this feature, because it was the primary basis for important and clearcut phenomenological criteria distinguishing hereditary from other influences. The most important of these criteria was homochrony; that is, the occurrence of the disease effect in the same period in the life cycle. Other criteria included the specificity of the disease and its incurability.

The distinction between "congenital" and "connate" causation was made by many authors in slightly different ways. This vocabulary is sometimes confusing, and as Pujol pointed out, even the Société's commissaires seem not to have been clear on the distinction themselves.[26] Typically, connate diseases were considered to be those acquired after the first formation of the fetus, by contact in the mother's womb with humors transferred through her blood. Some authors also included diseases and defects acquired by mechanical influences during pregnancy, like blows, and others believed that the mother's spirit, imagination, or state of mind could also exert influence and produce connate peculiarities as well. Congenital presences would, in contrast, be those integrated into the individual's constitution at the moment of his or her first formation via the parent's seminal contributions. But was there a real, important difference picked out by the distinction, or was it just a nicety with little substance? The main reason for confusion was, of course, the coexistence of different physiological explanations of disease in general, and of its transmission in particular, the most salient theoretical divide being that between solidist and humoralist physiologies. The issue of what sort of material influences to include in the congenital and what to leave out under the connate strongly depended on the previous commitment to one or the other theoretical standpoint, or to some eclectic combination of the two.

Under a solidist viewpoint the congenital influence was the only kind of truly hereditary character. Humoralists were more divided around this matter, some wanting also to restrict the hereditary to influences at the first formation, while others saw no reason for this restriction, considering any humoral cause communicated by any of the parents before the constitution of the infant becomes completely fixed (i.e., before the end of the first year) as worthy of being called hereditary. Under humoralism many authors did not see any reason not to include in the hereditary such influences as that exerted over a child's constitution by the mother's, or the nurse's, milk. Either way, the distinction between congenital and connate was difficult to carve out theoretically. The best participants in the 1790 contest can be said to have brought to it a renewed clarity. The solidists seemed to have had the upper hand in this, because they could argue for a simpler notion of hereditary transmission that reduced it to the sole seminal influence stemming from the parents and the state of the body once it was completely formed.

Nevertheless, humoral causes for hereditary diseases were defended by several of the contenders. Pujol, for instance, was especially annoyed that the Société Royale rejected his first essay and criticized it as a dogmatic upholding of solidism. For the second (1790) contest, he decided to write a whole new chapter trying to strengthen his humoralism. Pujol started by contending that "the transmission of virulent and humoral vices by the route of heritage will never be proven completely with speculative reasons."[27] Due to the close interdependence between humors and solids, Pujol claimed, it is never easy to isolate the original bearer of a morbific cause. Ill humors affect solid parts as much as diseased solid parts alter the fluid parts' normal composition.[28] To restrict the hereditary to solid to solid transmission is, according to him, to blatantly ignore most empirical evidence, both physiological and pathological. Different tainted humors can act over solid structures at different times; to call their effect hereditary, it is sufficient to show that some of them have acted on the conformation of the germ before or at the moment of its fecundation, or that they were incorporated into the bodily fluids at that moment by forming part of the semen. Many observations point in that direction, Pujol continued. The fact that some of these "ferments" (*levains*) can also be communicated nonhereditarily is no reason to deny this. Amoreux added in his contribution that strict solidism is untenable because many real hereditary diseases cannot be unambiguously classified as either humoral or solid.

The capacity that humoral vices have to produce wide-ranging effects on different organs at different times (their "proteism")—a feature cited by solidist opponents like Louis and Pagès as a main reason for discarding them as bearers of real hereditary influences—was in contrast seen by Pujol and Amoreux as a further reason for keeping them within the field. Tracing the "transformations" of humoral hereditary diseases from generation to generation within a family would eventually lead to a reduction of hereditary influences to a small set that could account for the observed diversity.[29] Hereditary causes, then, are not only reduced in number but they gain in "extension." Dispositions to different diseases need not each have a particular humoral cause, but perhaps only a particular effect on an organ or tissue. All hereditary diseases might still depend on constitutional flaws, but they would need a tainted humor as a vehicle for nonspecific transmission.[30]

The curability of some hereditary diseases was also at the root of the defense of humoral causes for them. It was widely believed among eighteenth-century physicians that humors were more easily affected by medical treatments than solids were. The former could be affected by nutritional, chemical, and some therapeutical means (like bloodletting), whereas most of the latter were incurable, and only the symptoms could be ameliorated. Contrary to the claims of both Pujol and Amoreux, however, their two-cause approach to the hereditary left too much room for all kinds of diseases to be considered, one way or another, as hereditary. This proliferative and unbounded character, as displayed in the rich classifications of the two pro-humoral authors,[31] was precisely what some judges were opposed to, and what was at the root of the Société Royale's quest for "Maladies vraiement héréditaires." For this reason, the humoralists were not favored in the competitions.

Disagreements between humoralists and solidists went beyond what kind of physiological interactions the different constituents of the body might undergo, to include which of them were capable of actual hereditary transmission. However, the obscurities surrounding generation, and the impossibility of tracing the details of physiological interactions, forced the decision of how to define the hereditary toward external evidence. This evidence could not in the end tip the balance one way or the other; it could only draw some limits or boundaries. As Silvia De Renzi argues in chapter 3, the implications of some physiological routes of transmission of resemblances within families went beyond medical issues. Paternal and maternal material influences on the child's body could become a legal matter. What was possible and what was not possible then became something socially and politically charged. Under humoralism all resemblances—moral, physical, and pathological—within a family or wider genealogical groups could be attributed to the similarities in the proportions of the different humors, affected by internal and external presences.[32] With the rise of mechanistic solidism, in the seventeenth century, family and group resemblance became linked with the explanation of the origin of solid parts during the first formation, and the posterior malleability under the action of humors was then held to be an open matter. Strict solidists believed that once the formation of solid parts was complete, external influences of a humoral or other character were of secondary importance.

Alternative generation theories had a different bearing on these issues. Preexistence could provide a basis for the radical questioning of the reality of any causal link between parents' and offspring's organizations—that is, of their resemblances. Constitution or solid organization is something that comes with the preexistent germ, and any predisposition to disease, or diathesis, is either already there or is acquired, but it is not inherited in any real sense. One possible exception was to extend the criteria and incorporate humoral causes into the hereditary, in which case the problem existed of separating those causes from other, external humoral influences. This position was adopted in Guillaume Rey's (1687–1756) treatise on hereditary diseases (1748), and also, with some variation, by many of the 1788 and 1790 competitors that were not prepared to abandon humoral hereditary influences. Those who argued for preexistence among the prize competitors, especially one Bésuchet (b. 1754), used Charles Bonnet's (1720–1793) ideas about the influence of seminal humors in the transformation of the germ during its development. One reason was that some of the traditionally hereditary diseases like scrofula seemed to be of a humoral nature. On the other hand—and under the assumption that lodged in different organs, one morbid humor would produce different diseases—the inclusion of humoral causes among the hereditary provided an economy of causes, given that the persistence in one family of only one kind of morbid humor could account for different affections in different individuals and generations. Many eighteenth-century medics seem to have believed this.

The strictest solidist in the 1790 competition was Pagès. Where others were prepared to accept that certain hereditary diseases had as their main cause a morbific humor (*levain, virus*), Pagès was inflexible in his refusal to allow that. He also related his solidism to the disputes over generation. Although he denied that generation theories should have a primal position in the discussion about hereditary disease,[33] he argued that dual seminal views, like those of Hippocrates or Maupertuis, made it easier to account for hereditary transmission of constitutional traits to offspring, from both the father and mother, and from ancestors. This dualism would also help to draw a line between real hereditary influences internally incorporated into solids at first formation and secondary or humoral, external ones. The obscurantist and abusive practice of many physicians in applying the adjective to any disease could be checked, Pagès believed, with clearer external criteria, derived from a clear definition. The principal criterion had to be the time of appearance of the disease. An essential character of hereditary dispositions, he wrote, is that they develop in the offspring at the same age at which they appeared in the parents.[34] In his view, only a small set of constitutional predispositions to certain diseases, specifically transmitted through generation by the semen, deserve the adjective *hereditary*. They are "epilepsie, hemoptysie, pthysie, manie, melancholie, hysteria, hypochondria, & apoplexie." I believe that Pagès' stern criteria and strong solidism appealed to Vicq D'Azyr more than the other authors' "eclectic" positions did, and that is why, although not given the prize by the Société Royale, it was this essay that Vic D'Azyr included in the *Dictionnaire de Médecine* of the *Encyclopèdie Méthodique*.

Pujol and Amoreux also used "homochrony" to separate humoral hereditary influences from nonhereditary ones. For them the transmission of the influence would occur through the semen or the mother's blood and it would then either affect the constitution during the first formation or remain in the body without effect until a later, determined stage, when it would produce its evil effects. As Amoreux wrote, hereditary diseases are usually not carried by the children at birth but are "developed" at a certain age because they are transmitted:

> At the moment of formation, by an heterogeneous mix in the prolific semen, the morbific principle enters, so to speak, into the germinal seed at the moment of the formation, and such principle with more or less strength is modified and altered during the growth of the foetus, of the child and the adult, and gives rise to a bad temperament, or to a disease, or finally to a simple disposition.[35]

Homochrony was thus agreed as the clearest of "external" criteria for the presence of hereditary transmission by both solidists and humoralists.

But for some it was only one manifestation of the main characteristic that hereditary causes had been seen to possess: latency, or the ability to hide for some time in a healthy body without any sign or symptom. Atavism, of course, was the other important manifestation of this property linked to hereditary transmission. In their attempt to establish the category of "maladies vraiement héréditaires," most competitors for the Société Royale's prize realized they needed to give some attention to the special status of hereditary causation. Louis and other skeptics had discarded as absurd any indirect or mute causation. Especially difficult to conceive of was a morbific presence, be it constitutional or humoral, that could remain hidden for an entire lifetime or several generations, get transmitted, and only manifest itself capriciously in a future generation. The transmission of predispositions rather than actual diseases was the answer.

The idea of the existence of latent causes was thus in need of a credible physiological description that gave it some substance in order to transcend its hypothetical status. Several entrants in the prize competitions seemed confident that the problem could be solved through a distinction between kinds of causes. The famous Fernel had provided the basic Aristotelian frame: that a hereditary, predisposing cause needed to be supplemented by a triggering, efficient cause, whose absence would leave the former one unnoticed. What most competitors tried to do was to elaborate on this scheme based on their physiological biases. Materialist approaches to causation were favored by most of them, both humoralists and solidists. Amoreux, for instance, stressed that an important thing about both humoral and solidist causes is that they avoid all plastic, immaterial efficient forces, *archées*, and so on, which he considered "empty" notions: "cette substitution des mots," he wrote, "n'a jamais donné une idée plus claire de la chose."[36] Under a humoral account, latency is best explained by the permanence in the body of the tainted fluid for an indefinite time without acting noxiously in any of its parts, thus amounting to a sort of predisposition. It is only at some stage of the development of the solid parts that one or several of them become vulnerable to the ill action and the disease develops. Other triggering causes can be external emotional, physical, climatic, or other influences.[37] The solidist explanation of latency is based on the acquistion at the moment of the first formation of a physical flaw or conformational mark that creates a predisposition to a given malfunction. It is only when a developmental or external triggering cause concurs with the latent predisposition that the disease develops.

In any case, secondary, triggering causes were given an important role in the manifestation of the hereditary disease. This opened a clear avenue of intervention for hygienist preoccupations of the period. These efficient causes, however, did not have a decisive role, as the variability of reactions within groups and families proved. "Everything being equal," wrote Amoreux, "secondary causes will act in the same way over individuals similarly disposed, and will act differently over individuals differently disposed, which explains why within a family not all the children are always afflicted by their parents' diseases."[38] Authors could thus cast doubt on the possibility that either morbid humors or constitutional defects could really remain hidden in a healthy individual for long periods of time. The idea of a predisposition itself, and its being hereditarily transmitted, could nonetheless be attacked on a priori grounds. But to most eighteenth-century French medical men, these ideas provided an excellent resource for picturing the "hereditary" which was among the most irregular and untamable set of observations they knew about.

In this case too, the analogy with normal life stages and characteristics helped reinforce their beliefs. Homochronic phenomena like dentition and puberty on the one hand, and latent hereditary transmissions like male baldness received from the mother on the other, could respond to a kind of latent causation analogous to the causation they were advocating. Amoreux addressed another important problem for hereditary transmission, the striking variation among children of the same family, which was traditionally ascribed to widely varying influences at the time of conception. This author argued that the relative weights of the various kinds of causal influences in shaping an individual's temperament are not equal. Primary, humoral, and solid hereditary causes by far outweigh the secondary, environmental ones. It seemed undeniable to Amoreux, as to many others, that peculiarities of temperament and constitution run in families. As Pujol wrote,

> If it is true that the color of the skin is hereditary among humans, why wouldn't the temperament be so, the color ordinarily being both the sign and the effect of it? . . . One cannot imagine how such a skillful man [Louis] could decide to deny such a noticeable and general fact. . . . Propagation of temperaments, through heritage and succession, is one of those general facts whose reality is easy to certify; one has only to examine with curiosity and detail different families among those gathered in the big cities.[39]

This general, holistic fact of inheritance not only affected the superficial qualities of color, height, weight, and form of the body; its influence extended to the internal constitution of tissues and organs. Alongside clinical observations—hinted at by Pujol by the phrase "big cities"—in the late eighteenth century, surgery provided a new observational window that few of the contenders failed to mention. Resemblance within families could be traced to the minor details of inner configuration, giving access to a multitude of new facts that increased both the number and the evidential strength of hereditary claims. This was especially the case when peculiar hidden defects began to turn up in autopsies. As Amoreux wrote, "Anatomists, when exploring dead bodies of several individuals of some families have sometimes noticed a given conformity of structure or of defective organisation between them; and the examples would undoubtedly be more frequent if they would follow more carefully this kind of research."[40]

The general fact of widespread resemblance within families is thus used analogically to justify the belief in "pathological" resemblance. The importance of this analogical argument cannot be exaggerated for the history of hereditarianism. As Georges Canguilhem has shown, the move from normality to pathology was not a natural one until well into the nineteenth century, and so it was also with respect to the hereditary.[41] Johann Friedrich Blumenbach (1752–1840), for instance, reacted strongly against this kind of argument before adopting such notions himself, which he then popularized.[42] Solidism and surgery tightened this analogical move among physicians. The emphasis on personal observation, less reliance on ancient reports, and the attention to structural detail reinforced their confidence in the reality of hereditary transmission of individual, idiosynchratic constitutional characters in general, and of the predispositions to diseases this process could entail. Whatever way the first "rudiment" or embryo came to be formed, many physicians were convinced there had to be a causal mechanism responsible for the impression on the embryo of some constitutional characteristics of each of its parents. For lack of a better model, which generation theories did not provide, Pujol described this causal mechanism with a metaphor: "The same hand that traces so scrupulously the child's physiognomy from those of the father and the mother has to move on to internal resemblances, and deliver with the same exactitude, organ after organ, entrail after entrail."[43]

This hand, this mechanism whose basic external manifestations these late eighteenth-century medical men were trying define, was to become, decades later, the physiological mechanism of l'hérédité. The notion of a general and unified explanation for hereditary transmission of both normal and pathological features was facilitated then by the strenghtening of solidism. Humoral morbific causes would always maintain a connotation of a poison in the form of alien, external disruptive influences that were somehow more easily eradicated than solid internal influences were. On the other hand, the solid to solid communication through a "normal" mechanism of conformational flaws provided a perfect frame for the analogical reasoning described above. This framework could encompass naturally all the biological phenomena that fell under the aegis of the adjective *hereditary*, normal or deviant. The hereditary domain extended from family resemblance to hybridization and from transmission of physical deformity to hereditary disease. This strengthening of the analogical domain, by unifying it under one kind of causal mechanism of transmission and not a diversity, opened the door to a further step from visible resemblances to invisible ones. Later, in the nineteenth century, this was to prove crucial for the leap from physical to moral inheritance.

Les Maladies Vraiment Héréditaires, as the French Société Royale de Médecine declared in 1791, a few months before its dissolution, was a category still in search of a precise definition. The stinging effect of Antoine Louis' skeptical arguments, together with the perceived stalemate in generation studies, had made the following facts evident to the best prepared members of the French medical community. If they were to preserve their authority with regard to hereditary transmission of all the mysterious diseases that seemed to appear spontaneously in some individuals within some families, they had to provide more clearcut characterizations of the diseases' main features and of their etiology and development. Chronic, constitutional diseases that once developed seemed impossible to eradicate, like gout, epilepsy, apoplexy, mania, or tuberculosis, were very difficult to account for except by a deep-rooted predisposition in the body. The category of a hereditary disease was one needed by the medical community, in part to account for its failure to prevent or cure this set of maladies,[44] in part to gain and maintain authority over matters of public medicine.

The discussion then had to be focused not so much on the reality of hereditary transmission as on the formation of the theoretical space in which to describe it. Including more phenomena than those truly necessary—in other words, creating a category of hereditary disease so loosely defined as to have room for all diseases with some familiar pattern—was absurd. Restrictive conditions based on the communication mechanism or route were needed, but the restrictions themselves had to be checked so that the category would not become impossible, as Louis had tried to argue it was. Pujol and Amoreux, the more enthusiastic defenders of the wide-ranging inclusion of diseases, saw the creation of some kind of gradient—or set of subclassifications based on kinds of causes and intensity of effects—as the solution to the riddle. Amoreux, for instance, proposed four orders of truly hereditary diseases. The first two are basically humoral, while the second two are solidist in cause: (1) those that are transmitted specifically, without a change of nature; (2) those nonspecific diseases that have a hereditary origin but change their nature; (3) essential bodily dispositions; and (4) bodily deformities that are transmitted and are indelible. As false hereditary diseases he considered all those products of accidental occurrences during pregnancy, nontransmissible bodily defects, like *taches*, and all those opportunistic diseases that emerge from a weak constitution but that are not determined in any way.[45]

This still left too much room for the demanding Société Royale judges, and only the solidist prudence of Rougemont and Pagès seems to have satisfied them on this point. The latters' strong restriction of the category of the hereditary diseases to only the most obviously constitutional and chronic ones leaves, however, a residual problem of how to account for the widespread occurrence of familial patterns that are not easily accountable by external contagions. The discussion around transmission through prenatal and postpartum nourishment through the mother's blood and the nurse's milk falls into this unstable domain.

Again, the lack of a clear physiological description of hereditary disease precludes the closure of the debate. The assault on the hereditary by those wishing to account for all its target phenomena through external causation was still an open possibility. However, the set of distinctions, like that between congenital and connate, of external observational criteria like homochrony, and of causal concepts like latency, that most French physicians had agreed on by the end of the eighteenth century seems to have given them a strong enough basis for belief in a kind of independently describable system of transmission of physical peculiarities from parents to offspring. In this depathologized sense, the whole system could be synthetically referred to as heredity.

5.3 Epilogue

In this chapter I have discussed European medical developments between the early seventeenth century and the early nineteenth century concerning the notion of hereditary disease. I have tried to show that in these developments we can find the first elements of theories of causal structure and empirical criteria that led to the conceptualization of heredity as a major biological cause of disease. I have focused on the outstanding theoretical analyses of the hereditary transmission of disease provided in the late eighteenth century by some French physicians responding to the challenge set by the Parisian Société Royale de Médecine.[46] I have shown how we can find surprising and quite revealing intimations of the theoretical labor that lay ahead. These early attempts reveal an awareness on the part of European physicians that they were dealing with a peculiar kind of natural mechanism that was not restricted to the transmission of pathological features and dispositions.

As I have discussed extensively elsewhere, the clarifications that physicians were able to make in this period laid the groundwork for the subsequent understanding of heredity.[47] The metaphor of the craftman's tracing hand used by Pujol in his 1790 prize essay clearly points toward this notion of physiological inheritance as a copying process. In the same period there is striking evidence of similarly advanced analyses of hereditary transmission in Britain, attributed to John Hunter (1728–1793) by his disciples. Hunter's thoughts on physiological inheritance are found in a series of paragraphs in his *Principles of Surgery*, which was published in 1835[48] and was based on the transcriptions by James F. Palmer of Hunter's 1786 lectures. That work maintains that there exists in nature a hereditary principle that "may be divided into two kinds: the transmission of natural properties, and the transmission of diseased, or what I shall call acquired or accidental properties."[49] It is an intriguing question as to what exactly was on Hunter's mind, but the evidence I have seen points toward a solidist analysis of the kind French physicians had advanced.[50] The opposition of solidist versus humoralist physiological accounts of hereditary transmission can be seen in eighteenth-century Britain by focusing on Hunter and Erasmus Darwin. Philip Wilson (chapter 6, this volume) provides a fine exposition of how Erasmus Darwin conceptualized the hereditary.[51]

By 1817 Antoine Petit, who I believe had access to the manuscripts from the competitions discussed in this chapter, had composed a small treatise on hereditary disease. This was immediately incorporated by the editors of the *Dictionnaire des Sciences Médicales* into the entry on "Héréditaire" (maladie) that can be taken to summarize the great lucidity that solidist French authors had achieved. This piece remained the most influential analysis published on the subject until the 1840s.[52] It was a clear and convincing attack on humoralist heredity. Echoes of Petit's precisely worded piece can be found in works written sixty or seventy years later. Petit summarized what he considered the main achievements that physicians had attained in the definition of the hereditary cause. Heredity, he asserted, has to be based on particular states of the bodily constitution communicated to children by parents. These states give an "organic disposition" to reproduce a given effect—for instance, a particular disease. He added that they can be localized states, or states of the whole *économie*. But he denied that general qualities of the constitution like weakness that establish in the body vague and indefinite tendencies to disease are to be seen as similarly hereditary. In heredity a specific, one-to-one connection must be shown to exist. Among other things Petit praised the ancient distinction between predisposing and efficient causes as the main analytic resource to deal with the hereditary. He summarized, with more clarity than any previous author, the determinant features of heredity. Latency, homochrony, atavism, all can be accounted for with a proper causal analysis. He upheld the importance of separating clearly congenital and connate influences, and accepted that only through the process of generation can real hereditary influence be transmitted.

Petit, however, joined previous authors in condemning attempts to solve the mystery of heredity by an even deeper and more unsolvable mystery of generation. According to Petit, hypothetical systems of generation only confuse the issue. It is far more likely, he noted, that proper observation of the patterns and nature of hereditary disease will illuminate the theorizing in generation, than the other way around. Although he was skeptical about the feasibility of any success, Petit left it to other specialists to determine the real, intimate nature of inherited dispositions. The good observer, however, can on occasion find visible, exterior characters that are linked to the disposition, before its effects are noticeable. Generally, however, this is not the case, and though there is an organic bases for hereditary causes, they usually remain hidden or latent until the predisposed time in the life pattern arrives for their expression.

This basically summarizes the rationale behind the adoption by a new generation of French physicians after the 1820s of the noun *hérédité* as the conceptual focus, or node, through which a set of widely ranging medical issues were processed. The articulation of the needs of a new kind of socialized, hygienic program, and a need for theoretical replacements for abandoned conceptual tools like temperament or constitution in their humoral version, made heredity an ideal explanatory and political resource. Laure Cartron's contribution in chapter 7 of this book builds on these proposals in the sense that in the postrevolutionary period heredity became an important disciplinary resource for the medical profession.[53] The growing importance this concept acquired in the natural and social sciences during the rest of the nineteenth century loses much of its mysterious abruptness once the developments in the medical scene are taken into account.

Elsewhere I have detailed the subsequent episodes of this story. I show how at some point in the early decades of the nineteenth century, French medical men and physiologists adopted the noun *hérédité* as the carrier of a structured set of meanings that outlined and unified an emerging biological concept.[54] The elements of this domain had previously been loosely connected by the metaphorical mirroring between physical resemblance among parents and offspring and the passing on of property and titles through the generations. Hereditary concepts furthermore found applications ranging from the medical to the zoological, from the agricultural to the ethnological. It was in those areas where, during the first half of the nineteenth century in several European countries, our modern concept of biological heredity was first adopted. However, our modern concept of biological heredity was first clearly introduced in a theoretical and practical setting by the generation of French physicians that were active between 1810 and 1830 and that were the direct inheritors of the physicians I have discussed in this chapter. From a traditional focus on hereditary transmission of disease, influential French medical men moved toward considering heredity a central concept for the conception of the human bodily frame, and its set of physical and moral dispositions. The notion of heredity as a natural force, with wide-ranging capabilities of transmitting differentially both fundamental and accidental characters, was generalized by that generation of physicians with the help of contemporary naturalists and physiologists. By 1830 the term *hérédité* was widespread, and it shared the explanatory and semantic qualities of traditional medical concepts like constitution and temperament.

Acknowledgment

The author wishes to acknowledge the substantial help he received from Staffan Müller-Wille while working on this chapter.

Notes

1. For a discussion of such theological and physiological frames and their relation to hereditary transmission, see López-Beltrán 2002; De Renzi, chapter 3, this volume. See also the relevant chapters in Roger 1973.

2. These long-term developments are described from another perspective by Sabean, chapter 2, this volume.

3. López-Beltrán 1994.

4. The issue of resemblance, particularly to the father, as being part of the hereditary in the early modern period has important legal consequences that are discussed in De Renzi's excellent chapter in this volume.

5. See van der Lugt, forthcoming; López-Beltrán, forthcoming.

6. See Olby 1994; Borie 1981; Morgentaler 1999; Malinas 1985.

7. For alternative explanations of the hereditary in the eighteenth century see López-Beltrán 2002. For the attribution of influence to the mother's imagination see Huet 1993; for the legal aspects of such attribution see De Renzi, chapter 3, this volume.

8. Montaigne 1582, book II, chap. xxxvii; English translation quoted from Montaigne 1999.

9. Like the Irish physician de Meara 1619.

10. For a discussion of the importance of the moment of conception for hereditary discussions see López-Beltrán 2002.

11. See López-Beltrán, forthcoming.

12. For close analysis of Louis 1749, see López-Beltrán 1995, 310–320; López-Beltrán 2002.

13. See De Renzi, chapter 3, this volume.

14. See Société Royale de Médécine 1786–1788, vol. 9, 1786–1787, 17.

15. See Williams 1994. For D'Azyr's views on the social role of medicine see Mandressi 2004.

16. Luis Mercado's treatise 1620 [1594] is a specific example of that.

17. *Minutes d'examen de mémoires sur maladies héréditaires*. See appendix 1 in López-Beltrán 1992.

18. Société Royale de Médecine 1786–1788, 18.

19. For a more detailed account see López-Beltrán 1995.

20. Other essays could be added to this list. For instance, those of Ladevère (1790) and Girard (1790) were excellent in some respects but were less well rounded.

21. "Dans ce genre les observations isolées considerées séparément, ne peuvent avoir qu'un degré d'utilité très-borné. Ce ne sera qu'en les réunissant & en les comparant, qu'on pourra leur donner de la valeur" (Société Royale de Médecine 1786–1788, vol. 9, 1787–1788, x–xi, Paris); my translation.

22. Louis 1749, 4.

23. Société Royale de Médecine 1786–1788, vol. 9, x–xi.

24. Steinau 1843; for biographical and bibliographical details on Rougemont see Dezeimeris, Ollivier, Raige-Delorme 1828–1839, vol. 4, 24–25. The same work also provides information on Alexis Pujol (vol. 3, 764–765) and Pierre-Joseph Amoreux (vol. 1, 111).

25. This distinction was already made with clarity by early modern authors like Mercado 1620 [1594] and de Meara (1619). Surprisingly, John Hunter and his followers claimed originality when they tried to argue for it, as in Adams 1814.

26. Pujol [1802] 1823, 231.

27. "La transmission des vices humoraux et virulens par voie d'héritage ne sera jamais prouvée complétement par des raisons speculatives" (Pujol [1802] 1823, 228); my translation.

28. "Un virus transmisse une organisation vicieuse dérangeront bientôt l'harmonie qui doit régner entre les parties solides et les fluides; de la des maladies organiques et des maladies humorales" (Amoreux 1790, 11).

29. Contrary to Pagès, both these authors accepted as hereditary a whole range of very different diseases, most of which adopted familial patterns. The idea of following these transformations over time was taken from some ancient authors, and was revisited later by several nineteenth-century hereditarians and advocates of degeneration. For discussion of these developments, see Dowbiggin 1991; Waller 2001.

30. Portal (1808) took this position to an extreme when he wanted all hereditary diseases to be caused by one proteic vice, the scrofular.

31. Amoreux (1790) considers a list of thirty-one different kinds of diseases, and to most of them he grants some "heritability." Pujol mentions at some point the existence of an "échelle d'hérédité" of diseases, in which all known ailments can be accommodated according to their feasibility of being inherited.

32. See López-Beltrán 2002.

33. Pagès 1798, 162.

34. Pagès 1798, 160. This author explains homochrony based on a peculiar physiological theory. He maintains that each organ of the body has a certain period, in the individual's life, in which it exerts its main influence. When it is "switched on" the period arrives in which the organ weaknesses and latent predispositions are revealed, in the form of ailments or disease. See also Pagès 1798, 163.

35. Amoreux 1790, 17–18.

36. Amoreux 1790, 13.

37. Amoreux 1790, 11.

38. Amoreux 1790, 12.

39. Pujol [1802] 1823, 248.

40. Amoreux 1790, 15. This confronts Albrecht von Haller's (1708–1777) denial of internal resemblance in Haller's critique of Buffon; Haller 1752.

41. Canguilhem 1972.

42. Blumenbach [1795] 1865. For discussion see López-Beltrán 1994.

43. Pujol [1802] 1823, 244.

44. As Ackerknecht (1967, 63) has rather sternly put it, "Heredity has always been the facile way of explaining the inexplicable." Waller (2002) extended this argument. For an excellent exposition of the role of predisposing causes in the medical debates of the early nineteenth century, see Hamlin 1992.

45. See Amoreux 1790, 27.

46. John Hunter's views on hereditary disease as well as the views of his followers in Britain could have provided us with an alternative case study. His position contrasts in an interesting way with the humoralist, proteic approach to hereditary transmission found in Erasmus Darwin. For the latter, see Wilson, chapter 6, this volume.

47. López-Beltrán 2002.

48. Hunter [1786] 1835–1837, vol. 1, 353–359.

49. Hunter [1786] 1835–1837, vol. 1, 353–354.

50. For a brief analysis of their views, see López-Beltrán 1992, chapter 3.

51. Wilson, chapter 6, this volume.

52. For an analysis of theories of hereditary transmission of disease in the *Dictionnaire des Sciences Médicales*, see López-Beltrán 2004a.

53. See Dowbiggin 1991; Waller 2001; López-Beltrán 2004b.

54. López-Beltrán 2004a.

References

Documents from the Archives de la Ancienne Société Royale de Médecine, at the Bibliothèque de l'Académie Nationale de Médecine, Paris

Amoreux, Pierre-Joseph. 1788. *Des Maladies Héréditaires.* 75 pp. In the classification system of the Bibliothèque de l'Académie Nationale, this is document 200-2-2.

Amoreux, Pierre-Joseph. 1790. *Des Maladies Héréditaires.* (10; 120-3-1.)

Bésuchet. n.d. *Sur les Maladies Héréditaires.* 59 pp. (E; 200-2-3.)

Gellei, Michel-Raphaël. 1788. *Morbi haereditarii.* (N; 200-2-6.)

Girard. 1790. *Recherches philosophiques et médicales sur les maladies héréditaires.* 80 pp. (12; 119-33-A.)

Ladevère. 1790. *Essai sur la question proposé . . . s'il existent vraiment des Maladies Héréditaires.* (119-30-5.)

Minutes d'examen de mémoires sur maladies héréditaires, 1788–1790. (181-23-1,5.)

Pujol, Alexis. 1788. *Essai sur les Maladies Héréditaires.* (H; 200-2-9.)

Pujol, Alexis. 1790. *Mémoire sur les maladies héréditaires.* (9; 120-3-5.)

Other Literature

Ackerknecht, Erwin H. 1967. *Medicine at the Paris Hospital 1794–1848.* Baltimore: Johns Hopkins University Press.

Adams, Joseph 1814. *A Treatise on the Hereditary Properties of Diseases.* London: Callow.

———. 1817. *Memoir of Life and Doctrine of the Late John Hunter.* London: Callow.

Blumenbach, Johann Friedrich. [1775, 1795] 1865. On the natural varieties of mankind. In *The Anthropological Treatises of Johann Friedrich Blumenbach.* Trans. T. Bendysche. London: Longman, Green, Roberts and Green.

Borie, Jean. 1981. *Les mythologies de l'hérédité au* XIXè *siècle.* Paris: Editions Galilée.

Burton, Robert. 1621. *Anatomy of Melancholy*. Oxford: Printed by John Lichfield and James Short, for Henry Cripps.

Cabanis, Pierre-Jean-Georges. [1802] 1956. Rapports du physique et du moral de l'homme. In Claude Lehec and Jean Cazeneuve, eds., *Oeuvres Philosophiques de Cabanis*, 2 vols. Paris: Presses Universitaires de France.

Canguilhem, Georges. 1972. *Le normal et le pathologique*. Paris: Presses Universitaires Françaises, coll. galien.

D'Azyr, Vicq, ed. 1787–1830. *Dictionnaire de Médecine de l'Encyclopédie Méthodique*. 13 vols. Paris: Panckouke.

Dezeimeris, Jean Eugène; Ollivier, Raige-Delorme, eds. 1828–1839. *Dictionnaire Historique de la Médecine Ancienne et Moderne*. 4 vols. Paris: Bechet Jeune.

Dowbiggin, Ian R. 1991. *Inheriting Madness: Professionalization and Psychiatric Knowledge in 19th-Century France*. Berkeley: University of California Press.

Fowler, Orson Squire. 1848. *Hereditary Descent*. New York: Fowler & Wells.

Haller, Albrecht von. 1752. *Reflexions sur le système de la génération de M. de Buffon*. Geneve: Barrilot et Fils.

Hamlin, Christopher. 1992. Predisposing causes and public health in early nineteenth century medical thought. *Social History of Medicine* 5 (1): 43–70.

Huet, Marie Hèléne. 1993. *Monstruous Imagination*. Cambridge, MA: Harvard University Press.

Hunter, John. [1786] 1835–1837. *Principles of Surgery*. Ed. James F. Palmer. London: Callow.

———. 1861. *Essays and Observations on Natural History*. Ed. R. Owen. 2 vols. London: Van Voorst.

López-Beltrán, Carlos. 1992. *Human Heredity 1750–1870: The Construction of a Domain*. Doctoral dissertation, King's College, London.

———. 1994. Forging heredity, from metaphor to cause: A reification story. *Studies in the History and Philosophy of Science* 25 (3): 211–235.

———. 1995. "Les maladies héréditaires": 18th-century disputes in France. *Revue d'Histoire des Sciences* 47 (3): 307–350.

———. 2002. Natural things and non-natural things: The boundaries of the hereditary in the 18th century. In *A Cultural History of Heredity I: 17th and 18th Centuries*, 67–68. Berlin: Max Planck Institute for the History of Science, Preprint 222.

———. 2004a. In the cradle of heredity: French physicians and "l'hérédité naturelle" in the early 19th century. *Journal of the History of Biology* 37:39–72.

———. 2004b. *El Sesgo Hereditario: Ámbitos Históricos del Concepto de Herencia Biológica*. Mexico City: Universidad Nacional Antonónoma de México, Estudios sobre la Ciencia.

———. Forthcoming. Les haereditarii morbi au debut de l'époque moderne. In Maaike van der Lugt and Charles de Miramon, eds., *L'hérédité à la fin du Moyen Âge*. Florence: Micrologus Library.

Louis, Antoine. 1749. *Dissertation envoyée à l'Académie des Sciences de Dijon, pour le prix de l'année 1748, sur la question . . . Comment se fait la transmission des maladies héréditaires?* Paris: Delaguet.

Lugt, Maaike van der. Forthcoming. Les maladies héréditaires dans la pensée scolastique (XIIè–XVIè siècle). In Maaike van der Lugt and Charles de Miramon, eds., *L'hérédité à la fin du Moyen Âge*. Florence: Micrologus Library.

Lyonnet, Robert. 1647. *Brevis Dissertatione de Morbis haereditariis*. Paris: Meturas.

Malinas, Yves. 1985. *Zola et les hérédités imaginaires*. Paris: Expansion Scientifique Française.

Mandressi, Rafael. 2004. *Félix Vicq D'Azyr, l'anatomie, l'Etat, la médecine*. Bibliothèque Interuniversitaire de médecine, Université Paris-5. http://www.bium.univ-paris5.fr/histmed/medica/vicq.htm.

Meara, Dermutius de 1619. *Pathologia Haereditaria generalis*. Dublin: Typis Deputatorum J. Franctoni.

Mercado, Luis (Ludovico Mercatus). [1593] 1620. De morbis haereditariis. In *Opera omnia medica & chirurgica*. Frankfurt: Typis Hartmanni Palthenij.

Montaigne, Michel. [1582] 1978. *Essais*. Bordeaux: S. Millange.

———. 1999. *Essays*. Trans. John Florio. Renascence editions, University of Oregon. http://darkwing.uoregon.edu/%7Erbear/montaigne/index.htm.

Morgentaler, Goldie. 1999. *Dickens and Heredity: When Like Begets Like*. London: Macmillan.

Olby, Robert C. 1994. Constitutional and hereditary disorders. In W. F. Bynum and R. Porter, eds., *Companion Encyclopedia of the History of Medicine*. London: Routledge.

Pagès, Jean-François. 1798. Héréditaires (maladies). In *Dictionnaire de Médecine*, vol. 7, 160–176. Paris: Agasse.

Petit, Antoine. 1817. *Essai sur les maladies héréditaires*. Paris: Gabon.

Portal, Antoine. 1808. *Considérations sur la nature et le traitement de quelques maladies héréditaires, ou de famille*. Paris: Baudouin.

Pujol de Castres, Alexis. [1802] 1823. Essai sur les maladies héréditaires. In *Oeuvres Médicales*. 4 vols. Reissued by Boisseau as *Oeuvres de Médecine Practique*. 2 vols. Paris: Bailliere & Bechet.

Rey, Guillaume. 1748. *Sur la transmission des Maladies Héréditaires, qui a balancé le Prix de l' Académie de Dijon en 1748*. Paris: Delaguette.

Roger, Jacques. 1963. *Les sciences de la vie dans la pensée française du xviiiè siècle*. Paris: Armand Colin.

———. 1964. Jean Fernel et les problèmes de la médecine de la Renaissance. In *Les conferences du Palais de la Découverte*, serié D, 70. Paris.

Rougemont J.-C. 1794. *Abhandlung über die erblichen Krankheiten*. Trans. Friedrich Gerhard Wegeler. Frankfurt: Johann Georg Fleischer.

Société Royale de Médecine. 1786–1788. *Histoire et Mémoires de la Société Royale de Médecine*. Vols. 9 and 10. Paris.

Steinau, Julius Henry. 1843. *Pathological & Philosophical Essay on Hereditary Diseases*. London: Simpkin, Marshall & Co.

Waller, John C. 2001. Ideas of heredity, reproduction and eugenics in Britain, 1800–1875. *Studies in the History and Philosophy of Biology and Biomedical Sciences* 32:457–489.

———. 2002. "The illusion of an explanation": The concept of hereditary disease, 1770–1870. *Journal of the History of Medicine and Allied Sciences* 57:410–448.

Williams, Elizabeth Ann. 1994. *The Physical and the Moral: Anthropology, Physiology, and Philosophical Medicine in France, 1750–1850*. Cambridge: Cambridge University Press.

Zirkle, Conway. 1946. The early history of the idea of the inheritance of acquired characters and of pangenesis. *Transactions of the American Philosophical Society* 38:91–151.

6 Erasmus Darwin and the "Noble" Disease (Gout): Conceptualizing Heredity and Disease in Enlightenment England

Philip K. Wilson

6.1 Introduction

The passage of hereditary traits has, as this book illuminates, intrigued natural philosophers and physicians for centuries. We have seen that the meaning of *heredity* itself was in flux during that time of considerable intellectual turbulence and societal discord that demarcates the shift from the Enlightenment to the Romantic era. Such discussion even occurred in provincial regions like England's West Midland county of Staffordshire, an area more commonly viewed as the seedbed of the European Industrial Revolution. There, one individual offered considerable written insight into contemporary thoughts on heredity, particularly in its relation to disease. This individual—the physician, industrialist, social reformer, educator of females, and in short, all around polymath—was Erasmus Darwin. Because he was better known among his contemporaries as "Dr. Darwin," I, too, will continue to use the sobriquet "Dr. Darwin," especially when distinguishing him from other family members.

Dr. Darwin has long been a favored study in pedigree analysis, particularly in relation to the intellectual predisposition of his grandson, Charles Darwin. Historians typically cite Dr. Darwin's section, "On Generation," from his *Zoonomia; or, The Laws of Organic Life* (1794–1796), to establish an intellectual background for his grandson's thoughts on adaptation, sexual selection, and evolution.[1] Such foreshadowing misses the opportunity to analyze Dr. Darwin's thoughts on heredity within their own Enlightenment and early Romantic era contexts.[2] Even the few Erasmus Darwin enthusiasts have neglected his views of human heredity in favor of his more voluminous and lusty botanical writings.[3]

This chapter concentrates on two of his writings that look particularly at human heredity and disease, namely, *Zoonomia* and the *Temple of Nature; or, The Origin of Society* (1804). In these works, Dr. Darwin argued that gout, consumption (i.e., tuberculosis), scrofula (the "King's Evil"), epilepsy, alcoholism, and insanity were hereditary. Consistent with what has been termed a "soft" hereditarianism view, Darwin viewed heredity as the result of a malleable admixture of nature and nurture causes.[4] After briefly introducing Dr. Darwin, his views, and those of several key contemporaries, this chapter will turn its attention to one disorder, gout—a disease thought by many to be as inherited as titles among the upper class. This focus allows us to examine his belief that hereditary disease, though predisposed, was not predestined. Preventing gout was achievable if one could learn how best to exert power over nature as well as to improve on nurture. This medical therapeutic aim was, as we will see, also intertwined with Dr. Darwin's critique of England's traditional social fabric that customarily enforced an aristocratic, class-controlled inheritance of property and power.

6.2 Dr. Darwin, Disease, and Heredity

Disease was not, to the industrial-minded Dr. Darwin, an entity that invaded the body. Rather, sharing the convictions of his Edinburgh medical school professors, he viewed disease more in constitutional terms as the result of a "malfunction" of the healthy motions within the body.[5] In particular, a disease typically expressed its mechanical defect as a "disturbance in one or more of the classes of fibrous activity." The physician's role, therefore, was "to apply remedies which would restore . . . [the body's] normal functioning," or even better, to prevent any malfunction in the first place.[6] This reliance on prevention was, to some extent, consistent with the concurrent revival of support for the "nonnaturals," the Galenic notion of keeping one's physical constitution healthy by moderating sleep, diet, drink, movement, and excretion as well as one's passions.[7]

But Dr. Darwin was also innovative and revolutionary within his medical thinking. Building on the Enlightenment projects of ordering and classifying information into new knowledge, he envisioned his *Zoonomia* as a Linnaean classification for disease.[8] "There is need of a theory in the medical profession," Darwin argued, "a theory founded upon nature, that should bind together the scattered facts of medical knowledge and converge into one point of view the laws of organic life."[9] The passion to reclassify medicine had also been the long-term goal of Edinburgh Professor of Chemistry and Medicine, and later Professor of the Institutes (i.e., theory) of Medicine, William Cullen. Like many during this reign of neurophysiological nosologists, Cullen averred that "almost the whole of diseases of the human body might be called *Nervous*."[10] However, he distinguished his 1769 *Synopsis Nosologiae Methodicae* from the taxonomies proposed by François Boissier de Sauvages (1763), Carl von Linné (Carolus Linnaeus) (1763), and Rudolph August Vogel (1764) by labeling the "Neuroses" or

Nervous Diseases a class of diseases by itself. Dr. Darwin's nosology, published a quarter of a century later, differed even further. Consistent with his view of society in general, he saw diseases more in terms of their dynamic, evolving qualities rather than as the static entity more typical of Enlightenment taxonomies.[11]

Within his medical writings, we find that more than many of his contemporaries, Dr. Darwin readily acknowledged a hereditary predisposition to disease. Ever prone to versify, Darwin stated his argument thusly:

E'en where unmix'd the breed, in sexual tribes
Parental taints the nascent babe imbibes;
Eternal war the Gout and Mania wage
With fierce unchecked hereditary rage;
Sad Beauty's form foul Scrofula surrounds
With bones distorted, and putrescent wounds;
And, fell Consumption! Thy unerring dart
Wets its broad wing in Youth's reluctant heart.[12]

Why did Darwin pay special attention to hereditary disease? In part, he had considerable experience with diseases that appeared in the same family—his own family. Throughout his life, Dr. Darwin suffered from stammering (i.e., stuttering). This "defect" was so pronounced that some said it to be "painful" to hear. Others, however, claimed that Darwin "repaid his auditors [i.e., attentive listeners] so well for making them wait for his wit or knowledge, that he seldom found them impatient."[13] When one young man questioned him as to whether he found this "deficit" to be inconvenient, Dr. Darwin replied, with typical wit, "No, sir, it gives me time for reflection, and saves me from asking impertinent questions."[14] Darwin's eldest son, Charles, was also afflicted with stammering. Darwin sent this son, when eighteen, to France, in hopes that "if he was not allowed to speak English for a time, he would be cured of . . . [stammering]. Charles returned a year later, spoke French fluently for the rest of his life, but continued to stammer in English."[15] For Dr. Darwin, this served as evidence early on in his career that at least some hereditary diseases could be modified in their expression through changing one's environment.

Turning again to his own family lineage, Darwin shared thoughts about intemperance (i.e., alcoholism) as a hereditary disease. In a letter to his son Robert Waring Darwin, Erasmus discussed Robert's uncle, Charles Howard. Howard, he claimed, was "a drunkard both in public and private—and when he went to London he became connected with a woman and lived a deba[u]ched life in respect to drink, hence he always had the Gout of which he died."[16] These effects also seem to have carried forth into the life of Mary Howard, Charles Howard's sister and Dr. Darwin's first wife. Mary

Darwin began to drink excessively early in their married life—though attempting to conceal this from her husband—in order to overcome episodes of "temporary delirium, or . . . insanity."[17]

In the words of his grandson, "No man ever inculcated more persistently and strongly the evil effects of intemperance than did Dr. Darwin."[18] If mankind was but temperate, Dr. Darwin argued, "men would still live their Hunderd [sic], and Methusalem [sic] would lose his Character; fewer [would be] banished from our Streets, limping Gout would fly the land, and Sedentary Stone would vanish into oblivion and death himself [would] be slain."[19] Although not restricted to the upper class, intemperance was, to Dr. Darwin, one sign of the gross overindulgence of that class. To broaden his appeal regarding intemperance particularly among the upper class, Dr. Darwin sought to persuade readers that intemperance was a problem far beyond drunken behavior. Indeed, he noted that much of the problem lay hidden from society. Similar to the other signs of lavishness and luxury among the aristocrats, their excesses in alcohol consumption was a "Parent" of disease.[20]

Moreover, Dr. Darwin reinforced the idea that intemperance laid the foundation for hereditary disease. All of the "hereditary diseases of [Britain]," he argued, originated as "the consequence of drinking much fermented or spiritous liquor."[21] According to Dr. Darwin's writings, all hereditary diseases were synonymous with what he called the "drunken diseases." Consequently, he argued that overcoming the hereditary tendency to intemperance was "perhaps . . . the most practical line of attack" against all "ill-health."[22]

Coupled with the evidence for hereditary disease garnered from his own family lineage, Dr. Darwin gathered additional evidence from many of the families for whom he offered medical care. "I have seen epilepsy," one of the "drunken diseases," produced "very often" in the same family, he said. As a treatment, he noted from his own success that "one sober generation" can cure epilepsy. Insanity, another "drunken disease," could be stopped in a similar manner. Indeed, Dr. Darwin claimed to have treated many people "who had insanity . . . [on] one side" of their family, yet whose "children now old people have no sign of it."[23]

Enlightenment medical authorities including William Heberden, William Hunter, and William Cullen—all of whom Dr. Darwin had studied with—as well as the physician Thomas Beddoes, founder of the Pneumatic Institute in Bristol, and the Birmingham physician and fellow Lunar Society man, William Withering, also viewed what were called "the drunken diseases" (including gout) as "hereditary in some degree."[24] The Scot-born gouty physician and literatus, Tobias Smollett, featured the hereditary passage of disease in his popular late eighteenth-century novels, *The Adventures of*

Peregrine Pickle (1751) and *The Adventures of Humphrey Clinker* (1771). Consistent with these authorities, Dr. Darwin claimed that it was the tendency, the diathesis, the predisposition to disease rather than the particular disease itself that was hereditary.[25] Contemporaries described "hereditary predisposition" as an "original conformation of the body, transmitted from the parent to the offspring" that, "when particular exciting causes are applied, a similar train of morbid phenomena takes place" in the child as was experienced by the parent.[26] Predisposition, the precise factor that was inherited, was identified as "the medium between health and disease."[27] Thus the susceptibility to be afflicted with a hereditary disorder like gout could be enhanced when an individual was exposed to certain triggering environments. Or, using the terminology of Dr. Darwin's day, the remote cause of disease (its physical inheritance) could be triggered by the proximate or "exciting" cause (cold, heat, some spasm or other debilitating disturbance in one's environment).[28] Again, we see Dr. Darwin's reliance on controlling the shaping influence of the "nonnaturals" as part of controlling disease:

The clime unkind, or noxious food instills
To embryon nerves hereditary ills;
The feeble births acquired diseases chase,
Till Death extinguish the degenerate race.[29]

Specifying the particular cause of disease was, to a nosologist like Darwin, of the utmost importance for the precision of his diagnoses. Hereditary disease, he argued, resulted from both inheritance (i.e., nature) and environment (i.e., nurture). Implicating the degree to which inheritance versus environment was viewed as the chief "cause" of a particular disorder helped Dr. Darwin target how best to prevent the further occurrence of this disease within a particular family. For example, if a patient's past pedigree suggested that he or she was the likely carrier of a hereditary propensity to disease, then it was the physician's role to prevent the patient from being exposed to the precipitating factors that would most likely induce the onset of disease. Neglecting to treat the symptoms of a reputed hereditary disease could also present a problem. For if "improperly treated," a hereditary disorder like gout may be "diverted from its proper course" such that "the miserable patient has a chance to be ever after tormented with head-achs, coughs, pains of the stomach and intestines."[30]

Dr. Darwin's view of heredity in relation to disease is consistent with general claims about the use of the label *hereditary* during this time period. His own practice demonstrated that diseases, like gout, though inheritable, were not inevitable. Consistent with the views of later historians of the Enlightenment, hereditary diseases could, at times, be disinherited.[31] In particular, hereditary predispositions "cease," Darwin claimed to have observed, "if one or two sober generations succeed; otherwise the family becomes extinct."[32] As evidence, he offered the claim that if the "drunken diseases" were not curable by imposing one sober generation upon the family, then, Darwin exclaimed, "there would not be a family in the kingdom without epileptic, gouty, or insane people in it."[33]

6.3 Dr. Darwin on Gout

Dr. Darwin also suffered from gout, a disease to which the upper class, the aristocracy, and the nobility were long thought to have been hereditarily predisposed. Indeed, it was an "early attack of gout" that turned Dr. Darwin into a "vehement advocate of temperance" for the rest of his life.[34] William Darwin, Erasmus's great-great-grandfather, had died in 1644, it was claimed, "from gout." It was "therefore, probable," claimed Erasmus's grandson, the gouty naturalist Charles Darwin, that Dr. Darwin, as well as many other members of the family, "inherited from this William, or some of his predecessors, their strong tendency to gout."[35] Dr. Darwin also gathered considerable case-study evidence from his wide-ranging practice, predominantly though not exclusively among the middle and upper ranks of society, that he interpreted as further proof of a hereditary transmission of gout.[36]

Darwin's thoughts on hereditary disease, especially gout, influenced several contemporaries, including the physician Thomas Trotter of Newcastle on Tyne. Trotter, whose mind was fixed on the nervous temperament as the exciting, proximate cause to all disease, also incorporated the predisposition to hereditary disease in his diagnoses.[37] Not everyone, however, shared Dr. Darwin's belief in the transmission of hereditary predispositions, or even in hereditary diseases themselves. The irascible Edinburgh physician, John Brown, whom Darwin admired for quantifying "excitation" and "stimulation" in his explanation of the actions of human physiology, disease, and medications, disapproved of applying the concept of "hereditary" to any disease. Brown's theory—known widely, especially throughout continental Europe, as Brunonian medicine—was based on the belief that every individual's temperament or constitution was the same. Drawing on Isaac Newton's revolutionary claim that "one single principle" governed all motion in "the whole planetary systems of the universe," Brown conjectured that the human form was similarly organized around one single human constitution.[38] Disease, according to Brown, an ex-pupil and extramural rival of William Cullen, resulted from either an excess or a deficiency of "excitability" (i.e., the capacity to react to external stimuli).[39] To claim that "a taint, transmitted from parents to their offspring" should be "celebrated under the appellation of hereditary" was, Brown argued, "a mere tale." For true Brunonians, "there is nothing in the fundamental part of . . . [the] doctrine" of hereditary disease.[40]

As evidence, Brown turned to gout, and he speculated that the "sons of the rich," who "succeed to their father's estate, succeed also to . . . [their] gout. Those [sons] who are excluded from the estate, escape the disease also, unless they bring it on by their own conduct." It is interesting that despite his adamant disbelief in the concept of hereditary disease, his recommendations for preventing such disorders were not that dissimilar from those of Dr. Darwin. In one of his rationalized case examples, Brown argued that although "Peter's father may have been affected with the gout, it does not follow that Peter must be affected; because, by a proper way of life, that is by adapting his excitement to his stamina, he may have learned to evade his father's disease." "If the same person," he continued, "who from his own fault and improper management, has fallen into the disease; afterwards, by a contrary management, and by taking good care of himself, prevents and removes the disease, . . . What then . . . [has] become of [the] hereditary taint?" Alas, there is nothing whatsoever hereditary about this or any other disease: "Whatever produces gout in one [individual], will produce it in another. And whatever cures it in any one [individual], cures it also in every other."[41]

Another contemporary, William Cadogan, physician to the Foundling Hospital in London, an institution formed on Enlightenment ideals, also questioned the claim that diseases were hereditary. Like Brown, he argued that if a disease like gout was truly hereditary, it would "necessarily be transmitted from father to son, and no man whose father had it could possibly be free from it." His own extensive practice and readings had provided him "many instances to the contrary." Cadogan explained further, "Our parents undoubtedly give us constitutions similar to their own, and, if we live in the same manner they did, we shall very probably be troubled with the same diseases; but this by no means proved them to be hereditary: it is what we do to ourselves that will either bring . . . [the diseases] on, or keep us free."[42]

Cadogan argued his case specifically in reference to gout.[43] This select paragraph exemplifies his numerical-based (i.e., statistical) critique:

> Those, who insist that the gout is hereditary, because they think they see it so sometimes, must argue very inconclusively; for if we compute the number of children who have it not, and women who have it not, together with all those active and temperate men who are free from it, though born of gouty parents; the proportion will be found at least a hundred to one against that opinion. And surely I have a greater right from all these instances to say that it is not hereditary, than they have from a few to contend that it is. What is all this, but to pronounce a disease hereditary, and prove it by saying that it is sometimes so, but oftener not so? Can there be a greater absurdity.[44]

Although many, including Darwin, concurred with Cadogan on temperance, they opposed his view regarding gout and heredity. William Falconer of Bristol rebuked Cadogan in a 1772 treatise, as did Aberdeen practitioner William Grant in an essay seven years later. Grant firmed up a general view of the relationship between constitution and hereditary disease by arguing that just as "constitutional diseases are often hereditary, . . . the hereditary diseases are always constitutional."[45] The Lichfield luminary, Samuel Johnson, praised Cadogan's work as "a good book in general" but criticized it as a "foolish one in particulars. . . . 'Tis foolish, as it says, the gout is not hereditary."[46] These distinctly different views expressed regarding the potential heritability of gout strongly suggest that the nature of hereditary disease itself was a matter of ongoing debate among Dr. Darwin and his contemporaries.

6.4 Darwin and Buchan: Progressive-Minded Reformers of Inheritance

Consistent with his deistic beliefs that development on the earth followed no fixed plan, Dr. Darwin argued that disease, too, was not predestined. Following the dissenter's practice of working to improve one's lot during his earthly existence, one could also work to overcome or prevent disease, including diseases to which one might be hereditarily predisposed. To achieve these aims, Darwin argued, one must learn how best to exert power over nature. Given his preoccupation with ideas concerning reproductive generation, it seems to have been a logical extension for him to seek societal reform through changing patterns of reproduction as a measure to induce progress in society over a mere few generations of offspring.

Dr. Darwin's objectives also represented a critique of England's traditional social fabric that enforced an aristocratic, class-controlled structure of the legal inheritance of property. England upheld a long tradition of patrimony and primogeniture that had, as Lawrence Stone has shown, "determined the life chances of every individual" in society.[47] Although particular patterns of property inheritance had been challenged since the early modern era, little had changed in this legal passage of power. Retaining societal control, in terms of property and power, required the regular oversight of elders within aristocratic families. In essence, they enforced a societal form of selective breeding between appropriate families that predictably solidified the transfer of current power and property through inheritance into the hands of future upper-class offspring.

By Dr. Darwin's era, some reform had been undertaken to modify this form of societal inheritance. For example, Strict Settlement laws had been enacted to broaden the distribution of inherited wealth. These laws provided for an "arranged inheritance" to be settled on before the birth of all children in a given family. In reality, this law focused primarily on offspring from the upper-class families who still wielded considerable power and had substantial wealth throughout England. Although the "claims of each child to some part of the inheritance were carefully protected," as Stone noted, no concomitant "decline in the emphasis on primogeniture" was immediately forthcoming.[48] However, by the time that Dr. Darwin's writings appeared, this method of arranging inheritance before birth represented part of a progressive-minded movement whereby, again in Stone's words, "primogeniture was successfully harmonized with individualism."[49] The time of this change regarding the control over property inheritance is consistent with the time frame that David Warren Sabean evocatively elaborates on regarding Europe in general in chapter 2 of this book.

Darwin's views about hereditary disease and the need to reform the aristocratic inheritance of power were in accord with the progressive-mindedness of his fellow Edinburgh-educated physician, William Buchan (1729–1805). Indeed, Buchan and Darwin were like-minded in many ways. They both supported the revolutionary action in the American colonies and in France; they both promoted the diffusion of useful knowledge; they were both avidly pro-temperance decades before any formal organization was to be found in Britain. In short, they were both liberal, progressive-minded reformers who exemplified that doctors, like politicians, could also be crusaders for social improvement.

Buchan, however, distinguished himself from Darwin through the aim of his medical writings. Unlike Darwin, whose chief medical writing, *Zoonomia*, was addressed to his fellow practitioners, Buchan, in the true spirit of the Enlightenment's view of the democratic power of nature, sought to put medical wisdom in the hands of the public. His principal work, *Domestic Medicine*, first appeared in 1769 and was reprinted throughout the nineteenth century. Like John Wesley's *Primitive Physic* (1747), and the anonymously written midwifery manual, *Aristotle's Masterpiece*, *Domestic Medicine* served many needs of the lay public. Some claim that Buchan's book partnered with the Bible in its popularity.[50] As a means of appreciating what much of the public in Darwin's Britain understood regarding hereditary disease, Buchan's *Domestic Medicine* continues its usefulness in serving us as a valuable guide as well.

Based on his own extensive practice and reading, Buchan claimed to have uncovered "manifold and decisive proofs" of the heritable transmission of disease. Although gout might not make its appearance until later in life, "the germs . . . must have lain in the system from the earliest periods of existence."[51] Indeed, the proclivity of such a disease for appearing among the upper classes was predictable, for once a disease is "contracted and riveted in the habit," it became—using the legal language of aristocratic inheritance—"entailed on posterity."[52]

Expanding beyond Darwin, Buchan suggested how to anticipate the future occurrence of hereditary disease. He informed readers that children were "particularly prone" to the diseases of the parent "to whom they bear the greatest personal similarity." Such direct linkage was not the sole key, however, for he warned that just as "we occasionally perceive the resemblance of some more remote ancestor break forth" in the personality and characteristics of a child, "so we shall [also] find the constitution and diseases of that child . . . in the nature of the progenitor whom it most resembles."[53]

Further noting that this "point of similarity between parents and children" regarding the hereditary disease had "not hitherto been sufficiently attended to," Buchan focused considerable attention on this familial connection in discussing diseases, including gout—a disorder that he also identified as having a particularly strong hereditary predisposition.[54] Familial disorders like gout, he went on, are "the result of a certain combination of habits, continued, perhaps, from one generation to another, combined with the peculiar circumstances in which the individual is placed. It is reasonable to suppose that, by altering the former, and counteracting the latter, the general constitution might be changed."[55] But just what did Buchan mean by "altering the former"—that is, nature's contribution—regarding the cure of hereditary disease?

Recalling Jean-Jacques Rousseau's analogies that were popular in England at this time, Buchan reminded readers of the farmers' inability to reap "a rich crop from a barren soil." Delicate women and men who pursue "irregular" habits may, he acknowledged, be able to bring forth children, but such children are "hardly . . . fit to live." The "first blast of disease" tends to "nip the tender [child] in the bud." Although some of these children may be able to "struggle through a few years' existence," their "feeble frame[s], shaken with convulsions from every trivial cause," leave them "unable to perform the common functions of life." In short, they "prove a burden to society."[56] This burden, Buchan was quick to note, was not only perpetuated among the lower classes. "How happy had it been," he charged, "for the heir of many a great estate had he been born a beggar, rather than to inherit his father's fortunes at the expense of inheriting disease."[57] Indeed, the lower classes might actually have some predilection against the onset of hereditary gout, which was claimed to be precipitated by the Epicurean living of England's upper class.

Hereditary aristocracy had long been based on the idea that the essence of nobility was transmissible. Retaining this essence, so the nobility customarily believed, required a purity of bloodlines without any interference through interclass marriage.

Yet by the time of Dr. Darwin's writings, England had undergone a change in its perspective on class-based family structure. Much of this change, according to economic historians, resulted from the rise of the middle class.[58] Many in society were enjoying an unprecedented upward social movement. This movement, to a progressive-minded thinker like Dr. Darwin—who, with his Lunar Society colleagues, was well known for adding material improvements *to* society—was evidence that improvements might be made *on* society as well.

Beyond the diagnostic significance of identifying hereditary predisposition, Buchan, perhaps more than any other contemporary, challenged readers to accept the responsibility they held, individually and collectively, in improving future generations. Most pointedly, Buchan proclaimed, it was in the name of progress that any "person labouring under any incurable [hereditary] malady ought not to marry." Doing so, he continued, not only "shortens his own life, but [also] transfers its misery to others." When "both parties are deeply tainted" with hereditary diseases, their offspring, he prognosticated, "must be very miserable indeed."[59]

Working to improve one's circumstances beyond what nature provided was consistent with industrial, progressive-minded, Enlightenment ideals. And by placing further responsibility for the future in the hands of the public, Buchan believed himself to be enhancing the progressive cause. "Want of attention to these things" in the past, he proclaimed, had "rooted out more families than plague, famine, or the sword." And unless individuals became more informed and active in suppressing the propagation of the diseased, the "evil" associated with such disorders was bound to continue.[60] Thus for Buchan, as for Dr. Darwin, the social reality of hereditary disease had significantly impeded human progress. To enhance a more positive evolution of humanity, he argued that people must assist and encourage nature to work in a progressive manner.

Believing that particular diseases like gout were introduced via marriage lines, Darwin, too, suggested that more attention should be directed to one's choice of marriage partner. The "art to improve the sexual progeny" in humans would, he argued, follow the choice of the "most perfect of both sexes." Fertility itself, Dr. Darwin claimed, was hereditary and subject to selection.[61] Noting that some aristocratic families had "become gradually extinct by hereditary diseases," he offered the cautionary and chastising note that "it is often hazardous to marry an heiress, as she is not infrequently the last of a diseased family."[62] Offspring would also be less liable to hereditary disease, Dr. Darwin argued, if they married into different families.

Such views clashed with traditional aristocratic family structure in which "like" tended to marry "like." Dr. Darwin's social progressive-mindedness was consistent with what Lawrence Stone (1979) and others have characterized as contemporary attempts to break from the traditional notions of fixed patterns of inheritance. Similar to and intertwined with his use of moral suasion to encourage his patients to modify their physical habits, Dr. Darwin framed his rhetoric as a suasion for change from the customary views of aristocracy. We find these views particularly resonant in his claims about the hereditary transmission of gout. By reinforcing that gout most typically owed its origin to noble ancestral lineage, he was critiquing the customary acceptance of the social structure of aristocracy. Conquering gout required, in part, the conquering of nature. The inheritance of gout could be overcome by breaking the lines of familial inheritance and marrying outside of the ancestral lineage.

These views, proposed in Dr. Darwin's last work, *The Temple of Nature,* were also foundational to what, a century later, became incorporated into programs of both positive eugenics (i.e., promoting the marriage and proliferation of "good stock") and negative eugenics (i.e., the prohibition of marriage and breeding between "defective stock"). Few physicians in Britain expanded more on the particulars of what might be termed the "eugenics" of this era than did William Buchan. Buchan noted that responsible actions were required of both women and men toward shaping the future in a progressive-minded manner. If women would "reflect on their own importance and lay it to heart, they would embrace every opportunity of informing themselves of the duties which they owe to their infant offspring." It was "their province," Buchan argued, "not only to form the body, but to give the mind its most early bias." Women "have it very much in their power" to make their children "healthy or valetudinary, useful in life or the pests of society."[63] "Were the time that is generally spent by females in the acquisition of trifling accomplishments employed in learning how to bring up their children . . . so as best to promote their growth and strength . . . , mankind would derive the greatest advantage from it." But until the education of females expands beyond "what relates to dress and public show, we have nothing to expect from them but ignorance" in important concerns such as their role in preventing the propagation of hereditary disease.[64]

Dr. Darwin also devoted considerable attention to improving the education of women. His chief work, *A Plan for the Conduct of Female Education in Boarding Schools,* written, ironically, for his two illegitimate daughters, the "Misses Parkers," was aimed at improving the minds and bodies of girls attending the boarding school that he had established at Ashbourne in Derbyshire. Progress in the body, as in society, was, in part, achievable through improving nurture.[65]

Buchan argued that men, too, must become more attentive to their role in improving their offspring. Why is it that a man is "not ashamed to give directions concerning the . . . [breeding of] dogs and horses, yet would blush were he . . . [to perform] the same office for that being who derived its existence from himself, who is the heir of his fortunes, and the future hope of his country!"[66] Dr. Darwin similarly noted that "those who breed animals for sale" had long known that "the sexual progenies of animals" were "less liable to hereditary diseases," if the breeding took place between families rather than within the same family.[67] For when both parents suffer from the same hereditary disease, he argued that the disease was more likely to descend to their posterity. Thus, another way to achieve progress among humanity resulted from improvements on nature. Both Drs. Darwin and Buchan popularized the proscription of marriage when hereditary disease was a family concern.

Buchan also charged physicians with negligence in showing attention to hereditary disease. This inattentiveness, he argued, was due largely to their unwillingness to attend to the sickness, management, and "preservation" of children. How striking the "labour and expense [that] are daily bestowed to prop up an old tottering carcase for a few years, while thousands of those who might be useful in life perish without being regarded! Mankind [is] too apt to value things according to their present, not their future usefulness."[68] Such characterization symbolizes a difficulty that had long plagued preventive thinking in medicine. Focusing on the here and now in medical practice had been (and continues to be) the norm, thereby leaving little energy (and funding) to expend on improving the lot of humanity for the future.[69]

Drs. Darwin, Buchan, and many of their contemporaries based their medical practices, in part, on altering one's physical constitution in order to enhance resistance to disease. Habits, characteristics—indeed physical constitutions—were thought to be subject to alteration. Subsequently, the altered or acquired forms of a constitution were heritable. Such a "transformation of descent"[70] exemplified what has been labeled "trans-generational progress."[71] In the minds of some, the altered constitutional makeup was thought to be passed along intact to the next generation. Thus Dr. Darwin anticipated what Gottfried Reinhold Treviranus and, more notably, Jean-Baptiste Lamarck would later argue about the passage of acquired characteristics. In his *Philosophie Zoologique* (1809), Lamarck provided considerably more evidence drawn from nature than did Dr. Darwin. Moreover, he argued for a directional development that Dr. Darwin had not done. Still, in the context of industrial revolutionaries, Dr. Darwin's arguments set forth in *Zoonomia* "biologised the concept of progress."[72] Nature may have provided a direction, but not a destiny; a potentiality, but not an inflexible future. The progressive developments as evidenced in the behavioral changes or adaptations of organic life were, according to Dr. Darwin, carried forth into the embryonic development of the next generation, or in other words, inherited. It might be said, contrary to the usual phrasing, that Lamarck was actually Darwinian in his thinking.[73]

6.5 Nineteenth-Century Heredity and Disease After Dr. Darwin

Dr. Darwin routinely had the ear of fellow Lunar Society men like Josiah Wedgwood, Matthew Boulton, and James Watt, industrialists whose success entailed unprecedented demands on the workforce. More than some, Wedgwood appreciated that his maximal financial gain depended in large part on maximizing the health of his workers. In part, through Dr. Darwin's efforts, public health and hygiene became critical concerns within the industrial Midlands. In the generation immediately following these Lunar luminaries, public health concerns were increasingly addressed in these heavily populated towns-turned-cities. The divide between the "haves" and the "have-nots" intensified. From an environmental perspective, the financially dependent were crowded into unsanitary areas that soon became known for their high morbidity and mortality. From a hereditary perspective, these rates were thought to increase even further, for these environments could easily excite those with predispositions to particular diseases into full-blown manifestations of their disorders.

It could be argued—though with sweeping generalizations—that concern about hereditary disease became relatively inconsequential in England during the decades immediately following Dr. Darwin. With the rise of industry, inheritance was no longer the only way to achieve substantial financial gain. The growth of industry also diminished the significance of inheritance within the minds of many medical reformers. Individuals including William Wilberforce, Lord Shaftsbury, Jeremy Bentham, and Edwin Chadwick focused much of their reform efforts on improving the environmental factors related to disease and public health. Although some claimed Dr. Darwin to be a pioneer of temperance reform, later temperance campaigns operated from the premise that drinking was more of a societal ill than an inherited one.[74]

The rise of bacteriology later in the nineteenth century further diverted the medical gaze, focusing more on the germ than the germ cell. With efforts aimed at cleansing society from germs, the concept of hereditary disease might appear to have lain dormant within British medical and popular writings throughout much of the latter half of the nineteenth century. According to standard historical accounts, heredity was excited into action only after another British polymath, Francis Galton, a grandson of Erasmus Darwin, drew attention to what he called the "study of agencies, under social control, that may improve or impair the social qualities of future generations either mentally or physically."[75] Encouraging a shift away from the primary focus on bacteriology, Galton proposed that massive efforts should be undertaken along the lines of "hard" hereditarianism to improve the human reproductive stock of society by giving "the more suitable races or strains of blood a better chance of prevailing speedily over the less suitable."[76]

Although some truth lies within such explanations, we are beginning to see more clearly through works of Carlos López-Beltrán (1994; chapter 5, this volume), John C. Waller (2002), and others in this book that support for hereditary explanations of disease remained strong though in flux as to the precise meaning of *heredity* throughout nineteenth-century Britain. Charles Darwin's son, Leonard Darwin—a key spokesperson for British eugenics in the early twentieth century—claimed that this interest in heredity, both as related to disease and to societal well-being, was itself a commonly noted trait passed along through family pedigrees. For his family, this may well have been true. However, our continually refined focus on the passage of ideas from of one generation of medical writers to succeeding generations will further illuminate beliefs about the possible connections between heredity, environment, and disease in the nineteenth century.

Notes

1. Many authors have forged connections on evolution between the two Darwins, including Glass, Temkin, and Straus (1959); Darlington (1961, especially, 7–13); King-Hele (1963, especially 63–96); Harrison (1971). One early notable exception is Bowler (1974), who carefully contextualizes eighteenth-century views of evolutionary thought. Porter (1989; 2000, especially 439–445) also addresses this point properly. Sheffield (2002) incorporates Dr. Darwin's evolutionary views into his historical science fiction.

2. For an overview of Erasmus Darwin and human reproductive generation, see Wilson 2005, 113–132. Hassler 1973 convincingly portrays Dr. Darwin as a bridge between Enlightenment and Romantic literati. Carlos López-Beltrán's foundational work in this book further elucidates the context of a growing number of Enlightenment medical authors who turned their attention to the hereditary transmission of disease.

3. Schiebinger 1993 surveys Erasmus Darwin's botanical writings in reference to gender.

4. López-Beltrán 1994 argues that this malleable view existed in the "soft hereditarianism" beliefs of the early nineteenth century in contrast to the more objective qualifications of a nature-based, "hard hereditarianism" later in the century.

5. Motions within the fibrous part of the constitution were the key to Dr. Darwin's understanding of physiology and pathology. Many scholars have pigeonholed Erasmus Darwin as a mechanist without equivocation. His mechanistic inclinations should not surprise us given all of the machines he contrived for his fellow Lunar Society members. However, from close scrutiny of

Zoonomia (E. Darwin [1794–1796] 1809), I discern his support of a deistic vitalism (or a vitalistic deism). In particular, I find that he proposed many vitalistic notions, albeit generally construed within a linguistic framework of mechanism. McNeil 1987 and King-Hele 1999 provide well-contextualized overviews of Dr. Darwin's industrial-mindedness.

6. Crum 1931, 124.

7. Emch-Dériaz 1992 discusses the nonnaturals in a late eighteenth-century context, though from the French perspective.

8. Uglow 2002, 266, depicts ordering and classifying as among the key projects of the Enlightenment. For a highly readable overview of eighteenth-century nosologies, including Dr. Darwin's, see King 1958.

9. E. Darwin [1794–1796] 1809, vol. I, viii.

10. Bynum 1993, 152.

11. See, for example, Bowler 1974, 166–179; McNeil 1987, 86–124.

12. E. Darwin 1804, Canto II: "Reproduction of Life."

13. Pearson 1930, 42.

14. Krause 1879, 40. For a historical overview of stammering, see Compton 1993.

15. Pearson 1930, 9–10.

16. E. Darwin, letter to Robert Waring Darwin, January 5, 1792, in E. Darwin [1792] 1958, 224. King-Hele 1981, 219, clarifies the previously mistaken identity of the Charles Howard in question.

17. Colp 1977, 119.

18. C. Darwin 1879, 56. As Pearson (1930, 27), notes, the novelist Maria Edgeworth, daughter of Lunar Man Richard Edgeworth, also testified that Dr. Darwin "believed that almost all the distempers of the higher classes of people arise from drinking . . . too much vinous spirit."

19. E. Darwin to his sister Susannah Darwin, March 1749, as reprinted in King-Hele 1981, 3. Written by Erasmus at the age of seventeen while a student at Chesterfield School, this letter attests that temperance was ingrained in his thinking from an early age.

20. E. Darwin to Susannah Darwin, March 1749, as reprinted in King-Hele 1981, 3.

21. E. Darwin 1804, 178.

22. C. Darwin 1879, 56.

23. E. Darwin [1792] 1958, 225.

24. E. Darwin [1792] 1958, 224. For complementary accounts of the Lunar Society, see Schofield 1963; Uglow 2002.

25. For an overview of diathesis, see Ackerknecht 1982, and for an elaboration on the importance of "constitution" in regard to hereditary disease, see Olby 1993.

26. Trotter 1808, 169.

27. Trotter 1808, 204.

28. Bynum 1994, 19.

29. E. Darwin 1804, Canto II: "Reproduction of Life."

30. Buchan 1828, 277.

31. Porter and Rousseau 1998, 117.

32. E. Darwin [1794–1796] 1809, vol. I, 414.

33. E. Darwin [1792] 1958, 225.

34. C. Darwin 1879, 1.

35. C. Darwin 1879, 1.

36. Although based in Lichfield and later Derby, Dr. Darwin's practice covered a large geographic area. He traveled extensively, either in his carriage or on horseback, to see patients. In 1766 alone, he calculated having traveled over 10,000 miles.

37. Trotter 1808, 172. Trotter also discusses the hereditary nature of tuberculosis, a disease more often associated with degeneracy than nobility. For descriptions of later views of hereditary tuberculosis, see Teller 1988; Worboys 2001; Wilson, 2006.

38. Brown 1803, 242–243.

39. Bynum 1994, 17. Yet, as Lawrence (1988) has insightfully pointed out, identifying the essential characteristic(s) of Brunonian medicine has been a difficult task for over 200 years.

40. Brown 1803, 396.

41. Brown 1803, 243.

42. Cadogan [1771] 1925, 70.

43. Porter and Rousseau (1998, 104) describe Cadogan's view on gout and heredity as the "paradigm-shift framing his entire theory."

44. Cadogan [1771] 1925, 70.

45. Porter and Rousseau 1998, 115.

46. Porter and Rousseau 1998, 81.

47. Stone 1979, 73.

48. Stone 1979, 412.

49. Stone 1979, 167. For an examination of this shifting pattern of inheritance from a legal-history perspective, see Spring 1993.

50. Porter 1992, 217.

51. Buchan 1828, 162.

52. Buchan 1828, 441–442. This legalistic language was commonly used by satirists of the period, as attested by the popular 1823 publication, *The Entail: or, The Lairds of Grippy*, written by the Scottish novelist and biographer John Galt. I am grateful to Ginna Pomata for drawing my attention to this work.

53. Buchan 1828, 162.

54. Buchan 1828, 162.

55. Buchan 1828, 163.

56. Buchan 1828, 441.

57. Buchan 1828, 441.

58. Harevan (1982) and Earle (1989) have substantiated E. A. Wrigley's claim that this rise preceded the Industrial Revolution rather than following it, as generations of historians once argued.

59. Buchan 1828, 442.

60. Buchan 1828, 442.

61. Darlington 1961, 11.

62. E. Darwin 1804, 179.

63. Buchan 1828, 440.

64. Buchan 1828, 440.

65. For further discussion of Erasmus Darwin's model of female education, see Schiebinger 1993.

66. Buchan 1828, 440.

67. E. Darwin 1804, 178.

68. Buchan 1828, 441.

69. Rosen 1977, 69–77.

70. Darlington 1961, 10.

71. McNeil 1987, 112.

72. McNeil 1987, 123.

73. Krause (1879, 133) made a similar comparison, rightly noting that it is "more proper" to view Lamarck as a "Darwinian of the older school" than to characterize Dr. Darwin as a Lamarckian. Julian M. Drachman (1930, 88) also addressed this point.

74. Pearson 1930, 27.

75. Galton [1909] 1984, 81.

76. Galton 1883, 24.

References

Ackerknecht, E. H. 1982. Diathesis: The word and the concept. *Bulletin of the History of Medicine* 56:317–325.

Bowler, Peter J. 1974. Evolutionism and the Enlightenment. *History of Science* 12:159–183.

Brown, John. 1803. *The Elements of Medicine.* Portsmouth, NH: William & Daniel Treadwell.

Buchan, William. 1828. *Domestic Medicine; or, A Treatise on the Prevention and Cure of Disease.* Exeter, NH: J. & B. Williams.

Bynum, W. F. 1993. Cullen and the nervous system. In A. Doig, J. P. S. Ferguson, I. A. Milne, and R. Passmore, eds., *William Cullen and the Eighteenth Century Medical World,* 152–162. Edinburgh: Edinburgh University Press.

———. 1994. *Science and the Practice of Medicine in the Nineteenth Century.* Cambridge: Cambridge University Press.

Cadogan, William. [1771] 1925. *A Dissertation on the Gout and All Chronical Diseases Jointly Considered.* Boston: Henry Knox. Reprinted in John Ruhräh. 1925. William Cadogan and His Essay on Gout. *Annals of Medical History* 7:64–92.

Colp, Ralph Jr. 1977. *To Be an Invalid: The Illness of Charles Darwin.* Chicago: University of Chicago Press.

Compton, David. 1993. *Stammering: Its Nature, History, Causes, and Cures.* London: Hodder and Stoughton.

Crum, Ralph B. 1931. *Scientific Thought in Poetry.* New York: Columbia University Press.

Darlington, C. D. 1961. *Darwin's Place in History.* New York: Macmillan.

Darwin, Charles. 1879. Preliminary notice. In Ernst Krause, *Erasmus Darwin,* trans. W. S. Dallas. London: John Murray.

Darwin, Erasmus. [1792] 1958. Letter to Robert Waring Darwin, January 5. In Nora Barlow, ed., *The Autobiography of Charles Darwin, 1809–1882.* New York: Norton.

———. 1797. *A Plan for the Conduct of Female Education in Boarding Schools*. Derby, Derbyshine: J. Johnson.

———. [1794–1796] 1809. *Zoonomia; or, The Laws of Organic Life. In Three Books*. Boston: Thomas and Andrews.

———. 1804. *The Temple of Nature; or, The Origin of Society*. New York: T. and J. Sword.

Drachman, Julian M. 1930. *Studies in the Literature of Natural Science*. New York: Macmillan.

Earle, Peter. 1989. *The Making of the English Middle Class: Business, Society and Family Life in London 1660–1730*. London: Methuen.

Emch-Dériaz, Antionette. 1992. The non-naturals made easy. In Roy Porter, ed., *The Popularization of Medicine 1650 –1850*, 134–159. London: Routledge.

Galt, John. 1823. *The Entail; or, The Lairds of Grippy*. London: Blackwood.

Galton, Francis. 1883. *Inquiries into Human Faculty and Its Development*. London: J. M. Dent.

———. [1909] 1984. Probability the foundation of eugenics. In *Essays on Eugenics*, 73–99. New York: Garland.

Glass, Bentley, Owsei Temkin, and William L. Straus Jr., eds. 1959. *Forerunners of Darwin, 1745–1859*. Baltimore: Johns Hopkins University Press.

Harevan, Tamara K. 1982. *Family Time and Industrial Time*. Cambridge: Cambridge University Press.

Harrison, James. 1971. Erasmus Darwin's view of evolution. *Journal of the History of Ideas* 32:247–261.

Hassler, Donald M. 1973. *Erasmus Darwin*. New York: Twayne.

King, Lester S. 1958. *The Medical World of the 18th Century*. Chicago: University of Chicago Press.

King-Hele, Desmond. 1963. *Erasmus Darwin*. New York: Scribner.

———. 1981. *The Letters of Erasmus Darwin*. Cambridge: Cambridge University Press.

———. 1999. *Erasmus Darwin: A Life of Unequalled Achievement*. London: Giles de la Mare.

Krause, Ernst. 1879. *Erasmus Darwin*. Trans. W. S. Dallas. London: John Murray.

Lamarck, Jean-Baptiste. 1809. *Philosophie Zoologique*. Paris: Dentu.

Lawrence, Christopher. 1988. Cullen, Brown, and the poverty of essentialism. *Medical History* suppl. no. 8:1–21.

López-Beltrán, Carlos. 1994. Forging heredity: From metaphor to cause, a reification story. *Studies in the History and Philosophy of Science* 25:211–235.

McNeil, Maureen. 1987. *Under the Banner of Science: Erasmus Darwin and His Age*. Manchester: Manchester University Press.

Olby, Robert C. 1993. Constitutional and hereditary disorders. In W. F. Bynum and R. Porter, eds., *Companion Encyclopedia to the History of Medicine*, 412–437. New York: Routledge.

Pearson, Hesketh. 1930. *Doctor Darwin.* London: J. M. Dent.

Porter, Roy. 1989. Erasmus Darwin: Doctor of evolution? In James R. Moore, ed., *History, Humanity and Evolution: Essays for John C. Greene*, 39–69. Cambridge: Cambridge University Press.

———. 1992. Medicine in Georgian England. In Roy Porter, ed., *The Popularization of Medicine 1650–1850*, 215–231. London: Routledge.

———. 2000. *Enlightenment: Britain and the Creation of the Modern World.* London: Penguin.

Porter, Roy, and G. S. Rousseau. 1998. *Gout: The Patrician Malady*. New Haven, CT: Yale University Press.

Rosen, George. 1977. *Preventive Medicine in the United States 1900–1975: Trends and Interpretations.* New York: Prodist.

Schiebinger, Londa. 1993. *Nature's Body: Gender in the Making of Modern Science*. Boston: Beacon Press.

Schofield, Robert E. 1963. *The Lunar Society of Birmingham.* Oxford: Oxford University Press.

Sheffield, Charles. 2002. *The Amazing Dr. Darwin: The Adventures of Charles Darwin's Grandfather—It's All in the Family*. Riverdale, NY: Baen Books.

Smollett, Tobias. 1751. *The Adventures of Peregrine Pickle*. London: Printed for the Author, Sold by D. Wilson.

Smollett, Tobias. 1771. *The Adventures of Humphrey Clinker*. London: W. Johnston and B. Collins.

Spring, Eileen. 1993. *Land, Law and Family: Aristocratic Inheritance in England, 1300–1800*. Chapel Hill: University of North Carolina Press.

Stone, Lawrence. 1979. *The Family, Sex and Marriage in England, 1500–1800*. Harmondsworth, Middlesex: Penguin.

Teller, Michael E. 1988. *The Tuberculosis Movement: A Public Health Campaign in the Progressive Era.* New York: Greenwood Press.

Trotter, Thomas. 1808. *A View of the Nervous Temperament; Being a Practical Inquiry into the Increasing Prevalence, Prevention, and Treatment of Those Diseases Commonly Called Nervous, Bilious, Stomach & Liver Complaints; Indigestion; Low Spirits, Gout, &c.* Troy, NY: Wright, Goodenow, & Stockwell.

Uglow, Jenny. 2002. *The Lunar Men: The Friends Who Made the Future*. London: Faber and Faber.

Waller, John C. 2002. "The illusion of an explanation": The concept of hereditary disease, 1770–1870. *Journal of the History of Medicine and Allied Sciences* 57:410–448.

Wilson, Philip K. 2005. Erasmus Darwin on human generation: Placing reproductive generation within historical and zoonomian contexts. In C. U. M. Smith and R. Arnott, eds., *The Genius of Erasmus Darwin*, 113–132. Aldershot, Hampshire: Ashgate.

———. 2006. Confronting "hereditary" disease: Eugenic attempts to eliminate tuberculosis in Progressive era America. *Journal of Medical Humanities* 27:19–37.

Worboys, Michael. 2001. From heredity to infection: Tuberculosis, 1870–1890. In Jean-Paul Gaudillière and Ilana Löwy, eds., *Heredity and Infection: The History of Disease Transmission*, 81–100. London: Routledge.

7 Degeneration and "Alienism" in Early Nineteenth-Century France

Laure Cartron

> Qu'elles (les mères) n'oublient jamais qu'elles portent un être susceptible de modifications et qui peut apporter des preuves d'un père ou d'une mère coupable.
> —Millot 1828

In this chapter 5, Carlos López-Beltrán shows how, up until the end of the eighteenth century, medical theories on hereditary transmission were by no means clearly formulated. Uncertainty over the physiological processes involved left doctors bemused by the mystery of fecundation.[1] This mystery was to be solved partly in 1847 when Prosper Lucas (1805–1885) published his *Traité philosophique et physiologique de l'hérédité naturelle dans les états de santé et de maladie du système nerveux*. Under the impetus of the work of accomplished physicians and naturalists like Alfred Louis Armand Velpeau (1795–1867), Karl Ernst von Baer (1792–1876), and Félix-Archimède Pouchet (1800–1872),[2] Lucas intended to discover the nature of heredity. In one of his first remarks on this subject, he said that understanding the procreative act was not enough to understand human reproduction. Hereditary transmission had to be studied as a problem in itself. In fact, an "enormous error had been made" in not having separated these "two sides of the problem."[3]

Long before Lucas, however, a number of physicians had already focussed on the problem of hereditary transmission, which occupied the borderland between public and private realms. One indication of their public interest in delineating hereditary relationships was that physicians began advocating the necessity for legislation governing marriages, unions whose settlement had traditionally been managed by families and clerics. In terms of familial inheritance, the most renowned physicians were quite sure that if children generally resemble their parents in outward form and in their mental capacities and dispositions, they also inherit the diseases of their parents. It was thus on the transmission of hereditary diseases that medical works would concentrate. In this context, degeneration and heredity were often considered virtually synonymous. This notion of hereditary transmission of diseases emerged at the point where clinical teaching converged with the use of medical statistics. Physicians were to take the hospital as their exemplary experimental site, where they found ample evidence of degeneration.

In the following pages, I would like to show that the vogue for theories of familial and national degeneration in France was already well underway by the mid-nineteenth century, when Bénédicte-Augustin Morel (1809–1873) wrote his famous *Traité des Dégénérescences* (1857). I will first outline the views of degeneration generally accepted in the eighteenth and first half of the nineteenth century. For this period, a distinct trend is recognizable, from degeneration as a consequence of accidental causes, to degeneration as a natural force. I will then try to relate this conceptual development to changes in the medical profession. In a first section, I will examine the situation of the medical profession after the Revolution and during the First Empire (1804–1815), when the medical field remained dominated by Old Regime physicians, Antoine Portal (1742–1832) in particular. Though pathological anatomy brought to the fore hereditary disposition or "diathesis" as a key concept in this period, hereditary transmission remained largely undistinguished from other, developmental and environmental processes. I will then turn to the period after Napoléon's fall when the medical profession tried to secure its standing by arguing for its status as a scientific discipline, which would further the interests of society. In particular, my discussion will focus on the work of the "alienists" (mental physicians) Philippe Pinel (1745–1826) and Jean-Etienne-Dominique Esquirol (1772–1840). I will argue that their use of statistical methods isolated hereditary diseases as being those transmitted in a number of families, running deeper and deeper in successive generations, and leading to the degeneration of the French nation as a whole.

7.1 The Fear of Degeneration

In ancient Greece and Rome, as in France until the fourteenth century, the verb *degenerate* (in French, *dégénérer*) meant to lose the qualities of one's lineage. The nouns *degeneration* (*dégénération*) and *degeneracy* (*dégénérescence*) appeared toward the end of eighteenth century. Both terms were used synonymously, with a tendency to reserve *degeneration* to stand for the process, and *degeneracy* for its result.[4] From the mid-eighteenth to the mid-nineteenth century, the term changed its meaning considerably, from an evil inflicted by luxury and civilization, to a natural tendency of pathological decline.

For eighteenth-century naturalists like Georges-Louis Leclerc Comte de Buffon (1707–1788), migrating populations degenerated due to abandonment of their original, ideal environment. The body was afflicted by the conditions of new milieus and new types of food.[5] This translated into the idea of human degeneration, assumed to be an evil of civilization as opposed to the condition of the "noble savage," who had been in vogue in the eighteenth century, particularly in the work of Jean-Jacques Rousseau (1712–1778). In this period, notions of degeneration were applied most often to the French idle class, with its luxurious style of living. In 1817, Julien-Joseph Virey (1775–1846), chief pharmacist at the military hospital Val-de-Grâce, maintained that degeneration was most common among civilized populations.[6] Only in the 1840s did physicians identify the main external cause of degeneracy in the population as the conditions of life of the working classes rather than idleness and luxury.[7]

Hence, it was rare in the eighteenth century to find philosophers or doctors who painted an apocalyptic picture of the French nation or the human race. As Ann Carol has stressed in her *Histoire de l'eugénisme en France* (1995), it was only in the early nineteenth century that such a picture began to emerge.[8] In the early years after the Revolution of 1789, a more optimistic picture dominated, one that, while acknowledging the past degeneration of the nation under the rule of the idle classes, also saw real opportunities for its regeneration. As exclaimed by Gabriel-Honoré Riquetti, Comte de Mirabeau (1749–1791), who was an elected representative of Tiers-Etat, which included those who were neither nobles nor clerics: "If only principles could be the seed of a much needed regeneration, one that is too necessary to be refused and too desired to be anything but inevitable."[9] However, regeneration was no longer conceived of as a return to a putatively original state of humanity. It concerned future generations, and targeted the nation as a whole rather than particular classes or sections of the population.

In part these new interests in degeneration and regeneration can be explained by changing social conditions. The twenty-six years of civil and foreign wars following the Revolution uprooted a large proportion of the country poor, who came to the cities to seek work.[10] This mobility and the collapse of religious authority led to the breakup of many regulatory traditions with respect to marriage and property devolution.[11] The specter of degeneracy therefore lurked in the minds of thinkers of the time with an ominous certainty. New political conditions allowed the possibility of combating this threat before too much damage was wrought among the population as a whole. The potential problem of national degeneracy called for an intense state engagement to make society healthier, stabilize a restless population, reinforce communications between the state and the population, and solidify public administration. Concern about degeneracy and interest in the possibility of state intervention were expressed in the literature of the postrevolutionary period. In Honoré de Balzac's (1799–1850) novel *Le Médecin de Campagne* (1833) an enlightened, rationalist country doctor is moved by a veritable sense of mission to eradicate cretinism in his country: "Marriages among these unfortunate creatures are not forbidden by law," he ponders and adds, "would it not do the country a great service to put a stop to this mental and physical contagion?"[12]

One opinion that held sway in scientific circles and among the public was that the number of mentally ill had been increasing since the time of the Revolution. Read at the *Institut de France* as early as 1796, and published in 1802, was an essay on the "relations of the physical and the moral" by Pierre-Jean-Georges Cabanis (1757–1808). Cabanis was a physician and then member of the Directoire, the first government following the French Revolution (1795–1799). His paper stimulated much discussion in medical circles.[13] Among others, it undoubtedly inspired the *Traité du goître et du crétinisme*, by François-Emmanuel Fodéré (1764–1835). In this book, published in 1800, Fodéré argued that "if children are born with goiters, if from father to son over two generations, men with goiters marry women so afflicted, and if the region is one where goiters are endemic, by the third generation, the child who is born will be not only with goiters, but also a cretin."[14] Thus medical opinion expressed the same concern about escalating mental weakness as evidenced in the literary work of Balzac.

Certainly, statistics showed that asylum populations were growing; in Paris the number of patients grew from 1,009 in 1786, to 2,000 in 1813, to almost 4,000 in 1839.[15] The fact was, however, that from the beginning of the nineteenth century, the state opened several hospitals. Here, for the first time, patients were interned in specialized establishments and census taking was made possible. In addition, there was a growing surveillance of the population in general, as France underwent rapid social and political change, and sequestered all the social deviants—unstable and asocial individuals alongside prostitutes—that before had been left on their own. Initially, socioeconomic explanations were sought for the apparent increase of the insane. To blame were moral laxity, literature, and unhealthy urban conditions, but the primary conclusion arrived at was that progress itself caused the problems. "A good many French heads have been led astray by ideas of liberty and reformation," wrote Étienne Esquirol (1772–1840) in his treatise *Des maladies mentales* in 1838.[16] Fighting these ills by hygienic measures was a task entrusted not only to physicians but also to engineers. The movement to engineer physical, moral, and mental hygiene started to gain ground at the end of the Enlightenment and had given rise to the Paris Council of Hygiene and Salubrity, responsible for streets, markets, and dwelling control.[17]

While illnesses were thus imputed above all to insalubrious living conditions, it was still thought that there had to be a natural predisposition to the illness for it to be actually contracted. From the 1830s on, the theme of degeneration thus selected new scourges: alcoholism and prostitution. Degeneration was increasingly seen by medical writers not as due to the social conditions of the poor, but as a self-reproducing force. In other words, degeneration came to be viewed as the cause rather than the effect of crime, moral destitution, and disease. The putative force of degeneration produced degeneracy in society.[18] Heredity thus reinforced the presumed dangers of moral degeneration. Mental physicians elaborated a system of thinking that made hereditary degeneration the center of moral concern. In La Sainte-Famille in 1845, Marx and Engels noted this powerful preoccupation with crime in European society, or, as they put it, with "the mystery of degeneracy in civilization."[19]

7.2 Survival of the Old Regime (1792–1815)

According to George Weisz in *The Medical Mandarins* (1995), until the early nineteenth century, there remained a kind of intellectual hierarchy in place in the French medical corps. Doctors who practiced their art in hospitals or in their private clinics were apparently despised by their peers. Only those who held academic or official positions could establish influential circles.[20] At the very top of these groups, we find a notorious and inescapable personality, Antoine Portal (1742–1832). He had begun his career in 1766 as professor of natural history to the Dauphin, and ended it in 1832 as First Physician to Louis XVIII and Charles X. From 1769, Portal held the teaching chair of medicine at the Collège de France and was appointed to the chair of anatomy at the Jardin du Roi in 1775. He kept this chair after the Revolution, when the Jardin du Roi became the Muséum d'Histoire Naturelle in 1793. During the Revolution, his life was threatened because of his close relationship to the royal family and other members of the court. Astutely curing some revolutionary politicians, he was "condemned to continue to treat humanity" in 1794. Portal became a member of the section of sciences morales et politiques of the Institut de France, and, finally, in 1820, was chosen by the government to become the president of the Académie de Médecine, which was created to replace the Académie Royale de Chirurgie and the Société Royale de Médecine, both dissolved in 1793. As president of the Académie de Médecine, Portal had to chair the committee that evaluated propositions made to the government by physicians. Playing the role of filter between government and his peers, Portal remained a key figure in French medicine for four decades after the Revolution, despite his links to the old regime, and despite frequent changes of government.

Portal was trained in Montpellier at the École de Médicine, where medicine and surgery had not been separated. He gave the first instructions in pathological anatomy in France. Based on case studies and personal observations, Portal argued for the heredity causes of diseases in his numerous medical writings, particularly in his *Considération sur la nature et le traitement des maladies héréditaires ou maladies de familles* (1808). Portal's medical philosophy, however, remained stamped by humoralism, which hampered the development of a notion of specific hereditary causes of disease, a point discussed by Carlos López-Beltrán and Philip K. Wilson in chapters 5 and 6, respectively, of this volume. The discovery of tubercles, for example, enabled Portal to draw together in one class some chronic diseases that had previously been considered distinct: scrofula, phthisis, and rickets. A variety of internal and external causes were proposed to account for the disease: food and drink, dirt, excreta and atmospheric influences (miasmas), lymphatic temperament, contagion or a syphilitic "virus," but also heredity. Portal proposed that the hereditary factor consisted in a "scrofulous virus" caused by an alteration of lymph. Contracted by parents during sexual intercourse, the "virus or poisoning principle" (*virus ou principe empoisonné*) spread slowly in every part of the bodily system and "sooner or later came out in the sperm or in the milk." During childhood, the virus first migrated to the lymphatic glands (causing scrofula). As the years passed this "virus" or "vice" continued to travel toward the limbs (causing phthisis) and became associated with a variety of other viruses, provoking rickets, gout, dropsy, venereal evil, and scurvy. Finally, by spreading to the nervous system, the scrofulous virus also became implicated in some mental disorders like epilepsy, mania, and hysteria.[21]

Due to Portal's fame the theory enjoyed a good reception in France. Anyone desirous of disputing Portal's theories had to be of considerable medical stature. This was true of Antoine Petit, Portal's former colleague at the Jardin du Roi, who had subsequently become professor of surgery and doctor at the Hôtel-Dieu in Paris. In his *Essai sur les maladies héréditaires*, Petit did not deny the existence of a virus, but he postulated its contagious rather than hereditary transmission. Yet Petit, too, tried to reconcile both theories by arguing that what was transmitted by generation was not the virus itself, but "a predisposing diathesis receptive to this virus." "Hereditary disease," in Petit's eyes, "was no more than the result or the product of an organic disposition" that children received from their parents.[22]

In his critique, Petit relied on the concept of hereditary disposition as it had been spelled out by François-Emmanuel Fodéré in his *Traité de Médecine légale et d'Hygiène publique ou de police de santé* in 1813. Fodéré was a member of the Marseilles Board of Health, a professor of medicine in the École de Strasbourg, and an expert on public health problems. During the Napoleonic wars he served as a health officer in the army. In the *Traité de Médecine légale*, Fodéré insisted on the transmission of a special condition, constitutional weakness. This could produce a variety of illnesses or disabilities, and was inherited at the moment of conception, at birth, or during weaning. As this new model of thinking developed and spread through the medical corps, an unprecedented notion took root: chronic diseases would appear only in those with an inherited constitutional weakness. From that moment on, every well-informed physician spoke about hereditary diseases in the key terms of "hereditary predisposition," "hereditary taint," or "predisposing factors."

Although their hypotheses about the mechanism of hereditary transmission diverged, virtually all physicians supported Portal's basic argument. Given that the transmission of external physical characteristics seemed to be constant, there could be no doubt that the same transmission process was also at work in determining the internal constitution of the body. Thus, whenever the same chronic disease affected members of the same family, transmission by heredity was readily assumed.[23] At the same time, an important, additional theme began to emerge. An external cause, even if unique and specific, was not enough to bring about a chronic disease; there also had to be a favorable ground, a "diathesis," for the disease in the afflicted individual. Focusing their attention on heritable, constitutional anomalies, medical practitioners showed that they had special competence in confronting a peril that, through generation, threatened to become entrenched in the population as a whole. Physicians thus developed ideas that had implications for the management of populations.

7.3 A New Generation of Physicians (1815–1830)

During the Restoration (1815–1830), physicians were putting increasing pressure on government to become involved in public health.[24] In part this was due to the large number of young physicians that had been discharged from Napoléon's armies in 1815, only to join a crowded medical profession. Like other middle-class professions in Western Europe between 1815 and 1848, medicine was suffering from the discrepancy between demand for services and supply, in this case of new doctors graduating from medical school. Paris, for example, was saturated, with only 3 percent of the national population but 13 percent of all doctors.[25] The result was bitter competition for patients and for salaried hospital positions.

Calls for a greater public role on the part of doctors predated the Restoration period. In the last decade of the eighteenth century, Cabanis in particular supported the medical corps' ambition to take a more active role in the governance of the people.

For Cabanis, if the moral could not be dissociated from the physical, then it followed that both medical treatment and political government had their origins in one and the same principle. They should accordingly be subject to the same laws, based on a science of man that included both physiology and the analysis of ideas and moral issues.[26] "It is up to medicine to try to seek direct ways [of improving the human race], of advancing future races, of designing the way of life of the human species," Cabanis said in 1799 in an address to the Conseil des Cinq Cents, one of the two executive power assemblies under the Directoire.[27]

Cabanis' arguments entailed the idea that doctors needed to prove to the French government and to the public that the new medicine was indeed a science, rather than a conjectural art, as it had been considered in the past. In earlier times, when all members of a family were in bad health, medical practitioners could account for this by referring to some hypothetical agent, be it contagious or hereditary. Now, with the official liberalism, something more was demanded to justify regulations, like quarantines or controls, with their associations with the Old Regime. Medicine had to become a positive science if it was to assume some social function. In particular, it had to become quantitative. Like other scientists, physicians were influenced by Pierre-Simon de Laplace (1749–1827), astronomer and mathematician and famous for his "calculus of probabilities." Laplace claimed that establishing the regularity of phenomena with mathematical rigor permitted the investigation of causes, and had himself asserted that this method could be extended to the human and moral sciences.[28]

One particular field that the new doctors investigated was mental medicine, or "alienism." At the same time, madness was attributed to so-called moral causes, passions like anger and pride, immoral habits, poverty, or unhealthy work. While the new doctors did not dispute the efficacy of such causes,[29] they raised protostatistical arguments that spoke for predispositional causes as well. Thus, Marie-André Ferrus (1784–1861), a surgeon in Napoléon's army, noticed that some soldiers went mad after intense battles, while others remained sane. So he rejected any role for accidental causes in the onset of madness. Because all soldiers had experienced the same events, but only some went mad, the only valid explanation lay in the inherent "sensitivity" of each individual, this "sensitivity" being a part of the peculiarities inherited by families.[30]

Until the end of the eighteenth century, the insane had been indiscriminately mixed in with vagabonds, beggars, abandoned children, criminals, prostitutes, and the poor in hospitals or hospices, which functioned simultaneously as orphanages, old age homes, and prisons.[31] Changes in the understanding of mental disorders took place when specialized asylums for the mentally ill were set up after the Revolution, which allowed for clinical observations. Yet, until the creation of the first chair of mental medicine in the Paris Faculty of Medicine in 1821, there was no particular instruction to prepare doctors to work in asylums. So, the "moral physicians," as they called themselves, had to acquire their specific skills in the course of their work. Although working conditions were poor, and although moral physicians saw themselves marginalized in the medical community, this was the first step toward the professionalization of one branch of medicine. With the return of the monarchy, the Catholic clergy was gaining in power again. The French state might indeed revoke the physicians' privileges and reinstate the religious orders that had been in charge of the hospitals, and hence the insane, before the Revolution.

In such a situation, the result was sharp competition for patients, official posts, prestige, and social authority within the profession. Mental physicians had to assert their specific knowledge of the origin of mental disease. One of the most influential writers in this respect was Philippe Pinel (1745–1826). In 1792, the Royal Society of Medicine organized a jury selection process and appointed Pinel to reorganize the Bicêtre Hospital, where male patients with mental disorders had been hospitalized. This was the first time that a trained physician was selected to supervise a mental asylum. In the same year, Pinel was chosen to teach hygiene at the new Paris School of Medicine. Three years later, in 1795, Pinel moved to La Salpêtrière to be in charge of a female population of curably insane that had been transferred from the Hôtel-Dieu. Pinel called for physicians to occupy the role of advisor that the Church had left vacant,[32] and he was the first to stress the importance of statistical tools in this context.

Jean-Baptiste Pussin (1745–1811), Pinel's predecessor at Bicêtre, had been a warden, not a physician, and the method of treatment he advocated accordingly became known as the *méthode des concierges*. This method was based on a belief in nature's self-healing power. Drugs and physical constraints were rejected. Instead, Pussin believed, patients would recover if they were only taken away from the unhealthy influences of their home environments to be isolated in specialized institutions. In the introduction to his *Traité médico-philosophique sur l'aliénation mentale ou la manie* of 1801, Pinel mentioned the *méthode des concierges* favorably and conceded that it had shown some evidence of success. However, he also stressed that the "calculus of probabilities" should be used to determine the effectiveness of various therapies by counting the number of times a treatment had a positive effect. Although Pinel alluded to the mid-eighteenth-century work on the "calculus of probabilities" by Daniel Bernoulli (1700–1782), he went no further than insisting on the compilation of figures.

It is only in the second edition of this *Traité* in 1809 that Pinel actually used statistics in a section titled "*Résultats d'observations et construction de tables pour servir à déterminer le degré de probabilité de la guérison des aliénés.*" This section was devoted to commenting on figures characterizing his patients in Bicêtre and La Salpêtrière and the circumstances of their illness. Pinel showed the importance of the patient's family background and his or her relationship with other patients. On this basis, Pinel defined two kinds of causes of madness: predispositional, or latent, and accidental. Starting his discussion of causes with heredity, he singled it out as a major predispositional cause: "It would be difficult not to admit the hereditary transmission of mania when we see everywhere that in certain families some members are affected by the disease over several successive generations."[33] At the same time Pinel abandoned all reference to the *méthode des concierges* in the second edition of his *Traité*.

7.4 Statistics and Alienism

According to Jan Goldstein, the first mental doctors' circle was formed around Pinel.[34] He trained several famous students, among them the already-mentioned Fodéré and Esquirol. It was the latter who played a crucial role in establishing mental medicine as a politically influential medical profession in France. Like Pinel, he had learned medicine at Montpellier. He replaced Pussin as Pinel's assistant in La Salpêtrière in 1811 and in 1825 he was appointed chief physician of the Hôpital de Charenton, an asylum for mentally ill men that had been built close to Paris in 1796. From 1802, Esquirol managed his own private clinic, the "Maison de traitement des aliénés," opposite the Salpêtrière. In contrast to Pinel, who considered himself a medical philosopher, Esquirol was a specialized practitioner. As a dedicated alienist, he argued publicly for mental medicine to consist of a system of specialized practitioners and particular institutions restricted to the mentally ill.[35] From the work of Esquirol originated a powerful group of moral physicians or "médecins aliénistes," their new name.

From 1817 on, Esquirol actively engaged in creating this circle. For example, he trained several dozen students who participated in plying the "specialty" of mental medicine to the exclusion of other types of medical practice. In addition, a private course for mental clinicians was opened in the dining room of La Salpêtrière where students came in flocks.[36] Furthermore, through visits to provincial asylums, Esquirol had been able to develop relations with the physicians charged with their administration. His patronage came to weigh heavily in selecting candidates for new positions. In this way, more or less all alienists who would go on to occupy the major positions in French asylums from the 1830s onward had been Esquirol's students. This is true of Jean-Pierre Falret (1794–1870), who started teaching the mentally disabled children at La Salpêtrière in 1821; Félix Voisin (1794–1872), who became senior doctor at Bicêtre asylum in 1836; Scipion Pinel (1795–1829), the son of Philippe Pinel; Etienne Georget (1795–1828), subsequently head physician at Esquirol's private clinique in Ivry; Ulysse Trélat, chief doctor at La Salpêtrière from 1840 to 1876; Achille-Louis Foville (1799–1878), chief doctor at Saint-Yon asylum at Rouen; Jacques Moreau de Tours (1804–1884), doctor at Bicêtre and subsequently at La Salpêtrière; and Jules Baillarger (1809–1890), Esquirol's private secretary, who in 1842 founded the first specialized journal in the field, the *Annales médico-psychologiques*.[37]

Esquirol was not the man to describe his works in wide theoretical synthesis. His *Des maladies mentales considérées sous les rapports médical, hygiénique et médico-légal*, published in 1838, was a collection of his previous works, published as contributions to specialized journals like the *Annales d'Hygiène Publique et de Médecine Légale*, which Esquirol founded with some students in 1829. This journal published work of medical statisticians and hygienists, and alienists contributed regularly to it with statistical studies on the etiology of madness. Several factors predisposed alienists to statistical studies. In contrast to general practitioners, who worked with a varied urban clientele, alienists were supposed to live and work in the asylums. Administered by the state since the Revolution, these hospitals kept careful registers of patient admission and discharge. The moral physicians were thus very well placed to collect data. Baillarger provided an overview of the situation:

> Consumptive, scrofulous and gout-afflicted persons are scattered here and there and one isolated observer would have much difficulty and spend much time collecting a sufficient amount of data. In contrast to this, the alienated are united by the hundreds in hospitals. Thus in La Salpêtrière and Bicêtre, there are more than two thousand patients with mental illness, and nothing is easier than to collect a great many observations in a few years.[38]

Additionally, the very close-knit circle surrounding Esquirol became solidly anchored in the state system by the 1830s. Esquirol's data, the first French assessment of the number of lunatics in the population, provided a basis for the 1838 Great Law on the Insane in France. Some of his disciples, Falret, Georget, and Scipion Pinel, sat on the commission in charge of preparing this law. The use Esquirol made of asylum registers went beyond concerns of the individual asylum, like calculating costs. In 1810, 1814, and 1817, Esquirol set off on visits to asylums throughout France to collect data. When the Académie de Médecine was reestablished in 1820, Esquirol was named a member of its Statistics Commission, and found himself right in the middle of a debate that developed around Pierre-Charles-Alexandre Louis (1787–1872).

Louis had ardently argued that only statistics could provide a sound basis for medicine in a number of treatises on phthisis, typhoid fever, and, most notoriously, bloodletting.[39] In particular, he thought it necessary to classify diseases exactly and to establish constant connections between diseases, their diagnosis, and patient therapy. Louis' work brought about one of the fiercest controversies at the Paris academies of science and medicine. Opponents like François-Victor Broussais (1772–1838), chief physician at the Paris military hospital and medical school, the Val de Grâce, argued that diseases could only be identified by observing the lesions on organs. Statistics could not teach anything about the causes of diseases and their treatment.[40]

Esquirol took Louis' side, and favored statistical methods to determine the relative importance of various causes of madness. In the *Annales d'hygiène publique et de médecine légale* he wrote in 1835, at the height of the debate:

> What is experience if not observation of facts, often repeated and stored in memory? But sometimes memory is unreliable. Statistics record and don't forget. Before a physician puts forward a prognosis, he has done a mental calculus of probability and solved a statistical problem. Notice that he has observed the same symptoms, ten, thirty, a hundred times in the same circumstances and from this, he draws his conclusion. If medicine had paid attention to this tool of progress, it would have acquired a greater number of positive truths. It would not be taxed with being a vague and conjectural science lacking strong principles.[41]

Esquirol's method of statistical reasoning relied essentially on comparative studies of the insane. Instead of just compiling data from individual asylum registers, he intended to compare them with respect to variables of social context, climate, and conditions of observation. "Tabular forms of statistics," he wrote in the same 1835 article quoted above, "established with daily notes taken down during several years, on a great number of aliens in the same conditions, will give comparative terms with other tabular forms obtained from observations of sick living in opposed climates and subjected to different morals, laws, and treatments." From such studies, physicians could gain knowledge of the causal factors involved in mental disease. As Esquirol confidently added: "What invaluable results could be had for understanding madness and its causes!"[42] Establishing the frequencies of madness with respect to age, gender, season, and so on, it would be possible to orient mental medicine toward causal factors in the social and natural environment of the sick. "The consideration of professions and ways of life brings us to the study of manners [*mœurs*] and their relationship to madness, which among all diseases is most manifestly dependent on private and public manners," he hypothesized.[43]

Now, if members of a particular population basically shared the same living conditions, why did some nevertheless become mad and others not? It was in such cases that Esquirol resorted to hereditary predisposition as a possible cause by turning to family histories. On the basis of a three-year survey (1811–1814) conducted at La

Salpêtrière, Esquirol thus found hereditary predisposition to be the cause of mental illness in 105 out of 466 cases. This impressive proportion, almost 23 percent, was apparently not enough for him to draw the desired conclusions, and he made assertions that his figures did not in fact support: "I am sure that heredity predisposes to mental illness more frequently than my figures show here."[44] Comparing the figures from La Salpêtrière to those from his private clinic, he concluded that the proportion of hereditary to nonhereditary mental disease was greater in poor people than in rich people. He determined that one in two were cases of hereditary madness among the rich patients, as compared to only one in four among the poor. However, Esquirol did allow that the proportion of hereditary madness among the poor would in reality be greater if further facts were known. At La Salpêtrière, for example, poor female patients "often did not even know their parents' name." In the private clinic where the rich were treated, on the other hand, doctors had more background information on the patients' families.

The spread of the idea that mental illness was hereditary can in large measure be attributed to Esquirol. From 1820 on, Esquirol's figures were repeated in many works such as in *De la folie*, published in that year by Georget. In 1826, Félix Voisin in *Des causes morales et physiques des maladies mentales* confidently asserted with respect to Esquirol's work that the role of heredity in madness had become a "mathematical truth."[45] Whereas earlier, only a few doctors had considered insanity to be exclusively inherited, mental illness now became the hereditary disease par excellence.[46] From 1830 on, the medical discourse became alarmist and the transmission of individual characters seemed to tend only in one direction: toward degeneration. The notion of hereditary predisposition to mental illness had not changed, but the pronouncements concerning the danger it presented to individuals became more and more deterministic. Thus Esquirol wrote in 1838: "This harmful transmission paints itself on the face, in the conformation, ideas, passions, habits and instincts of those who are bound to be its victims. Alarmed by one of these symptoms, I sometimes arrived at predicting the onset of madness several years ahead."[47] The notion of degeneration had been transformed from a vague idea into convincing data expressed in figures.

7.5 Conclusion: From Individual to Family and from Family to Species

With the identification of heredity as a source of madness, mental physicians in the early nineteenth century turned to the study of the consequences of hereditary transmission. Whereas before, the administrative perspective sought information in order to ensure a reasonable management of lunacy in a world considered otherwise stable, now a rationalizing perspective took over, aiming at defining effective modes of action in an industrializing and increasingly complex world. Since the causal processes involved in human generation were beyond their immediate grasp, physicians created a series of theories on the basis of statistical reasoning that made hereditary transmission of diseases the center of moral concern. The first Constitution after the deposition of Louis XVI had already elevated the responsibility of one generation toward the next to the status of law: "One generation does not have the right to subjugate a future generation to its laws."[48] The nineteenth century took up this notion of generation, as Ohad Parnes shows in chapter 14 of this book.

Such notions took root among the bourgeoisie in particular. While the nobility had its aristocratic bloodlines—long ancestry and a high value placed on the proper marriages—the bourgeoisie looked in the other direction for longevity. As a class of prosperous and upwardly mobile families, they were concerned with the body's health and the character of one's descendants. The concern with genealogy was replaced with a preoccupation with heredity.[49] In marriages, it became necessary to consider not only the rules of social homogeneity and the wedding patrimony, but the prospects and risks of heredity. The question was no longer whether the parents had transmitted a disease or disability to their children but, more importantly, whether these children in turn would transmit the disease or disability to their offspring.[50] Whereas the nobility looked to its past, the bourgeoisie was oriented toward the future.

This redirection in general social perspective toward the future accentuated the interest that the eighteenth century had already begun to take in children. In 1799, Fodéré, in *Les Lois éclairées par les sciences physiques*, argued that the Republic needed people to bear and raise healthy children and added that people in poor shape should be compelled to remain single. Demands that marriage be regulated became more and more frequent until the end of the Empire.[51] It was time, many argued, to participate in a regenerative project.[52] The child, presented as a privileged object of susceptibility, was in the process of becoming the central preoccupation of collective responsibility. It was no longer perceived as the result of the combination of both parents, but as the threatened realization of suffering or shame that already resided, perhaps latently, in adult physiologies and mentalities.

The analysis of hereditary transmission thus portrayed matrimonial unions as having "biological responsibility" for the species. Mental physicians therefore subscribed to the principle that the species had a pathological capital. Not only could ill-assorted marriages make the partners unhappy but they could, if not controlled, transmit disease or create it in future generations. The "degeneracy" theory explained how a family history of diverse illnesses, by destroying individuals, led to the gradual contamination of entire populations. This can be seen in the work of Philippe-Joseph-Benjamin Buchez (1796–1866). In his youth, Buchez was an adherent of Saint-Simon's social physiology, and in the 1820s he became involved in the founding of the French Carbonarie movement, a revolutionary sect dedicated to the overthrow of the Restoration. Buchez assumed that everybody, present, past, and future, was bound up in a larger social collectivity. It was in this context that the hereditary mental illness posed a threat to society as a whole: "All men, all generations, all peoples, of whatever time, be it present, past, or future, are united by the bonds of reciprocal dependence, as too by a community of function and responsibility; whence as a result . . . they form a single society embracing the century."[53]

Notes

1. For examples, see Petit 1817; Millot 1828, 342.

2. Lucas 1847–1850, xv–xvi.

3. Lucas 1847–1850, xv–xvi. Unless otherwise noted, all translations are my own.

4. See the entries "Dégénération" and "Dégénérescence" in *Dictionnaire des sciences médicales*, Laënnec 1823; Adelon et al. 1821–1828, vol. 4, 390; Audouin et al. 1822–1831, vol. 5, 378–383.

5. Buffon [1766] 1868, vol. 4, 128.

6. Virey 1817. See also Fodéré 1800b; Jeune 1802; Petit 1817.

7. For example, see Devay 1846.

8. Carol 1995, 17–37.

9. Quoted in Barnave 1988, 116. See also Guizot 1858–1867, vol. 3, 150.

10. See Chevalier [1958] 2002.

11. See Vedder, chapter 4, this volume.

12. Balzac [1833] 1999, 82.

13. Cabanis 1802.

14. Fodéré 1800a.

15. Esquirol 1838, 55.

16. Esquirol 1838, 43.

17. Williams 1994.

18. Pick 1989, 1–36.

19. Marx and Engels [1845] 1956.

20. Weisz 1995.

21. Portal 1808, 25.

22. Petit 1817, 56.

23. For example, see Pujol 1801, 285, with respect to leprosy.

24. For example, see Trélat 1828, 64.

25. Dowbiggin 1993, 27.

26. Cabanis 1802.

27. Cabanis 1989, 554.

28. See Lecuyer 1994, 135–152.

29. For example, see Esquirol 1805, 7.

30. Ferrus 1834.

31. Foucault [1963] 1997.

32. Pinel in a treatise titled *Mémoire sur la manie périodique ou intermittent* (1797), included in Pinel 1801.

33. Pinel 1809, 13.

34. Goldstein 1997, 103.

35. See the entry "Maison d'aliénés "in *Dictionnaire des sciences médicales*, Esquirol 1818, vol. 30, 47–95; Esquirol 1819.

36. Goldstein 1997, 181.

37. See Goldstein 1997, 169–208.

38. Baillarger 1844.

39. See Matthews 1995.

40. See Peisse 1857, 134.

41. Esquirol 1835, XIII 5–192.

42. Esquirol 1835, 148.

43. Esquirol 1838, 48.

44. Esquirol 1838, 64.

45. Voisin 1828, 287.

46. For example, see Aubanel and Thore 1841, 71–72.

47. Esquirol 1838, 65.

48. Article 30, Constitution de l'An I.

49. Foucault 1976, 140–147.

50. For example, see Buchez 1840.

51. For example, see Mahon 1801; Esquirol 1838, vol. 1, 50, the entries in *Dictionnaire des sciences médicales* "copulation" by Marc 1813, vol. 6, 285; "héréditaire" by Petit 1817, vol. 20, 77; "mariage" by Fodéré 1819, vol. 31, 34; as well as the entry "génération" in *Dictionnaire de Médecine ou Répertoire général des sciences médicales considérées sous le rapport théorique et pratique* 1824, vol. 1, 88–89.

52. Vigarello 2001.

53. Buchez, *Introduction to Science* 1842; quoted in Pick 1989.

References

Adelon, Nicolas Philibert et al. 1821–1828. *Dictionnaire de Médecine.* 21 vol. Paris: Béchet Jeune.

Adelon, Béclard, Bérard et al. 1832–1846. *Dictionnaire de Médecine ou Répertoire général des sciences médicales considérées sous le rapport théorique et pratique.* 2nd edition. 30 vol. Paris: Béchet Jeune et Labé.

Aubanel, Honoré, and Ange Marie Thore. 1841. *Recherches statistiques sur l'aliénation mentale faites à l'hospice de Bicêtre.* Paris: J. Rouvier.

Audouin, Jean Victor et al., eds. 1822–1831. *Dictionnaire classique d'histoire naturelle.* 17 vols. Paris: Rey et Gravier.

Baillarger, Jules Gabriel François. 1844. Recherches statistiques sur l'hérédité de la folie. *Annales médico-psychologiques* 3:122–173.

Balzac, Honoré de. [1833] 1999. *Le Médecin de Campagne.* Paris: Le Livre de Poche.

Barnave, Antoine. 1988. *De la révolution à la constitution.* Grenoble: Presses Universitaires de Grenoble.

Buchez, Philippe. 1840. *Essai d'un traité complet de philosophie,* Paris: Eveillard.

Buffon, Georges-Louis Leclerc Comte de. [1766] 1868. De la Dégénération des animaux. In *Oeuvres,* vol. 4. Paris: Parent-Desbarres.

Cabanis, Pierre-Jean Georges. 1802. *Rapports du physique et moral de l'homme.* Paris: Crapart, Caille et Ravier.

———. 1989. Rapport fait au Conseil des Cinq Cents, sur l'organisation des Ecoles de Médecine, séance du 29 brumaire an 7. In *Du degré de Certitude de la Médecine*, rev. ed. Paris: Slatkine.

Carol, Anne. 1995. *Histoire de l'eugénisme en France, les médecins et la procréation XIX–XXe siècle.* Paris: Seuil.

Chevalier, Louis. [1958] 2002. *Classes laborieuses et classes dangereuses.* Paris: Perrin.

Devay, Francis. 1846. *Hygiène des familles.* Paris: Labé.

Dowbiggin, Ian. 1993. *La folie héréditaire ou comment la psychiatrie française s'est constituée en un corps de savoir et de pouvoir dans la seconde moitié du XIXe siècle.* Paris: E.P.EL.

Esquirol, Jean Etienne-Dominique. 1805. *Des passions considérées comme causes, symptôme's et moyens curatifs de l'aliénation mentale: Thèse de médecine.* Paris.

———. 1819. *Des Etablissements des Aliénés en France et des moyens d'améliorer le sort de ces infortunés: Mémoire présenté à son Excellence le Ministre de l'Intérieur en septembre 1818.* Paris: Imprimerie de Madame Huzard.

———.1835. Mémoire historique et statistique sur la Maison Royale de Charenton. *Annales d'hygiène publique et de médecine légale* 13:5–192.

———. 1838. *Des maladies mentales considérées sous les rapports médical, hygiénique et médico-légal.* Paris: Baillière.

Ferrus, Marie-André. 1834. *Des Aliénés.* Paris: Imprimerie de Madame Huzard.

Fodéré, François-Emmanuel. 1800a. *Traité du gôitre et du crétinisme.* Paris: Bernard.

———. 1800b. *Les lois éclairées par la science physique, ou traité de médecine légale et d'hygiène publique.* Paris: Croulebois.

———. 1813. *Traité de médecine légale et d'hygiène publique, tome V, 3e partie. Police médicale et hygiène publique.* Paris: Mame.

Foucault, Michel. [1963] 1997. *Naissance de la clinique.* 5th ed. Paris: PUF Quadrige.

———. 1976. *Histoire des la sexualité I: La volonté de savoir.* Paris: Éditions Gallimard.

Goldstein, Jan. 1997. *Consoler et classifier, l'essor de la psychiatrie française.* Paris: Synthelabo.

Guizot, François. 1858–1867. *Mémoires pour servir l'histoire de mon temps.* Paris: Michel Levy.

Jeune, Robert le. 1802. *De l'influence de la Révolution française sur la population.* 2 vols. Paris: Allut et Crochard.

Lecuyer, Bernard-Pierre. 1994. Probabilistic thinking, the natural sciences and the social sciences: Changing configurations (1800–1850). In L. Bernard Cohen, ed., *The Natural Sciences and the Social Sciences.* Dordrecht: Kluwer Academic Publishers.

Lucas, Prosper. 1847–1850. *Traité philosophique et physiologique de l'hérédité naturelle dans les états de santé et de maladie du système nerveux, avec l'application méthodique des lois de la procréation au traitement général des affections dont elle est le principe.* 2 vols. Paris: Baillière.

Mahon, Paul-Augustin-Olivier. 1801. *Médecine légale et policer médicale.* Paris: Buisson; Rouen: Robert.

Marc, Charles Chrétien Henri. 1840. *De la folie considérée dans ses rapports avec les questions médico-judiciaires.* Paris: J.-B. Baillière.

Marx, Karl, and Engels, Friedrich. 1845. *La Sainte-Famille ou Critique de la Critique critique.* Frankfurt: J. Rütten.

Matthews, J. Rosser. 1995. *Quantification and the Quest for Medical Certainty.* Princeton, NJ: Princeton University Press.

Millot, Jacques A. 1828. *L'art de procréer les sexes à volonté ou Histoire physiologique de la génération humaine.* Paris: Béchet.

Panckoucke, Charles-Louis-Fleury, ed. 1812–1822. *Dictionnaire des sciences Médicales.* Paris: C. L. F. Panckoucke.

Morel, Bénédicte-Augustin. 1857. *Traité des dégénérescences physiques, intellectuelles et morales de l'espèce humaine.* Paris: J. B. Baillière.

Peisse, Louis. 1857. *La médecine et les médecins, philosophie, doctrines, institution, etc. critiques, mœurs, et biographies médicales.* Paris: J.-B. Baillière.

Petit, Antoine. 1817. *Essai sur les maladies héréditaires.* Paris: Gabon.

Pick, Daniel. 1989. *Faces of Degeneration: A European Disorder, c. 1848–1918.* New York: Cambridge University Press.

Pinel, Philippe. 1798. *Nosographie philosophique ou la méthode de l'analyse appliquée à la médecine.* Paris: De Crapelet.

———. 1801. *Traité médico-philosophique sur l'aliénation mentale ou la manie.* Paris: Richard, Caille et Ravier.

———. 1809. *Traité médico philosophique de l'aliénation mentale.* 2nd edn. Paris: Brosson.

Portal, Antoine. 1808. *Considération sur la nature et le traitement de quelques maladies héréditaires ou de famille.* Paris: Beaudouin.

Pujol, Alexis. 1801. Essai sur les maladies héréditaires, Séance publique de la Société Royale de Médecine 31 août 1790. In *Oeuvres de médecine pratique.* Paris: Baillière & Bechet.

Trélat, Ulysse. 1828. *De la constitution du corps des médecins et de l'enseignement médical, des réformes qu'elle devrait subir dans l'intérêt de la science et de la morale publique.* Paris: Villeret et Cie.

Vigarello, Georges. 2001. *Le corps redressé.* Rev. ed. Paris: Armand Colin.

Virey, Julien Joseph. 1817. *Recherches medico-philosophiques sur la nature et les facultés de l'homme*. Paris: Panckoucke.

Voisin, Félix. 1826. *Des causes morales et physiques des maladies mentales*. Paris: J. B. Baillière.

Weisz, George. 1980. *The Organization of Science and Technology in France, 1808–1914*. Cambridge: Cambridge University Press.

Weisz, George. 1995. *The Medical Mandarins: The French Academy of Medicine in the 19th and early 20th centuries*. Oxford: Oxford University Press.

Williams, Elizabeth A. 1994. *The Physical and the Moral: Anthropology, Physiology, and Philosophical Medicine in France, 1750–1850*. Cambridge: Cambridge University Press.

III Natural History, Breeding, and Hybridization

8 Figures of Inheritance, 1650–1850

Staffan Müller-Wille

8.1 Genealogy and Heredity

As emphasized in several contributions to this book, it was not before the first half of the nineteenth century that phenomena connected with the generation of living beings were addressed in terms of inheritance.[1] This makes it difficult to trace the developments in the premodern life sciences that eventually gave rise to the concept of heredity. In this chapter I want to get around this difficulty by attending to genealogical terms and figures that naturalists used to address generation. After all, inheritance follows paths outlined by kinship terms and rules of succession, and both belong to the domain of genealogy.

In a sense, genealogy is the oldest logic. Relations of affinity and descent are used universally, in myths, philosophy, and the sciences, to describe structures of the social and the natural world. The key concepts of ancient logic, genus and species, have genealogical connotations, and the relationship of these concepts was also modeled on a genealogical relationship. A species, it was said, relates to its genus in the same way "as Agamemnon is an Atride, a Pelopide, a Tantalide, and finally [a son] of Zeus."[2] Some identity persists in descending genealogies, and properties can thus be inferred for individuals by retracing their origin or *principium*—again a term with genealogical connotations—through the chain of ancestors.[3]

And yet, despite this belief in persistence, the premodern world was a world full of transmutations, monstrous births, and bizarre couples. Although Aristotle, for instance, maintained, that "a man is generated by a man,"[4] he equally conceded the possibility that "two animals different in species produce offspring which differs in species; for instance, a dog differs in species from a lion, and the offspring of a male dog and a female lion is different in species."[5] And still in 1690, no less a figure than John Locke claimed in his *Essay Concerning Human Understanding* that he "once saw a Creature that was the Issue of a Cat and a Rat, and had the plain marks of both about it."[6] As fantastic as this promiscuity of nature appears in hindsight, it nonetheless rested on rational grounds. The foundation for similarities between parents and offspring was provided by the fact that similar conditions prevail, as a rule, during procreation and development. Conversely, this meant that any deviation from the ordinary course of things—for example, *mesalliances* like the ones Aristotle and Locke referred to—would produce as deviant results. "All things are governed by law" is the conventional translation of the opening sentence of the Hippocratic tract *De genitura* (On seed).[7] Yet, it is worthwhile to consult its Renaissance Latin translation—"Law strengthens everything"—with "law (νομος)," as the translator Girolamo Mercuriale (1530–1606) noted, meaning "customs, pasture, region, tribe (*instituta, pascua, regionem, classem*)."[8] Law, in *De genitura*, did not refer to universal laws of nature, but to the persistence of local tradition and circumstance.

In this premodern perspective—which prevailed, as we will see, well into the seventeenth century—specific and individual similarities between ancestors and descendants result from similarities in particular constellations of climatic, economic, political, and social factors, from the persistence of a "fabric," as Claude Lévi-Strauss once formulated it, "in which warp and filling yarn correspond to localities and tribes."[9] The phenomenon of heredity—that "like begets like"—was thus as trivial as it was precarious: it was trivial insofar as it was stabilized and reinforced by municipal rules; it was precarious insofar as it always remained open to transgressions against such rules.[10]

Against this background, the doctrine of the constancy of species, which so dominated eighteenth-century natural history, appears as a radical departure. It shared the age-old intuition that the number of organic forms is limited, and that these forms are immutable. At the same time, however, it connected these intuitions to the assumption that organic reproduction is governed by universal laws.[11] The doctrine of the constancy of species implied that organisms *under all circumstances* reproduce within the bounds of species—that is, without changing into other species. An individual, even if different in this or that respect from its parents, would always belong to the same species as its parent. Genealogical relationships, like common descent or the production of fertile offspring, thus took precedence over relations of similarity in the definition of species.[12] As Ernst Mayr once explained, naturalists in the eighteenth century began to employ a species concept whose "essence" did not consist in some "intrinsic property" but in "the relationship of two coexisting natural populations." The word *species* became "equivalent to like, let us say, the word *brother*, which also has a meaning only with respect to a second phenomenon. An individual is a brother only with respect to someone else."[13]

In the following I want to argue that the epochal shift described by Mayr provided the context in which heredity eventually acquired its biological meaning. I will do so in two steps. In the first three sections, I will describe tropes or figures of genealogy used by William Harvey (1578–1657), Carolus Linnaeus (1707–1778), and Charles Darwin (1809–1882) in order to add more detail to the conceptual shift Mayr pointed at. My central claim will be that it involved a shift of emphasis from vertical relations of descent to horizontal relations of affinity. Metaphors of inheritance became inserted in the latter context to account specifically for the persistence of differences among organisms affiliated by descent. In the fourth section, I will then identify the social correlate of this conceptual development. I will argue that it was the institutionalization of a system of specimen exchange between naturalists that provided this correlate, a system whose growth resonated with broader cultural developments like the development of long-distance trade and colonialism, the growth of markets and industrial production, as well as the increasing dependence of the bourgeois class on mobile property.

8.2 William Harvey: Generation, Patriarchy, and Emancipation

William Harvey's explanation of blood circulation, contained in his *Anatomical Disputation Concerning the Movement of the Heart and Blood in Living Creatures* (1628), is considered the starting point of modern physiology, based as it was on experiment and quantification. Much less historical influence is accorded to Harvey's *Anatomical Exercises on the Generation of Animals*, which appeared more than twenty years later in 1651.[14] Yet the concept of circulation (*circulatio*) unites both works to such an extent that Harvey has been called a "lifelong thinker" of circulation.[15]

Indeed, both the source for the movement of blood and the beginning of a new life cycle coincide in what Harvey referred to as the *punctum saliens* or "pulsating point."[16] Thus, Harvey did not distinguish between the development of an individual organism and the transmission of properties or dispositions from one generation to the next. To explain generation was, for Harvey, to explain the procreation of an individual being. This perspective also dominated the way Harvey explained similarities between parents and offspring:

> Now, I maintain that the offspring is of a mixed nature, inasmuch as a mixture of both parents appears plainly in it, in the form and lineaments, and each particular part of its body, in its color, mother-marks, disposition to diseases, and other accidents. In mental constitution, also, and its manifestations, such as manners, docility, voice, and gait, a similar temperament is discoverable. For as we say of a certain mixture, that it is composed of elements, because their qualities or

virtues, such as heat, cold, dryness, and moisture, are there discovered associated in a certain similar compound body, so, in like manner, the work of the father and mother is to be discerned both in the body and mental character of the offspring, and in all else that follows or accompanies temperament.[17]

Generation is here portrayed as an individual act in which parents are actually *making* their children, employing the same substances that they themselves are made of. Similarities result trivially under this view. The view also implied a fundamental problem, however, which since antiquity had not only dominated discussions about generation, but discussions about the movement of projectiles and the origin of surplus value as well.[18] In a nutshell, this problem consisted of the following. The process of generation is characterized by the fact that something—the male seed in higher animals—is displaced from its original source of motion and yet remains active in its new place, the female body. How could this be possible if one presupposed, as Harvey did, an understanding of causation according to which cause and effect are contiguous? As Harvey formulated the problem: "The knot therefore remains untied . . . , namely: how the semen of . . . the cock forms a pullet from an egg . . . especially when it is neither present in, nor in contact with, nor added to the egg."[19]

Harvey tried to find a solution to this problem, as Aristotle had done before him, by restricting the role of the parents to the formation only of the first rudiments of the embryo. The subsequent growth and differentiation of the embryo could then be explained by assuming that, once procreated, it was endowed with the capacity to fashion itself out of the nourishment provided by the maternal body. In explaining this, Harvey employed a genealogical metaphor that he borrowed, again, from Aristotle:

What Aristotle says of the generation of the more perfect animals, is confirmed and made manifest by all that passes in the [chicken] egg, viz.: that all the parts are not formed at once and together, but in succession, one after another; and that there first exists a particular genital particle, in virtue of which, as from a beginning, all the other parts proceed. . . . And this particle is like a son emancipated, placed independently, a principle existing of itself, from whence the series of members is subsequently thrown out, and to which belongs all that is to conduce to the perfection of the future animal.[20]

The genealogical figure used in this passage is that of patrilineal succession, as indicated by the expression "emancipated son" (*filius emancipatus* in Harvey's Latin original).[21] The metaphor evinces that the embryo is initially formed by the direct action of its progenitor, just as a father forms his son by nourishing and educating him. Each link in the patrilineal chain is thus initially determined by its particular progenitor.

Once procreated, however, the embryo continues its life independently. It is now endowed with a capacity for self-perfection and produces the different parts of its own body successively, just as a son will provide for himself on release from patriarchal authority.

Harvey also granted an effective role to the female in generation, in contrast to Aristotle, and in accordance with Galenic tradition. But even the female contribution ends with conception according to Harvey. During pregnancy, the role of the maternal body is reduced to providing the substances and instruments, the "household," so to speak, for the embryo's own perfection. This becomes particularly evident when Harvey compares the embryo to an ancestral household deity, the *lar famliaris* of ancient Rome, "whence life proceeds to the body in general, and to each of its parts in particular."[22] The embryo, after conception, enjoys autonomy and leads a life of its own.

Aristotle held that the starting point for the independent life of the embryo was an organic body, namely the heart, structured by the interaction of male seed and female blood. Harvey's embryological studies can be seen as an attempt to determine this starting point more precisely, by retracing the "series of members" empirically through the anatomical study of fertilized chicken eggs at different stages of development.[23] Not surprisingly, this led him further back than Aristotle. For Harvey, it was not the heart, but the already-mentioned "pulsating point (*punctum saliens*)" from which the independent life of the embryo proceeded: a minuscule pulsating droplet of blood, as yet without any discernible internal structure, but endowed with the potential to produce structure. In its inconspicuousness it occupied the threshold between life and death.[24]

Harvey thus radicalized Aristotle's solution of the problem of generation. The embryo has an independent life of its own from its very first, still undeveloped, unstructured, and ultimately invisible beginnings. Ironically, this only reinforced the original problem. It became impossible for Harvey to see an immediate, physical connection between the embryo and any concrete product of the sexual act. Harvey faced the "paradox of the empty uterus."[25] The coherence of the lineage was thus endangered, because there seemed to be no physical connection between ancestor and descendant. Independent from its very beginning, nothing in the embryo indicated that it was actually made by its parents.

To account for the way parents acted on their offspring Harvey took recourse to action at a distance, comparing generation with contagious disease, magnetic attraction, and astral influence.[26] This was a dangerous move, however. If generation was subject to causes acting at a distance, then it was difficult to see how there should be any limit to their action. Indeed, Harvey acknowledged the reality of equivocal generation, where living beings are procreated from beings different in species, or even nonliving matter.[27] How was it then possible to uphold the figure of a patrilineal chain, in which like was supposed to beget like?[28]

Harvey rescued himself from this aporia by placing generation into the context of a cosmological model. The skies, the weather, the polity, with the king as its "heart" and "sun," and matter itself, in distillation, are all subject to cyclical movements that bring about a regular renovation of beings.[29] And parents represented instruments only in this cosmic system of cyclical movements:

> The male and female, therefore, will come to be regarded as merely the efficient instruments [of generation], subservient in all respects to the Supreme Creator, or father of all things. In this sense, consequently, it is well said that the sun and moon engender man; because, with the advent and secession of the sun, come spring and autumn, seasons which mostly correspond with the degeneration and decay of animated beings . . . ; for, if future generation and corruption are to be eternal, it is necessary that something likewise move eternally, that interchanges do not fail, that of the two actions one only do not occur.[30]

According to this picture, each individual being is ultimately determined by the place it occupies in relation to local circumstances, especially its place with respect to celestial bodies like the sun, the "climate" in its original meaning.[31] Generations always result from singular constellations of causes, and order is only upheld by the tendency for similar constellations to reappear due to the regular movement of celestial bodies. Harvey refers to the resulting cycles of generation and corruption as "interchanges" (*mutationes* in the Latin original).[32] And indeed, they can be seen as exchanges, albeit fundamentally asymmetric ones. Each generated being will take the place of its particular progenitor only. Harvey reinstates the figure of a patrilineal chain, threatened by the autonomy he had to ascribe to the individual being in the last analysis, by projecting patriarchal relations onto nature at large.

8.3 Carolus Linnaeus: Species and Laws of Generation

Late seventeenth- and eighteenth-century speculations about generation became increasingly dominated by theories assuming that living beings preexisted in their germs. Some of these theories went so far as to maintain that the germ contained a miniature representation of the future organism, including representations of the germs of all its future offspring. All links of the genealogical chain were thus believed to be present in the first organisms, which had been created by God directly, and the succession of beings in time consisted in nothing but their successive unfolding or "evolution." The only open question seemed to be which sex contained the series of germs, the male or the female, thus defining the alternative positions of animalculists and ovists.[33]

This extreme preformationism was actually rarely defended. All preexistence theories shared the assumption, however, that germs had come into existence by an act of divine creation. In a sense, then, they were not simply opposed to the "epigenetic" theory of generation proposed by Harvey. They rather hypotatized two aspects of that theory, by maintaining on the one hand an absolute independence of the offspring from its parents, while maintaining on the other their complete dependence on a "divine watchmaker." The result was that the relationship between ancestors and descendants changed from a relationship of physical procreation to a formal relationship of mere succession. Parents were not creators, but rather containers of their offspring, and the latter developed independently from their parents, subject to universal laws of motion only.[34]

One radical version of this reduction of generation to formal relationships of succession has so far escaped the attention of historians. In 1746, the Swedish naturalist Carolus Linnaeus published a short essay titled *Sponsalia plantarum* (Marriages of Plants).[35] The aim of this essay was to prove that "plants live no less than animals."[36] By far the greatest part of this proof consisted in establishing that plants, as animals, reproduce sexually. Interestingly, Linnaeus rejected both ancient, epigenetic theories according to which the embryo resulted from "effervescences and precipitations" brought about by the interaction of male and female generative substances, and preexistence theories in their ovist and animalculist versions. Instead he simply stated that "we do not know how generation occurs."[37]

But how, if there is nothing to say about its physiology, can anything be said about generation at all? The answer Linnaeus offered in *Sponsalia plantarum* consisted in drawing attention away from causal connections and focusing it exclusively on what he called the "process of generation (*processus generationis*)"—that is, on the regular series of changes and movements that plants undergo in reproduction. The descriptions Linnaeus compiled on these processes border on pornography. The reader learns that the plant world knows of "voluptuously gaping vulvae" and of "males ejaculating their prolific substance."[38] It is no wonder that Caspar Friedrich Wolff (1733–1794) could level the following charge against Linnaeus some twenty years later: "He may well know how to copulate, but has no understanding of the theory of generation."[39]

Indeed, the detailed descriptions Linnaeus offers on the generation of plants are totally devoid of any physiological content. They do allow, however, certain empirical generalizations about the process of generation. Thus, "the flower always appears before the fruit"; "anthers and pollen always appear before the fruit"; "stigmas always flower at the same time as the anthers."[40] These generalizations add up to the Linnaean dictum that "all life proceeds from an egg" (*omne vivum ex ovo*), and provided the basis for his highly idiosyncratic theory of creation. This theory had already been laid out in the first three aphorisms of Linnaeus's famous *Systema naturae* (1735). In the first aphorism he maintained that "every single living being is propagated from an egg," that "each egg produces offspring similar to the parents," and that therefore "no new species are produced nowadays." This was not so much a denial of change in general, but more precisely a denial of equivocal generation and species transmutation. In the second aphorism Linnaeus stated that "individuals are multiplied by generation," such that "there is at this time a greater number of individuals in each species than there was originally." The third aphorism then carried out a "backward" calculation on the basis of the two previous aphorisms: "If we count the multiplication of individuals in each species backwards . . . the series will end in one parental being, be it that this consists in an hermaphroditic individual (as is common in plants), or that it consists in two individuals, the male namely, and the female (as in most animals)."[41]

The argument Linnaeus makes here is based on a peculiar inversion of a common representation of ancestry. In so-called ancestor tables, the full ancestry of a particular individual is depicted by retracing lines of descent from that individual to its two parents, from those two parents to the four grandparents, and so on. Because the number of ancestors grows exponentially with each generation, it is easy to see that a generation would quickly be reached that encompassed the entire human species, and thus presented no unique ancestry anymore. However, insofar as ancestor tables stop retracing descent after four to five generations, they can represent an individual as being determined by a unique set of ancestors, which it shares, if at all, only with its siblings.

Linnaeus, on the contrary, proceeded in his argument from the undifferentiated mass of individuals that constitutes each species, including the human, at the present, and reduced this mass by "counting backward"—that is, by retracing relations of ancestry to a concrete individual or pair of individuals. In contrast to the construction of ancestor tables, which aims at determining an individual by its particular ancestry, Linnaeus's procedure represents each and every individual member of a species as being essentially the same, and as being essentially different from members of other species, no matter what particular ancestry gave rise to it. The only thing that changes in descent, despite the myriad causes that can play a role in the generation of individual beings, is the number of individuals. Generation is reduced to a strictly formal relation of descent, constituting the species as a set of individuals determined in essentially the same, uniform way by a "procreative unit" instituted by God at the beginning of time.[42]

Linnaeus thus radically separated two aspects of genealogy: the aspect of production, and the aspect of classification of descendants. This separation found its full expression in the influential definitions that Linnaeus gave of the concepts of species and variety. According to these definitions, varieties are distinguished by differences that are due to particular circumstances accompanying generation—"place or accident" as Linnaeus put it. Properties distinguishing species—"specific differences (*differentia specifica*)" in the technical language of the time—remain unchanged in generation, obeying universal "laws of generation (*leges generationis*)."[43] Transmutations and equivocal generations are thus not only excluded as rare events, but in principle. This was a radical departure from traditional theories of generation—and it was such a departure precisely to the extent to which Linnaeus's theory of generation did not deal with the procreation of individual beings as Harvey's theory had done for example, but with the reproduction of species.

There is a concomitant departure from tradition to be detected in the way Linnaeus conceptualized "exchange." In Harvey "exchange" consisted in each generated being taking the place of its particular progenitor. In Linnaeus, this diachronic exchange relation extended to synchronic relations as well. Because all members of a particular species are subject to "laws of generation," they are essentially equal and can also stand in for each other. In the "economy of nature," therefore, two dimensions of exchange emerge. One is the "continued series (*series continuata*)" in which descendants replace their ancestors, retaining their essential identity, and being assigned a preordained place in the economy of nature. The other dimension consists in the "interconnection (*nexus inter se*)" of organisms—that is, in the network of predator-prey relationships that connect species with each other. Because members of a species are essentially the same by virtue of descent, and because they "multiply" continually, a constant abundance of supply and demand exists, constituting a perfect equilibrium. As Linnaeus put it, "All natural things should contribute and lend a helping hand to preserve every species . . . [so that] the death and destruction of one thing should always be subservient to the restitution of another."[44] Prey serves its predators for food, but this service is immediately compensated. By feeding on them, predators restrain the number of prey, and thus contribute toward preserving the latter's own resources.[45] Members of different species offer one another products and services in the form of their own, prolific nature, and the economy of nature thus constitutes a system of circulation characterized by symmetric relationships of mutual benevolence and indulgence, rather than asymmetric relationships of domination and servitude.

In a powerful image, Linnaeus compared this system with a "weekly market," where "at first one only sees how a great mass of people spreads out in this or that direction, while nevertheless each of them has his home (*domicilium*), from where he approached and to which he will proceed."[46]

8.4 Charles Darwin: Community of Descent and the Struggle for Life

The shift in emphasis from diachronic ancestor-descendant relationships to synchronic relationships among members of a collective entity, which Harvey's and Linnaeus's theories of generation exemplify, was used by Michel Foucault in his late work to mark a major political transition around 1800. According to Foucault, it was around 1800 that a new technology of power, the "dispositif of sexuality," came to complement the "principle of alliance" that had regulated power relations in premodern societies. Foucault defined the "principle of alliance" as the system of marriage, kinship, and transmission of names and goods, which lies at the foundation of every society and serves the reproduction of power relations. The dispositif of sexuality, on the contrary, did not aim at the mere reproduction of particular power relations. It went beyond that by aiming for a global control of populations through the multiplication, renovation, and combination of individual bodies.[47]

Insofar as this diagnosis of Foucault can be applied at all to the history of the life sciences, it seems to miss one of the most fundamental developments of nineteenth-century biology. Was it not the nineteenth century that saw the emergence of evolutionary theories? Clearly, such theories attempted to account for the present order of nature by assuming a "common descent" of species.[48] And they would thus, it seems, deploy principles analogous to the "principle of alliance," rather than a "dispositif of sexuality." Darwin himself argued in the *Origin of Species* that the meaning of what naturalists called the "natural system" of species could easily be explicated by assuming their descent from common ancestors.[49] As he put it in the summary of the *Origin*: "The natural system is a genealogical arrangement, in which we have to discover the lines of descent."[50]

Darwin illustrated this idea with the help of the only figure that was included in the *Origin of Species* (figure 8.1). The diagram shows a complicated, branching structure, corresponding to Darwin's assertion that "the natural system is genealogical in its arrangement, like a pedigree."[51] A closer look at the diagram and the text passages that comment on it reveals some interesting details, however.[52] Most important, the drawing has a bushy rather than treelike appearance. Unlike usual pedigrees, the branches do not issue from a single stem. This reflects Darwin's dislike of speculation

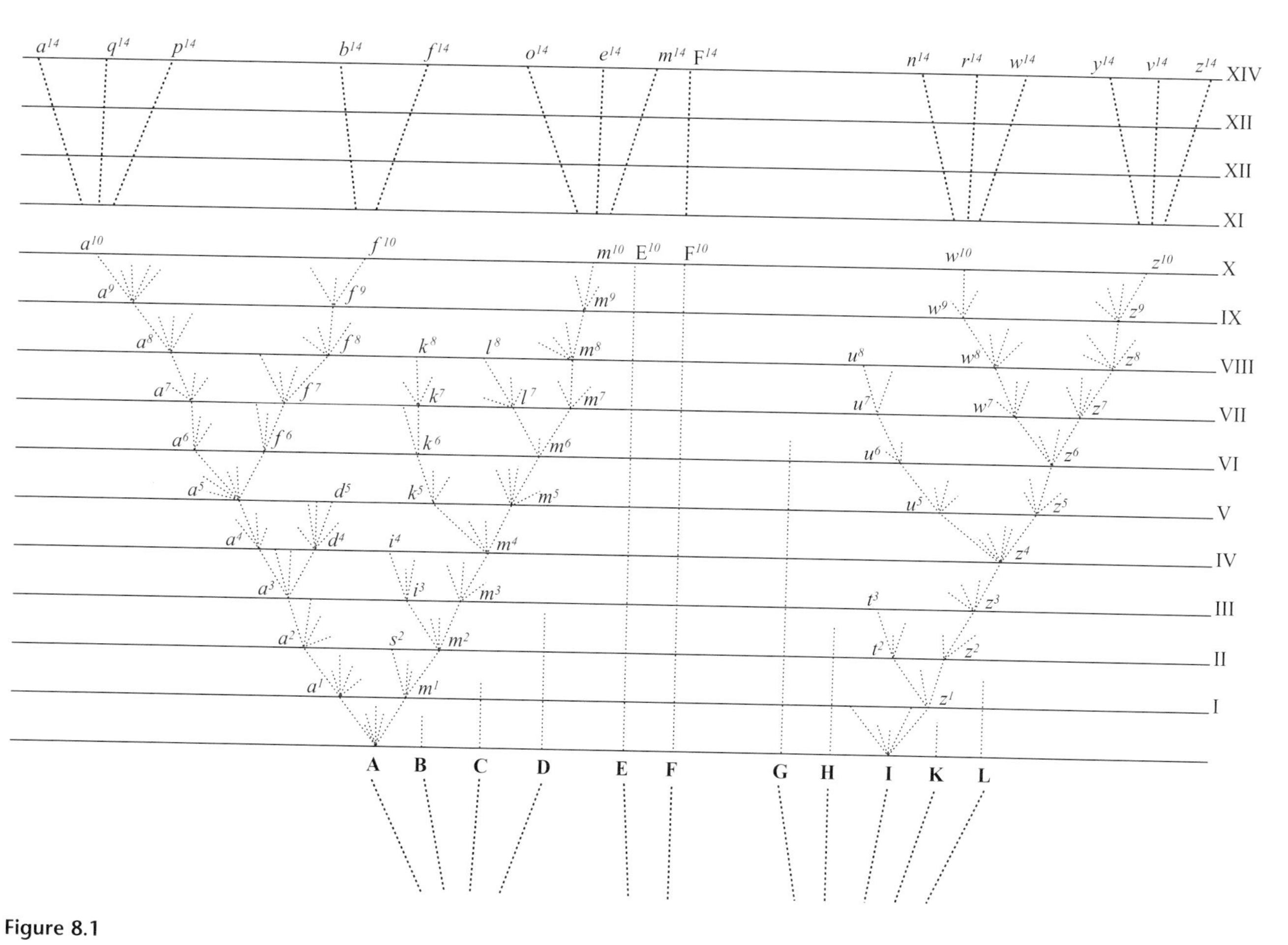

Figure 8.1
Diagram from Charles Darwin, *Origin of Species* (1859). See text for explanations.

on absolute origins, especially the origin of life.[53] "Community of descent" does not refer to descent from a primordial source that determines the shape and direction of all that originates from it. It rather refers to the mere fact that a number of divergent lines start from a common, contingent point. A metaphor that Darwin employed in explaining his diagram is revealing in this respect: "All these modified descendants from a single species [a14 to z14 in Darwin's diagram], are represented as related in blood or descent to the same degree; they may metaphorically be called cousins to the same millionth degree." The metaphor of "cousins" emphasizes some distant degree of affinity, rather than identity by virtue of descent, and indeed, Darwin immediately went on to point out that affiliated species may "differ widely and in different degrees from each other."[54]

What Darwin had in mind with these qualifications becomes clear if we follow some further explanations he gave of his diagram. First, species may undergo modifications in descent on a geological time scale that make them appear as belonging to a different genus or even family than the parental species they descend from. The descendants of A in the diagram, for instance, belong to three distinct families. Descendants may even become "so much modified as to have more or less completely lost traces of their parentage."[55] Second, the lines of descent do not fan out in any regular fashion. They may remain straight like the series of forms emerging from F and leading up to F14. Or they may follow meandering pathways, like the one leading from A to f10. The latter extinct form is apparently intermediate between the extant forms a14 and m14, both descending from A as well. But it has reached this position through a series of ancestral forms (a2–a5 and f6–f9) that had already occupied much more eccentric positions in relation to A.[56] "Extinct forms are seldom directly intermediate between existing forms; but are intermediate only by a long and circuitous course through many extinct and very different forms," as Darwin summarized this point.[57] This reflected Darwin's general conviction that there was no "fixed law of development"[58]—neither a law guaranteeing the stability of species, as in Linnaeus, nor a law constituting progress toward forms of higher complexity, as in the theories of species transformation put forward by Jean-Baptiste Lamarck (1744–1829) or Darwin's grandfather Erasmus Darwin (1731–1802).[59] Darwin's theory was not one of "common descent" but, as he himself called it, a "theory of descent with modification."[60]

It was precisely in the context of elaborating one of the main principles of this theory, the "principle of divergence of character," that Darwin made use of the metaphor of inheritance in a crucially peculiar way. Relying on "Dr. Prosper Lucas's treatise, in two large volumes" as well as the experiences of breeders, Darwin defended inheritance against "doubts thrown on this principle by theoretical writers" in the following way:

> When a deviation appears not unfrequently, and we see it in the father and child, we cannot tell whether it may not be due to the same original cause acting on both; but when amongst individuals, apparently exposed to the same conditions, any very rare deviation . . . appears in the parent . . . and it reappears in the child, the mere doctrine of chances almost compels us to attribute its reappearance to inheritance.[61]

Inheritance, as is evident from this passage, was for Darwin not a process that simply accounted for similarities between parent and offspring. Such similarities could easily be explained by assuming that similar causes remained active in generation. Much more specifically, inheritance comprised cases in which a difference was reproduced under essentially the same conditions of life. Inheritance and its counterpiece, variation, thus turn out to be capricious, not necessarily adaptive "tendencies." If varieties reproduce under essentially the same conditions—if varieties reproduce, for instance, side by side in the same, closely confined locality—then there is no reason to assume that the difference is caused by different conditions of life, nor that the varieties are adapted to the same degree to these conditions.[62]

The two sources that Darwin drew on, Prosper Lucas's (1805–1885) *Traité philosophique et physiologique de l'hérédité naturelle* (1847), epitomizing a long medical tradition of dealing with hereditary diseases, and breeders' knowledge in general, make this point abundantly clear.[63] In both cases, inherited differences are conceived of as individual deviations from a norm, in the former case as a nonadaptive, in the latter case as an adaptive one. "Divergence of character," brought about by the combined effects of variation, inheritance, and natural selection, therefore can but must not happen. This divergence depends entirely on the degree to which a new variety is indeed better adapted to the prevailing conditions of life. "We must not . . . assume that divergence of character is a necessary contingency," as Darwin put it; "it depends solely on the descendants from a species being thus enabled to seize on many and different places in the economy of nature."[64]

Darwin's conception of "divergence of character" can account for a further peculiarity of his diagram. In its vertical dimension forms replace each other that are not related by direct filiation. Replacement occurs due to the fierce competition among affiliated, but different varieties or species. "The modified offspring from the later and more highly improved branches in the lines of descent, will . . . often take the place of, and so destroy, the earlier and less improved branches," as Darwin explained the diagram; "whole collateral lines of descent . . . will be conquered by later and improved lines of descent," he said. Step by step, for instance, descendants of the original species

A and I occupy places that before had been occupied by other species affiliated with but not immediately related to them (B–E, G–H, K–L).[65] The relationship of affiliated forms is a violent one, subject to conquest, extrusion, and elimination.

Darwin emphasized on several occasions that the relationships among species resulting from this "struggle for life" were actually too complex to be represented in a two-dimensional diagram.[66] But what remains of the order of nature, if the shape that organisms take on, and the place they occupy in nature, only stands in an indirect relation to their descent? Darwin's diagram contains a further element of order—horizontal lines—that he referred to as "stages of descent,"[67] and that cut across the branching lines of filiation. The intervals between the lines represented two things: on the one hand, the time of organic reproduction measured by "ten thousand generations,"[68] and on the other hand, geological time in terms of "geological formations."[69] It is these formations, defined by two unrelated time frames—generations and geological strata—that provided the scene for the antagonistic forces that Darwin's theory of natural selection called on.[70] The biological phenomena that engaged Darwin's lifelong interest—the "associated life" in colonial organisms like bryozoans; the growth of coral reefs; the propagation of fruit trees by seedlings; the "life" and "death" of species on the scale of earth history—were precisely defined by occupying such formations.[71]

Darwin's theory of descent with modification can thus indeed be seen as an expression of Foucault's "dispositif of sexuality."[72] In the framework of this theory, it is indeed of minor significance along which lines of descent a certain property had been reproduced in the past. It is rather the reservoir of inheritable dispositions, passed down from ancestors along a multiplicity of lines to a given generation as a whole, and the degree to which these dispositions confer some competitive advantage to individuals under present conditions of life, that determine future success in reproduction.[73] This is the deeper sense of Darwin's famous remark that "each [geological] formation does not mark a new and complete act of creation, but only an occasional scene, taken almost at hazard, in a slowly changing drama."[74]

8.5 Natural History and Heredity

The three figures of genealogy explored in the preceding sections mark a transition from thinking in terms of vertical to thinking in terms of horizontal relationships of parentage, and this accords well with the general development of European kinship systems from "clan" to "kindred" between 1650 and 1850.[75] Is this to imply that Harvey's, Linnaeus's, and Darwin's theories of generation should be seen as mere expressions of a grand, underlying ideological trend denouncing patriarchy and endorsing the ideals of civic society? All three theories bear the clear signs of political issues like autonomy, equality, and progress. Yet it is difficult to see how they came into play in the material culture of the life sciences, which, after all, gave rise to the three theories discussed in the first place.

In proposing a solution to this problem, I would like to start with the protagonist that might strike the reader as the odd one in the triad I have chosen: Linnaeus. His place in the history of the life sciences, in contrast to that of Harvey and Darwin, was certainly not secured through "ideas" he contributed. His theory of creation was far too naive and idiosyncratic to be palatable to the *philosophes* of his time, and the theory of organic reproduction put forward by his contemporary and rival Georges-Louis Leclerc Comte de Buffon (1707–1788) was certainly more influential.[76] But there is another level—other than that of ideas—on which Linnaeus's work played a pivotal role. The rules of classification and nomenclature he laid down in his *Philosophia botanica* (1751) were intended to achieve a "total reform" of the practice of natural history, and there is no doubt that they did so on the more mundane level of amateurs and local practitioners of natural history.[77]

At the core of Linnaeus's reform stood his distinction between species and variety, which was thoroughly based on his theory of generation. In distinguishing between species, Linnaeus advised his fellow naturalists, one should rely exclusively on "constant" characters—that is, not on characters that varied with external conditions like climate or nutrition, but on characters that reproduced in offspring under various external circumstances.[78] Two things are notable about this distinction. First, both "constant" and "accidental" characters were defined as differences. And second, "constant" and "accidental" characters could be separated on empirical grounds. Two procedures existed to achieve this separation. One could either subject organisms possessing a certain character to different conditions of life; if they then exhibited differences with respect to that character, it could not be trusted as a "constant" one. Or one could subject organisms differing with respect to a certain character to the same conditions of life; if the difference vanished, the character could equally not be trusted as a constant one.[79] Interestingly, it is precisely the latter constellation that Darwin invoked to make a case for the reality of inheritance.

Now, the important point about the Linnaean distinction between species and variety was that it did not represent an idea only but regulated an ongoing practice, at least in botany. Since the Renaissance, botanical gardens had grown into veritable institutions of "big science," collecting plants from the most diverse climates and locations and bringing them under essentially homogeneous conditions controlled by horticultural techniques. Linnaeus's distinction between species and variety mirrored these institutions—"gardens of paradise," as he liked to call them—by claiming that all accidental variability, insofar as it was caused by external factors only, would vanish under these homogeneous conditions. "Botanists, who saw that soil and sky make so many varieties," as Linnaeus explained, "understood also, that soil and sky would reduce them [to their species], and therefore sowed them out in botanical gardens."[80]

Moreover, Linnaeus made his distinction with an eye to the social workings of the community of naturalists—that is, with an eye to what he liked to refer to as the "republic" or "marketplace" of natural history.[81] Naturalists, in distinguishing species, relied heavily on the recognition of their distinctions by peers. The specific differences they pointed out to each other therefore had to prove to be reproducible intersubjectively. A system of specimen exchange connected the institutions of natural history for this purpose. Linnaeus's species concept entailed clear criteria for deciding whether a specific difference was indeed reproducible, and thus could count as a specific difference, or not. It was thus not on abstract metaphysical grounds, but on the basis of a constant flow of specimens—including living specimens—from one local context to another that Linnaeus instituted the distinction of species and variety.[82]

Linnaeus was not the only one to work out such distinctions in the eighteenth century. Of equal influence, though geared much more toward zoology, was the criterion introduced by Buffon—against a theoretical background that, like that of Linnaeus, emphasized reproduction. According to Buffon's criterion, species can be separated along lines of intersterility. Two organisms, even if strikingly similar, who fail to produce fertile offspring belong to two different species, and two organisms, even if strikingly dissimilar, who produce fertile offspring belong to one and the same species.[83]

Both Linnaeus and Buffon, as well as the great many naturalists who had reached similar conclusions or followed in their footsteps during the eighteenth century, indicate a profound transformation in the culture of natural history. In preceding centuries natural history had largely been a descriptive and encyclopedic science, based on the tradition of authoritative knowledge. With the growth of a system of institutions engaged in a global exchange of knowledge rather than in the preservation of scholarly and professional traditions, natural history became a science where the political issues of autonomy, equality, and progress were of immediate relevance for the workings of that science itself—not as issues of general political relevance, that is, but much more specifically as issues that pertained to the social standing of those who produced knowledge. To exchange knowledge, rather than pass it down, presupposed relative autonomy and equality of knowledge producers, as well as criteria to decide on intersubjective grounds when progress was made—in other words, when it was justified to replace "old" knowledge with "new" knowledge.

Linnaeus's and Buffon's species concepts provided such criteria by replacing relationships of personal authority, which secured the transmission of knowledge from teacher to disciple, with relationships among peers mediated by the objects of knowledge themselves. Transplantations and hybridizations were supposed to decide if a difference observed among organisms was indeed a difference that could count as a specific difference. This is the core of what Mayr called the "nondimensional" species concept. The significance of a property did not grow from the regularity with which it recurred as a stable form through time. It was rather measured by the degree to which that property was reproduced as a difference within a changing network of relationships among organisms and their respective environments. As a consequence, naturalists became the nodes in what Carl Friedrich Kielmeyer (1765–1844) called "that greater system of effects, which may be called the life of the species."[84]

The institutions of natural history, virtually representing the reproduction of species through the far-flung network of transplantations and hybridizations that connected them, thus became sites of inadvertent experimentation; they became laboratories of the life of species, so to speak. This is indicated by a peculiar dynamic that Linnaeus's and Buffon's conceptual creations became subject to. The nice dichotomies that they established between species and varieties soon gave way to a much more complex picture. Linnaeus himself discovered a number of plant forms that could count as varieties belonging to the same species, but whose distinctive characteristics nevertheless bred true and thus appeared not to be caused by some environmental factor. He called these varieties "constant varieties," and speculated about their origin through hybridization.[85] Buffon, on the other hand, hybridized varieties of domestic animals adapted for widely different purposes—dogs, horses, donkeys, and mules, as well as various kinds of ruminants—and was led to speculations about the origin of species through processes of environmentally caused degeneration.[86] Both Linnaeus and Buffon inspired an experimental research tradition that can be followed right up to Gregor Mendel (1822–1884), and in which ever more sophisticated species concepts flourished.[87] The results achieved in the hybridization and acclimatization experiments carried out in this tradition were deeply troubling, because it turned out that the reproduction of organic forms obeyed laws that stood in no relation whatsoever to the places they occupied in the "economy of nature."

On a much grander scale, evidence accumulated by eighteenth-century naturalists on the biogeographic and stratigraphic distribution of diverse organic forms pointed in the same direction. There simply was no clearcut correlation between conditions of life and the organic forms that reproduced under these conditions. Widely different forms occupied what seemed to be essentially the same places in the "economy of nature," and closely related forms seemed to prosper under widely different conditions of life. In the case of fossils, this noncorrelation could even be used to construct temporal scales for columns of otherwise homogeneous sediment.[88] As François Jacob in particular has forcefully argued, it was these developments that forced naturalists to take recourse to history in the emphatic, modern sense of that word. What was at stake was no longer the task of explaining the adaptedness of organic forms, but of accounting for the processes through which those forms, whose reproduction was governed by autonomous laws, became adapted to their physical environments.[89]

It was precisely in this context that naturalists began to apply metaphors of inheritance. Gärtner, the most influential of the hybridists, referred to "hereditary disposition" as the source of the "limitless" variability of forms in hybrid offspring.[90] Immanuel Kant (1724–1804) made innovative use of the German verb *vererben* and its countless derivatives in his anthropological treatises to explain the curious fact that racial differences reproduced even outside of the climates in which they could be regarded as adaptations.[91] And Darwin, as we saw, introduced the "tendency to inheritance" to account for the "geological succession of organic beings." In natural history, that is, inheritance became an issue when the intimate relationship between property, place, and ancestry, which organisms seemed to enjoy in their "natural places," was upset by transplantations and hybridizations effected by a culture of exchange.

8.6 Conclusion

Natural history did not undergo the large-scale and long-term developments sketched out in the last section in splendid isolation. The activities of naturalists were rather deeply entangled, and partly coextensive, with activities in other social arenas, especially colonial trade, horticulture, and animal breeding. The first botanical gardens at Padua and Pisa, for instance, were not only resources for local universities teaching pharmacology to medical students, but also served as reference collections for long-distance trade in spices and drugs.[92] Ship captains and surgeons sailing for the Dutch East and West India Companies were asked to collect specimens and keep journals recording plants and animals they observed in foreign countries. In the seventeenth century, more and more botanical gardens were not only established all over Europe, but in the European colonies and trading posts as well, serving plant and animal trafficking as well as acclimatization experiments in both directions.[93] Animal breeders and horticulturalists modeled their practices on those of naturalists and contributed massively to the growing literature in natural history,[94] while naturalists, on the other hand, learned from their trade and engaged in sophisticated acclimatization and hybridization experiments.[95]

I have tried to show in this chapter that metaphors of inheritance in natural history were introduced to cover phenomena that pointed toward an underdetermination of organic reproduction by local conditions. Heredity was introduced into biology as a tendency of a fortuitous nature, with no immediate cause or immediate purpose, although with determinate effects. The astonishing clarity with which political themes like the autonomy of the individual, equality before the law, and the legitimacy of progress emerge from the genealogical tropes underlying Harvey's, Linnaeus's, and Darwin's theories of generation respectively is a sign of the degree to which modernity's culture of exchange has placed difference and affinity over identity and descent.

Notes

1. See the contributions to this book by Müller-Wille and Rheinberger (chapter 1), López-Beltrán (chapter 5), and Terrall (chapter 11).

2. Porphyrius 1887, 6.

3. See Durkheim [1912] 1995, 441; Heinrich 1981, 98–100; Gayon and Wunenburger 1995, 8.

4. For example, Aristotle *De gen. anim.*, 735a20; see Lesky 1950, 139, and Glass [1959] 1968a, 31, for further references.

5. Aristotle *De gen. anim.*, 747b33–36; see Zirkle 1935, 15–17.

6. Locke [1690] 1975, 451; for a rich account of premodern "hybrids" see Zirkle 1935.

7. Hippocrates 1978, 317; see Stubbe 1965, 18–21, for a discussion of this important Hippocratic tract.

8. Hippocrates 1588, 10, 15: "Lex quidem omnia corroborat."

9. Lèvi-Strauss [1962] 1985, 111.

10. See chapter 2 on the "logic of totemic classifications" in Lévi-Strauss 1962, 48–99. Lévi-Strauss explicitly extends his analysis to "the naturalists and hermetics of antiquity and the middle ages: Galen, Plinius, Hermes Trismegistos, Albert the Great" (p. 57). For an in-depth discussion of medieval notions of heredity see the contributions in van der Lugt and Miramon, forthcoming.

11. See Zirkle 1959; Glass [1959] 1968b; Roger [1963] 1993, 206–223; Jacob [1970] 1993, chapter 1.

12. Müller-Wille, 2006.

13. Mayr 1957, 15; see Lefèvre 2001.

14. Bayon 1947, 52.

15. Pagel 1976, 153; see also Jucovy 1976.

16. Harvey [1651] 1847, 374; cf. Harvey [1628] 1976, 45–47. See White 1986 on this aspect of Harvey's theory.

17. Harvey [1651] 1847, 363; see White 1986, 513–514.

18. See Wolff 1978, 184–191, on the relationship of these problems. Harvey referred to the movement of projectiles in his *Exercises*; see Harvey [1651] 1847, 365. The way the three sets of phenomena resonate with each other is nicely captured in a simile that the thirteenth-century English lawyer Henry de Bracton used to define inheritance (*hereditas*): "There is another *causa* for acquiring dominion called succession . . . which ought to descend to nearer heirs, male and female, in the right or transverse line, for the right descends, like a weight downwards" (Bracton [1240] 1968–1977, vol. 2, 184).

19. Harvey [1651] 1847, 354.

20. Harvey [1651] 1847, 372–373; translation slightly altered (the original translates *filius* as "child"). The passage contains paraphrases from Aristotle, *De gen. anim.*, 734a17 and 740b35–a6.

21. Harvey [1651] 1766, 389.

22. Harvey [1651] 1847, 373; see Keller 1999 on gender in Harvey and the background in contemporary challenges to patriarchal, especially royal, authority.

23. Roger [1963] 1993, 114.

24. Harvey [1651] 1847, 374; see Harvey [1628] 1976, 47.

25. Goltz 1986.

26. Bayon 1947; Goltz 1986.

27. Foote 1969.

28. For eighteenth-century discussions of this problem, see Terrall, chapter 11, this volume.

29. Duchesneau 1998, 218–220; Cohen 1994; Di Meo 1995.

30. Harvey [1651] 1847, 367.

31. Gregory 2001.

32. Harvey [1651] 1766, 291.

33. For systematic overviews of preexistence theories see Roger [1963] 1993, 37–131; Roe 1981, 2–9; McLaughlin 1990, 10–16; Terrall, chapter 11, this volume.

34. Adelmann 1966, vol. 2, 872; Müller-Sievers 1997, 26–30.

35. Linnaeus [1746] 1749. To my knowledge, this essay was never translated into English for reasons that will presently become clear. Translations in the following are therefore my own. Some flavor of Linnaeus's language, and its resonance with contemporary political issues of gender and civil society, can be gained from Erasmus Darwin's poem "The Loves of the Plants," however; see Browne 1989 and Schiebinger 1991 for interesting interpretations.

36. Linnaeus [1746] 1749, 327.

37. Linnaeus [1746] 1749, 347–349.

38. Linnaeus [1746] 1749, 359.

39. Wolff [1764] 1966, 25.

40. Linnaeus [1746] 1749, 344, 351, 354, 358.

41. Linnaeus 1735, *Observationes in regna III. naturae*, aph. 1–3 (unpaginated).

42. Linnaeus 1735, *Observationes in regna III, naturae*, aph. 4 (unpaginated).

43. Linnaeus 1737b, *Ratio operis*, aph. 5 (unpaginated).

44. Linnaeus [1749] 1762, 40.

45. See Limoges 1972 on Linnaeus's conception of an economy of nature. On its background in Linnaeus's peculiar brand of theology see Rausing 2003.

46. Linnaeus [1760] 1764, 18.

47. Foucault 1976, 140–141.

48. Mayr 1982, 436–439.

49. Darwin [1859] 1964, 413–420.

50. Darwin [1859] 1964, 479.

51. Darwin [1859] 1964, 422.

52. See also Voss 2004.

53. Schweber 1977.

54. Darwin [1859] 1964, 420–421.

55. Darwin [1859] 1964, 421.

56. Darwin [1859] 1964, 421–422.

57. Darwin [1859] 1964, 345.

58. Darwin [1859] 1964, 314; see also 147, 318, 351.

59. On Lamarck's theory of species transformation and its contemporary background see Corsi 1988; on Erasmus Darwin's medical and industrial progressivism see McNeil 1987 and Wilson, chapter 6, this volume.

60. Darwin [1859] 1964, 331.

61. Darwin [1859] 1964, 12–13. On the kinds of doubts that Darwin may have had in mind here see López-Beltrán, chapter 5, this volume.

62. On the Darwinian concepts of variation and inheritance see Winther 2000. Winther's thorough analysis provides two caveats. First, Darwin gave more and more room to the inheritance of acquired characters in his later work. Second, transmission remained entirely contingent on development for Darwin, although he identified it as a separate problem.

63. On inheritance in medicine and the significance of Lucas see López-Beltrán 2004.

64. Darwin [1859] 1964, 331.

65. Darwin [1859] 1964, 119–121.

66. Darwin [1859] 1964, 118, 331, 333, 422.

67. Darwin [1859] 1964, 412.

68. Darwin [1859] 1964, 117.

69. Darwin [1859] 1964, 331.

70. Darwin [1859] 1964, 315.

71. Hodge 1985.

72. See Foucault 1991.

73. See Parnes, chapter 14, this volume, on the new meaning that the term *generation* acquired in the first half of the nineteenth century as a consequence.

74. Darwin [1859] 1964, 315.

75. Sabean, chapter 2, this volume.

76. See Terrall, chapter 11, this volume.

77. Stafleu 1971; see McOuat 1996 on their codification into mandatory rules of "good" taxonomic practice in the early nineteenth century, in which Darwin participated.

78. Linnaeus 1737b, *Ratio operis*, par. 5 (unpaginated). With his recommendation Linnaeus was generalizing a point that had already been made by John Ray (1627–1705); see Ray 1686, 40.

79. Linnaeus 1751, 225, 247–247. Linnaeus summarized his views on these procedures by stating: "Culture is the mother of so many varieties, that it is also their best examinator (*examinatrix*)."

80. Müller-Wille 2001.

81. Linnaeus 1737a, 204.

82. Müller-Wille 2003b; on exchange practices in mid-eighteenth-century French natural history see Spary 2000, 61–78.

83. Farber 1972.

84. Kielmeyer [1793] 1993, 5.

85. Müller-Wille 2003a.

86. Sloan 1979.

87. Olby 1985.

88. Jacob [1970] 1993, chapter 3; for an overview of these developments in the eighteenth century see Larson 1994; on biogeography in particular see Browne 1983; on paleontology see Rudwick 1985.

89. Jacob [1970] 1993, chapter 3; see also Lefèvre 2003.

90. Gärtner 1849, 321.

91. See McLaughlin, chapter 12, this volume.

92. Savoia 1995, 181.

93. Schiebinger and Swan 2005.

94. See Wood, chapter 10, this volume; Ratcliff, chapter 9, this volume. Ratcliff shows that the boundary work between botany and horticulture was fraught with potential conflict, but that the resulting boundaries became increasingly porous in the eighteenth century.

95. Osborne 1994; Koerner 1999.

References

Adelmann, Howard B. 1966. *Marcello Malpighi and the Evolution of Embryology*. 5 vols. Ithaca, NY: Cornell University Press.

Bayon, H. P. 1947. William Harvey (1578–1657): His application of biological experiment, clinical observation, and comparative anatomy to the problems of generation. *History of Medicine and Allied Sciences* 2:51–96.

Bracton, Henry de. [1240] 1968–1977. *On the Laws and Customs of England*. Ed. G. E. Woodbridge; trans. S. L. Thorne. 4 vols. Cambridge, MA: Belknap.

Browne, Janet. 1983. *The Secular Arch: Studies in the History of Biogeography*. New Haven, CT: Yale University Press.

———. 1989. Botany for gentlemen: Erasmus Darwin and "The Loves of the Plants." *Isis* 80 (4): 592–621.

Cohen, I. Bernhard. 1994. Harrington and Harvey: A theory of the state based on the new physiology. *Journal of the History of Ideas* 55 (2): 187–210.

Corsi, Pietro. 1988. *The Age of Lamarck: Evolutionary Theories in France, 1790–1830*. Los Angeles: University of California Press.

Darwin, Charles. [1859] 1964. *On the Origin of Species: A Facsimile*. Introduced by E. Mayr. Cambridge, MA: Harvard University Press.

Di Meo, Anotonio. 1995. Il concetto di "circulazione": Storia di una rivoluzione transdisziplinaria. In G. Cimino and B. Fantini, eds, *Le rivoluzioni nelle scienze della vita*, 31–81. Florence: Leo S. Olschki.

Duchesneau, François. 1998. *Les modèles du vivant de Descartes à Leibniz*. Paris: Vrin.

Durkheim, Émile. [1912] 1995. *The Elementary Forms of Religious Life*. New York: Free Press.

Farber, Paul L. 1972. Buffon and the concept of species. *Journal of the History of Biology* 5:259–284.

Foote, Edward T. 1969. Harvey: Spontaneous generation and the egg. *Annals of Science* 25 (2): 139–163.

Foucault, Michel. 1976. *Histoire de la sexualité 1: La volonté de savoir*. Paris: Gallimard.

———. 1991. Faire vivre et laisser mourir: La naissance de racisme. *Les temps modernes* 535:37–61.

Gärtner, Carl Friedrich. 1849. *Versuche und Beobachtungen über die Bastarderzeugung im Pflanzenreich*. Stuttgart: Auf Kosten des Verfassers.

Gayon, Jean, and Jean-Jacques Wunenburger. 1995. Présentation. In J. Gayon and J.-J. Wunenburger, eds, *Le paradigme de la filiation*, 7–11. Paris: L'Harmattan.

Glass, Bentley. [1959] 1968a. The germination of the idea of biological species. In B. Glass, O. Temkin, and W. L. Straus, eds, *Forerunners of Darwin: 1745–1859*, 30–48. Baltimore: Johns Hopkins Press.

———. [1959] 1968b. Heredity and variation in the eighteenth century concept of the species. In B. Glass, O. Temkin, and W. L. Straus, eds, *Forerunners of Darwin: 1745–1859*, 144–172. Baltimore: Johns Hopkins Press.

Goltz, Dietlinde. 1986. Der leere Uterus: Zum Einfluß von Harveys "De generatione animalium" auf die Lehren von der Konzeption. *Medizinhistorisches Journal* 21:242–268.

Gregory, Andrew. 2001. Harvey, Aristotle and the Weather Cycle. *Studies in the History and Philosophy of Biological and Biomedical Sciences* 32 (1): 153–168.

Harvey, William. [1651] 1766. Exercitationes de generatione animalium. In *Guilielmi Harveii Opera omnia, a collegio medicorum londinensi edita*, 160–540. London: Guilielmus Bowyer.

———. [1651] 1847. Anatomical exercises on the generation of animals. In R. Willis, ed. and trans., *The Works of William Harvey*, 143–518. London: Sydenham Society.

———. [1628] 1976. *An Anatomical Disputation Concerning the Movement of the Heart and Blood in Living Creatures*. Trans. G. Whitteridge. Oxford: Blackwell.

Heinrich, Klaus. 1981. *Tertium datur: Eine religionsphilosophische Einführung in die Logik*. Vol. 1, Dahlemer Vorlesungen. Basel and Frankfurt/M.: Stroemfeld/Roter Stern.

Hippocrates. 1588. De genitura. In *Opera Hippocratis Coi qvae Graece et latine extant*. Venetiis: Apud iuntas.

———. 1978. *Hippocratic Writings*. Ed. G. E. R. Lloyd. Harmondsworth: Penguin Books.

Hodge, Jonathan. 1985. Darwin as a lifelong generation theorist. In D. Kohn, ed., *The Darwinian Heritage*, 207–243. Princeton, NJ: Princeton University Press.

Jacob, François. [1970] 1993. *The Logic of Life: A History of Heredity*. Princeton, NJ: Princeton University Press.

Jucovy, Peter M. 1976. Circle and circulation: The language and imagery of William Harvey's discovery. *Perspectives in Biology and Medicine* 20:92–107.

Keller, Eve. 1999. Making up for losses. The workings of gender in William Harvey's "De generatione animalium." In S. C. Greenfield and C. Barash, eds, *Inventing Maternity: Politics, Science, and Literature, 1650–1865*, 34–56. Lexington: University of Kentucky Press.

Kielmeyer, Carl Friedrich. [1793] 1993. *Ueber die Verhältnisse der organischen Kräfte unter einander in der Reihe der verschiedenen Organisationen*. Reprint, introduced by K. T. Kanz. Marburg: Basiliskenpresse.

Koerner, Lisbet. 1999. *Linnaeus: Nature and Nation*. Cambridge, MA: Harvard University Press.

Larson, James F. 1994. *Interpreting Nature: The Science of Living Form from Linnaeus to Kant*. Baltimore: Johns Hopkins University Press.

Lefèvre, Wolfgang. 2001. Natural or artificial systems? The 18th-century controversy on classification of animals and plants and its philosophical context. In W. Lefèvre, ed., *Between Leibniz, Newton, and Kant: Philosophy and Science in the Eighteenth Century*, 191–209. Dordrecht: Kluwer.

———. 2003. Inheritance of acquired characters in Lamarck's and Geoffroy Saint Hilaire's zoology. In *Conference: A Cultural History of Heredity II: 18th and 19th Centuries*, Preprint, vol. 247, 93–107. Berlin: Max-Planck-Institute for the History of Science.

Lesky, E. 1950. *Die Zeugungs- und Vererbungslehren der Antike und ihr Nachwirken*. Vol. 19, Abhandlungen der Geistes- und Sozialwissenschaftlichen Klasse der Akademie der Wissenschaften und der Literatur in Mainz, Jahrgang 1950. Wiesbaden: Franz Steiner.

Lévi-Strauss, Claude. 1962. *La pensée sauvage*. Paris: Plon.

———. [1962] 1985. *Le totémisme aujourd'hui*. 6th ed. Paris: Presses universitaires de France.

Limoges, Camille. 1972. Introduction. In *C. Linné: L'équilibre de la nature*, trans. B. Jasmin. Paris: Vrin.

Linnaeus, Carl. 1735. *Systema naturae, sive, Regna tria naturae systematice proposita per classes, ordines, genera, & species*. Leiden: de Groot.

———. 1737a. *Critica botanica in quo nomina plantarum generica, specifica, & variantia examini subjicuntur. . . .* Leyden: Wishoff.

———. 1737b. *Genera plantarum eorumque characteres naturales secundum numerum, figuram, situm, proportionem omnium fructificationis partium*. Leyden: Wishoff.

———. [1746] 1749. Sponsalia plantarum. In vol. 1 of *Caroli Linnaei Ammoenitates academicae, seu Dissertationes variae Physicae, Medicae, Botanicae antehac seorsim editae*, 327–380. Stockholm and Leipzig: Kiesewetter.

Linnaeus, Carolus. 1751. *Philosophia botanica in qua explicantur Fundamenta Botanica cum definitionibus partium, exemplis terminorum, observationibus rariorum*. Stockholm: Kiesewetter.

———. [1749] 1762. The Oeconomy of Nature. In *Miscellaneous Tracts Relating to Natural History, Husbandry, and Physick*, trans. B. Stillingfleet, 2nd ed., 3–129. London: R. & J. Dodsley.

———. [1760] 1764. Politia naturae. In vol. 6 of *Caroli Linnaei Ammoenitates academicae, seu Dissertationes variae Physicae, Medicae, Botanicae antehac seorsim editae*, 17–39. Stockholm: Salvius.

Locke, John. [1690] 1975. *An Essay Concerning Human Understanding*. Ed. P. H. Nidditch. Oxford: Clarendon Press.

López-Beltrán, Carlos. 2004. In the cradle of heredity: French physicians and l'hérédité naturelle in the early 19th century. *Journal of the History of Biology* 37 (1): 39–72.

Lugt, Maaike van der, and Charles de Miramon, eds. Forthcoming. *L'hérédité à la fin du Moyen Age*. Micrologus Library. Florence: SISMEL–Edizioni del Galluzzo.

Mayr, Ernst. 1957. Species concepts and definitions. In E. Mayr, ed., *The Species Problem*. Washington, DC: American Association for the Advancement of Science.

———. 1982. *The Growth of Biological Thought: Diversity, Evolution, and Inheritance*. Cambridge, MA: Belknap.

McLaughlin, Peter. 1990. *Kant's Critique of Teleology in Biological Explanation: Antinomy and Teleology*. Lewiston, NY: Edwin Mellen Press.

McNeil, Maureen. 1987. *Under the Banner of Science: Erasmus Darwin and His Age*. Manchester: Manchester University Press.

McOuat, Gordon R. 1996. Species, rules and meaning: The politics of language and the ends of definitions in 19th century natural history. *Studies in History and Philosophy of Science* 27 (4): 473–519.

Müller-Sievers, Helmut. 1997. *Self-Generation: Biology, Philosophy, and Literature around 1800*. Stanford, CA: Stanford University Press.

Müller-Wille, Staffan. 2001. Gardens of paradise. *Endeavour* 25 (2): 49–54.

———. 2003a. Characters written with invisible ink: Elements of Hybridism 1751–1875. In *Conference: A Cultural History of Heredity II: 18th and 19th Centuries*, Preprint, vol. 247, 47–59. Berlin: Max-Planck-Institute for the History of Science.

———. 2003b. Nature as a marketplace: The political economy of Linnaean botany. *History of Political Economy* 35 (suppl.): 154–172.

———. 2006. La science de Bacon en action: la place de Linné dans l'histoire de la taxonomie. In T. Hoquet, ed., *Les fondements de la botanique: Linné et la classification des plantes*. Paris: Vuibert.

Olby, Robert C. 1985. *Origins of Mendelism*. 2nd ed. Chicago: University of Chicago Press.

Osborne, Michael. 1994. *Nature, the Exotic, and the Science of French Colonialism*. Bloomington: Indiana University Press.

Pagel, Walter. 1976. The philosophy of circles: Cesalpino–Harvey. *Journal of the History of Medicine and Allied Sciences* 12:140–157.

Porphyrius. 1887. *Isagoge et in Aristotelis Categoria commentarium. Übers. von Boethius*. Ed. A. Busse. Vol. 4.1 of *Commentaria in Aristotelem Graeca*. Berlin: Reimer.

Rausing, Lisbet. 2003. Underwriting the Oeconomy: Linnaeus on nature and mind. *History of Political Economy* 35 (suppl.): 173–203.

Ray, John. 1686. *Historia plantarum Species hactenus editas aliasque insuper multas noviter inventas & descriptas complectens*. . . . 3 vols. London: Clark and Faithorne.

Roe, Shirley. 1981. *Matter, Life, and Generation: 18th-Century Embryology and the Haller-Wolff Debate*. Cambridge: Cambridge University Press.

Roger, Jaques. [1963] 1993. *Les sciences de la vie dans la pensée française du XVIIIe siècle*. Paris: Albin Michel.

Rudwick, Martin. 1985. *The Meaning of Fossils: Episodes in the History of Palaeontology*. Chicago: University of Chicago Press.

Savoia, Andrea Ubriszy. 1995. The Botanical Garden of Padua in Guildano's days. In A. Minelli, ed., *The Botanical Garden of Padua: 1545–1995*. Venice: Marsilio.

Schiebinger, Londa. 1991. The private life of plants: Sexual politics in Carl Linnaeus and Erasmus Darwin. In M. Benjamin, ed., *Science and Sensibility: Gender and Scientific Enquiry, 1780–1945*, 121–143. Oxford: Blackwell.

Schiebinger, Londa, and Claudia Swan, eds. 2005. *Colonial Botany: Science, Commerce, and Politics in the Early Modern World*. Philadelphia: University of Pennsylvania Press.

Schweber, Sylvan. 1977. The origin of the Origin revisited. *Journal of the History of Biology* 10:241–255.

Sloan, Phillip R. 1979. Buffon, German biology, and the historical interpretation of biological species. *British Journal for the History of Science* 12 (2): 109–153.

Spary, Emma. 2000. *Utopia's Garden: French Natural History from Old Regime to Revolution*. Chicago: University of Chicago Press.

Stafleu, Frans A. 1971. *Linnaeus and the Linnaeans: The Spreading of Their Ideas in Systematic Botany, 1735–1789*. Utrecht: Oosthoek.

Stubbe, Hans. 1965. *Kurze Geschichte der Genetik bis zur Wiederentdeckung der Vererbungsregeln Gregor Mendels*. Vol. 1, Genetik. Grundlagen, Ergebnisse und Probleme in Einzeldarstellungen, H. Stubbe, ed., 2nd ed. Beitrag Jena: VEB Fischer.

Voss, Julia. 2004. Das erste Bild der Evolution: Wie Charles Darwin die Unordnung der Naturgeschichte zeichnete und was daraus wurde. In B. Kleeberg, W. Lefèvre, and J. Voss, eds, *Der Darwinismusstreit*, Preprint, vol. 272, 51–89. Berlin: Max-Planck-Institute for the History of Science.

White, John S. 1986. William Harvey and the primacy of the blood. *Annals of Science* 43:239–255.

Winther, Rasmus. 2000. Darwin on variation and heredity. *Journal of the History of Biology* 33:425–455.

Wolff, Caspar Friedrich. [1764] 1966. *Theorie von der Generation in zwei Abhandlungen erklärt und bewiesen*. Hildesheim: Olms.

Wolff, Michael. 1978. *Geschichte der Impetustheorie. Untersuchungen Zum Ursprung der klassihen Mechauik*. Frankfurt/M.: Suhrkamp.

Zirkle, Conway. 1935. *The Beginnings of Plant Hybridization*. Philadelphia: University of Pennsylvania Press.

———. 1959. Species before Darwin. *Proceedings of the American Philosophical Society* 103 (5): 636–644.

9 Duchesne's Strawberries: Between Growers' Practices and Academic Knowledge

Marc J. Ratcliff

9.1 A Discovery and Its Interpretations

In July 1764, a young Versailles botanist,[1] Antoine-Nicolas Duchesne (1747–1827), aged seventeen, who had recently published a small handbook of botany, "personally presented King Louis XV with a pot of strawberries."[2] Thanks to a particular hybridization, the result was spectacular and the King decided to patronize him, authorizing Duchesne "to raise more F. chiloensis in the royal kitchen garden at Versailles and to collect all varieties of strawberries known in Europe for the Trianon garden."[3] The King's patronage opened many doors for Duchesne.[4] Tutored by Bernard de Jussieu (1699–1777), who had helped him from the beginning, he wrote to Albrecht von Haller (1708–1777), Carolus Linnaeus (1707–1778), and other French and European naturalists to ask for seeds and specimens.

A year earlier in 1763, Duchesne had discovered a new kind of strawberry in his Versailles garden that had not yet been described. It had monophylla leaves, with only one lobe, while leaves of common strawberries possess three lobes (figure 9.1).

Now, in 1764, after he had reproduced this plant from budding, Duchesne collected the seeds obtained, and sowed them out. Two weeks later, the first strawberries appeared and he was astonished to find that the plant, which had not yet been described, reproduced from seeds. The latter feature characterized a stable species. Duchesne dispatched the seeds obtained and asked colleagues to sow them in order to repeat the experiment. Sowed in other places, the seeds stubbornly reproduced the new plant. On these grounds, Duchesne launched a three-year inquiry that led him to publish in 1766 a *Histoire naturelle des fraisiers*. Still nonextant for the community of botanists, the plant might be baptized at last. Linnaeus, who had recently been elected a foreign member of the Paris Académie des Sciences,[5] had prescribed strict rules in his *Philosophia botanica* (1751) that enabled only "orthodox Botanists" to give the "right names" to plants.[6] This secured the authority of Linnaeus and a few other

Figure 9.1
Fragaria Monophylla by Sydenham Teast Edwards (1768–1819). From *The Botanical Magazine* 2 (1788).

privileged botanists over amateurs, by reserving the power of naming plants to professionals. In the 1767 twelfth edition of his *Systema naturae*,[7] Linnaeus thus named the new plant *F. monophylla*, a name later widely used.

Although Duchesne drew hundreds of figures of strawberries to illustrate *Histoire naturelle des fraisiers*,[8] the only plate published in 1766 illustrated a "genealogy of strawberries" (figure 9.2) in which the author arranged ten kinds of strawberries in a genealogical tree.[9] Based on this illustration, certain scholars have regarded Duchesne as an early representative of mutationism and evolutionism, while others think his use of a genealogical tree does not prove such inclinations. To historians of breeding, Duchesne's work is the starting point of the modern strawberry that contrasts with the wild strawberry.[10] Scientists have discussed the importance of *F. monophylla* in relation to the issue of mutation and new species, and consider Duchesne to have anticipated modern evolutionism with his genealogical tree.[11] This interpretation of Duchesne became topical in Emile Guyénot's 1941 history of biology.[12] Nowadays, historians tend to reject the category of precursor, but they disagree on the meaning of Duchesne's genealogical tree. For Pascal Tassy, Duchesne was the first to make use of a temporal genealogy to classify living organisms.[13] Yet, for Giulio Barsanti, Duchesne's tree is not a genealogical tree at all, because it has no boughs, no bifurcations, and no intermediary species.[14]

In this chapter I will develop Tassy's argument and analyze how Duchesne conceived of evolution through time genealogically. This new conception emerged because he worked at the intersection of several fields previously separated: the grower's know-how, botanical theory, and aristocratic conceptions of descent. By breaking down the boundaries between these fields, he established the significance of his discovery.

9.2 Growers' and Breeders' Practices in Relation to Morphological Changes in Plants

In Europe since the Renaissance, practices that bore on the morphological changes in plants provided a background to the work of cultivators, florists, breeders, and cattle farmers. Several treatises on husbandry, gardening, and agriculture allowed for the progressive improvement of these practices. In the second half of the eighteenth century, the leading publication that provided information on empirical practices in France was the *Journal économique*. A monthly journal created in 1751, it was partly controlled by the Académie des Sciences through the censorship of Jean-Etienne Guettard (1715–1786). The *Journal économique* excluded nonapplied scientific research, and

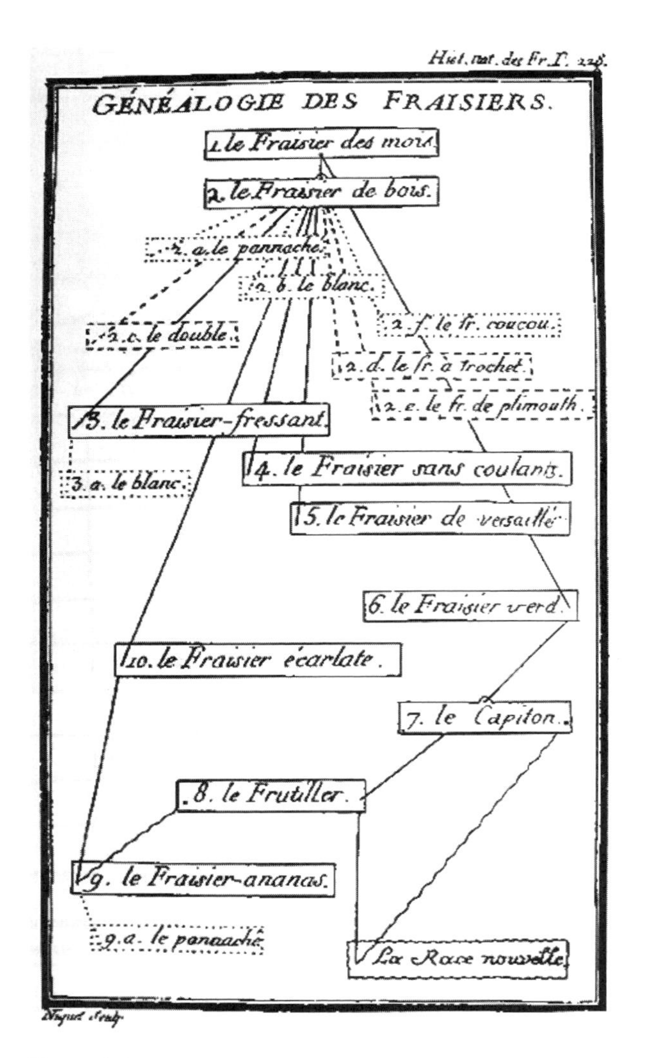

Figure 9.2
Genealogy of the strawberries according to Duchesne. From Antoine-Nicolas Duchesne, *Histoire naturelle des Fraisiers* (1766). Courtesy Bibliothèque Publique et Universitaire of Geneva.

served as a link between applied knowledge and civil society, for subjects such as natural history, domestic economy, law, public health, navy, geography, technology, gardening, growing, and breeding.[15]

The practitioners of that period distinguished between several techniques that caused morphological changes in plants or animals. The transformation of simple into double flowers, for instance, was achieved by these methods, resulting in an increase in value of the flowers. In the seventeenth century, the price of certain rare tulips was so high that the Dutch government published a law regulating the market, to avoid excessive expenditure on flowers as luxury goods. In the late 1630s, Dutch tulip speculation was so significant and potentially ruinous that it even served to alert investors to the dangers of market speculation.[16] The business was profitable and German and Dutch florists distributed catalogs of their best specimens of carnations and other flowers all over Europe.

To transform a flower from simple into double, florists transplanted it from one soil to another. This method resembled that of livestock farmers. When a flock of sheep was seen as "degenerating," being too meager and small, farmers imported a strong ram from another region to impregnate the failing female sheep. Farmers from the Limousin reported that thanks to this method, they "multiplied by three the income of the wool."[17] Other livestock farmers experimented on the selection of domestic animals with the aim of "improving the species," as discussed in chapter 10 of this book. Similarly, growers and florists learned to transform the shape of flowers, and horticulturists wondered if it was possible to "treat the young trees in the same manner as the flowers, in order to change the structure of their organs, and create, so to speak, double species."[18] It was commonly admitted that if a double flower ceased to be cared for, it returned to the state of a simple flower.

Another method of effecting changes in plants pertained to the inheritance of colors in tulips and other flowers: "When cutting all the stamens of a red tulip before the emission of dust, and when powdering the stigma of this same tulip with the dust of the stamen of a white tulip . . . , the seeds obtained were red for a part, white for another part and red and white for the last part."[19] The French botanist Michel Adanson (1727–1806) reported this example of hybridization, but he made no attempt at quantifying the results. This example was known, especially after Jean-Paul de Rome d'Ardène's (1689–1769) *Traité des tulipes* published in 1760, yet careful experiments on hybridization were not frequently reported in the *Journal économique*, contrary to the importance given to them in academic *Mémoires*, by such botanists as Joseph Gottlieb Koelreuter (1733–1806) and Lazzaro Spallanzani (1729–1799).[20] Practitioners used hybridization as a routine to a large extent based on tacit knowledge, and did not often resort to technical terms. For instance, d'Ardène ventured to call "mulâtres" and not "hybrids" the interbreeding product of two tulips.[21] Except for the word *degeneration*, no other particular names (such as *atavism*, *reversion*, or *heredity* used in the nineteenth century) referred to the return to the natural condition of the specimen.[22]

A shared model by which florists, cultivators, and livestock farmers understood their practices of transforming sorts and species was the influence of human art on nature.[23] The practitioners wanted to push nature beyond its limits, including morphology and calendar rhythms. They modified the color and morphological properties of plants, and also their blooming season. The *Journal économique* reported on several methods to make flowers—hyacinths and others—bloom at a precise date, to display at private parties and Christian festivals.[24] Grafting was used to modify the morphology or the color of particular species—for instance, an anonymous grower produced "green or yellow roses," thanks to a graft.[25] In other papers, authors reported various grafting experiments, creating a horned cock, a pear-apple, or seedless cherries.[26] Although the authors discussed the constancy of these "species," they usually regarded them as monstrous productions with no possibility of sexual reproduction. The influence of art on nature was limited by teratological barriers.

Thanks to his education by the gardener Louis Claude Richard (1754–1821) and Bernard de Jussieu, who were both involved in the development of the Trianon Garden, Duchesne knew a great deal of the breeders', florists', and growers' crafts.[27] A frequent visitor to Montreuil near Paris, a town known as a marketplace for cultivators and florists, he was skilled in breeding experiments. The Montreuil community of growers was famous for its products supported by horticultural and agricultural experiments. They were especially known for their practices of producing new kinds of fruits, notably peaches and strawberries, which dated back to the first part of the seventeenth century.[28] Duchesne had acquired much of the information for writing the domestic economy section in *Histoire naturelle des fraisiers* from "the practices of the grower traders of the town of le Bois and of Montreuil."[29]

One can nevertheless hardly identify a coherent tradition of gardening and breeding during the Ancien Régime. The know-how varied according to country, and according to the kind of plant or animal treated. Silkworms were cultivated in Italy and France since the Renaissance, while flower breeding was concentrated in the Dutch Republic and in certain areas of Germany. New trends developed for gardening that expanded in the realm of English home gardening between 1750 and 1850.[30] Growers' and breeders' practices were locally determined, in contrast to the monolithic and supposedly universal system of the botanical tradition. Practitioners nevertheless demonstrated a unity when confined to a particular locus. Until the middle of the nineteenth century, many small towns that surrounded Paris produced goods and especially vegetables to feed the capital through *les halles* (the marketplaces). Written just before the Revolution, Abbé Rozier's (1734–1793) *Comprehensive Course of Agriculture* reported that young florists apprenticed to masters from many of these towns specialized in producing vegetables. Montreuil produced peaches in particular, while Montmorency, Bagnolet, Vincennes, and Charonne specialized in other fruits.[31] This fertile environment was grown on the bedrock of the localized tradition of practitioners, as opposed to the international tradition of botany. The regional tradition was, however, destroyed before the end of the nineteenth century with the growth of Paris and the railways that facilitated the transport of goods.

9.3 Managing Varieties and New Species between Botanists and Practitioners

Criticized by Linnaeus since the 1730s as a "tower of Babel," the botanical tradition demonstrated a different kind of cohesion than the practitioners who were organized in local guilds. Botanists worked within an international network. Shared rules and skills as well as the use of Latin enabled botanists distributed over Europe and the world to collect plants, to give them a place in their classifications, and to spread botanical information everywhere.[32] Yet, while the botanical tradition perpetuated its collective memory and piled up lists of names, memory was of minor importance to practitioners. They wrote papers in journals, dictionaries of gardening, agronomic treatises, and other works, but except for a few heroes such as Thomas Fairchild (1667–1729), Olivier de Serres (1539–1619), and Philip Miller (1691–1771), the vast majority remained anonymous.

It is known that Linnaeus's specific project was to unify his "army of botanists" thanks to institutional means as well as guideline books such as *Critica botanica* and *Philosophia botanica*.[33] In these books he formalized the botanical rules used for nomenclature and defined the morphological language employed by botanists. Historians have underlined both the sociological model supplied by Linnaeus for botanists and his social impact on the formation of a new type of scientific society at the end of the eighteenth century.[34] To unify botany, Linnaeus also diminished its complexity, and reduced the number of species by eliminating varieties from the botanists' realm. While John Ray (1627–1704) had described about 18,500 species, Linnaeus reduced this number to 7,000. Eliminating varieties from botany laid the practice of variations in the hands of practitioners, a break that strengthened the demarcation between the two professions.[35] Thus, Linnaeus's impetus for demarcation also strengthened the botanists' professional identity. Yet, excluding varieties had a price, which was that it prevented botanists from examining the sources of variation. Rejecting varieties was thus a political and theological act that eliminated variation from the botanists' cabinet, which performed the role of strengthening the myth of the fixity of species and genera. Therefore, the world of practitioners who made their living in part from the production of varieties and hybrids was clearly set apart from the botanists' realm.

Four events, however, disturbed the clear demarcation of species to which Linnaeus and other botanists aspired. First, the definition of species that Ray and Linnaeus had given was a sort of genealogical definition. Second, the fixist system was seriously challenged by disobedient plants that did not comply with the botanists' rules, such as *Mercurialis* from Jean Marchant (1650–1738), *Peloria* from Linnaeus himself, and *Delphinium* from Johann Georg Gmelin (1709–1755).[36] Third, botanists such as Adanson did not neglect varieties and called for a sort of "evolutionary" theory.[37] Fourth, Duchesne's strawberries and similar productions made the practitioners heard. They advanced the exploration of this no-person's-land of variation and hybridization, which were the objects of many practices yet neglected by theoreticians. One important borderline problem that appeared in the process was the question of constancy. Botanists worked with species that were considered constant and rejected varieties that did not reproduce constantly from seeds. As Chrétien Guillaume Lamoignon de Malesherbes (1721–1794) said in the early 1750s, "All these alleged species [pears, peaches, carnations, tulips] degenerate when one wants to multiply them from seeds; consequently they are only varieties for botanists."[38]

It remains an open question whether the disciplinary boundaries of botany were strong enough to make those such as Adanson or Duchesne consider themselves botanists even when working on varieties. Although a practitioner, Duchesne was also a botanist, and he demonstrated a comprehensive use of botanical skills and knowledge: use of Latin names, morphological description of species, use of classification, historical knowledge, quotation of botanists, and correction of "bad descriptions." He had already used all these skills in his *Manuel de botanique*.[39] In describing species, Duchesne followed a method taken right from the botanical tradition, dealing with, in the following order, the systematic and historical survey of the species, its name, description, morphological characters, differences, reproduction, cultivation, synonymy, authors, geography, and so on. Recognized by the community of botanists, Duchesne corresponded with Linnaeus, Haller in Bern, Giuseppe Monti (1682–1760) in Bologna, and Ludovico Allioni (1728–1804) in Turin. He also knew many French botanists personally.[40]

The attitudes of practitioners and botanists toward variety and variation reveal their beliefs concerning the stability of species. Adanson remarked that "modern botanists do not agree with these changes, which are, to speak properly, only varieties more characterized."[41] He nonetheless noticed that there existed varieties that became almost constant, "being fixed during several generations."[42] Adanson aimed at reintroducing varieties into botany, because some varieties could become constant, or in other words, transmute into species. Facing this problem, botanists had to take a stand on the fixity of species. This was a problem pertaining to the delimitation of the botanical field, and thus had to be either rejected out of hand, or integrated into the field. Perhaps this borderline situation was the reason Linnaeus changed his mind many times regarding new species, and cast doubt on the constancy of the species from the 1750s onward.[43]

Practitioners, however, had not been waiting for the *Philosophia botanica* to begin working on varieties and on methods of generation and hybridization. On the local level, traditions of practitioners were unified enough to establish certain routines in dealing with varieties. Utilitarian methods of hybridization defined one of the borders between botanists and practitioners, but there were no systematic attempts to establish a theory of varieties produced by hybridization among practitioners. Duchesne tried to bring about a change in this milieu while treating many of these problems in a separate volume added to his *Histoire naturelle des fraisiers* under the title *Remarques particulières*.

9.4 Duchesne's Enterprise: Culture, History, and Genealogy

In the first volume of his *Histoire naturelle des fraisiers*, Duchesne surveyed the history of strawberries, including changes in their morphology due to climate and other causes. He described the various species of strawberries, and compared them with other genera placed within the family Rosacea. Then he studied each particular variety, "race" as he called it, within respective species, including their morphological differences and history, "notifying at the same time what race seemed the older, and indicating with regard to new ones those among which they were born."[44] This was followed by a summary of the characters particular to each race together with their genealogy. A section on the domestic economy of strawberries followed, including reflections on art, ornamental aspects, gardening, growers' practices, harmful insects, recipes, and remedies.[45] In the 100 pages of the *Remarques particulières* added to the first volume, Duchesne developed six specific topics.[46] They dealt with constancy, the definition of variety, species and genus, the existence of hermaphroditic strawberries, a comparison between hybrids in plants and animals, and synonymy.[47]

9.4.1 Of Species, Races, and Constancy

Duchesne begins his discussion by addressing the influence of the method of generation (sexual or budding) on constancy. Commenting on a passage from Philip Miller's *Gardeners Dictionary*,[48] he emphasized sexual reproduction as the true basis of species, while budding and even grafting never constituted species. Plants obtained by the latter methods were only an extension of the same specimen and did not usually produce seeds.[49] This distinction—grounded in propagation methods—was shared by botanists, and established a clear limit between fixed species and varieties. But the new strawberry was a previously undescribed variety, originating in hybridization and yet reproduced *constantly* from seeds, thus showing features of both variety and species: "Is it a species?" asked Duchesne. "Then there are new species. Is it a variety? Then how many varieties, which are regarded as species, do other genera include?"[50] So, while using the current botanical categories, one ended with a conundrum: either new species existed, or many genera actually included varieties and not species.

By asking these questions, Duchesne opened the door to a concept that was much used in animal breeding, but less so in plant breeding, and that could bring the concept of fixity into harmony with that of variability: the concept of *race*. Florists and gardeners used to call constant varieties either "species" or "sorts," but not "races," at least until the 1770s. Guyénot has maintained that Duchesne failed to distinguish race, species, and genus, confusing "race with the meaning of species, and the term species with that of genus."[51] But Duchesne analyzed the botanical definitions of his time and developed a precise line of argument: "The confusion was mainly caused by the use by several authors of the same words applied to ideas totally different."[52] Both practitioners and botanists were the focus of his criticism, and he detected three major sources of confusion. Florists threw language into confusion when calling species precisely what botanists called variety. For instance, the well-known florist d'Ardène claimed to use "the language of florists . . . [and] call species the plant which is termed variety in Botany."[53] Once this (first) linguistic confusion was eliminated by using botanical names, however, a second problem emerged—that of specifying the meaning of constancy with respect to species and varieties respectively. Constancy was a property of the species, so how could one deal with constant variety?

It was the discovery of *F. Monophylla* that pushed Duchesne to rethink the relationship between species and variety and led him to write that "cultivation and other accidental causes do not produce new species, but they cause, in certain individuals, some changes which, being persistent through their posterity, make new races."[54] Once it was admitted that new varieties could originate from stable species, there remained two questions. First, how could one secure the distinction between variety and species, if both were constant? Second, what was the relationship between species and variety? For the first problem, crossbreeding of various species allowed the determination of which ones were the true species. If different plant forms could be crossbred, they belonged to one and the same species; if not, to two different species. But, according to Duchesne, this method was too time consuming and difficult. Instead, he reverted to the investigation of the "true" characters determining species. For Duchesne, the cause of the second confusion between species and race was that many specific characters were not "true ones," and plants had not been well determined. This attempt to specify the "true" characters of species was influenced by Jussieu's natural method in botany.[55] Instead of looking, like Linnaeus, to the number, figure, proportion, and situation of the external parts of the plant, one should look at the "position and internal shape" of the parts for "constant characters . . . , as in animals."[56]

Linnaeus had regarded hybridization as a model to understand the origin of new species. Duchesne attempted to shape a consistent theory of genealogical species, based on internal characters, which could explain the natural relationship between species and constant varieties. He used the experience of the practitioners to understand the origins of constancy and to map the genealogical aspects of hybridization. Like Linnaeus, livestock farmers employed the category of "constant variety," which the latter usually termed *race*. Georges-Louis Leclerc Comte de Buffon (1707–1788) had utilized the term *race* in his *Histoire naturelle des animaux* with regard to varieties of humans and domestic animals.[57] According to Duchesne, this term needed "to be introduced into the history of the vegetables,"[58] to designate objects in the territory between botanists and cultivators. He considered *race* to be the better word for "constant variety," which was an "improper denomination" in his view.[59] As Buffon stated in 1779, "in animal species, races are simply constant varieties that propagate through generation."[60] Yet, as we have seen, botanists held apart species (on which they worked) and varieties (a problem set aside for cultivators). They thus ended up with a third conceptual confusion: "Following Ray's axiom, since one cannot label a constant race a variety, they call them species."[61] Duchesne's claim was that, although both races and species were constant, races were not the same as species. In other words, the lack of the term *race* in botany had confusing consequences: botanists were unable to differentiate between varieties that interbreed and vary, and races that interbreed but do not vary.

To sum up, Duchesne criticized three sorts of confusion emerging at the border between botanists and practitioners. First, practitioners fed a linguistic confusion, calling all their varieties "species," while not meeting the botanical species criterion of reproduction through seeds. Second, botanists created confusion in the determination of species by not using "true" characters to determine species. The use of "true" characters would have been particularly sensible when a new species emerged that forced botanists to either change the specific characters, or to neglect the principle of the constancy of the species.[62] Third, while following Ray's axiom, botanists fostered a conceptual confusion because they called species what were actually races—that is, varieties stabilized over time. Along with the concept of race, both time and genealogy were material to explaining what Linnaeus himself had acknowledged as a "singular event" in the history of botany: the appearance of the Versailles strawberry.[63]

Duchesne's genealogical system can be outlined as follows. There are three constant kinds of strawberries, regarded as species. It was empirically known that one of them, the Versailles strawberry, came from the common wood strawberry, which many clues showed was the only kind known to the ancients. Clearing up the three confusions, the following picture emerged. First, the various sorts of strawberries came from only one primitive kind of strawberry. Second, taking all these sorts together, strawberries represent but one species containing several kinds, each of these kinds being a constant variety, or, properly named, a race of its own. To understand their relationship, one needed genealogy.[64] Genealogical classification allowed botanists to retain a modified principle of the constancy of species. It was helpful to consider a species as a totality over time, an expression of all constant and interbreeding races, which a "primitive race" could yield in posterity.[65] "From there came all the various races" of strawberry.[66] To explain the spreading of varieties, Duchesne used his genealogical model to conceive of the constancy of the species as relative to time.

9.4.2 A New Reading of History?

One of the learned customs of botanists was frequent reference to their three-century-old tradition. An eighteenth-century botanist had a particular relationship to memory and, going back to the Renaissance, referred to names and morphological descriptions of plants recorded by previous botanists. This relationship to history in the botanical tradition was very distinct from that of another important tradition, the experimental tradition, which took root in the second half of the seventeenth century.[67] And indeed, when one looks for the historical sensibilities demonstrated by eighteenth-century experimentalists, one finds that they do not usually refer to works before 1660. The line that separated the "ancients" from the "moderns" was placed in the mid-seventeenth century, corresponding with the birth of the academies and the famous quarrel of the ancients and moderns.

Such a line of demarcation did not exist in the botanical and natural historical traditions. Although there were controversies on the boundaries between the herbalist tradition and the botanical tradition,[68] eighteenth-century botanists tended to consider much of the documented knowledge about plants as part of the tradition. Linnaeus's classification of botanists and Haller's *Bibliotheca botanica* (1771–1772) left spaces for the inclusion of Renaissance authors. Similarly when Adanson discussed botanical iconography, he began his inventory with publications by Jean de Corbichon in 1482 and Adam Lonitzer in 1491.[69] Using a typical botanist's expertise, Adanson systematically surveyed ancient literature with critical and cumulative targets. He wanted to distinguish creators from copyists. For instance, according to him, only 10,000 out of 70,000 engravings of plants were original; the others were copies.[70] The aim was also to establish a cumulative selection of data to avoid losing information on authors, plants, figures, names, descriptions, classifications, and other important aspects of the tradition.

While the reading of history was a duty in the botanical tradition, the mission of botany was conceived of as a systematic description of fixed plants existing since the Creation. Instead of using history to survey the description of all plants, Duchesne regarded history as a temporal locus wherein new plants could emerge. Following a Linnaean procedure,[71] he was very precise when dealing with historical material:

> *In 1665*, Colbert . . . asked for the catalogue of the King's garden. . . . They included the double strawberry. *At around the same time*, Morison also describes it. . . . Father Barrelier, *who died in 1673*, drew it . . . it has been engraved *since*, and published *in 1714*. . . . Zanoni, *at the same time*, drew it also . . . ; it is in his botanical history published *in 1675*. . . . It appears in Furetière's dictionary, *from 1690*. . . . M. Haller quotes a description of it . . . in the memoir of Breslau, *July 1722*.[72]

Much of Duchesne's reading of the history of strawberries focused on contexts, dates, and places, and he frequently used temporal connectors. Time could produce new forms of constancy, while for some botanists time did not allow for this creative dimension.

This temporal approach supplied Duchesne with the historical foundation for his genealogical conception, as shown in the following example: "From Plymouth where our strawberry seems to be born, it arrived in Leiden, then in Paris, and eventually in Bologna, where Zanoni saw it in 1675. It also stayed in England, its native country, for Ray said in 1686, that he had cultivated it during several years in Cambridge. But since that time no one has seen it."[73] Duchesne's reading of history pushed him to reshape the relationships among races into a classification that would include the dimension of time. The genealogical tree was the main and probably the most efficient model available for the job.

9.4.3 From Genealogy to Experiments on New Races

Duchesne's use of genealogy was further enriched by sociocultural imagery consisting of the use of aristocratic conceptions and language. Before him, as already mentioned, the term *race* was rarely used by florists, gardeners, and horticulturists even in the *Journal économique*,[74] which usually preferred *sort*. Similarly, the *Encyclopédie* used *race* only for human and noble domestic animals, such as horses.[75] *Souche*—that is, stock—was used by one Marquis de Puismarais with reference to sheep; he also employed *alliance*.[76] D'Ardène used the latter term as well.[77] Duchesne used many words derived from the language of noble genealogies, such as "offspring" (*postérité*), "head of the tree," "branches of the same house," and "genealogical tree."[78] Three reasons account for this usage. First, the awareness of a true genealogical origin of new races was shared knowledge in the breeding milieu, although it did not reflect an evolutionary conception. Second, the sullying connotations of the process of hybridization were overcome by the attribution of a noble name, a successful tactic for the marketing of many flowers.[79] Third, Antoine-Nicolas Duchesne himself came from a noble bourgeois family in Paris. His great-grandfather, Denis, had been locksmith to King Louis XIII; his grandfather, Nicolas, became *Prévot des bâtiments* under King Louis XIV, a head-property-manager office that Duchesne's father, Antoine, inherited.[80]

Duchesne's genealogical tree (figure 9.2) illustrated the descent from a primitive race of strawberries to another and contained its own logic. It was organized according to certain rules and particular symbols. Constant races, represented by bigger rectangles, were separated from derived varieties or monstrous specimens, represented by smaller and dotted-line rectangles. The genealogical descent stopped with varieties, but could in theory endure with the formation of constant species. Another rule located the common head of all races on the top of the tree, in a way similar to that used for noble genealogical trees. Younger races were on the bottom. The third rule dealt with the proximity of races. The three races 3, 4, and 5, close to each other, had "several common characters," which made possible their reduction to a common root. The places of the other strawberries were organized according to morphological and ecological similarities, and of course, genealogical descent. The seventh comes from the sixth, the tenth from the second, and the sixth from the first. The lines also represented differences in the way kinds of strawberries came into being. The normal line showed a descent without hybridization, and the wavy line depicted a hybrid. An example of the latter was the ninth, *F. Ananassa*, which was suspected to come from a crossbreeding of 8 and 10.[81] A new race was also expected to result from the crossbreeding of 7 and 8.

However, if it established a temporal relationship from the origin (top) to the latest races (bottom), the time scale of the genealogical tree was not quantified precisely, even if Duchesne laid many hints in his historical reconstruction of strawberries that helped in this regard. He was able to date the appearance of certain races, and established which were the older races described by the ancients. Thus he could establish *Fragaria semperflorens* as the head of the line.

In addressing the questions of genealogy and new constant races, Duchesne also proved to have been consistent with his methodology and his numerous appeals for naturalists to use empirical techniques. Indeed, his genealogical investigations found an important counterpart in his experimentation. He carried out hybridization experiments on the issue of new constant races, and the *Histoire naturelle des fraisiers* could thus discuss two new races, not just one. The first race was an accidental discovery, and he named it *Versailles strawberry* (later to be named *F. monophylla* by Linnaeus). But the second race was the result of an experiment. Having received a female *Frutiller*, Duchesne placed it next to the male *Capiton*, which it resembled. A week later, the *Frutiller* gave a new fruit. Duchesne repeated the experiment, and, thanks to the help of people from the *Jardin du Roi*, they obtained four new *Frutiller* pollinated by the *Capiton*. The seeds were given to several botanists, such as Jussieu and Richard, who sowed them at the time the printing press was turning out Duchesne's book. Other scholars, such as Le Monnier and Le Normand had tried to cross the *Frutiller* with *F. Silvestris*, but with no success.

On this point, the book disappointed its readers, for the results of the latest hybridization were not reported. However, the practitioners preserved the memory of these experiments.[82] Thirty years later, Abbé Rozier's *Cours complet d'agriculture* reproduced, under the entry "Strawberries," the genealogical tree of Duchesne, now containing the new race, however, which had by then proved to be undoubtedly constant (figure 9.3) and had been identified with the Scarlet strawberry from Bath.[83] Rozier's copy of Duchesne's genealogy contained the new name for race 11, while another race was added (12). Duchesne's experiment seems to have left, at least in the late eighteenth century, some offspring.

In the 1780s, after Duchesne's discovery was diffused, practitioners and botanists nevertheless maintained their differences of opinion on the concept of species. Rozier remarked that the botanists distinguished three species of *Fragaria* (*semperflorens, muri-*

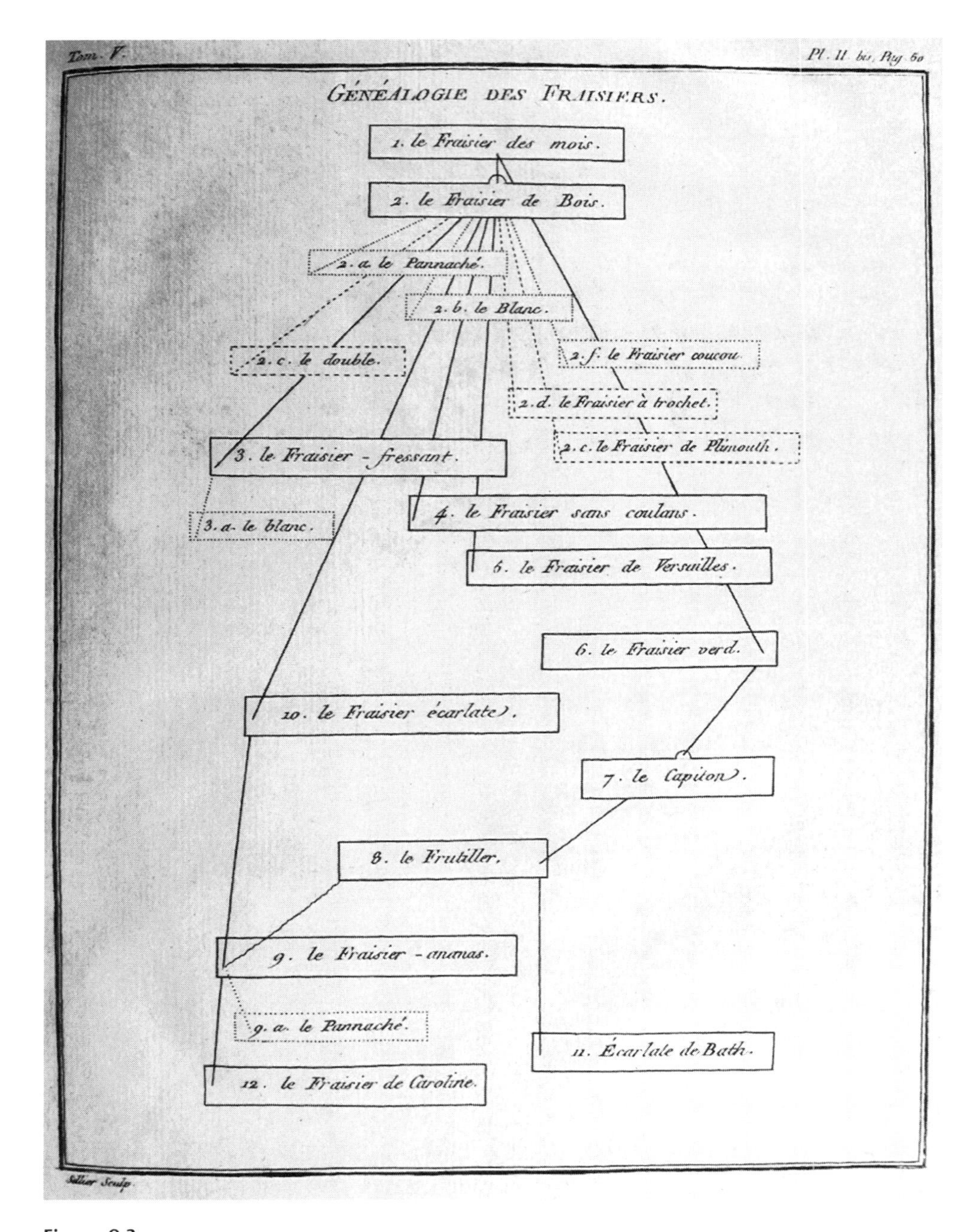

Figure 9.3

Genealogy of the strawberries according to Rozier. From François Rozier, *Cours complet d'agriculture . . . ou Dictionnaire universel d'agriculture,* vol. 5 (1787). Courtesy Bibliothèque Publique et Universitaire of Geneva.

cata, and *sterilis*), while Duchesne had attempted to unify them under one species. His work sketched out some of the links between natural method, genealogy, and heredity, and, as Roger L. Williams phrased it, "The probable descent of species from a common ancestor was already in the minds of the earliest practitioners of the natural method."[84] In spite of his work, practitioners still considered races to be species, and labeled them as such. Two models for the species coexisted in France at the end of the eighteenth century, which served the rather different interests of two distinct communities, the practitioners and the botanists. Yet the genealogical model had emerged at the border of these two communities, and, while drawing on noble imagery, met with some diffusion.

9.5 Conclusion

Various influences encouraged Duchesne to use the genealogical tree. In addition to the empirical approach to hybridization used by scholars such as Marchant, Linnaeus, Buffon, Koelreuter, Adanson, and Gmelin, Duchesne developed his own method. To explore the no-person's-land of the "constant races" between botany and the practitioners, he systematically examined the history of botany and gardening, looking for the appearance of new sorts. He cultivated a temporal history as opposed to the systematic history in use by the botanical tradition, and discussed the issue of the common ancestor for one species considered as a whole. Accounting of nature according to temporal and genealogical schemes was later used by Charles Darwin (1809–1882), Ernst Haeckel (1834–1919), and Hugo de Vries (1848–1935), but with a universal history in mind that was totally absent in the work of Duchesne.

This story has shown us the entrance of a skilled practitioner into the fields left unexplored between two traditions—botany and breeding—and their hybridization, resulting in a new field spanning the border between them. Practices of heredity took shape through know-how and hybridizations cultivated by breeders. The double education of Duchesne led him to look for a theoretical and schematic account of novelty while proposing a reform of the current language and concepts used by breeders and botanists. Relevant to his problem, a noble custom of illustrating genealogy was available, which he drew on in order to design one of the first genealogical accounts of species in transformation, based on experiments. This conception of the genealogical tree had little immediate impact on the scientific realm. Scholars could reproduce a particular experiment, but they were not able to reproduce Duchesne's contextual configuration of scholarly botany, practical breeding, and genealogical reconstruction. The impervious barriers between botany and practical know-how, along with the Ancien Régime's hermetic separation of practitioners from the ruling classes, did not help the diffusion of this new thought. An emerging scientific discourse on genealogy, hybrids, and heredity called for the hybridization of the knowledge of practitioners with that of theoreticians, which occurred, this time definitely, during the nineteenth century.

Notes

1. Duchesne was born into a noble family in Versailles in April 1747. He studied law, and since his father was administrator of the Trianon Garden, Antoine-Nicolas could freely work and study botany along with Bernard de Jussieu and his friend and gardener Richard. For a biographical notice, see Bernier 1898.

2. Darrow 1966, chapter 5. (I quote Darrow from an Internet version without pagination.)

3. Darrow 1966, chapter 5.

4. Duchesne's direct patron was Abel-François Poisson de Vandières, Marquis de Marigny (1727–1781), the brother of Mme de Pompadour. See Duchesne 1764, *dédicace*.

5. Larson 1994, 17.

6. Linnaeus 1751, §211. On orthodox botanists, see Linnaeus 1751, §26.

7. Linnaeus 1767, 349.

8. This information was provided by Bernier 1898, 5. Staudt (2003) has published sixty of Duchesne's plates.

9. Duchesne 1766a, 228, plate I.

10. Darrow 1966, chapter 6.

11. De Candolle 1882, 8–10; De Vries 1909, 378; Blaringhem 1911, 8.

12. Guyénot 1941, 375–376.

13. "Avec la publication d'une véritable généalogie d'êtres vivants qui se veut comme telle, à la manière des fraisiers de Duchesne, apparaît la notion du temps" Tassy 1991, 26.

14. Barsanti 1992, 84.

15. On the history of the relationship between the Paris Académie des Sciences and French societies directed to utilitarian practices, see Briggs 1991.

16. Garber 1989, 535. Tulip prices were up to 2,000 guilders between 1625 and 1637 for a rare bulb.

17. Puismarais 1754, 74.

18. Anonymous I, 1761, 172.

19. Adanson 1763, cxiii.

20. On Spallanzani's hybridization experiments, see Plantefol 1987.

21. d'Ardène [1760] 1765, 123.

22. On the early history of hybridization, see Zirkle 1935.

23. Anonymous F, 1758, 500.

24. Anonymous G, 1760, 335.

25. Anonymous E, 1758, 170.

26. Anonymous A, B, C, D, H, I.

27. On the influence of Richard and Jussieu see Williams 2001, 32–34.

28. Schabol 1758, 75–76; Rozier 1787, 56.

29. Duchesne 1766a, x.

30. See Lustig 2000.

31. Roger 1755.

32. On this feature of the botanical tradition around Linnaeus's network, see Müller-Wille 2003.

33. Stafleu 1971.

34. See Stafleu 1971; Duris 1993.

35. Linnaeus 1751, §306.

36. On early hybridization, see Zirkle 1935.

37. Adanson (1763) laid down his evolutionary conception in his 1763 *Familles des Plantes*. He admitted the possibility of the creation of new species: "One could perhaps apply these examples to a number of insects, shellfish and worms, which would demonstrate the possibility of the mutations or of the creation of new species in animals, as it seems proved that there are new species in plants showing to be not immutable" (p. clxii). See also Atran 1993, 306; Larson 1994.

38. Malesherbes [1798] 1971, 31. Malesherbes's book was written in 1750 as a quick reaction to Buffon's *Histoire naturelle* (1749). Yet Malesherbes's contribution remained in manuscript form until 1798. A state minister, Malesherbes was among the few who belonged to the three main French academies, and he was a skilled botanist.

39. Duchesne 1764, xv.

40. Duchesne 1766a, vii–viii.

41. Adanson 1763, clxii.

42. Adanson 1763, clxiii.

43. See Müller-Wille, chapter 8, this volume.

44. My translation. Duchesne 1766a, ix: "Avertissant en même temps quelles races paroissent les plus anciennes, et indiquant pour les nouvelles celles dont elles ont pris naissance."

45. Duchesne's plan is set out on pages viii–x.

46. Duchesne 1766b, 1.

47. Duchesne 1766a, xi–xii.

48. Miller [1724] 1759; see *Fragaria*.

49. Duchesne 1766b, 5–6, 8.

50. Duchesne 1766b, 13.

51. Guyénot 1941, 376n.

52. Duchesne 1766b, 13–14.

53. d'Ardène [1760] 1765, 78.

54. d'Ardène [1760] 1765, 21.

55. On the influence of the Jussieu natural method on Duchesne, see Bernier 1898, 4; Williams 2001, 36–38.

56. Duchesne 1766b, 25.

57. See for instance Buffon 1755, *Histoire du chien*, 196.

58. Duchesne 1766b, 18.

59. Duchesne 1766b, 18.

60. Buffon 1778, 252.

61. Buffon 1778, 26.

62. In Duchesne 1766b, 22–25, Duchesne criticized Linnaeus's views on hybrids.

63. Letter from Linnaeus to Duchesne, December 23, 1764, quoted in Duchesne 1766b, 13.

64. Duchesne 1766b, 14.

65. Duchesne 1766a, 7–8. Duchesne was certainly also inspired by Buffon (1755), who considered all dogs as one species (p. 193), spoke of one "primitive race" (pp. 197, 201), and discussed genealogy (p. 225). Yet Buffon did not witness the birth of a new race, was highly speculative

on the descent of dogs, and his figure was geographic and not genealogical. See on this Hoquet 2005, 605–614.

66. Duchesne 1766a, 9.

67. See Shapin and Schaffer 1985; Licoppe 1994.

68. See Atran 1993.

69. Adanson 1763, lxxxi, cxlii.

70. Adanson 1763, cxlii.

71. On this, see Müller-Wille 2001.

72. Duchesne 1766a, 77–79. My emphasis.

73. Duchesne 1766a, 104.

74. Anonymous H, 1761, 122.

75. Diderot and d'Alembert, 1778, t. 28, 202.

76. Puismarais 1754, 70; also used in Duchesne 1766a, 224.

77. d'Ardène 1746, 232.

78. Duchesne 1766a, 214, 221; Duchesne 1766b, 23, 47, 63. *Famille*, a botanical term often used by Adanson and Jussieu, is also present in Duchesne's work.

79. d'Ardène's *Traité des oeillets* (Carnation) (1762, 243–251) mentioned a list of 300 sorts of carnations, among which one finds *Bâton royal*, *Duc de Milan*, *Cléopâtre*, *Roi d'Alger*, and so on.

80. Bernier 1898, 3. For Duchesne's genealogy, see Bernier 1989, 9.

81. Darrow (1966, chapter 6) believed Duchesne "must be credited as the first to identify F. chiloensis x F. virginiana as the origin of the modern strawberry." In the plate, F. chiloensis is *frutiller* (8), F. virginiana is *fraisier écarlate* (10), and the modern strawberry (F. ananassa) is *fraisier-ananas* (9).

82. Certain botanists also adopted views similar to Duchesne's on the acquisition of stability in the offspring of a race; see Williams 2001, 105.

83. Rozier 1787, 50, plate II.

84. Williams 2001, 38.

References

Adanson, Michel. 1763. *Familles des plantes*. Paris: Vincent.

Anonymous A. 1752. Pour faire venir sur un même pied des fleurs de la même espèce et de différentes couleurs. *Journal économique* (June): 25–28.

Anonymous B. 1756. Méthode facile pour avoir des pommes et des poires dont les quartiers soient de différentes espèces. *Journal économique* (April): 75–76.

Anonymous C. 1756. Observations sur la manière de greffer par M. *Journal économique* (April): 83–95.

Anonymous D. 1757. Manière d'élever les oies et les canards, et de se procurer des canards mulets. *Journal économique* (December): 44–57.

Anonymous E. 1758. Manière de faire venir des roses vertes et des jaunes. *Journal économique* (April): 170.

Anonymous F. 1758. Observations économiques sur les brebis, et la manière de les élever. *Journal économique* (November): 498–505.

Anonymous G. 1760. Méthode pour avoir en hiver des fleurs naturelles nouvelles épanouies le jour que l'on veut. *Journal économique* (July): 335.

Anonymous H. 1761. Des coqs cornus, et de la manière de leur faire venir des cornes sur la tête. *Journal économique* (March): 120–122.

Anonymous I. 1761. Moyen d'avoir des cerises sans noyau. *Journal économique* (April): 171–172.

Atran, Scott. 1993. *Cognitive Foundations of Natural History*. Cambridge: Cambridge University Press.

Barsanti, Giulio. 1992. *La Scala, la Mappa, l'Albero Immagini e classificazioni della nature fra Sei et Ottocento*. Florence: Sansoni.

Bernier, Paul-Dominique. 1898. Notice sur Antoine Duchesne et sur Antoine-Nicolas son fils. In P.-D. Bernier, ed., *Voyage de Antoine-Nicolas Duchesne au Havre et en Haute Normandie, 1762*, 3–9. *Mélanges de la Société d'histoire de Normandie*, 4th series.

Blaringhem, Louis. 1911. *Les transformations brusques des êtres vivants*. Paris: Flammarion.

Briggs, Robin. 1991. The *Académie Royale des Sciences* and the pursuit of utility. *Past and Present* 131:38–88.

Buffon, Georges-Louis Leclerc, Comte de. 1755. *Histoire naturelle, générale et particulière*. Vol. 5. Paris: Imprimerie Royale.

———. 1778. *Des époques de la nature*. Paris: Imprimerie Royale.

d'Ardène, Jean-Paul de Rome. 1746. *Traité des renoncules*. Paris: Lottin.

———. [1760] 1765. *Traité des tulipes*. Avignon: Chambeau.

———. 1762. *Traité des oeillets*. Avignon: Chambeau.

Darrow, G. M. 1966. *The Strawberry: History, Breeding, and Physiology*. New York: Holt, Rinehart and Winston.

De Candolle, Alphonse. 1882. *Darwin considéré au point de vue des causes de son succès et de l'importance de ses travaux*. Geneva: Georg.

De Vries, Hugo. 1909. *Espèces et variétés*. Paris: Alcan.

Diderot Denis, and Jean le Rond, d'Alembert, eds. 1778. *Encyclopédie ou dictionnaire raisonné des sciences, des lettres et des arts*. Vol. 28. Geneva: Pellet.

Duchesne, Antoine-Nicolas. 1764. *Manuel de botanique*. Paris: Didot.

———. 1766a. *Histoire naturelle des fraisiers*. Paris.

———. 1766b. *Remarques particulières*. Second and separate section in *Histoire naturelle des fraisiers*. Paris.

Duris, Pascal. 1993. *Linné et la France (1780–1850)*. Geneva: Droz.

Garber, Peter M. 1989. Tulipmania. *Journal of Political Economy* 97 (3): 535–560.

Guyénot, Emile. 1941. *Les sciences de la vie aux XVII et XVIIIe siècles, l'idée d'évolution*. Paris: Albin Michel.

Hoquet, Thierry. 2005. *Buffon: histoire naturelle et philosophie*. Paris: Honoré Champion.

Larson, James L. 1994. *Interpreting Nature: The Science of Living Forms from Linnaeus to Kant*. Baltimore: Johns Hopkins Press.

Licoppe, Christian. 1994. The crystallization of a new narrative form in experimental reports (1660–1690): The experimental evidence as a transaction between philosophical knowledge and aristocratic power. *Science in Context* 7 (2): 205–244.

Linnaeus, Carolus. 1751. *Philosophia botanica*. Stockholm: Kiesewetter.

———. 1767. *Systema naturae*. Vol. 2. Stockholm: Laurentius Salvius.

Lustig, Abigail J. 2000. Cultivating knowledge in nineteenth-century English gardens. *Science in Context* 13:155–181.

Malesherbes, Chrétien-Guillaume de Lamoignon de. [1798] 1971. *Observations sur l'histoire naturelle, générale et particulière de Buffon et Daubenton*. Vol. 1. Geneva: Slatkine reprints.

Miller, Philip. [1724] 1759. *The Gardeners Dictionary*. 7th ed. London: Printed for the author.

Müller-Wille, Staffan. 2001. La storia raddoppiata: La sintesi dei fatti nella storia naturale di Linneo. *Quaderni storici* 108 (3): 823–842.

———. 2003. Joining Lapland and the Topinambes in flourishing Holland: Center and periphery in Linnaean botany. *Science in Context* 16 (4): 416–488.

Plantefol, Lucien. 1987. Spallanzani botaniste. *History and Philosophy of Life Science* 9:37–56.

Puismarais, Marquis de. 1754. Mémoire sur les moyens de bonnifier les laines dans les provinces du royaume. *Journal économique* (August): 68–77.

Roger, Abbé. 1755. Sur les villages de Montreuil, Bagnolet, Vincennes, Charonne et villages adjacents à deux lieues ou environ de Paris, au sujet de la culture des végétaux, avec une idée de leur méthode de traiter les arbres, surtout les pêchers. *Journal économique* (February): 44–79.

Rozier, François. 1787. De la culture des fraisiers. *Cours complet d'agriculture*, vol. 5, 56–59. Paris: Serpente.

Schabol, Roger, Abbé. 1758. De la culture des fraisiers. *Journal économique* (April): 248–253.

Shapin, Steven, and Schaffer, Simon. 1985. *Leviathan and the Air-Pump: Hobbes, Boyle, and the Experimental Life*. Princeton, NJ: Princeton University Press.

Stafleu, Franz. 1971. *Linnaeus and the Linnaeans: The Spreading of Their Ideas in Systematic Botany, 1735–1789*. Utrecht: International Association for Plant Taxonomy.

Staudt, G. 2003. *Les dessins d'Antoine Nicolas Duchesne pour son Histoire naturelle des fraisiers*. Paris: MNHN, Ciref.

Tassy, Pascal. 1991. *L'arbre à remonter le temps*. Paris: Christian Bourgois Editeur.

Williams, Roger L. 2001. *Botanophilia in Eighteenth-Century France: The Spirit of Enlightenment*. Dordrecht: Kluwer.

Zirkle, Conway. 1935. *The Beginnings of Plant Hybridization*. Philadelphia: The American Philosophical Society.

10 The Sheep Breeders' View of Heredity Before and After 1800

Roger J. Wood

What can be revealed from horse breeding within a century can be achieved in sheep breeding within a decade.
—Weckherlin 1846, original in German

Breeding sheep for desired characteristics has a long history. The ingenuity of generations of farmers ensured that the best of breeds would become valued objects of trade, war booty, and gifts between monarchs. By the eighteenth century when advances in selective breeding by pioneer British and Western European farmers had become associated with improved rearing conditions and carefully designed feeding experiments, progress in breed improvement was seen to accelerate. It was then that artists began to bring the evidence of novelty and transformation to a fascinated public. Meanwhile an expanding army of agricultural writers tried to explain how the various breeding successes had come about. By the mid-nineteenth century, principles of breeding had been established that would serve farmers for years to come. Attempts by then to discover patterns of relationship between generations, with the intention of predicting the results of breeding, had met with gratifying success in skilled hands although never, it must be stressed, to the point of providing a functional explanation of biological inheritance. Theory was restricted to practice-based axioms and breeding rules, and it evolved with technological advance. It is these technology-driven changes in the concept of heredity[1] that I will be reviewing.

10.1 The Background

10.1.1 "Like Begets Like"

Animal breeders had long accepted that the property of living organisms by which offspring receive the nature of their parents or ancestors must be associated, in some mysterious way, with an animal's blood, the essence of life. The idea rested on the assumption of blood being transformed into semen (seed), the basis of new life.[2] As the scripture said, each new being is "fashioned in flesh . . . , being compacted in blood of the seed."[3] By the eighteenth century, authorities were increasingly ready to postulate that the embryo was formed by the "fermentation" in the womb of seminal particles, produced by the blood of both sexes. Inheritance from both sexes fitted with the farmer's own experience. Differences between generations and within families were explained ultimately by reference to "accidents of development" caused by a variety of supposed influences. These might be fluctuations of climate, vegetation or mode of life, or even dreams and passions that deflected the generative process from its established course, that were *degenerative*.[4] Farmers recognized no greater achievement than to regulate the generative process, so as to preserve or create an ideal type. It was a result to be brought about by a combination of selective breeding, environmental management, and other aspects of good husbandry, requiring an expert knowledge of the relative influence of the blood compared with other forces for change.

10.1.2 Blood and Locality

Especially to be considered was the observed association between an animal's racial (breed) characteristics and its traditional environment. Local differences between races led to the conviction that blood was linked with locality—that is, that an animal's own nature and "external" nature were somehow interconnected. "Every soil has its own stock," wrote an Englishman after surveying the county of Berkshire.[5] The domestic animal inherited its environment just as truly as it inherited its blood. Successful breeds were those in harmony with their environment, and problems with degeneration[6] were expected when a breed was transported outside its familiar territory: "We have, at present, through time and the industry of our ancestors, various breeds; some of them adapted, though not perfectly, yet in very considerable degree, to the soil they are upon, and the purpose to which they are wanted."[7]

Transportation of sheep breeds between localities, a practice becoming more and more widespread during the period under review, allowed the theory to be tested. With experience, the intelligent farmer could establish the limits of natural environmental influence on heredity. Evaluating the connection between blood and locality was to prove a key issue in selective breeding, opening the way, step by step, to a greater understanding both of heredity and of adaptation to the outside world. An explanation was required about how external influences, interacting with the animal's own ancestral nature, might bring about continuity within a race in one circumstance but lead to changes in another. The final conclusion drawn from these "experiments" would prove to be reassuring, that when mating was controlled, adaptation (degeneration) generally took place much more slowly and gradually than the pessimists predicted.[8] Practical experience of moving breeds over long distances into new locations led to a shift of emphasis in dealing with heredity, encouraging a "harder" view of it. The earliest and most dramatic example to be documented in sheep is Jonas Ahlströmer's transportation of merinos from Spain to Sweden in 1723, moving them into a most unsuitably cold, damp climate during much of the year. Despite early losses of stock, he had the faith to build up Swedish fine-wool production into a thriving business, under royal patronage.[9] On the basis of further imports and improved husbandry, under carefully controlled conditions of housing and nutrition, the Swedish merino breed gained an early reputation.

The most direct way for breeders to obtain domestic animals of higher quality than those they already possessed was to breed exclusively from blood stock of both sexes obtained from wherever the best could be found. If this course of action proved impractical, the alternative was to introduce superior blood by importing only males, to be crossed with native females and to their progeny for several successive generations. With each generation of such crossing, the proportion of superior ("noble") blood would increase: one-half in the first generation, three-fourths in the second, 7/8 in the third, 15/16 (93.75 per cent) in the fourth, rising to more than 99 per cent in the seventh. This technique, known as "grading" or "grading up," was based on a *proportionate* concept of heredity, as a fraction of the blood. It probably had a long history but gained special significance in Sweden, where hybrid vigor seems to have proved particularly advantageous. As time passed, Swedish merinos, infused with a degree of local blood, grew larger and stronger than the original Spanish imports but still yielded wool of excellent quality, equalling or even excelling that from Spain itself.[10]

Merino experience from Sweden inspired other countries to follow Ahlströmer's example, leading to the development of specialized merino varieties in Prussia, Saxony, Austria/Hungary, and France, each with its own characteristics. For grading Swedish sheep, Ahlströmer had recommended that none but merino rams should be used for three generations, advice to be repeated many times by writers of various nationalities in the coming years. Eventually, however, the number of generations the experts considered necessary to make a suitable transformation rose from three to four and then to five.[11] Even after five generations it proved necessary to practice selective breeding if stability was to be achieved and the risk of degeneration avoided. Experience thus gained convinced J. H. Fink (1730–1807), the King of Prussia's head bailiff, that fineness of wool was attributable more to breeding than to environmental influences, an opinion he stated in a published letter to Sir John Sinclair, president of the Board of Agriculture in London,[12] which was endorsed by the English farmer George

Culley in the fourth edition of his book *Observations on Livestock*.[13] The supremacy of the blood in controlling the characteristics of sheep was something that he and his mentor Robert Bakewell had learned to accept from their own experience. They were in full agreement that even the best ancestry was not in itself a guarantee of quality. Breeding stock always had to be selected most carefully.

10.2 Bakewell's Revelation

Under the influence of a succession of breeders, it came to be recognized that heredity could be directed by selection, trait by trait. Taking the lead for a time was Robert Bakewell (1725–1795), who farmed at Dishley, near Loughborough in Leicestershire. By precision in defining an animal's traits, he was able to regulate the hereditary process in a manner that made selective breeding more predictable, allowing him to demonstrate publicly how rapidly a transformation could take place. Among his achievements was the creation of a new breed of sheep, known as the Dishley or "New Leicester," derived from rare individuals ("accidental varieties") able to transmit the particular characteristics he desired (figure 10.1).

10.2.1 Male and Female

No successful breeder from Bakewell's time onward was in doubt that both sexes contributed to inheritance,[14] but less certainty was expressed about whether the two parents contributed equally,[15] it being easier to evaluate a good male than a good female because of the much larger number of progeny he could produce. The livestock expert W. Youatt spoke for many, however, when he stipulated, with direct reference to sheep, that "no certain degree of excellence can be attained unless the female possesses an equal degree of blood with the male."[16] Bakewell himself believed that "by bringing together males and females possessing the same valuable properties, he should ensure their presence in their offspring, probably to an increased degree, they being inherited from both parents."[17] He was among those who had established by experience that both sexes contributed to inheritance of the same characters.[18]

10.2.2 Individual Trait Selection, Inbreeding, and Progeny Testing

Bakewell had startled the farming world in the 1780s with the inflated prices he could demand for his breeding stock. What made his highly selected Dishley sheep so prized was their rapid growth rate, associated with a unique body shape, designed to minimize unprofitable parts of the carcass and maximize joints of meat that would gain the highest prices. The difference in their appearance from traditional breeds was

Figure 10.1
A Dishley ram drawn by Samuel Howitt (1756–1822). From Rev. W. Bingley, *Memoirs of British Quadrupeds* (1809).

remarked on with astonishment. It seemed that he worked like an artist who created an image in his mind and then transformed it into reality.[19] This startling achievement depended on more than a proportionate, "holistic" view of heredity such as underpinned the principle of grading. His was a "trait-based" approach, one in which defined characteristics could be followed in families, sometimes independently, sometimes in association with one another.[20] By taking practical advantage of hereditary interactions—for example, between small leg bones and rapid maturation[21]—he could push his selection programs forward without delay.

As a natural accompaniment to selective breeding for growth and body shape, Bakewell carried out feeding experiments. After comparing different breeds side by side, he discovered the best regime to maximize productivity at minimum cost. Because he was selecting intensively he was also inbreeding very closely, father to daughter, mother to son, and brother to sister. His technique, given the name "breeding in-and-in,"[22] had to be used with great care, for it ran the risk of concentrating deleterious traits as well as favorable ones. Just as the latter were to be preserved, the former had to be rigorously excluded. Bakewell's attitude on the matter was well known; only animals in robust good health, as well as of the highest quality, judged by touch as well as eye, were considered suitable for breeding at this level. To minimize the potentially injurious effects of breeding in-and-in, Bakewell developed an early form of progeny testing in cooperation with fellow farmers who formed the Dishley Society in 1783. The original membership extended over seventeen counties, from Devon to the Scottish border.[23] Through the interchange of stock among the Society's members, Bakewell and his friends were able to test their rams against a much wider variety of ewes than could possibly have been kept on their individual farms.[24] Bakewell's skill in defining economically significant traits, and employing a broad raft of techniques to enhance them, above all making maximum use of inbreeding and progeny testing, as well as ensuring optimum housing and diet, earned him the title of "Prince of Breeders."[25] His Dishley sheep stock was used for crossing into almost every British breed, and many abroad, to introduce its quick-fattening quality.[26]

10.2.3 Heredity as a Changing Concept

When fellow farmers were asked about Bakewell's breeding philosophy, they would frequently stress his great faith in the old adage that "like produces like."[27] By observing successive generations of his breeding stock under controlled conditions, he determined which traits were most strongly inherited. The result, as reported by William Marshall, a prominent agricultural writer, was that "a number of traits were found, in some considerable degree at least, to be hereditary."[28] Bakewell's confidence in heredity extended to less readily defined traits, including resistance to bad weather, tolerance of poor food ("hard fare"), and even propensities to certain disorders. Visible traits of no economic value in themselves ("nicks") could be used as inherited markers of useful properties and propensities.[29]

10.2.4 Population Thinking

Bakewell's actions tell us that his aim was to produce a breed in which the same characteristic traits would appear in every individual. To be "well bred" from good parentage was not enough. Breeding stock had to be selected on the basis of "form" (i.e., the expression of desired traits under prevailing conditions), as well as "blood" (i.e., ancestry). Selection had to be applied in every generation, supported by consistent husbandry practice, aimed at "concentrating" the desired traits within the breed and avoiding detrimental ones.[30] Furthermore, not only was the form and ancestry of an animal of concern to Bakewell but also, even more importantly, the form of its already-existing progeny. It was on this broadest possible basis that every animal used for breeding had to be selected most carefully.[31] Successful breeders like Bakewell had come to appreciate the significance of the whole flock as the ultimate target for improvement. Their attitude toward breeding exemplifies what Ernst Mayr has defined as "population thinking." He believes that breeders were the first group to gain an understanding of the concept.[32] Bakewell's own attitude was revealed in a conversation with Young, who quoted him as saying: "The merit of a breed cannot be supposed to depend on a few individuals of singular beauty: it is the larger number that must stamp their character on the whole mass: if the breed, by means of that greater number, is not able to establish itself, most assuredly it cannot be established by a few specimens."[33]

To "establish itself," in Bakewell's terms, meant that the breed not only had to compete successfully with the one it replaced locally but ideally also with breeds elsewhere. In each locality, as Marshall pointed out, a breed had to be adapted to "the farmer's *climature, soil* and *system of management.*" Otherwise, as he stated, "if we reason from analogy, the improver appears to be setting himself up against nature, a powerful opponent."[34] How was opposition to nature to be avoided? Marshall had no doubt what Bakewell and his friends would answer: by carefully selecting all breeding stock against a particular environmental background. Only then could the highest levels of breeding excellence be achieved. "By this process . . . , the term *blood* became distinctively applied. When reference could be made to a number of ancestors of distinguished excellence the term *blood* was admitted."[35]

Dishley Society membership ensured that this was possible, and Bakewell was their model, a man with plenty to say but wishing to be judged above all by his actions: "We are best of all led to Men's principles by what they do." This quotation can be found in one of a series of letters to George Culley,[36] dated December 8, 1786. Letters to Culley, Arthur Young, and one or two others are almost all that is known from the pen of Bakewell himself. Fortunately there were many reporters ready to write about him, recording both his words and actions, whose accounts can be checked against one another to reveal his methods, and to provide an interesting picture of his lifestyle and personal philosophy. A major source of information on his methods is Culley's book *Observations on Livestock*, in its four editions (1786–1807). Culley revealed in the introduction that he did not feel he was the best person to write the book, being convinced that that there was someone else more qualified to take it on. He wrote of "one whom it is not necessary here to name, for whose abilities I have the highest respect, whose whole life has been employed in breeding and improving stock, and has carried it to a very great perfection." It was only after being unsuccessful in persuading Bakewell to do it that he reluctantly took on the task himself, producing a most influential book, a work of which Bakewell strongly approved, as he wrote in one of his letters to Culley (April 11, 1786).

10.3 After Bakewell

10.3.1 Races and Species

When breeders, following Bakewell's example, began to compare races/varieties native to different counties or countries, side by side at a single location but isolated reproductively, they noted the capacity of stock to remain distinct for as many generations as there was time to observe them. It seemed obvious then to conclude that any environmental effect on heredity must be slow and gradual, which raised the significant question of how races/varieties, so different from one another, had ever come about. One possibility was the persistent influence of human selection over many centuries; another rested on the idea that some races/varieties were natural subdivisions of species, created separately and, for this reason, remaining separate.[37] William Marshall felt bound to consider the problem: "Whether in the Animal Kingdom VARIETIES are altogether *accidental* or *artificial,* or whether they are not, or have been originally *natural subdivisions* of SPECIES, would, with respect to DOMESTIC ANIMALS be now difficult to determine."[38]

In his own opinion, varieties arose "by climature [sic], soil, accident or art, under the guidance of reason or fashion, during a succession of centuries." In thus supporting the idea of a species being a potentially "splitable" entity, he was reacting against the widely held opinion by naturalists that members of a species differed only in superficial, nonessential characters.[39]

Dr. James Anderson, a Scottish farmer and scholar, a friend of Culley's who had business dealings with Bakewell,[40] believed it useful to make a distinction between *accidental varieties* that were "discriminated by certain minor qualities," and *permanent varieties* that could be very different from one another and stable, except when altered by "intermixture with other breeds." Anderson did not accept that any one permanent variety could be "transmuted" into another "without any intermixture of blood, purely by a change of circumstances only," as the "celebrated Buffon" claimed. Nor did he see such a transmutation coming about by controlled breeding from particular individuals,[41] an opinion seemingly brought into question by Sir John Saunders Sebright in a published letter to Sinclair: "The alterations that can be made to any breed of animals by selection, can hardly be conceived by those who have not paid some attention to this subject; they attribute every improvement to a cross, when it is merely the effect of judicious selection."[42]

Sebright draws no boundary between "improvement" and "transmutation," although he recognizes the "degenerative" effect of extreme selection when associated with an "injudicious" degree of inbreeding. Expert opinion on the complex matter of selection was divided.

10.3.2 Inbreeding and Prepotency

Farmers recalled Bakewell's own attitude to inbreeding, by quoting his supposed dictum that "inbreeding gives prepotency and refinement." The term *prepotency* described the ability of particular individuals to pass on their traits with extra certainty. It was seen to result from a "concentration of the blood" that allowed selective improvement, also known as refinement, to be carried out more rapidly. The concept of blood in a hereditary sense demanded an explanation from science. In a pamphlet published in 1812 the surgeon John Hunt, a supporter of the "Dishley System," deplored the idea of blood being the actual vehicle of heredity as "far exceeding the laws of nature." He agreed with the merino breeder, Dr. Parry of Bath, who had written earlier that "the word blood is nothing more than an abstract term expressive of certain external and visible forms which from experiment we infer to be separately connected with those excellencies which we most covet."[43] Similar sentiments were expressed by other breeding experts at the time. All that could be said for sure was that blood, in the hereditary sense, was divided between the parents in proportion. Thus a son mated to his mother would cause her to produce lambs with six parts of herself and only two of his father.[44]

By 1800 two traditions in animal breeding had thus been established: (1) the grading technique with its proportionate (blood fraction) view of heredity; (2) selective breeding from individuals (accidental varieties) showing desired traits, isolated from the progeny of controlled matings. On the basis of experience, breeders were guided to success by a number of breeding maxims, but even the deepest thinkers could suggest no causal explanation for heredity. Anderson tried to face the issue in a published essay on the nature of varieties but had to confess it to be a mystery.[45] For practical purposes, the priority for a breeder in the Bakewell tradition was to possess breeding stock with a recognized capacity to transmit desirable traits, as surely and certainly as possible, through either sex. The appearance of accidental varieties, "arising from a cause which is entirely unknown to us," as Anderson noted, could not be predicted.

10.3.3 The Bakewell Approach to Breeding Is Evaluated

A number of breeders in Britain benefited from Bakewell's example, although by no means everyone who wished to. For those with only a shadow of his judgment, and without the equivalent of a Dishley Society to evaluate their breeding stock, inbreeding could bring drastic penalties. As more and more breeders jumped onto the inbreeding bandwagon, British experts like Sebright and Sinclair recommended caution.[46] Too many breeders were finding that close inbreeding "impairs the constitution and affects the procreative powers."[47] The care Bakewell had taken to avoid the various weaknesses and disorders, by evaluating rams and exchanging stock with other Dishley breeders, was widely recognized.

Bakewell's original and unusually systematic approach to breeding created interest beyond the British Isles. On the other side of the channel his combination of techniques was being taken seriously in relation to fine-wool production. By the end of the century the technical term "breeding in-and-in" had entered the French, Austrian, and German breeding literature, either in English or in translation.[48] In a German textbook on animal improvement, published in 1785, the author, Christian Baumann (1739–1803), a Cistercian monk, wrote in favorable terms about a fellow Bavarian who was evaluating his rams in conscious imitation of the famous Englishman. He was referring to the "talented economist" von Bori (Borie) who farmed at Neuhaus, close to Bad Neustadt: "Mr Bori hired his rams to his neighbours to evaluate their offspring, following the example of Bakewell."[49]

Toward the end of his life, Baumann moved to Moravia, where he produced a new textbook on agriculture published in 1803.[50] Describing fine-wool production in Central Europe, he picked out for special mention the estate of Ferdinand Geisslern at Hoštice in the Hapsberg province of Moravia.[51] Baron Geisslern made a practice of exchanging rams with neighboring landowners, mating ram with ewe in carefully controlled matings.[52] By that time Geisslern had attracted sufficient notoriety to be called the "Moravian Bakewell."[53]

10.4 Sheep Breeding in Moravia

An interesting amount is known about sheep breeding after 1814 from the proceedings of an association for sheep breeders, created in Brno, Moravia. "The Association of Friends, Experts and Supporters of Sheep Breeding" (SBA) was set up to discover "incontrovertible principles to ensure favourable results in sheep improvement."[54] A regular exchange of rams between stockowners was a feature of the SBA from the beginning. Like Bakewell and his collaborators, the best breeders were ready to adopt "population thinking" when dealing with selected stock, female as well as male, in order to produce the highly uniform product the cloth manufacturer demanded. In a practical sheep-breeding manual written by an SBA member, we read: "No animal should be perceptively better or worse than the others, particularly in wool. In this way one should work from the beginning of the improvement programme."[55]

10.4.1 The Brno Inbreeding Debate

By 1817 an area of disagreement was developing within the SBA about the value of inbreeding as a means of "fixing" traits, as a route to more constant inheritance. Despite reassurances by the SBA president, E. Bartenstein,[56] doubt remained among some members that pairing of nearest blood relatives must disrupt "the principal plasma of the animal's organization" (*Hauptplasma der thierischen Organisation*).[57] In support of Bartenstein, it was argued that the potential for weakness was present within a race even before inbreeding took place.[58] A similar divergence of opinion and practice had been taking place in England.[59] A Hungarian member of the SBA, E. Festetics, summarized his own fifteen years of experience of inbreeding under the heading "genetic laws of nature" (*genetische Gesetze der Natur*) in four parts.[60] Here he (1) associated pure "noble" breeding with good health, (2) appreciated that inherited traits could be recessive for one or more generations, (3) recognized the extent of variation, even within seemingly pure breeds, and (4) stipulated that the precondition for applying inbreeding safely must be scrupulous selection of stock animals. The final point was acknowledged to be the most critical.

Reflecting the successes achieved by Geisslern and his followers, the SBA continued to recommend close inbreeding for the totality of its existence (1814–1845). Constancy of inheritance could only be maintained by matching the best rams to their close female relatives, each ram forming "a sire's family."[61] Even so, as Bakewell's followers had shown earlier, inbreeding had its dangers. Furthermore, racial stability could never be absolute. The Moravian experience confirmed that selective breeding was required even in the "pure noble race," obtained directly from Spain. Even an apparently "genetically fixed race" (*genetisch befestigte Rasse*) was expected to degenerate. It could not be doubted that eventually a chance deviation would appear and multiply, and would have to be removed to regain the original "principal racial form" (*Hauptsgeschlechtsform*).[62] Hence the firm rejection by the SBA of the popular idea of inherent racial constancy, proposed by Justinus (1815) with reference to horses, that the noble race stays forever noble when the purity of the blood is maintained. The concept was simply not justified by the experience of Moravian sheep breeders, who recognized the need for persistent selective breeding.

10.4.2 Improvement of the Pure Race

Sometimes a chance deviation could be advantageous, by providing an opportunity to make an improvement to a race, to add to its economic value. Selection would then be directed toward making such favorable "accidental varieties" endure (*zufällige Varietäten bleiben zu machen*).[63] It was a strategy that had its roots with Bakewell and his followers in the British Isles. Marshall had referred to Bakewell as having used this technique as the basis of his Dishley breed of longhorn cattle: "solicitously seizing the superior accidental varieties produced."[64] The possibility of exploiting favorable variants, of which there appeared to be no lack, implied that there could be no fixed limit to the improvement of a race. Even in the case of original merinos from Spain, breeders were taking into account their "predispositions for a higher perfection" (*die Anlagen für einer hoher Vollkommenheit*).[65] To increase the efficiency of selective breeding, the SBA recommended that all animals in a flock should be numbered, with all parents and their progeny recorded, as a standard procedure.[66] Improvements were becoming evident all over Europe where different forms of so-called pure merino stock were established, varying in size, degree of skin wrinkling, and length of staple.

10.4.3 "Inner and Outer Organisation"

By the late 1830s the SBA debates were partly taken up with considerations of breeding theory, to which some interesting contributions were made by Cyrill Napp, abbot of the Augustinian monastery in Brno, which Mendel would join in 1843. At the 1836 meeting Napp commented that, when pure breeding was the aim, it was important not only that parents should be of the same *type*, but that they should correspond in "both inner and outer organisation":

> Napp asserts that, according to his view, heredity of characteristics from the "engenderer" (*Erzeuger*) to the "engendered" (*Erzeugten*) consists above all in the "mutual elective affinity" (*gegenseitige Wahlverwandschaft*)[67] of paired animals. As a result of this, a ram chosen for the ewe should correspond to it in both inner and outer organisation. This process deserves to be the subject of a serious physiological study.[68]

Napp's concept of "inner organisation" brings to mind a similar idea expressed by B. Petri, who attached particular significance to selection from animal crosses—that is, from blending (*Vermischung*), "where the inner cohesion of the external formation of the individual in its different varieties has to remain hidden from the eye."[69] A matter of key importance to the breeder was whether an animal's inner organization/cohesion could be reliably determined by examining its outward appearance. Experience showed that selection directed toward producing a pure breeding strain made this more likely, but what happened in crosses was still a mystery. As Napp pointed out, the relationship between inner and outer organization was a problem requiring "serious physiological study."

10.4.4 Inheritance Capacity

Among the mysteries of heredity to be investigated was the variability observed in the capacity of individual animals to transmit their traits, discussed in English literature in terms of their "prepotency." The phenomenon was discussed in the SBA with reference to an animal's "inheritance capacity" (*Vererbungsfähigkeit*),[70] which was found to differ among individuals of both sexes and also with respect to the particular traits and combinations of traits. Experience showed that its value in any particular case depended on purity of stock, being enhanced by selective improvement,[71] which implied enhancement by inbreeding.[72] The SBA members, led by Bartenstein, were bringing up to date Bakewell's supposed saying that "inbreeding gives prepotency and refinement." Napp's conviction about the significance of inheritance capacity and its physiological basis led him, at the 1837 meeting, to ask the question "what is inherited and how?,"[73] to which no answer was then forthcoming.

10.4.5 The Physiological Basis of Heredity

By 1843 when Mendel entered the monastery, it was accepted in Moravia that heredity was brought about by the meeting of two parental "germs," located in the egg of the female and the semen or pollen of the male.[74] The subject was still wide open for investigation if a way could be found to do this investigation. It continued to puzzle Napp that "nothing certain can be said in advance as to why production through artificial fertilization remains a lengthy, troublesome and random affair." He remarked on a problem that all breeders faced when attempting to produce new varieties by artificial fertilization, the significance to be attached to chance.[75] The chance element demanded to be understood and, if possible, controlled. Napp's statement of 1836 that the process of heredity "deserves to be the subject of a serious physiological study"[76] has obvious significance in light of further events, instigated by his protégé Mendel. It seemed that although heredity could be manipulated by selective breeding, following certain defined procedures, there was no escape from the element of chance invisible in the germ.

Acknowledgments

I am greatly indebted to my friend and collaborator Vítezlav Orel for his enthusiasm, encouragement, and wise counsel during the past thirty years, and for his generosity in providing me with photocopies of early published material from Moravia, unobtainable in England. This chapter was presented in a fuller form at the MPI Workshop "A Cultural History of Heredity II: 18th and 19th Centuries," 2003, 21–46.

Notes

1. The noun *heredity* (*hérédité*) was little used in a biological sense before the mid-nineteenth century. Its use here acknowledges that the "conceptual space" it would later occupy was already recognized by adjectival and verbal forms. See López-Beltrán 2003, 7. Traits that distinguished one variety from another were referred to as "hereditary," and thus "inherited." See Marshall 1790, vol. 1, 419.

2. Wood and Orel 2001, 47.

3. *Wisdom of Solomon* 7, vol. 1, 2.

4. For the basis of this idea see Roger 1997, 460f.

5. Pearce 1794, 46.

6. In modern terms, immediate "degeneration" is explicable in terms of genotype-environment interaction, and subsequent "degeneration" as a result of natural selection or genetic drift.

7. Youatt 1837, 494.

8. Anderson 1800a, 172–174; 1800b, 325–326; 1800c, 170–171.

9. Schulzenheim 1797, 314–317; Wood and Orel 2001, 126–127.

10. Schulzenheim 1797, 315–316; Lasteyrie 1802.

11. Hastfer 1752; Daubenton 1782; Stumpf 1785; Fink 1799, 48; Parry 1806, 339; Tessier 1811.

12. Fink 1797, 278; 1799, 54.

13. Culley 1807, 260.

14. Berry 1826; Lincolnshire Grazier 1833, 171; Sinclair [1817] 1832, 97–98.

15. Boswell [1825] 1829; Youatt 1834, 523–524; Sinclair [1817] 1832, 97.

16. Lincolnshire Grazier 1833, 238.

17. Berry 1826, quoted by Youatt 1834, 522.

18. Marshall 1790, vol. 1, 481.

19. Pitt 1809, 249; Somerville 1806; Wood and Orel 2001, 69. See White, chapter 16, this volume, for his comments on the reactions of later writers, including Charles Darwin, to the claim of the breeder as an artist.

20. Culley 1786, 186; Marshall 1790, vol. 1, 298; Wood and Orel 2001, 6–7.

21. Young 1771, vol. 1, 111.

22. Marshall 1790, vol. 1, 300–301.

23. Pawson 1957, 75–76.

24. Pitt 1809, 256ff.; Wood and Orel 2001, 82–85.

25. Young 1791.

26. Young 1791; Trow-Smith 1959, 66–69, 269–274; Walton 1983; Wood and Orel 2001, 85–88, 145–146, 148.

27. Anon 1800; Berry 1826, quoted by Youatt 1834, 522.

28. Marshall 1790, vol. 1, 419.

29. Culley 1786, 186; Wood and Orel 2001, 78.

30. Marshall 1790, vol. 1, 464–465.

31. Lawrence [1805] 1809, 25.

32. Mayr 1971.

33. Young 1791, 570.

34. Marshall 1790, vol. 1, 464–465. The "analogy" referred to is with the natural world, in which it was believed, as expressed in British Protestantism, that the divine plan ensured perfect harmony, based on the perfection in adaptation of all structures and organic interactions. See Mayr 1982, 372; Wood and Orel 2001, 39.

35. Berry 1826, quoted by Youatt 1834, 52 (Berry's emphasis). It was understood that in the hands of another skilled breeder in another environment, the same race might appear slightly different. This is because it would have become adapted naturally to achieve compatibility with that environment, being "transformed" in order to maintain health and fertility in new conditions of nutrition and climate. By definition, however, a superior race would never lose the essential qualities that made it superior (i.e., in an economic sense), provided that selection for those qualities was maintained. See also White, chapter 16, this volume, for further reference to the idea of "essential quality" in other contexts.

36. A series of letters to Culley, written between April 1786 and August 1792, printed, with some other Bakewell letters, by Pawson (1957, 101–177).

37. Home 1776, 309; Anderson 1799, 53–55.

38. Marshall 1790, vol. 1, 462; his emphasis.

39. Mayr 1972.

40. Pawson 1957, 106.

41. Anderson 1799, 51, 62.

42. Sebright 1809, 26.

43. Parry 1806.

44. Sebright 1809.

45. Anderson 1799, 87, also 63; see also Marshall 1818, vol. 1, 43.

46. Sebright 1809; Sinclair [1817] 1832, 93–95.

47. Berry 1826, quoted in a footnote to Youatt 1834, 526; Berry 1829.

48. Fink 1799, 73; Bourde 1953, 145; Klemm and Meyer 1968, 139, 179; C. C. André 1804; Thaer 1804; Wood and Orel 2001, 140, 157, 164, 167, 212, 215, 231.

49. Baumann 1785, 217, 273.

50. Wood and Orel 2001, 157–158, 211–212.

51. Baumann 1803, 706.

52. Köcker 1809; "K in Mähren" 1811; C. C. André 1812; R. André 1816; Wood and Orel 2001, 231–232.

53. Köcker 1809.

54. R. André 1816; Wood and Orel 2001, 229f. In German, the SBA was called "Verein der Freunde, Kenner und Beförderer der Schaftzucht, zur noch höheren, gründlichen Emporhebung dieses Oekonomie-Zweiges und der darauf gegründeten, wichtigen Wollindustrie in Fabrikation und Händel."

55. R. André 1816, 37, original in German; see also Wood and Orel 2001, 201–202.

56. Bartenstein 1818.

57. Ehrenfels 1817.

58. Festetics 1819a.

59. Sebright 1809, 11.

60. Festetics 1819b; Orel and Wood 1998; Wood and Orel 2001, 237–238.

61. R. André 1816, 6–7; C. C. André 1804; Petersburg 1815; Nestler 1836; Stieber 1842, 41–44; d'Elvert 1870, vol. 2, 48–49; Wood and Orel 2001, 195–196, 244–245, 268.

62. "Irtep" 1812; Wood and Orel 2001, 224–225.

63. "Irtep" 1812; R. André 1816, 94–96; Nestler 1829; Wood and Orel 2001, 224, 243.

64. Marshall 1790, vol. 1, 383–384; see also Anderson 1799, 23, referring to variations "accidentally produced" being inherited, as well as the "general characteristics of the parental breed."

65. R. André 1816, 95; Ehrenfels 1837; Wood and Orel 2001, 257.

66. Köcker 1809; R. André 1816, 7, 35; Petersburg 1816, 113; Wood and Orel 2001, 194, 200–202, 232–233.

67. This term was originally coined in the dissertation *De attractionibus electivis* by Torben Olof Bergman in 1775, based on a study of chemical affinities, as between acids and alkalis. Used in this sense it was translated into English as "elective affinity" or "elective attraction." Later it was used in a figurative sense by Goethe, in his novel *Die Wahlverwandtschaften* (1809), to characterize human sexual attraction. His choice for the title was widely assumed to argue for a chemical origin of sexual attraction. Exactly what Napp had in mind is not evident, although he surely appreciated that parents most likely to correspond in both internal and external features would be those most closely related.

68. Teindl, Hirsch, and Lauer 1836, quoting Napp, original in German.

69. Petri 1838.

70. Teindl, Hirsch, and Lauer 1836, original in German; see also Bartenstein 1837; Bartenstein et al. 1837; Wood and Orel 2001, 246–247.

71. Bartenstein et al. 1837, 204–205.

72. Nestler 1839.

73. Bartenstein et al. 1837, 227: "*was verebt und wie?*"

74. Purkinje 1834.

75. Nestler 1841, 337, quoting Napp.

76. Teindl, Hirsch, and Lauer 1836.

References

Abbreviations

Folia Mendeliana Published annually by the Moravian Museum in Brno since 1966, issues 1–9 appeared as special volumes. Issues 10–20 were published as a supplement to *Acta Musei Moraviae*, sc. nat. with the pagination of these volumes. The part *Folia Mendeliana* was reprinted separately. *Folia Mendeliana* vol. 21 and further volumes are being published again as special volumes with their own pagination.

Hesperus *Belehrung und Unterhaltung für Bewohner des österreichisches Staates–Hesperus*, Prague.

Mittheilungen *Mittheilungen der k.k. Mährisch-Schlesischen Gesellschaft zur Beförderung des Ackerbaues, der Natur- und Landeskunde in Brünn.*

ONV *Oekonomische Neuigkeiten und Verhandlungen*, Prague.

PTB *Patriotisches Tagesblatt oder öffentliches Corrensspondenz- und Anzeiger-Blatt für sämtliche Bewohner aller kais. köng. Erbländer über wichtige, interessierende, lehrreiche oder vergnügende Gegenstände zur Beförderung des Patriotismus*, Brünn.

Anderson, J. 1799. An inquiry into the nature of that department of natural history which is called varieties. *Recreations in Natural History* 1:49–100.

———. 1800a. On the varieties of the sheep kind. *Recreations in Natural History* 2:81–94, 161–175, 241–252.

———. 1800b. On the varieties of the goat kind. *Recreations in Natural History* 2:321–329.

———. 1800c. Practical remarks on the management of the dairy. *Recreations in Natural History* 3:161–176, 241–256, 321–347, 401–419.

André, C. C. 1804. vol. 2, Cited in: K in Mähren (Köller, K), 1811, and republished by d'Elvert 1870, vol. 2, 145–152.

———. 1812. Anerbieten, Gutbesitzern auf dem kürzesten und sichersten Wege zur höchsten Veredlung ihrer Schafherden behülflich zu seyn. *ONV*: 181–183.

André, R. 1816. *Anleitung zur Veredelung des Schafviehes. Nach Grundsätzen, die sich auf Natur und Erfahrung stützen*. Prague: J.G. Calvé.

Anon (J. L.). 1800. Robert Bakewell in *The Annual Necrology for 1797–1798*, 191–207. London: R. Phillips.

Bajema, C. J. 1982. *Artificial Selection and Development of Evolutionary Theory*. Stroudsburg, PA: Hutchinson Ross Co.

Bartenstein, E. 1818. Bericht des Herrn Präfes Baron Bartensteins an die k.k. Ackerbaugesellschaft. *ONV, ausserordentliche Beilage*: 81–84.

———. 1837. Äusserungen von Bartenstein über das von dem Herrn Professor Nestler bei dem Schaf-Züchter-Verein im Jahre 1836 aufgestellte und debatirte Thema der Vererbungsfähigkeit edler Stammthiere. *Mittheilungen*: 9–10.

Bartenstein, E., F. Teindl, J. Hirsch, and C. Lauer. 1837. Protokol über die Verhandlungen bei der Schafzüchter-Versammlung in Brünn in 1837. *Mittheilungen*: 201–205, 225–231, 233–238.

Baumann, C. 1785. *Nothwendige Anstalten zur Vermehrung, Verbesserung und Verschönerung der Pferd- Rindvieh- Schaf- Geiss- und anderer Thierzuchten ohne Ausartung*. Frankfurt-Leipzig.

———. 1803. *Der Kern und das Wesentliche entdeckter Geheimnisse der Land- und Hauswirtschaft, zur bequemern Uebersicht und zum ausgebreitetern Gebrauch, mit der neunsten bewährten Versuchen und Nahrungsquellen, Liebhabern zum Handbuch gewidmet*. Brünn: F.K. Siedler.

Berry, H. 1826. Whether the breed of live stock connected with agriculture be susceptible of the greatest improvement, from the qualities conspicuous in the male or from those conspicuous in the female parent. *British Farmer's Magazine* 1:28–36.

———. 1829. Prize essay on "Whether the breed of live stock connected with agriculture be susceptible of the greatest improvement, from the qualities conspicuous in the male or from those conspicuous in the female parent?" *Transactions of the Highland Society of Scotland* 7 (new series 1): 39–42. (Shorter, revised version of Berry 1826.)

Boswell, J. [1825] 1829. Prize essay on "Whether the breed of live stock connected with agriculture be susceptible of the greatest improvement, from the qualities conspicuous in the male or from those conspicuous in the female parent?" *Transactions of the Highland Society of Scotland* 7 (new series 1): 17–39.

Bourde, A. J. 1953. *The Influence of England on the French Agronomes 1750–1789*. London: Cambridge University Press.

Culley, G. 1786. *Observations on Livestock; Containing Hints for Chusing and Improving the Best Breeds of the Most Useful Kinds of Domestic Animals*. (2nd ed., 1794.) London: G.G. & J. Robinson. (German translation of 2nd ed., Berlin: F. Maurer, 1794.)

———. 1807. *Observations on Livestock; Containing Hints for Chusing and Improving the Best Breeds of the Most Useful Kinds of Domestic Animals*. 4th ed. including appendix. London: G. Wilkie, J. Robinson, J. Walker, G. Robinson.

Daubenton, L. J. M. 1782. *Instructions pour les bergers et pour les propriétaires des troupeaux*. Paris, Dijon. (German translation published in Leipzig, 1784.)

Ehrenfels, J. M. 1817. Ueber die höhere Schafzucht in Bezug auf die bekannte Ehrenfelsische Race. Belegt mit Wollmustern, welche bei dem Herausgeber in Brünn zu sehen sind. *ONV*: 81–85, 89–94.

———. 1837. Schriftlicher Nachtrag zu den Verhandlungen der Schafzüchter-Versammlung in Brünn, am 10. Mai 1836. *Mittheilungen*: 2–4.

d'Elvert, C. 1870. *Geschichte der k.k. mähr. schles. Gesellschaft zur Beförderung des Ackerbaues, der Natur- und Landes-kunde, mit Rücksicht auf die bezüglichen Cultur-Verhältnisse Mährens und Oestrr. Schlessiens*. Brünn: M.R. Rohrer.

Festetics, E. 1819a. Erklärung des Herrn Grafen Emmerich von Festetics. *ONV, ausserordentliche Beilage*: 9–12, 18–20, 26–27.

———. 1819b. Weitere Erklärungen des Herrn Grafen Emerich Festetics über Inzucht. *ONV, ausserordentliche Beilage*: 169–170.

Fink, J. H. 1797. Answers to questions posed by Sir John Sinclair, concerning the breeding of sheep, particularly in Upper Saxony and neighbouring provinces. In A. young, ed., *Communication to the Board of Agriculture*, vol 1. (2nd ed., 1804.) London: W. Bulmer and Co.

———. 1799. *Verschiedene Schriften und Beantwortungen betreffend die Schafzucht in Deutschland und Verbesserungen der groben Wolle, aus eigener Erfahrung und Thathandlung, zusammengetragen in Frühjahr 1799*. Halle.

Hastfer, F. W. 1752. *Ütförlig och omständig underrättelse om fullgoda fä6rs ans och skjötsel, til det all männas tjänst sammanfaltad af Fried. W. Hastfer.* Stockholm: J. Merckell. (German translation published in Leipzig, 1752; French translation published in Dijon, 1756.)

Home, H., Lord Kames. 1776. *The Gentleman Farmer.* Edinburgh: W. Creech & T. Cadell.

Hunt, J. 1812. *Agricultural Memoirs or History of the Dishley System, in Answer to Sir John Saunders Sebright.* Nottingham.

Irtep (Petri). 1812. Ansichten über die Schafzucht nach Erfahrung und gesunder Theorie. *ONV*: 1–5, 9–16, 21–23, 27–28, 45–48, 60–61, 81–85, 91–92, 106–107.

Justinus, J. C. 1815. *Allgemeine Grundsätze zur Vervollkommung der Pferdezucht, anwendbar auf die übrigen Hausthierzuchten.* Wien and Trieste.

"K in Mähren" (Köller, M). 1811. Ist es nothwendig, zur Erhaltung einer edlen Schafherde stets fremde Original-Widder nachzuschaffen, und artet sie aus, wenn sich das verwandte Blut vermischet? *ONV*: 294–298.

Klemm, V., and Meyer, G. 1968. *Albrecht Daniel Thaer.* Halle: UEB Max Niemeyer.

Köcker, M. 1809. Auszüge aus Briefen des Herrn Oekonom Köcker, auf den fürst. Salmischen Herrschaft Raiz in Herbst 1808 an den Herausgeber des letzten. *Hesperus*: 277–303.

Lasteyrie, C. P. 1802. *Histoire de l'introduction des moutons à laine fine d'Espagne dans les divers états de l'Europe, et au Cap du Bonne-Espérance.* Paris. (German translation published by G. Fleischer, Leipzig, 1804; English translation published by Benjamin Thompson, London, 1810.)

Lawrence, J. [1805] 1809. *A General Treatise on Cattle, the Ox, the Sheep and the Swine. Comprehending Their Breeding, Management, Improvement and Diseases. Dedicated to the Rt. Hon. Lord Somerville.* 2nd ed. London.

Lincolnshire Grazier (William Youatt). 1833. *The Complete Grazier.* 6th ed. London: Baldwin and Craddock.

López-Beltrán, C. 2003. Heredity old and new: French physicians and l'hérédité naturelle in early 19th century. In *A Cultural History of Heredity II: 18th and 19th Centuries*, 7–19. Berlin: Max Planck Institut für Wissenschaftsgeschichte.

Marshall, W. 1790. *The Rural Economy of the Midland Counties.* 2 vols. London: G. Nicol.

———. 1818. *The Review and Abstract of the County Reports to the Board of Agriculture.* 5 vols. (Volumes published separately between 1808 and 1817; reissued in 1818.) York: T. Wilson. (Reprinted by David and Charles Newton Abbot about 1968.)

Mayr, E. 1971. Open problems of Darwin research (essay review). *Studies in the History and Philosophy of Science* 2 (3): 273–280.

———. 1972. The nature of the Darwinian revolution. *Science* 176:981–989.

———. 1982. *The Growth of Biological Though: Diversity, Evolution and Inheritance.* Cambridge, MA; Harvard University Press.

Nestler, J. K. 1829. Ueber den Einfluss der Zeugung auf die Eigenschaften der Nachcommen. *Mittheilungen*: 369–373, 377–380, 394–398, 401–404.

———. 1836. Ueber die Andeutung zur Veredelung der Schafe, von Herrn Rudolf v. Löwenfeld. *Mittheilungen*: 153–155, 163–164, 173–176, 185–189, 205–208.

———. 1839. Ueber Innzucht. *Mittheilungen*: 121–128.

———. 1841. *Amts-Bericht des Vorstandes über die vierte, zu Brünn von 20. bis 28. September 1840 abgehaltene Versammlung der deutschen Land- und Forstwirthe.* Olmütz: A. Skarnitze.

Orel, V., and Wood, R. J. 1998. Empirical genetic laws published in Brno before Mendel was born. *Journal of Heredity* 89:79–82.

———. 2000. Essence and origin of Mendel's discovery. Comptes Rendus de l'Académie des Sciences *Paris/Life Sciences* 323:1037–1041.

Parry, C. H. 1806. An essay on the nature, produce, origin and extension of the Merino breed of sheep. *Communications to the Board of Agriculture* 5, Part 1, (XVIII): 337–541.

Pawson, H. C. 1957. *Robert Bakewell, Pioneer Livestock Breeder.* London: Crosby Lockwood & Son.

Pearce, W. 1794. *General View of the Agriculture of the Country of Berkshire.* Drawn up for consideration by the Board of Agriculture and Internal Improvement, London. London: Nicol.

Petersburg, J. 1815. Veredlung des Schafviehs in der Blutverwandschaft betreffend. *ONV*: 1–4.

———. 1816. Äusserung des Repräsentanten für den Olmützer-Kreis Herrn Wirtschaftsraths Petersburg über die acht ausgestelleten Haupt-punkte. *ONV*: 113–115.

Petri, G. 1838. Ueber die Inzucht. *Versammlung der Naturforscher und Aerzte zu Prag vom 18. bis 26. September 1837. Sektion für Landwirtschaft, Pomologie, Technologie und Mechanik.*

Pitt, W. 1809. *General View of the Agriculture of the County of Leicester.* Drawn up for consideration by the Board of Agriculture and Internal Improvement, London. London: Nicol.

Purkinje, J. E. 1834. Erzeugung (generatio, genesis, procreatio). *Encyclopedisches Wörterbuch der medicinischen Wissenschaften, Berlin* 11:515–549.

Roger, J. [1963, in French] 1997. *The Life Sciences in 18th Century French Thought.* Ed. K. R. Benson. Stanford, CA: Stanford University Press.

Schulzenheim, Baron D. Schulz de. 1797. Observations on sheep, particularly those of Sweden. *Communications to the Board of Agriculture* 1:306–3204. (2nd ed., 1804.)

Sebright, J.S. 1809. *The Art of Improving the Breeds of Domestic Animals.* London: John Harding. (Reprinted in Bajema 1982, 93–122.)

Sinclair, J. [1817] 1832. *The Code of Agriculture*. 5th ed. London: Sherwood, Gilbert & Piper. (The 1st ed., 1817, was reviewed in *Möglin Annalen*, the review being reproduced in *ONV* 1820:201.)

Somerville, J. 1803. *Facts and Observations Relative to Sheep, Wools, Ploughs and Oxen: In Which the Importance of Improving the Short Wooled Breeds by a Mixture of the Merino Blood Is Deduced from Actual Practice. Together with Some Remarks on the Advantages, Which Have Been Derived from the Use of Salt*. London: W. Miller. (2nd ed., 1806; 3rd ed., 1809.)

Stieber, F. 1842. In welchem Alter und unter welchen Lebensverhältnissen zeigt sich die Vererbung des Schafbockes und der Schafmutter am kräftigsten und sichersten? *Mittheilungen*: 41–44.

Stumpf, J. G. 1785. *Versuch einer pragmatischen Geschichte der Schäfereien in Spanien, und der Spanischen in Sachsen, Anhalt-Dessau etc*. Leipzig.

Teindl, F. J., J. Hirsch, and J. G. Lauer. 1836. Protokol über die Verhandlungen bei der Schafzüchter-Versammlung in Brünn am 9. und 10. Mai 1836. *Mittheilungen*: 303–309, 311–317.

Tessier, H. A. 1811. *Instruction sur les bêtes à laine et particulièrement sur la race des mérinos*. 2nd ed., augmented. Paris. (German translation: *Über die Schafzucht, insbesondere über die Race der Merinos*. Trans. W. Witte. Berlin, 1811.)

Thaer, A. 1804. *Einleitung zur Kenntnisse der englischen Landwirtschaft*. Vol. 3. Hannover: Gebrüder Halm. (The last of three volumes on English agriculture, 1798–1804.)

Trow-Smith, R. 1959. *A History of British Livestock Husbandry 1700–1900*. London: Routledge & Kegan Paul.

Walton, J. R. 1983. The diffusion of improved sheep breeds in eighteenth and nineteenth century Oxfordshire. *Journal of historical geography* 9:115–155.

Weckherlin, A. von. 1846. *Die landwirtschaftliche Thierproduktion*. Vol. 1. Stuttgart and Tübingen: J. G. Gotta. (Later editions in 1851, 1857, 1865.)

Wilson, J. 1912. *The Principles of Stock Breeding*. London: Vinton & Co. Ltd.

Winstanley, W. 1679. *The Countryman's Guide*. London.

Wood, R. J., and V. Orel. 2001. *Genetic Prehistory in Selective Breeding: A Prelude to Mendel*. Oxford: Oxford University Press.

Youatt, W. 1834. *Cattle, Their Breeds, Management and Diseases*. London: Robert Baldwin.

———. 1837. *Sheep: Their Breeds, Management and Diseases to Which Is Added the Mountain Shepherd's Manual*. London: Baldwin and Cradock.

Young, A. 1771. *The Farmer's Tour through the East of England*. London.

———. ("The Editor"). 1791. A month's tour to Northamptonshire, Leicestershire, etc. *Annals of Agriculture* 16:480–607.

IV Theories of Generation and Evolution

11 Speculation and Experiment in Enlightenment Life Sciences

Mary Terrall

In his work on medical discourse about hereditary diseases, Carlos López-Beltrán has argued convincingly that heredity did not emerge as a coherent theoretical concept until the 1830s. He has also pointed out, however, that evidence for the hereditary transmission of characteristics was a crucial part of eighteenth-century debates about generation.[1] Well before the era of biology, and the formulation of the modern concept of heredity, naturalists and philosophers drew on hereditary phenomena to investigate the nature of life through the problem of generation. Their practice was eclectic, to be sure, but many of them were struggling to define methods for understanding the organic world, even as they were arguing about competing theoretical explanations for organization and transmission of traits. The formation of the individual organism and the forces behind organization made up a core problematic for the investigations of living beings spread across the subject areas of natural history, medicine, botany, physiology, and philosophy. The concerns of eighteenth-century writers on these subjects were rather different in focus and nuance than those of nineteenth-century biologists. This is perhaps an obvious point, but it bears emphasis because it reminds us that reading the earlier period in light of the later may obscure debates and insights rather than illuminating them. Experiment and speculation (or conjecture, in contemporary parlance) combined in investigations of the formation of organized beings; hereditary phenomena were observed and manipulated to support this project. Breeding experiments and reflections on hybrids made significant contributions to this venture. The practice of breeding animals and collecting evidence from human families played a positive part in articulating the role of the two sexes in reproduction, as well as in theories about the forces driving the process.

Regardless of the absence of a principle or law of heredity in this period, the transmission of particular traits across generations entered into a wide range of discussions about the nature and organization of living matter in the Enlightenment. Since the scientific community was not yet strictly parsed by discipline or profession, it is not surprising that these discussions easily crossed lines that later would become less permeable. In addition to the writings of physicians on hereditary diseases, phenomena from plant and animal breeding and from observations of human families made their way into works of various genres that addressed the vexed question of generation. Precisely because biology was not yet constituted as a discipline, this topic was open to interpretation, speculation, and experimentation by philosophically minded materialists as well as physicians, anatomists, and naturalists. Inheritance of traits, whether abnormal or not, was one kind of evidence brought to bear on the fraught topic of generation. Indeed, many writers mingled empirical evidence of various sorts with their conjectures about unseen forces and hidden mysteries. Ideas and experiments related to the passing of traits from parent to offspring and the relative roles of the two sexes thus belong in a cultural context characterized by a shifting and contentious discourse about the nature of life.

The philosophers and naturalists and physicians who proposed and defended theories of generation were equally concerned about the methods or modes of reasoning appropriate for investigating the mysterious world inaccessible to the senses, where organization or life begins. The *Encyclopédie* article on generation, for example, first describes the mechanics of copulation, and then proceeds to reflect that the really interesting part of the process is not mechanical, but "physical." Mechanism implies intelligibility; that is the virtue of mechanical explanations, whether or not they correspond to reality. In this context, the physical occupies a more fundamental level than the mechanical, and "nature employs the most secret means, the least available to the senses, to put fertilization into operation." Ironically, the most sensual of phenomena masks a process, the conception and formation of the fetus, beyond the reach of the senses:

> This mystery has always excited the curiosity of physicists [*physiciens*] and has led them to conduct so many investigations in order to penetrate it, so many experiments in an effort to take nature in the act; . . . they have imagined so many different systems, which have destroyed each other in succession, . . . without shedding more light on the subject. On the contrary, it seems that the veil behind which nature hides herself is essentially impenetrable to the eyes of the subtlest mind, and that the cause of the formation of animals must be ranked among first causes, like that of motion and gravity, of which we will never be able to know anything but the effects.[2]

If this is the case, what hope is there for a physical science of life? We find the same fascination with the ineffable in Diderot, who referred to the ordered succession of generations by which species preserve themselves as "the greatest marvel." He went on to rhapsodize, "The machine is finished, and the hours strike under the eye of the clockmaker. But among the sequences of the mechanism, we must admit that this faculty of animals and vegetables to produce their kind, . . . this procreative power that operates perpetually . . . is for us . . . a mystery whose depths it seems we will not be allowed to sound."[3] Here Diderot articulates the tension between predictable clockwork mechanism and the "procreative power" essential to life. The sense of mystery, and the recalcitrance of the problem, pervaded investigations of generation and hereditary phenomena in this period. If we are to understand the efforts of eighteenth-century thinkers to explore this mystery, we need to keep this in view. Otherwise we risk reading these writers as simply groping toward nineteenth-century biology, an approach that ignores the complexity of their efforts and the meanings attached to them by their readers.

The suspicion that the "first causes" of animal organization might remain beyond the reach of human understanding did not keep people from trying to sound the depths of the mystery. By the mid-eighteenth century, whenever the question of how to understand the generation of living forms came up, it brought along related philosophical, theological, and methodological freight. Empirical evidence came from anatomical observations going back to Harvey, Malpighi, Leeuwenhoek, and others working in the previous century. By the time the volumes of the *Encyclopédie* started to appear, in 1751, the problem of generation had attracted the attention of various medical men, academic anatomists, experimenters, naturalists, and philosophers. Some were willing, even eager, to peer behind the veil of the visible to speculate about fundamental forces and principles; others recoiled with something like repugnance. The interplay between eager speculation and prudent restraint characterized this discursive terrain, and gave an edge of danger, a whiff of suspicion, to theories of generation that went beyond the mechanical. I look here at competing theories and interpretations in the context of related debates about method, and especially about how speculation was interconnected with experiment and observation of organisms. Defining a science of life—setting parameters for how to go about investigating organic phenomena like generation—was inseparable from questions about the nature of life itself, and this inevitably impinged upon claims with moral or religious overtones. The investigation of inherited traits also entailed reflection on the properties of matter, not to mention the role of God in the universe. In the spectrum of discourse about generation, and the boundary between living and inert, we find conflict not just about theories, or interpretations of experiments, but about what is thinkable, what is dangerous, what is threatening or liberating.

11.1 Theories of Generation

Debates about animal generation centered on the fundamental opposition between preexistence (sometimes called preformation) and epigenesis (or development).[4] Preexistence theories flourished in the aftermath of late seventeenth-century microscopy, which revealed intricate organization in the reproductive organs of many kinds of animals. Though hardly rigorous, this kind of evidence suggested the possibility of fully organized creatures in the germ; opinions differed as to the location of the preformed being in egg or sperm. The organism, or the form of the future organism, was assumed to reside in submicroscopic particles, only just beyond the range of microscope-enhanced vision. The origin of organization was thus referred back to God, and therefore not open to scientific investigation. The "development" of an individual was nothing more than the mechanical unfolding of a form already organized. This mechanism was the butt of Diderot's objections, quoted above, to reducing life to the mechanical.

Any alternative to preexistence had to assign a cause (whether physical or metaphysical) for the organization of material particles into a functioning animal. The structure, and the life of the individual, emerge over time from the "chaos" of undifferentiated matter. Epigenesis, then, implied a successive process of organization following conception. Most versions of epigenesis accounted for this successive organization through properties or forces associated with elements of organic matter drawn from the bodies of both parents. Both parents contributed symmetrically to the offspring, which then formed from heterogeneous mixtures of matter, rather than unfolding or expanding according to a preordained plan. The problem was how to make such a nonmechanical process intelligible. Recent developments in the study of gravity, electricity, and chemistry (especially the mapping of the relative strength of affinities) made it conceivable to associate forces with matter, and these analogies were exploited regularly by proponents of epigenesis.

In spite of the obvious conceptual difficulties (all future organisms had to be present either in Eve or in Adam, or else seeds have to be created directly by God in each generation), preexistence was compelling because it avoided self-acting matter or spontaneous generation. Order had been imposed by God at the creation, and that was that. Of course, some people had no objections to active matter or to what was sometimes called "equivocal generation." In any case, there was a remarkable lack of consensus about how to explain even the most mundane phenomena of inheritance, like the resemblance of children and parents, or the transmission of abnormalities.[5] No single experiment could resolve the controversy, but starting in the 1740s new empirical evidence came into play, and certain remarkable phenomena became touchstones for the debate. None of these phenomena were transparent to interpretation, which kept the arguments going for decades.

Perhaps the most famous of these was the freshwater polyp, or hydra. This tiny organism's capacity to regenerate whole organisms from pieces of itself, or to grow several heads where one had been, captured the imagination of naturalists, philosophers, and their readers alike. It made an appearance in virtually every text that touched on the origin of life, the relation of soul to body, or generation. The polyp became emblematic of the mysterious capacities of organized matter, and by virtue of its stranger-than-fiction quality, inspired speculation about the capacity of matter to direct its own organization. Abraham Trembley, who first noticed the remarkable properties of these "little organized bodies," was perhaps the least inclined of any of his contemporaries to speculate about the metaphysical significance of his discovery. Instead, he undertook to show that, however unusual they might appear, these creatures followed normal patterns just like any other species—they were not monsters or sports of nature. He identified their feeding habits, their normal mode of reproduction by budding, and their response to being cut in all conceivable ways.[6] Through focused, almost obsessive, investigations, he tried to contain the startling nature of his discovery, and to deny any possible implications for the properties of organic matter, and the relation between body and soul. Others were not so timid. Georges-Louis Leclerc de Buffon, for example, drew out the very implications Trembley had hoped to avoid. To Buffon's eyes, "each part [of the polyp] contains a whole," and he went on to claim that they imply the existence of a multitude of ubiquitous organic particles, which come together to form the germs of all organisms. From the uncontested evidence of regeneration in these simple "insects," Buffon leapt to a conjecture about organization across the spectrum of complexity. In polyps, organization means a simple repetition of the same "form"; in more complex organisms, the process entails more complex combinations of organized parts.[7]

Maupertuis also took off from the startling behavior of the polyp—"a Hydra more marvelous than that of the fable"—to speculate about more profound questions. His quizzical mode captures the tenor of contemporary responses to this phenomenon:

> What are we to think of this strange kind of generation? of this principle of life spread throughout each part of the animal? Are these animals anything other than collections of embryos just ready to develop, as soon as they are allowed? Or do they reproduce by unknown means all that the mutilated parts are missing? Might Nature, which in all other animals attached pleasure to the act by which they multiply, have caused these [creatures] to feel some kind of sensual delight when they are cut into pieces?[8]

One might understand this "principle of life spread throughout each part" as the organizing force guiding the process of generation and regeneration alike. In higher organisms, Maupertuis speculated, parts of the body send organic elements to the reproductive organs, where they collect and combine under the direction of selective organizing forces, analogous to chemical affinities. The mixing of male and female fluids is essential, as evidenced by the resemblance of children to both parents; neither sex is privileged. At the elemental level, on this view, matter must have the capacity to organize itself. Whatever the forces are, they must be intimately associated with material elements.[9]

Looking at the response provoked by strange phenomena like polyps, we see uncertainty, not only about how to interpret unexpected observations, but about whether they are even open to interpretation. Where were the limits of intelligibility? And just how far could speculation safely go? Diderot, to take one example, drifted quite comfortably from the evidence of natural history to a vision of nature's activity going well beyond what might be seen through a microscope: "It seems that nature varies the same mechanism in an infinity of different ways. . . . When we consider the animal kingdom, . . . would we not willingly believe that there never has been more than one original animal, prototype of all animals, in which nature only lengthened, shortened, transformed, multiplied, or obliterated certain organs?"[10] The prototype is not simply an ideal—it is an ancestor, dating back to a moment before natural forms proliferated into the variety evident today. The intermediate steps are missing in Diderot's reflections about nature's power to transform organs; he offers no mechanism for the transmission of the altered structures to subsequent generations. The notion of a prototype animal does, however, link the present to the deep past. Diderot calls his suggestion a "philosophical conjecture," bringing the imagination to bear on solid evidence from natural history. From the indefinite past, and the malleable form of the original prototype, he ventured into the submicroscopic realm of organic molecules and proposed "muffled sensibility [*sensibilité sourde*]" as a property of all matter, the property that made all those variations in form possible. According to Diderot's conjecture, this force drives molecules to seek situations of stable equilibrium, through an "automatic restlessness [*inquiétude automate*]."[11]

Diderot's primary interest was not in a theory of generation as such, but a theory of matter, which he tied in turn to a theory of life. He embedded the theory in reflections on method, arguing for a combination of exploration and synthesis, experiment and interpretation. The philosopher seeks new knowledge in the same way that molecules seek their places, by a kind of restless touching and retouching, trial and error, a "*tâtonnement*" like that of a blind man's stick. The ideal method involves moving back and forth from sense impressions to reflection, from experiment to theory, in a kind of oscillatory exploration.[12] Any position, whether material or philosophical, is always either in flux or in a state of dynamic equilibrium, and potentially unstable. In his *Pensées sur l'interprétation de la nature* (1754) Diderot vigorously defended "conjecture" as a productive method; the interpreter of nature starts where the senses leave off: "He draws, from the order of things, abstract and general conclusions that are for him just as evident as sensible and particular truths."[13]

Diderot developed his radical defense of "philosophical conjecture" or "*l'esprit de divination*" in dialogue with the books and experiments of Buffon and Maupertuis. Toward the beginning of his investigation of reproduction, Buffon declared that "the living and animate, instead of being a metaphysical aspect [*un degré métaphysique*] of creatures, is a physical property of matter."[14] To understand the persistence, if not permanence, of forms, or the transmission of traits, the naturalist must understand these properties of matter. Thus Buffon linked the theory of organization and organic matter inextricably to suggestions about the means for passing forms from one generation to the next. Organic molecules are not yet organisms—they cannot reproduce themselves—but they are the material of life, the matter from which organisms build themselves. In collaboration with the English microscopist John Turberville Needham, who had earlier found teeming animalcules in the milt of squid, Buffon examined fluids taken from the reproductive organs of dogs and other animals. The microscope revealed "moving bodies" in both male and female fluids, and they decided that these apparently ubiquitous entities must be the organic elements of "germs" that in turn combined to generate new organisms. From the modern point of view, these experiments involved considerable confusion about what was actually seen.[15] But for Buffon and Needham themselves, these observations implied first of all that mammalian eggs did not exist, which they used as evidence against the preexistence of organized germs, and second that male and female contributed symmetrically to generation. After further experiments with infusions of seeds and meat, in which moving particles appeared spontaneously, they concluded that organic particles could be found throughout nature, and not just in reproductive organs. They claimed to have actually seen the building blocks of living organisms through their microscope.

The next question was how to understand the organization of these elements into functioning, living organisms. Buffon's organic molecules come together through the action of "penetrating forces" that guide them into "internal molds" where they take on the appropriate structure and form of body parts: "In the same way that we can make molds by which we give to the exterior of bodies whatever shape we please, let us suppose that Nature can make molds by which she gives not only the external shape, but also the internal form. Would this not be a means by which reproduction could be effected?"[16] It is worth noting the interrogative form of this suggestion, for which Buffon never claimed anything like certainty. The notoriously obscure formulation, ridiculed and misunderstood from its inception, bears closer examination for what it reveals about Buffon's search for an analytic tool for talking about "hidden mysteries" without retreating into agnosticism. I want to emphasize the interpretive power of analogy for Buffon. He did not claim that physical, three-dimensional molds hover in the sex organs, waiting to receive matter exactly as metal molds do: "It is possible that nature has these internal molds, which we [humans] will never have, just as she has the qualities of gravity, which in effect penetrate to the interior; the supposition of these molds is therefore founded on good analogies."[17] If our senses were not limited to the surfaces of things, we might be able to conceive of internal molds more immediately. As it is, Buffon tells us, we have no reason to assume that Nature cannot employ means that are simply beyond the reach of our direct perception. This does not mean that the naturalist should disregard empirical evidence, of course. But Buffon wrestled with the crucial question of how to use such evidence to explore natural forces that cannot be revealed directly by experiment. In fact, this may be the only way to know about the most profound aspect of natural processes, and Buffon invoked the authority of Newton to support the analogy between gravity and internal molds:

> I have admitted in my explanation of development and reproduction first the accepted principles of mechanics, then that of the penetrating force of gravity which we must accept, and by analogy I thought I could say that there were other penetrating forces that act on organic bodies, as experience assures us. I have proved by facts that matter tends to organize itself, and that there are an infinite number of organic particles. I have thus done nothing but generalize from observations, without having advanced anything contrary to mechanical principles.[18]

These mechanical principles resemble Newton's active principles more closely than the impact mechanics of Descartes. Buffon's internal molds are conceptually slippery because they sometimes operate as forces, by analogy to gravity, and sometimes as constraining structures, by analogy to a sculptor's molds. Buffon struggled to combine a quasi-mechanical explanation of forces, shapes, and motions with a notion of active organic matter that resisted this kind of explanation. He tried to escape the limitations of the human senses, while simultaneously drawing on empirical evidence. Experiment and observation were essential components of his theory of life, but he insisted that the limitations of the human senses meant that complex phenomena could only be understood by other means. Opponents attacked him at precisely this point, unwilling to make the analogical leap; his most outspoken defender, Charles

Joseph Panckoucke, explained that we would need a sixth or even a seventh sense, to conceive of internal molds more immediately.[19]

Following his collaboration with Buffon, Needham continued his microscopical work. His experiments with infusions of animal and plant material, as well as observations of seminal fluids of animals, led him to insist on the existence of "a new Class of Beings," creatures he saw come to life in his preparations. Infusions of seeds and meat gravy gave rise to swarms of microscopic bodies after sitting for a few days. Macerated wheat, for example, produced tangles of filaments; these "would swell from an interior Force so active, and so productive, that even before they resolved into, or shed any moving Globules, they were perfect Zoophytes teeming with life, and self-moving." Needham saw these microscopic creatures as the key to generation. Based on many observations, he decided that all organic substances, animal and plant alike, had "real vegetating Force residing in every microscopical Point."[20] Needham was quite explicit about transferring the locus of nature's admirable consistency from the macroscopic world of animals and plants to the microscopic realm of "exuberating ductile Matter, actuated with a vegetative Force."[21] Needham's experiments with infusions, though wildly controversial, provided another touchstone in the ongoing controversy about generation and the line dividing living from brute matter. For better or worse, infusoria captured the imagination of Needham's contemporaries and raised the stakes, both experimentally and philosophically.[22] Voltaire railed against the implications of spontaneous generation, which seemed to him to imply a disturbing lack of order: "If animals were born without a [preexisting] seed, there would no longer be a cause of generation: a man could be born from a clod of earth just like an eel from a bit of flour paste. Besides, this ridiculous system would obviously lead to atheism. . . . Needham's microscope appears to be the laboratory of atheists."[23]

Needham's microscopic creatures impressed Maupertuis, too, though he did not buy Needham's musings about epigenesis through the action of vegetative forces. Writing to Buffon, Maupertuis commented, "I have just been reading Needham's book. What disorder, what a reasoner! But it also contains plenty of marvels. It's too bad that such a man wants to make systems, and that a maker of bricks wants to be an architect."[24] The microscope had opened up the prospect of a "new world," but it was not obvious how to make sense of it. The ultimate question remained unresolved, regardless of the status of Buffon's and Needham's microscopic bodies: "Even if material organic parts of the bodies of animals had been found, it [still] would not have fundamentally explained generation: because the formation of these organic particles would [also] need to be explained. It is too bad that those who do the experiments hardly attempt these speculations, and those who do the speculating are devoid of experiments."[25]

Maupertuis attempted to bridge this gap between speculation and experiment, by opening up the boundaries of both categories. He used evidence from microscopy, animal breeding, and inheritance of human traits to argue that matter must have other properties in addition to impenetrability, extension, and inertia. While gravity and chemical affinities could account for many kinds of phenomena, organization required properties "of another order than those we call physical. . . . We must have recourse to some principle of intelligence, to something similar to what we call desire, aversion, memory."[26] This "principle of intelligence" resides in matter, down to its smallest parts. In effect, the elements of matter have organic affinities modeled on psychic categories: desires, aversions, perceptions, habits, and memories. In the course of generation, both male and female contribute material to the embryo, which forms from a mixture of the two seminal fluids. Desire and aversion direct the organic elements to their places in the embryo; memory links the elements to comparable particles in the parent organism. Each one "retains a kind of memory [*souvenir*] of its previous situation and will resume it whenever it can, in order to form the same part in the fetus."[27] This is hardly a principle of heredity, in the modern sense, but it does engage the question of the transmission of hereditary characters. The notion of memory provides for the possibility of forgetting as well, and so gives a rudimentary explanation for mistakes in the process of transmission. Even if abnormalities originated accidentally, or by chance, the mechanism of remembering can serve to perpetuate them in subsequent generations. Once a new trait becomes established, "the particles become accustomed to their locations, which makes them place themselves similarly [in succeeding generations]."[28]

11.2 Reactions to Epigenesis

The works of Buffon, Maupertuis, and Needham were widely read, cited, admired, and challenged. A number of writers regarded these theories as morally and theologically suspect, as well as anatomically questionable (especially on the matter of whether male and female fluids actually mix in the uterus). Albrecht von Haller, for example, objected vociferously to Buffon's organic molecules and seminal fluids on anatomical grounds, although he clearly rejected active matter for theological reasons as well. He argued, point by point, that Buffon simply was not a competent anatomist, and incompetent anatomy could not be the basis for correct theoretical conclusions. "I have opened, without preconception or prejudice, the bodies of hundreds and hundreds of women, old and young," he boasted, but he had rarely encountered the glandular bodies where Buffon had found the female seminal fluid. Internal molds, being invisible by their nature, were even more problematic for Haller; he rejected them not only as philosophically unintelligible, but as anatomically incorrect as well. "Anatomy has taught me the troublesome truth" that no two individuals have exactly the same layout of arteries, nerves, veins, muscles, and bones. Therefore, since no two individuals truly resemble each other, to build a model of inheritance on resemblance is to start from false premises.[29] Haller was perhaps disingenuous here, in denying continuities between parent and offspring, but he objected forcefully to the way Buffon moved from his microscopic examination of glandular bodies in dogs to conjectures about active matter.[30] Haller's own work on chick embryos subsequently led him to the conviction that generation actually involves the mechanical unfolding of a preformed germ.[31]

Just when Buffon was preparing the first volumes of his *Histoire naturelle* for the press in 1749, the naturalist Réaumur published a book on poultry breeding, where he allowed himself a digression on the perils of attributing activity to matter. In general, Réaumur refrained from theorizing, and saw no virtue in speculating about unseen causes and forces. Maupertuis's conjectures, however, provoked Réaumur to articulate his objections to epigenesis. Although he never proposed a fully articulated theory of generation, he took the preexistence of the germ for granted. Without such a starting point for organic development, blind chance would have free rein. But even if we were to suppose that male and female fluids mix to form the germ, how could we imagine that order will emerge from the chaos of the mixture? "Who is the agent which is to disentangle and clear this chaos," Réaumur wondered, "to sort the several parts which are to come together, to construct organs with them, . . . in short, to finish that germ . . . ? We must not expect . . . that the bare action of a gentle heat can ever be capable of producing such a work, a work infinitely more complicated than any repeating watch can possibly be."[32] Design and knowledge could not reside in matter, however fashionable attractive forces might be: "In order to arrive at the formation of so complicated a piece of work, it is not enough to have multiplied and varied the laws of attraction at pleasure; one must besides attribute the most complete stock of knowledge to that attraction. . . . How will attractions give to such and such a mass the form and structure of a heart, to another that of a stomach, to another that of an eye . . . !"[33] Clearly, theological considerations were at the root of Réaumur's repugnance for self-organizing matter. But rejecting the concept he also rejected the method of imposing "fashionable" but imaginary innovations, like attractive forces, on the naturalist's subject matter. The notion of the productive use of conjecture was anathema to Réaumur, a consummate observer always reluctant to go beyond the evidence of his senses.

Réaumur's protégé Lelarge de Lignac, a Jesuit enthusiast for natural history, went further and openly accused the epigenesists of materialism. "Everything happens fortuitously" in Buffon's theory, where "animals make themselves from what he calls living elements, equally appropriate for entering into the formation of animals and vegetables."[34] Referring to the organic molecules as "little men," Lignac took Buffon to be saying that all parts of the body must be made up of miniature bodies.[35] As for the internal molds, Lignac refused to see them as anything other than literal molds, a reading that allowed him to trivialize his opponent: "In the end, these internal molds are the ingenious plastic forms of the peripatetic philosophers, under another name."[36] Lignac, working under Réaumur's guidance and sometimes in his company, also set out to challenge with his own observations the results of Buffon and Needham on infusoria, concluding that the claims of spontaneous generation were spurious, much like the deceptive illusions of a conjuror. Lignac and Réaumur agreed that the germs of the infusoria must have been present in the broths before sealing, just as the germ of the chicken preexists in the egg, and the structure of the butterfly is hidden within the body of the caterpillar. Lignac was incensed by every aspect of Buffon's book, from its content to its style, pointing to its subversion of faith and morality along with its absurd and untenable use of attractive forces.

The Genevan naturalist Charles Bonnet, an admirer and correspondent of Réaumur, also rejected epigenesis, though he found Lignac's critique heavy-handed and misguided. Widely respected as an observer of insect behavior, Bonnet's eyesight failed him in middle age. In this period, starting in the 1760s, he developed a sophisticated, and speculative, version of preformation theory to counter the dangers of active matter and downplay the effects of chance in nature: "I am not undertaking in any way to explain how the animal is formed: I am assuming it to be preformed from the outset, and my assumption rests upon facts that have been well observed."[37] A brief summary cannot do justice to the subtleties of this theory, which evolved through several different versions and incorporated the latest experimental results in embryology and microscopy. Bonnet came to the problem of generation from his work on parthenogenesis in aphids and on regeneration in worms—just the sort of phenomena that, in other hands, fed conjectures about active matter. In scaling up from insects, he had to address the problem of the resemblance of offspring to the father, since this was the strongest evidence for the mixing of contributions from both parents, and hence for epigenesis. Bonnet decided that the semen serves to bring the circulatory system of the germ to life, initiating its sequential development. Along the way, this development submitted to "modifications" caused by the action of the semen: "It imprints upon the germ points of resemblance with different parts of the [father]."[38] Just how this imprinting works remained unclear, but Bonnet conceived the seminal fluid as a mix of particles corresponding to different parts of the paternal body, which affect the corresponding parts of the germ. The particular qualities of the individual had somehow to be transmitted against the background of the constancy of the preformed germ, while submitting to the influence of the action of the semen and the nourishment provided by the womb. "We must not believe," Bonnet cautioned his readers, "that the germ possesses in miniature all the features that characterize the mother as an individual. The germ bears the original imprint of the species, and not that of individuality. It is, in miniature, a man, a horse, a bull, etc. . . . , but not a *certain* man, a *certain* horse, a *certain* bull."[39]

11.3 Animal Breeding

Microscopic observations of infusions raised more questions than they answered, about the nature of life and the nature of matter. On the macroscopic scale, these open-ended questions were explored from another angle through breeding animals and speculating about hybrids. Farmers and pet fanciers had bred domesticated animals for selected characteristics for centuries. These breeders did not worry about what caused certain traits to be inherited with a certain frequency. But in the eighteenth-century hybrids and domestic animals entered the discourse about generation alongside Needham's microscopic animals, and the inheritance of visible traits became evidence for the mixing of contributions from male and female parents. In breeding practices, and in thought experiments about hypothetical crosses, we can see how the specialist discourse of life science (coming out of anatomy and microscopy) intersected with polite culture, specifically in the cultivation of pets and other domestic or farm animals, like horses and fowl, and also in literature written for this same genteel readership.

Maupertuis himself kept quite a menagerie of pets (as did Buffon and Réaumur), some of which he bred systematically, looking for hereditary patterns. When he came across a rare combination of colors in the coat of a female dog, for example, he attempted to perpetuate the trait; after four litters a male puppy with identical markings was born.[40] This dog eventually went on to father a puppy with the distinctive coloring, showing that the trait was not sex linked, and easily perpetuated. Another example came from a human family in Berlin, some of whom were born with extra fingers and toes: Maupertuis recorded the genealogy of this family for three generations with names of individuals, their spouses, and their children, and the occurrence of extra digits among them. The trait occurred often enough to show that it was transmitted bilaterally, and that marriage to a five-fingered spouse affected the frequency of its appearance in the offspring. Although he did not try to predict this frequency with any precision, it seemed to him not only definitive evidence of mixing of material elements from both parents, but of an orderly sort of mixing quite unlike randomly occurring monstrosities. These examples opened up questions about the meaning of individual variations for a science of life—especially the analytic value of examining populations over time—and about the relation of individuals to their species (or races). To show that the sixth finger in this Berlin family was inherited rather than accidental, Maupertuis estimated the probability of its appearing at random in consecutive generations of one family. The probability decreases by the same factor for each generation, to the point where the chances of three generations of the same family randomly producing such individuals would be impossibly high: "numbers so large that the certainty of the best-demonstrated things in Physics does not approach these probabilities."[41]

Breeding also raised questions about the malleability of living forms and about the implications of human efforts to breed new types of animals. Maupertuis understood the perpetuation of naturally occurring variations, like albinism, by analogy to selective breeding practices: "Whether one takes this whiteness [of the albino's skin] for an illness or accident, it will never be anything but a hereditary variation that reinforces itself or erases itself over the course of generations."[42] That is, the organism has the capacity to pass on its traits, though rare traits like albinism may disappear in the offspring without reinforcement from the other parent. Similarly for polydactyly: "By such repeated marriages [with five-fingered mates], it would probably die out, and [conversely] it would perpetuate itself through marriages where the trait was common to both sexes."[43] As animal breeders knew very well, variations that appeared unexpectedly could be perpetuated through cultivation. Over time, such selective practices could cause varieties to solidify into a separate species stable enough to perpetuate itself. Variations became grist for the mill of human intervention, where the breeder's choices operated just like the natural forces that might give rise to new heritable attributes.

In his most speculative mode, Maupertuis suggested that naturalists might use the collections of animals in menageries to explore heredity through hybridization, even to the point of crossing animals that would never mate in nature: "The efforts of a hardworking and enlightened naturalist would result in plenty of curiosities of this type, by causing animals of different species to lose their natural repugnance for each other, through education, habit and need." And for animals that could not be pressured to mate, he imagined the enlightened naturalist developing techniques for artificial insemination to create new "marvels": "We might see from these unions plenty of monsters, of new animals, perhaps even entire species that nature has not yet produced."[44] These suggestions were meant to push the limits of current knowledge into the unknown, the possible, and even the marvelous, not just by looking at nature, but by manipulating it. Maupertuis implied that much of what we do not know is only mysterious because it has not yet been carefully (or imaginatively) investigated. Reflecting on the regenerative capacity of polyps and lizards to regrow severed tails, he asked, "Is it probable that this marvelous property belongs only to the small number of animals we know about? . . . Perhaps it only depends on the method for separating the parts of other animals to see them reproduce themselves [in the same way]."[45]

In exactly the period when Maupertuis was breeding his dogs and reflecting on hereditary variations, Réaumur was developing techniques to improve poultry breeding, especially through artificial incubation of eggs. When he wrote up his results for publication, he included his objections to epigenesis (discussed above) as well as some elliptical comments about what might be learned by crossing different varieties of chickens. In one of his lighter moments, Réaumur described the "philosophical amusements of the barnyard," regaling his readers with the amorous adventures of a duck and a rooster and, even more outlandish, of a rabbit and a hen. These animals, which Réaumur had kept under observation in his house for a time, had entertained "all of Paris" and prompted, naturally enough, speculation about the outcome of the interspecies union: "It was the general wish, as well as my own, that it might have procured us chickens covered with hair, or rabbits clothed with feathers."[46] We must read this wish as a rhetorical flourish, since a furry chicken or a feathered rabbit might have forced Réaumur to question his commitment to preformed germs. He did, however, outline a variety of observations that might stimulate the efforts of "those who love to find in their poultry-yards amusements conducive to the progress of natural knowledge."[47] He himself pursued an extensive program of selective breeding of chickens for readily observable variations like extra claws, especially to see if they were transmitted exclusively by one or the other sex, though he did not disclose his results in print.[48] He designed chicken coops to allow for the isolation of hens and roosters with different traits, and used his incubators to hatch the carefully marked eggs. His unpublished notes show that he did obtain crosses that challenged his expectations, though he adopted ad hoc explanations for why these characters were passed on by both hens and roosters. He also knew that hybrid crosses of birds could produce a mixing of traits, as in the goldfinch-canary cross whose song resembles that of neither parent, but a pleasant mixture of the two.[49] This kind of mixing, familiar to bird fanciers and poultry farmers alike, remained without any theoretical explanation. In the second edition of his book on poultry (1751), Réaumur added a detailed genealogy of a six-digited family sent to the Paris Academy by a correspondent in Malta. This example corroborated exactly Maupertuis's account of the Berlin family, but again, was presented without interpretation. Réaumur made no comment other than "it does not seem favorable to the preexistence of germs."[50] Réaumur's impulse to collect stories and specimens of hybrids, including the fabulous North African beast of burden supposed to be a cross between a cow and an ass, alongside his own breeding experiments, brought him up against recalcitrant evidence for bilateral inheritance. He managed to avoid troubling conclusions by emphasizing descriptions of the curiosities rather than explanations.

Buffon also devoted substantial effort to breeding across species (wolves and dogs, for example, or donkeys and horses, or goats and sheep). Crossing a mare with a donkey, for instance, yielded a different kind of hybrid from that produced by a stallion crossed with a female donkey. "Both," he says, "look more like the mother than the father, not only in size, but also in the shape of the body. . . . But for the shape and dimensions of the head, they look more like the father than the mother."[51] He also noted that males predominated in the results of such crosses: "The male thus influences the offspring more in general than the female, since he gives his sex to the larger number, and this number of males becomes greater as the species are more distantly related."[52] Buffon saw hybrids as the key to tracing the contributions of each sex, as well as to understanding limits of fertility among species, and to mapping how closely different species might be related to each other. His programs for breeding agricultural animals (within species) were intended to counter what he saw as the natural degeneration of animal forms if left to themselves: "In order to have beautiful horses, good dogs, etc., it is necessary to give foreign males to the native females, and reciprocally to the native males, foreign females; failing that, animals will degenerate. . . . In mixing the races, and above all in renewing them constantly with foreign races, the form seems to perfect itself, and Nature seems to revive herself."[53] Buffon argued that people can use the forces of nature to shape animals to their purposes. Breeding for specific traits (as in pet dogs, for example) required constant attention, however, to avoid reversion to the original forms. Nature is flexible, with the potential to be molded by environmental factors, including human intervention.

The Paris physician Charles Augustin Vandermonde incorporated Buffon's theory of organic molecules into a system for improving the human species through application of principles of nutrition, education, hygiene—and breeding. Vandermonde played on contemporary concerns about declining population and degeneration by prescribing healthy practices for everything from the choice of mate to engendering vigorous offspring to the care of pregnant women and the feeding and education of children. Since, following Buffon, both male and female contribute equally to the offspring, they must be well matched to complement each other. Just as animal stock is strengthened by crossing with natives of other regions, he says, humans benefit from mating with people from different climates, who will necessarily have different constitutions. The goal of rational mating should be to make beauty and strength "hereditary among men."[54] When people move from the provinces to the city and intermarry with urban dwellers, the result is improvement in strength and health of the city stock. Further, he implies that human agency could easily be put to work improving its own species, by analogy to breeders who "create new races" of animals: "We can easily see that one could perfect animals by varying them in different ways. Why should we not work also on the human species? It would be just as possible, . . . in combining all our rules, to embellish men, as it is routine for an able sculptor to cause a model of beautiful nature to emerge from a block of marble."[55] The breeder, or the advising physician, effectively puts the forces of matter to work to improve physical and constitutional inheritance. Vandermonde's book reads as a handbook for healthy living; he assumes his readers will take an interest in shaping their own progeny. While he does not indulge in fantasies about cross-species mating, his rules for choosing suitable partners from other climates dictate a kind of subtle hybridization, or at least mixing. The program rests on a mechanics of inheritance inspired by Buffon's organic molecules, whereby specific traits materialize in oscillating spiral-shaped particles that wind together selectively to make up the germ of the offspring.[56] These operations at the submicroscopic level, well beyond the reach of direct observation, locate the origin of human traits firmly in the stuff of organic matter with its resilience, activity, vibrations, and self-propulsion. Vandermonde adapted Buffon's version of active matter to his own medical vision of prescriptions for human betterment, with heredity playing a starring role.

The fascination with hybridization evident in Buffon and Vandermonde also emerged in contemporary satire. An anonymous pamphlet published in Paris in 1772 tells the "true and marvelous story" of a "lynx-girl" who could see into solid objects. The author discusses the girl's odd trait by analogy to hydroscopes, people with the ability to find water underground by seeing through the earth. The lynx-girl (herself presumably a sport of nature) could look into bodies to see the circulation of blood, the nerves, and all the organs, including the pineal gland, where she found the soul: "She reads there very distinctly the thoughts of the individual, so that she will tell you your plans, your reflections, even your first ideas, with the precision and in the order that they come into being and order themselves." The remarkable trait should be heritable, and therefore its perpetuation ought to be subject to the breeder's art. Inspired by Maupertuis's speculations about crossbreeding, the satirist goes on to suggest that the Royal Society of London and the Paris Academy of Sciences really ought to preside over (and pay the expenses of) the marriage of this girl to a hydroscopic boy, in order to produce more such gifted lynx-people. The benefits are obvious: "There is no need to mention what advantages would result from a lynx race, for the good of humanity; what light [*lumières*], what vision, what insight, these living telescopes, born in the sanctuary and under the auspices of physics, could communicate to savants, the authors and the cause of their existence!"[57] He went on to calculate how long it would take for the trait to multiply in subsequent generations, and reflected on the utility of these academy-bred people for police work, for uncovering court intrigues, and so on. The calculations might be spurious, but their appearance here indicates how this way of thinking about inherited traits, and the manipulation thereof, could be taken for granted in this literary context.

11.4 Conclusion

I have laid out the landscape of discourse about heredity and generation to highlight resonances with other claims—about active matter, about the origin of organization, about the change of organic forms over time, and also about method. Experiments played an essential role, often as provocation for conjectural interpretation and rarely as conclusive or indisputable. While Haller claimed that everything could be sorted out by multiplying careful observations of embryos, a number of French writers at midcentury addressed the question of how to accommodate method to mystery, and to the limits of human senses. Their attempts to lay the foundations of a science of life included the use of hypothetical experiments, as well as conjectures and queries; we have seen how these theoretical modes interacted with actual experiments and observations. From this point of view, Réaumur's five-clawed chickens and Maupertuis's hypothetical hybrids between exotic African jungle animals and Vandermonde's perfected humans belong in the same discursive space. These speculative experiments, along with breeding programs that traced abnormalities through generations of dogs or birds or people, aimed to get below the surface to general laws of nature and of life. Maupertuis articulated this desideratum in his critique of natural history (and here he had Réaumur in mind as the contemporary exemplar): "All our treatises on animals, even the most methodical, form nothing but paintings pleasing to the eye. To make natural history into a true science [*véritable science*] we would have to apply ourselves to researches that would allow us to know, not the particular figure of such and such an animal, but the general processes of nature in her production and preservation."[58] Experiments must be designed to ask the right questions, he suggests, but the naturalist must also be willing to interpret the results, whatever they might be. For many thinkers at midcentury, the complementary use of conjecture and experiment seemed the only viable method for getting at what Buffon called "the hidden means that nature might be employing for the generation of creatures."

Nature's "hidden means" could be interpreted as general laws or mechanisms; these were discernible only vaguely to theorists of epigenesis in the eighteenth century. Generation in the first analysis refers simply to the formation of individuals as the result of a procreative act. As we have seen, challenges to the preexistence of germs brought the nature of matter, and possibly the definition of life itself, into the discussion. Hereditary phenomena—transmission of abnormalities or other variations, resemblances across generations—became integrated into the discourse of generation as theorists looked for new kinds of evidence to support their claims about matter and organization. The evidence of inheritance opened up possibilities for thinking across generations, for linking the present and the future with the past. When Maupertuis recorded the genealogy of a family and estimated the frequency of a trait in a population, or when Réaumur traced the occurrence of distinctive traits in his chickens, they were putting pressure, however tentatively, on the limits of explanation.

Notes

1. López-Beltrán 1994. On hereditary disease, see also López-Beltrán 1995 and chapter 5, this volume.

2. Aumont 1757, 568.

3. Diderot 1751.

4. The standard source is Roger 1971. I use *epigenesis* in its eighteenth-century sense to mean the postconception organization of material elements into the embryo. See also Roe 1981.

5. See López-Beltrán, chapter 5, this volume.

6. Trembley 1744. See also Ratcliff 2004; Dawson 1987; Vartanian 1950.

7. Buffon 1954, 239.

8. Maupertuis 1980, 107–108.

9. For a full discussion of Maupertuis's theory of generation, see Terrall 2002, chapters 7 and 10.

10. Diderot [1754] 1981, Pensée XII, 36–37.

11. Diderot [1754] 1981, Pensée LI, 84–85.

12. Diderot [1754] 1981, 34.

13. Diderot [1754] 1981, Pensée LVI, 88.

14. Buffon, *Histoire Naturelle*, vol. 2, in Buffon 1954, 238.

15. On Needham, see Roger 1971, 424–520; Roe 1983. For the experiments on seminal fluids and the microscope used by Buffon and Needham, see Roger 1997, 140–145 and Sloan 1992. Sloan concludes that Buffon was very likely seeing bacteria and cell fragments in Brownian motion. Ratcliff (2001) disputes this interpretation.

16. Buffon 1954, 243.

17. Buffon 1954, 244.

18. Buffon 1954, 254.

19. Panckoucke 1761.

20. Needham 1748, 645.

21. Needham 1748, 658–659.

22. Mazzolini and Roe 1986.

23. Voltaire, pamphlet against Needham, 1765, cited in Mazzolini and Roe 1986, 82.

24. Maupertuis to Buffon, September 1, 1750, Saint-Malo Municipal Archives.

25. Maupertuis to La Condamine, August 24, 1750, Saint-Malo Municipal Archives.

26. Maupertuis, *Système de la nature*, in Maupertuis 1756, vol. 2, 147.

27. Maupertuis, *Système de la nature*, in Maupertuis 1756, vol. 2, 158.

28. Maupertuis, *Système de la nature*, in Maupertuis 1756, vol. 2, 160–161.

29. Haller 1751, 48, 32.

30. See López-Beltrán 1995.

31. Roe 1981.

32. Réaumur 1751, 457.

33. Réaumur 1751, 463.

34. Lelarge de Lignac 1751, vol. 1, 7.

35. Lelarge de Lignac 1751, vol. 2, 26.

36. Lelarge de Lignac 1751, vol. 2, 64.

37. Bonnet 1762, vol. 2, 27–28.

38. Bonnet 1762, vol. 2, 243.

39. Bonnet 1762, vol. 2, 256.

40. The dog-breeding experiments are described in "Maupertuis: Sur la génération des animaux," *Lettres de M. de Maupertuis,* in Maupertuis 1756, vol. 2, 310–311. See also Hoffheimer 1982. For a fuller account of Maupertuis's work on generation and heredity, see Terrall 2002, especially chapter 10.

41. Maupertuis 1756, vol. 2, 310.

42. Maupertuis 1980, 138.

43. Maupertuis 1756, vol. 2, 308.

44. Maupertuis, *Lettre sur le progrès des sciences* (Berlin 1752), in Maupertuis 1756, vol. 2, 420–421.

45. Maupertuis, *Lettre sur le progrès des sciences* (Berlin 1752), in Maupertuis 1756, vol. 2, 421–422.

46. Réaumur 1751, 457.

47. Réaumur 1751, 467.

48. Réaumur's notes on his crosses, from roughly 1747 to 1750, are in the Archives of Académie des Sciences, Paris, Fonds Réaumur.

49. Réaumur 1751, 337.

50. Réaumur 1751, 376.

51. Buffon 1777, 2–3. For crosses between dogs and wolves, see "Le chien," in Buffon 1749–1767, vol. 5.

52. Buffon 1777, 15.

53. Buffon, "Le cheval," in Buffon 1749–1766, vol. 4, 215–217.

54. Vandermonde 1756, viii.

55. Vandermonde 1756, 155.

56. On Vandermonde's theory, see Wellman 2002.

57. *Histoire véritable et merveilleuse d'une jeune angloise . . .* (1772, 58). I thank Emma Spary for the reference to this pamphlet.

58. *Lettre sur le progrès des sciences,* in Maupertuis 1756, vol. 2, 418.

References

Aumont, Arnulfe d'. 1757. Génération (physiologie). In Denis Diderot and Jean Le Rond d'Alembert, eds., *Encyclopédie,* vol. 7, 559–574. Paris: Briasson.

Bonnet, Charles. 1762. *Considérations sur les corps organisés*. 2 vols. Amsterdam: M.-M. Rey.

Buffon, Georges-Louis Leclerc comte de. 1749–1767. *Histoire naturelle, générale et particulière*. 15 vols. Paris: Imprimerie royale.

———. 1776. *Histoire naturelle, générale et particulière. Supplément*. Vol. 3. Paris: Imprimerie royale.

———. 1954. *Oeuvres philosophiques*. Ed. Jean Piveteau. Paris: Presses Universitaires de France.

Dawson, Virginia P. 1987. *Nature's Enigma: The Problem of the Polyp in the Letters of Bonnet, Trembley, and Réaumur*. Philadelphia: American Philosophical Society.

Diderot, Denis. 1751. Animal. In Denis Diderot and Jean Le Rond d'Alembert, eds., *Encyclopédie*, vol. 1, 468–474. Paris: Briasson.

———. 1964. *Oeuvres philosophiques*. Vol. 9. Ed. Paul Vernière. Paris: Garnier.

———. [1754] 1981. *Pensées sur l'interprétation de la nature*. In Jean Varloot, ed., *Oeuvres complètes*. Paris: Hermann.

Diderot, Denis, and Jean Le Rond d'Alembert. 1751–1765. *Encyclopédie; ou Dictionnaire raisonné des sciences, des arts et des metiers*. 17 vols. Paris: Briasson.

Formey, Jean-Henri-Samuel. 1789. *Souvenirs d'un citoyen*. 2 vols. Berlin: Lagarde.

Haller, Albrecht von. 1751. *Réflexions sur le système de la génération de M. de Buffon*. Geneve: Chez Barrillot.

Hoffheimer, Michael H. 1982. Maupertuis and the eighteenth-century critique of preexistence. *Journal of the History of Biology* 15 (1): 119–144.

Lelarge de Lignac, Joseph Adrien. 1751. *Lettres à un Amériquain sur l'Histoire naturelle, générale et particulière de Monsieur de Buffon*. 2 vols. Hamburg.

López-Beltrán, Carlos. 1994. Forging heredity: From metaphor to cause, a reification story. *Studies in the History and Philosophy of Science* 25:211–235.

———. 1995. "Les maladies héréditaires": Eighteenth-century disputes in France. *Revue d'histoire des sciences* 48:307–350.

Maupertuis, P.-L. M. de. 1756. *Oeuvres*. 4 vols. Lyons: Bruyset.

———. 1980. *Vénus physique*. Ed. Patrick Tort. Paris: Aubier Montaigne.

Mazzolini, Renato, and Shirley Roe. 1986. *Science against the Unbelievers: The Correspondence of Bonnet and Needham, 1760–1780*. Oxford: Voltaire Foundation.

Needham, John Turberville. 1748. A Summary of some late observations upon the generation, composition, and decomposition of animal and vegetable substances. *Philosophical Transactions* 45:615–666.

Panckoucke, Charles Joseph. 1761. *De l'homme, et de la reproduction des différens individus. Ouvrage qui peut servir d'introduction & de défense à l'Histoire naturelle des animaux par m. de Buffon*. Paris.

Ratcliff, Marc J. 2001. *Europe and the Microscope in the Enlightenment*. Doctoral dissertation, University College London.

———. 2004. Abraham Trembley's strategy of generosity and the scope of celebrity in the mid-eighteenth century. *Isis* 95:555–575.

Réaumur, René-Antoine Ferchault de. 1751. *On the Hatching and Breeding of Domestick Fowls*. London. (Translation of *L'art de faire éclorre . . . des oiseaux domestiques*. 1st ed., 1749.)

Roe, Shirley. 1981. *Matter, Life and Generation: Eighteenth-Century Embryology and the Haller-Wolff Debate*. Cambridge: Cambridge University Press.

———. 1983. John Turberville Needham and the generation of living organisms. *Isis* 74:159–184.

Roger, Jacques. 1971. *Les sciences de la vie dans la pensée française du XVIIIième siècle*. 2nd ed. Paris: Armand Colin.

———. 1997. *Buffon: A Life in Natural History*. Trans. Sarah Bonnefoi. Ithaca, NY: Cornell University Press.

Sloan, Philip. 1992. Organic molecules revisited. *Buffon 88: Actes du Colloque international pour le bicentenaire de la mort de Buffon*. Paris & Lyon: J. Vrin.

Spary, Emma. C. 1999. Codes of passion: Natural history specimens as a polite language in late eighteenth-century France. In H. E. Bödeker, P. H. Reill, and J. Schlumbohm, eds., *Wissenschaft als kulturelle Praxis, 1750–1900*, 105–135. Göttingen: Vanderhoek and Ruprecht.

———. 2000. *Utopia's Garden: French Natural History from Old Regime to Revolution*. Chicago: University of Chicago Press.

Terrall, Mary. 2002. *The Man Who Flattened the Earth: Maupertuis and the Sciences in the Enlightenment*. Chicago: University of Chicago Press.

Trembley, Abraham. 1744. *Mémoires pour servir a l'histoire d'un genre de polypes d'eau douce*. Leiden.

Vandermonde, Charles-Augustin. 1756. *Essai sur la manière de perfectionner l'espèce humain*. 2 vols. Paris: Vincent.

Vartanian, Aram. 1950. Trembley's polyp, La Mettrie, and eighteenth-century French materialism. *Journal of the History of Ideas* 11:259–286.

Voltaire, François Marie Arouet de. 1752. "*Les Oeuvres de M. de Maupertuis*," *Bibliothèque raisonnée* 49:158–172.

Wellman, Kathleen. 2002. Physicians and philosophes: Physiology and sexual morality in the French Enlightenment. *Eighteenth-Century Studies* 35:267–277.

12 Kant on Heredity and Adaptation

Peter McLaughlin

The consideration of heredity in the latter eighteenth century seems to have focused primarily on the transmission of defects, especially hereditary disease. Immanuel Kant, too, uses the concept of *Vererbung* in this context: he analyzes the discussion of the transmission of defects as carried on in the three "higher" university faculties (law, medicine, and theology) with their notions of *Erbschuld*, *Erbkrankheit*, and *Erbsünde* (hereditary debt, hereditary disease, and hereditary, i.e. original, sin). However, Kant also deals with the problem of *Vererbung* in biology, where it is located in an area with decidedly positive connotations: adaptation. This chapter will be concerned with the relation of heredity to adaptation in the biological writings of Kant. Individual adaptations—as opposed to the appropriateness of the species form for its place in nature—did not constitute a central theme in eighteenth-century biological thought. And even for Lamarck they are of only secondary interest. Kant's position may be an important point of departure for the study of adaptive heredity and for the quite different relation of nineteenth-century thought to the problem of adaptation.

12.1 Heredity and Adaptation

Neither the heredity nor the adaptation of individual traits was a primary concern of eighteenth-century biology. The fit between organism and environment was most often conceptualized as an intrinsic aptitude (of the species form) for life accompanied by a contingent and superficial adaptation of individual characters to aspects of the environment. The realm of the regular, the lawlike, the vitally significant was the basic species form—and this at the time had little to do with heredity. It was not unusual to celebrate the purposiveness of organic structure, but it was unusual to speak of individual traits as adaptations—in fact, the first use of the word *adaptation* in this sense mentioned by the *Oxford English Dictionary* occurs in Darwin's *Origin of Species* (1859).

Seventeenth- and eighteenth-century science inherited from classical antiquity a threefold classification of kinds of intergenerational similarity. *Individual peculiarities*, *species form*, and *sex* mark off different ways progeny can be similar to parents. The distinction between the first two forms—the lawlike transmission of species form and the contingent disturbance or supplementation of the results of this transmission through the transmission of individual traits—structured early modern theorizing. For instance, the spermatic parts representing all the organs and tissues of the body could by pangenesis form an outline that is then fleshed out by the sanguineous parts (Everard); or the inherited species soul could direct the pangenesis of the individual body (Gassendi); or the pangenesis of the somewhat ethereal species form out of "spiritual" or "formal" atoms might guide and govern the pangenesis of the body by the "material atoms" (Highmore); or the species form is encased in the germ while individual peculiarities are transmitted pangenetically with the first nourishment (Bourguet).[1]

As a rule the two levels were kept distinct and the species form was what was scientifically interesting. For instance, the *moule interieur* introduced by Georges-Louis Leclerc Comte de Buffon guaranteed the unity of the species and the boundary lines between different species in spite of all the individual peculiarities that can arise and might be accumulated. Even Lamarck's later transformation theory still displays this distinction in the two basic processes of organic change: the progressive tendency to increased complexity of organic structure—so to speak a great escalator of being—and the collective adaptation of the individuals to changing environmental conditions that distorts the otherwise lawlike linear progression of forms. In all these theories the species form is fixed either by divine preformation or by natural law; and pangenesis provides a simple mechanism for the transmission of acquired characters: either directly to the newly formed germ or indirectly to a preformed germ's first nourishment. But the individual peculiarities that may genuinely be said to be inherited are just that—individual and peculiar. They are important not for theoretical science but for the application of science to individual cases in medicine and agriculture. The species characters, on the other hand, are not in any sense products of a contingent process of heredity.

The individual hereditary traits studied were also normally not viewed as being particularly beneficial to their bearers. The best-documented cases in the species that interests us most were the transmission of sixth digits and of hereditary diseases.[2] The adaptation, purposiveness, fitness of an organism for its environment, and so on were properties of the species form. And here, too, there was a certain tension—at least in the materialistic theories that dominated the second half of the eighteenth century. Whereas deistic preformation theories could imagine that there were innumerable possible organic forms that might have been actualized and that God elected only some of these (presumably the better ones), the materialistic theories of Buffon, Blumenbach, and others were committed to the natural necessity of all actually existing species. Their emphasis was on *viability*, not *adaptedness*—on living, not on living well. According to Buffon, every kind of organism that *can* exist *does* exist. The sloth, for instance, is a borderline case; it has just barely everything that it needs for life, but if it had just one more defect, it would become extinct.[3] In such a scheme the nature of matter determines species form and its continuity over time: if a catastrophe were to occur and wipe out some or all species, they would return again by spontaneous generation as soon as the environment returned to normal. There is no serious role for heredity or adaptation in such a scheme.[4]

We might want to look for the beginnings of adaptationism in physicotheology, inasmuch as the purported influence of William Paley's "argument from design" on Darwin's theorizing is widely cited. And it is true that this kind of argument did increasingly put single adaptive traits at the focus of attention—though there were limits. While the detailed fit of particular traits to aspects of the environment might evoke amazement at the skill of the technician, it was the greater, harmonious system of the world, not the anatomy of insects, that inspired awe at the grandeur of the Creator. There were thus two versions of the design argument, which differed in the object typically studied—but also in the kind of design appealed to. The divine artisan normally has two things in mind when he creates something, and both were called "design": there was first the "design" or plan (of the world system) and then there was the "design" or purpose of particular organs. The first version, the argument from the design of the world system, appealed to a *formal cause* of the world in the plan of a divine architect; this plan could be inferred from the order and harmony of the system without speculation as to the ultimate purpose of the clockwork universe. However, it was only the second version, the argument from the *final cause* of particular traits, that to some extent survived Hume's critique of the design argument.[5] This argument inferred the existence of a designer from the direct perception of the aims or purposes of useful or vital organs. Even Hume admitted that he "perceived" that legs were for walking. But here, too, it is *species* characters that give occasion to speculation about the intentions of the Creator. Even physicotheologians had difficulty with individual peculiarities. Albrecht von Haller at one point in his preface to the German translation of Buffon's *Histoire naturelle* has God's general plan produce the species, while his special providence produces the individual details and differences. At other times (in the same text), however, he leaves the details to chance and only the species form to God. What remains constant in Haller's exposition is the distinction between the two levels of transmission.[6] There was little theological mileage to be gotten out of the individual peculiarities that might be passed on. Since the physicotheologically-minded tended to be preformationists even well into the nineteenth century, and since preformation excludes the species form from the domain of the hereditary, we should not expect much in the way of inheritance of adaptations from this field.

12.2 *Vererbung*

The concept of heredity seems to have entered German science in the same breath as the concept of race—in the writings of Immanuel Kant. Kant lectured on physical geography once every year for half his life, and in these lectures the concept of race played an important role. In 1775—announcing what seems to have been his *twentieth* run through the material—he published a short brochure on human races and their origin: *On the Different Human Races*.[7] Ten years later he published a paper in a Berlin journal trying to defend and clarify his concept of a human race: "Determination of the Concept of a Race of Humans."[8] When this paper occasioned severe criticism by Georg Forster, Kant returned again to the subject in a third essay titled "On the Use of Teleological Principles in Philosophy," published in early 1788 (*Teutscher Merkur*). In these papers Kant is systematically dealing with the problems of heredity and grasping for a terminology that fits the phenomena under investigation. In the third paper he uses the German term *vererben* for the first time (in the verb form).[9] There is some reason to believe that his use of the term is prototypical—at least I know of no previous biological uses outside of medicine. In the *Critique of Judgment* of 1790, which contains his philosophy of biology, Kant says little or nothing on the subject of either race or heredity. Finally, in a slightly later work titled *Religion within the Bounds of Mere Reason* (1793), Kant also uses the concept *Vererbung* (in the noun form) in connection with the transmission of hereditary defects—that is, in a context in which it was already somewhat familiar.

Much of the discourse on heredity in late eighteenth-century France, as López-Beltrán (chapter 5, this volume) has shown in great detail, was focused primarily on the transmission of defects, especially hereditary disease. In *Religion within the Bounds of Mere Reason* Kant, too, takes up this question. In the course of an analysis of the origin of moral evil and its transmission, he discusses the explanations of this phenomenon offered by the three so-called higher faculties of the university (law, medicine, and theology). Each of these faculties, he tells us, has developed a notion of the inheritance of defects: the concepts of *Erbschuld, Erbkrankheit*, and *Erbsünde* (hereditary debt, hereditary disease, and original sin).[10] But in this work the question of biological mechanisms of heredity is only obliquely raised in long and speculative footnotes on how various theories of generation might handle the hereditary transmission of original sin and how they might explain the lack of transmission of this defect to Jesus. Some elaborate manuscript drafts of these passages also exist where Kant weighs the merits of epigenesis and the ovist and spermist forms of preexistence on this question, but it is unclear how seriously all this should be taken—it all does seem to be somewhat facetious.[11] What is important is that with Kant the notion of the hereditary moves from the higher or applied faculties to the lower, philosophical or theoretical faculty and that it is not confined to or even primarily oriented toward *defects*.

The German term for heredity (or inheritance), as I have mentioned, is *Vererbung*, which is slightly different in meaning from the English *inheritance*, inasmuch as it takes the perspective of the donor, not the recipient: it should actually be rendered as "bequeathment," but this is seldom done. In the mid-eighteenth century the concept *Vererbung* was a somewhat complicated notion in German property law. An heir to property had a calendar year in which to accept or reject the inheritance definitively—this was particularly relevant if the inheritance contained significant debts. Problems could arise if the prospective heir were to die during this grace period. In such cases the tentatively inherited property was passed on to his or her heirs. It was specifically this procedure that was called *Vererbung*. Thus, use of the legal term *Vererbung* was restricted exclusively to situations in which one person inherited from another something that person had received by inheritance. Zedler's *Lexicon* (1746) defined the term as follows:

> *Vererbung* . . . is when an heir, before he accepts or rejects the inheritance accrued to him, dies and that inheritance which he neither really accepted nor rejected is transmitted and further bequeathed . . . to his heirs. . . . And this occurs first of all when someone receives an inheritance from his parents or from elsewhere . . . and dies within the year's period after acquiring notice of the inheritance . . . in such a case this kind of transmission is called in Latin *transmissio ex jure deliberandi*.[12]

So conceived, there can, for instance, be no *Vererbung erworbener Eigentümer*, no bequeathment of acquired properties, because only property acquired by birth falls under the concept as defined. Now this is precisely the meaning that Kant wanted for the biological concept he was groping for in the race papers.[13]

In the race papers Kant introduces the concept of heredity into the philosophical faculty—into biology—where he uses it with decidedly positive connotations, in connection with adaptation, and he applies it to traits that are neither species-typical nor merely *individual* peculiarities. These considerations are much more interesting than the religious speculations in other works that also invoke the concept of heredity. Although he develops his ideas on heredity and adaptation on the example of *human races* and all the context of eighteenth-century anthropological discourse is relevant to a comprehensive historical explanation of Kant's theorizing here, I will be dealing only with one small aspect of his racial thought, an aspect that applies just as well to dogs and pigs. Races of dogs for Buffon, races of pigs for Blumenbach, and races of humans for Kant were also occasions to discuss the question of *subspecific* variations that breed true. I will stick to the narrower biological context.[14]

12.3 Hereditary Adaptation

In the three papers on race Kant is primarily concerned with introducing and justifying a distinction between various kinds of *subspecific* classes: in particular between races and varieties. The theoretical background of the distinction is "Buffon's rule" that animals that can produce fertile progeny together belong to the same species. The ostensible empirical phenomenon that Kant takes as the point of departure for his reflections is the experience of the Portuguese colonists in Africa and of black Africans transported to Europe. Kant not only consumed massive amounts of anecdotal travel literature, he also followed the more serious literature on the stability of racial traits and on the results of intermixture very closely—even to the extent of taking a position on the Spanish crown's policies on interracial marriages in Mexico.[15] While he draws conclusions about other races as well, the central question is always about blacks and whites.

12.3.1 Theoretical Background

From Buffon's breeding criterion and the ability of all humans to interbreed, Kant draws the conclusion that all humans whatever their morphological differences should be viewed as descendants of one common stock and thus not only as belonging to the same species but as constituting one family. Kant wants to distinguish between two kinds of disciplines: a purely descriptive and classificatory discipline—which he calls *Naturbeschreibung*—and a causal explanatory discipline—which he calls *Naturgeschichte*. (Linnaeus and Buffon provide the prototypes of these two disciplines.) The first gives us a school system for our memory, the second a physical system for our understanding. But, as we will see, Kant's causal, physical natural history also has something contingently historical about it. The classificatory traits he is interested in are not explained by natural law alone but also by the contingent family history of man.

12.3.2 Empirical Phenomenon: Skin Color

The direct effect of the climate in Africa according to Kant is to darken the skin, but the Portuguese who had been living in tropical Africa for many generations were just as white at birth as their countrymen in Lisbon. And the blacks transported to Europe did not seem to be turning white—there was much literature from Paris about blacks and one of the Hessian Dukes had his own little imported colony of blacks established near Kassel. Furthermore, it is assumed that dark skin is *beneficial* in the African climate and white skin in the European—that is, that skin pigmentation is adaptive. The basic questions that arose were: Why don't the later generations of Europeans in Africa turn black? Why don't the Africans in Europe turn white (again)? And how did the blacks get black in the first place? The easy way out of course would be just to deny the phenomenon and assume that the time spans needed to adapt skin color are very very great,[16] thus expecting that the Portuguese in West Africa would eventually turn black and that the Africans in Paris would turn white. But Kant takes the occasion to ask a principled question: How can something purposive like a *new* adaptive trait arise and become heritable—in fact become heritable in such a way that it cannot be removed?

Kant's first step is to distinguish at least four kinds of subspecific taxa, which can be arranged on a sort of scale according to how invariably (*unausbleiblich*) the identifying characters are transmitted (figure 12.1).[17]

The most important difference is the difference between races and the other three categories—varieties, sports, and strains—inasmuch as this difference also correlates more or less with that between natural history and natural description. The possession of a character defines a *race* only if it is *invariably* bequeathed to the next generation even when the organisms are transplanted to a new environment and even when they are crossed with individuals from another race. In the latter case all progeny are

Kant on the Permanence of Defining Characters

low heritability	Strain (*Schlag*)	Variety (*Varietät*)	Sport (*Spielart*)	Race (*Rasse*)	**high heritability**
	Variable, dependent on environment	Generally transmitted	Invariably transmitted in incrosses	Invariably transmitted even in outcrosses	

Figure 12.1
Schematic representation of Immanuel Kant's concepts for subspecific varieties. For explanations see text.

half-breed (Kant's term is *halbschlächtig*)—that is, they express a mixture of the two invariable racial characters. In *sports* some traits, like blond hair and blue eyes, are passed on invariably and independent of change in environment, but sports do not produce half-breed forms when crossed with other types, say, when blue-eyed and brown-eyed individuals are crossed. If the trait sometimes disappears in outcrosses, it signals not a race but a sport. In *varieties* the characters are generally but not always passed on in incrosses but are at least independent of the specific environment; not all children of brown-eyed pairs have brown eyes. In what Kant calls a *strain* the relevant trait is environmentally induced and disappears if the organisms are transferred to a different environment. Even if the trait is quite stable and is dependably passed on in the same environment and even if half-breed traits always result when the bearers are crossed with other forms, a trait signals only a strain as long as it is environmentally dependent.

Kant then tells a speculative story about how the original (and presumably white) humans spread out from their original habitat to all parts of the globe. On the way they adapted to the local climate. The problem for Kant is how to explain this adaptation in such a way that it is compatible with his other ideas about the explanation of organisms. Kant doubts that the laws of mechanics can explain the first origin of a purposive organic form—though they might be able to explain its replication—and thus he is committed to some kind of original organization in animals and plants. If the environment could directly change organisms *to their advantage* in such a way that the changes could be passed on, preserving or even improving the purposiveness of their structure, there is no reason why it should not in time also be able to change the organism to any arbitrary extent (even beyond the species boundary) or, for that matter, why it could not have produced them in the first place. There would be no limit to the bizarre forms that creatures might take and it might become impossible to reconstruct the original species form.[18] But Kant claims that it is impossible to imagine how environmental conditions could be able systematically to *cause* beneficial changes that can be passed on. Thus, he insists that all beneficial and hereditary traits must have been included in the original purposive organization. All adaptations to any particular environment must have been part of the original species equipment; they were just not yet expressed—or, in Kant's terminology, they had not yet been "unfolded." Kant asserts that the lineage must have contained "germs" or "natural dispositions" for any adaptive trait that arises and then breeds true. The *germs* unfold into particular organs or traits; the natural *dispositions* are responsible for new relations among given parts. Once brought to expression they are permanent, and thus alternative forms that were possible in the beginning are excluded for the future: the unfolding germs for black skin, for instance, close off the developmental possibilities for other colors. But only beneficial traits already contained potentially in the species form will breed true. The environmental, climatic circumstances cannot produce the new traits; they can only trigger their release. Everything passed on invariably in generation is itself inherited. In somewhat anachronistic terms: every hereditary adaptation is a preadaptation. As Kant puts it in the first race paper,

> Accident or universal mechanical laws cannot produce such fitnesses [*Zusammenpassungen*]. Therefore we must consider such occasional unfoldings [*Auswickelungen*] as *preformed*. However even where nothing purposive is displayed, the mere capacity to propagate its particular assumed character is itself proof enough that a special germ or natural disposition was to be found in the organic creature. For external things can be occasional causes but not productive causes of what necessarily is passed on and breeds true [*anerbet und nachartet*]. Just as little as accident or physical-mechanical causes can produce an organic body can they add something to its generative force, that is, effect something which propagates itself, if it is a particular shape or relation of the parts.[19]

The particular adaptive traits that Kant is interested in here do not occur in every current individual of the species; they also did not occur at all in the original ancestral form: they are historically new, but now permanent. But they are also not mere individual peculiarities. These are not the adaptations of one Immanuel Kant to the cold winter in Königsberg of 1774. It would have been fairly implausible to assume millions upon millions of little germs for this or that possibly useful adaptation. The skin is seen to be the major adaptive interface between organism and environment, and it is skin color and texture that make up the primary racial character.[20] As it turns out Kant assumes only *four* racially relevant sets of germs in humans. He maintains that the human species possessed germs and dispositions for traits preadapted to four basic types of climate. Linnaeus had explained his four races by means of the distribution of four Galenic humors, so that the dominance of, say, sanguineous or phlegmatic aspects might even be taken to give a causal explanation of the racial differences between whites and blacks. Kant on the other hand goes back even further to the presocratic doctrine of elements and the qualities derived from their combinations—cold and wet, hot and wet, hot and dry, cold and dry—and uses them to give a teleological explanation of races and their differences. Europe is of course cold and wet; Africa hot and wet; and the other two races are somehow assigned to the other two climates. As humans radiated out of the Garden of Eden—or whatever it was that Kant thought was located between the 31st and the 32nd parallels in the Old World[21]—the germs and dispositions appropriate to the new climates they first encountered were unfolded and their skins began to assume the appropriate color and structure. Kant also stipulates that this is a historically irreversible choice. The chosen germs gradually shut off the others—that is why blacks cannot become white again. The process of adaptation is irreversible even if the end product, due for instance to later transplantation, should turn out to be detrimental.

12.4 Recapitulation and the Terminology of Heredity

To recapitulate briefly, the relevant assertions of Kant's are that

1. All humans belong to one species.
2. Racial differences are primarily due to adaptation to different climates.
3. Any trait that is dependably passed on in generation is presumed to be adaptive, to belong to the purposive basic organization of the species.
4. Even traits that only arise at some particular point in time, insofar as they afterward breed true, must have been contained as potencies in the original equipment of the species.

In trying to deal with these questions Kant introduces a complicated and basically untranslatable jungle of special terms to distinguish various aspects, permutations, and combinations of hereditary phenomena. (And thus it is not surprising that the term that later came to be used for heredity is among the crowd.) Here is a list of the major entries, arranged according to the roots *erb* and *art*:

erben—to inherit (an *organism* inherits a trait)

erblich—hereditary (*traits* are hereditary)

anerben (intrans.)—to continue or breed true (a *trait* breeds true)[22]

ererben—to inherit (an *organism* inherits a trait)

forterben (intrans.)—to be continued (a *trait* is continued)

vererben—to pass on (an *organism* passes on a trait)

sich vererben—to pass oneself on (a *trait* passes itself on)

abarten—to degenerate, deviate from the original form (an *organism* deviates)

anarten (*Anartung*) (intrans.)—to be propagated (a *trait* is propagated)

anarten (*Anartung*) (with dat. object or with preposition *an* + dat./acc.)—to adapt to (an *organism* adapts to a climate)

ausarten (*Ausartung*)—to deviate beyond the possibility of reversal (literally *to speciate*, but used here for change within the species that establishes a race)[23]

einarten (with preposition *in*)—acclimatize (an *organism* fits into an environment)

nacharten (*Nachartung*) (trans.)—to take after (an F_2 *trait* takes after an F_1 *trait*) *nacharten* (*Nachartung*) (intrans.)—to continue or breed true (a *trait* breeds true)

Verartung—diversification (of a *species* into geographic varieties)[24]

What is important in this terminological morass is that Kant is systematically groping for conceptual tools to deal with phenomena we conceptualize using the notions of heredity and adaptation. For the purposes of this analysis the following results of Kant's speculations seem to be historically relevant.

First, the characters focused on are not *species* characters determined by divine preformation or physical natural law. The germs and dispositions do still fit the deterministic eighteenth-century model of pregiven potentialities, but at least the fact of their empirical expression as traits is a product of contingent history.

Second, some of the contingent traits that an organism possesses solely because its parents happened to possess them are not just accidental modifications but rather binding choices among species-given possibilities. They are thus perhaps susceptible to rules and may be explainable by the same kinds of mechanisms as specieswide traits.

Third, the realm of the hereditary is not confined to defects or mere peculiarities but also includes beneficial traits—and, as pursued by Kant in the "philosophical" faculty, it includes only beneficial traits.

Fourth, hereditary characters of the kind discussed must themselves have been inherited. Environmentally induced characters are not hereditary.

Now the question arises, whether Kant's verbal calisthenics with the German roots *erb* and *art* is more than just one of many similar Enlightenment gropings for concepts that later generations were to invent. Does it actually lead anywhere? It is true that for Kant historical contingency remains within the limits of the species lineage and to that extent his gropings remain firmly within the boundaries set by Enlightenment paradigms, but Kantian races have all the characteristics of traditional species except the sterility of hybrids. They provide examples of lawlike purposiveness conceptualized as improvements on a given species form relative to contingent actualizations of basic climatic possibilities. The racial characters are neither completely determined by species membership nor are they merely individual, arbitrary, or accidental. They are phenomena that do not fit and thus pose a problem and a challenge. They partake of both worlds, the lawlike and the contingent, and thus they must at some point lead to inconsistent theory—as long as the lawlike is confined to the species. This threatened inconsistency need not be unproductive. The contingent but hereditary racial characters present an occasion for the application of available concepts and theories to situations for which they were not originally devised or intended, and the inconsistency or instability that arises may lead to some change or development in these concepts and theories. Kant's speculation on the contingency and necessity of heredity and adaptation does indeed presuppose the traditional species concept, but in doing so it undermines it.

Notes

1. Everard 1661; Gassendi 1658, 260–262, both excerpted and translated by Howard B. Adelmann 1966; Highmore 1651; Bourguet 1729, 154–156.

2. On some of the theories of generation see Terrall, chapter 11, this volume; on the study of hereditary diseases see López-Beltrán, chapter 5, this volume, and the literature cited there.

3. "We formerly remarked that every thing that possibly could be, really did exist; of which the sloths are a striking example. They constitute the last term of existence in the order of animals endowed with flesh and blood. One other defect added to the number would have totally prevented their existence" (Buffon 1749–1767, vol. 13 (1765), 40; Buffon 1780, "The Sloth," vol. 9, 7–8).

4. Buffon 1774–1789, vol. 2, 496–515, vol. 4, 364; Blumenbach 1806, 19.

5. Hume [1779] 1990, parts 2 and 3. See McLaughlin 2005.

6. "If nature were not the hand of creative wisdom, then there would be differences just as much in the basic constitution as in the small and numerous parts of the structure, but nonetheless the latter occur constantly and the former never at all" (Haller 1752, unpaginated).

7. *Von den verschiedenen Rassen der Menschen*, 1775. This was Kant's only publication during his so-called silent decade preceding the *Critique of Pure Reason*. It was revised and republished in *Der Philosoph für die Welt*, ed. J. J. Engel (Leipzig, 1777); here: Kant 1956–1964, vol. 6, 9–30.

8. "Bestimmung des Begriffs einer Menschenrasse" (*Berlinische Monatsschrift*, November 1785), Kant 1956–1964, 63–82.

9. The term *Vererbung* also occurs once in connection with hereditary vice and disease in the *Physische Geographie*, published by Rink in 1801 and based on Kant's earlier lectures (Kant 1900, vol. 9, 456).

10. *Die Religion innerhalb der Grenzen der bloßen Vernunft* (Kant 1956–1964, vol. 4, 689).

11. "For, according to the hypothesis of epigenesis, the mother, who is descended from her parents through *natural* generation, would still be tainted with this moral blemish, and would bequeath [*vererben*] it to her child, at least half of it, even in case of a supernatural generation. Consequently, in order for this not to be the case, the system of preexistence of the germs in the parents must be assumed, but not the [system] of development on the female side as well (because through this the above consequence is not avoided) but only on the male side (not the [system] of *ova* but of *spermatic animalcules*); which side is omitted in case of a supernatural pregnancy, and thus this way of representing [generation] can be defended theoretically appropriate to that

idea [i.e., virgin birth]" (Kant, 1956–1964, vol. 4, 736). See also the drafts in Kant 1900, vol. 23, 105–108).

12. "*Vererbung*, oder *Verfällung auf einen Erben*, sonst auch die *Transmission* oder *Versendung einer Erbschaft* genannt, Lat. *Transmissio* oder *Transmissio haereditatis* und *Tranmissio ad Haeredes*, ist, wenn ein Erbe, ehe er die ihm angefallene Erbschaft antritt, oder ausschlägt, verstirbet, und solche von ihm weder würcklich angetretene, noch ausgeschlagene Erbschaft in denen durch die Rechte nachgelassenen Fällen auf dessen Erben transmittiret und weiter verfället wird. Daher es denn sonst auch das *Transmissions-* oder *Versend-Recht*, Lat. *Jus Transmissionis* oder *Jus transmittendi* genennet wird. Und zwar geschiehet solches erstlich, wenn jemanden von seinen Eltern, oder anders woher, von Verwandten oder Fremden, aus einem Testament, oder auch ohne Testament, eine Erbschaft anfällt, und derselbige verstirbet innerhalb Jahres-Frist nach erlangter Wissenschaft, als der einem Erben in denen Rechten vergönnten Bedenck-Zeit, ob er nehmlich die ihm solcher Gestalt zugefallene Erbschaft annehmen wolle, so heist diese Art der Transmission auf Lateinisch *Transmissio ex Jure deliberandi*" (Zedler 1732–1750, vol. 47, 510–511).

13. As we know, the ploy did not quite work. It may be a contradiction in terms to speak of the bequeathment of acquired properties in a mid-eighteenth-century German courtroom, but in the real world it is hard to impose the meaning constraints of technical terms out of their technical context. Kant himself also uses the term *Vererbung* later in the more general sense of any transmission of property (e.g., "Über den Gemeinspruch"; Kant 1956–1964, vol. 6, 148), and in his in his own last will and testament he speaks of "bequeathing" his household goods (*von der Vererbung meines übrigen Hausgerätes*), few of which presumably were acquired by inheritance (Kant 1900, vol. 12, 384).

14. For the larger context see Moebus 1977; Barkhaus 1993; Bernasconi 2002.

15. For instance, Kant 1900, vol. 23, 456; on the literature on racial mixing in Mexico see Mazzolini, chapter 15, this volume.

16. In the second race paper Kant (1956–1964, vol. 5, 155) asks rhetorically whether we should wait "12 times 12 generations" to see if the color goes away, asserting that this would be "to string along the naturalist with dilatory answers."

17. Kant 1956–1964, vol. 5, 12.

18. Not being able to reconstruct the original form would compromise Kant's project of developing methods for reconstructing the past in order to pursue natural *history* as opposed to mere natural description.

19. Kant 1956–1964, vol. 6, 18.

20. In the second race paper "Bestimmung des Begriffs" (§3) Kant insists explicitly that skin color is the only trait that is invariably inherited and thus can be used to characterize a race—implying that there are no other hereditary physical or moral racial characters—but his other writings are not consistent in this regard. See Kant 1956–1964, vol. 6, 68–69.

21. The Academy Edition (Kant 1900, vol. 2, 441) changes the text without indication to say 52nd parallel instead of 32nd, so as to include more or less all of Europe. Jerusalem is located at 31° 47′N.

22. Also used as a passive participle: *angeerbt*, "inherited" (an inherited *trait*).

23. Note that *ausarten* here does not refer to the sterility of hybrids but only to the fixation of the racial character. In the third paper on human races, "On the Use of Teleological Principles in Philosophy," Kant distinguishes between *Abartung* (*progenies classificata*) and *Ausartung* (*progenies specifica*), the latter of which he excludes (Kant 1956–1964, vol. 5, 144–145).

24. Used only in Kant's review (1785) of Herder's *Ideen zu einer philosophischen Geschichte der Menschheit* (Kant 1956–1964, vol. 6, 798).

References

Adelmann, Howard B. 1966. *Marcello Malpighi and the Evolution of Embryology*. Vol. 2. Ithaca, NY: Cornell University Press.

Barkhaus, Annette. 1993. *Rasse: Zur Konstruktion des Begriffs im anthropologischen Diskurs der Aufklärung*. Doctoral dissertation, University of Konstanz.

Bernasconi, Robert. 2002. Kant as an unfamiliar source of racism. In J. K. Ward and T. L. Lott, eds., *Philosophers on Race*. Oxford: Blackwell.

Blumenbach, Johann Friedrich. 1806. *Beyträge zur Naturgeschichte*. Göttingen: Dieterich.

Bourguet, Louis. 1729. *Lettres philosophiques sur la formation des sels et des crystaux et sur la génération & le mechanisme organique des plantes et des animaux*. Amsterdam: L'Honoré.

Buffon, Georges-Louis Leclerc Comte de. 1749–1767. *Histoire naturelle, générale et particulière, avec la description du Cabinet du Roi*. 15 vols. Paris: Imprimerie royale.

———. 1774–1789. *Histoire naturelle, Supplément*. 7 vols. Paris: Imprimerie royale.

———. 1780. *Natural history, general and particular*. 9 vols. Trans. William Smellie. Edinburgh: William Creech.

Everard, Anton. 1661. *Novus et genuinis hominis brutique animalis exortus*. Middleburg.

Gassendi, Pierre. 1658. *Syntagma philosophicum*. Vol. 2 of *Opera omnia*. Lyon.

Haller, Albrecht von. 1752. Preface to *Allgemeine Historie der Natur*, vol. 2, by G. L. L. de Buffon. Hamburg: Grund und Holle.

Highmore, Nathaniel. 1651. *A History of Generation*. London: Printed by R. N. for John Martin.

Hume, David. [1779] 1990. *Dialogues Concerning Natural Religion*. Ed. M. Bell. London: Penguin.

Kant, Immanuel. 1900. *Gesammelte Schriften*. Berlin: De Gruyter.

———. 1956–1964. *Werke*. Studienausgabe, ed. Wilhelm Weischedel. Darmstadt: Wissenschaftliche Buchgesellschaft.

Lagier, Raphaël. 2004. *Les Races humaines selon Kant*. Paris: Presses Universitaires de France.

McLaughlin, Peter. 2005. Lehren was man selber nicht weiß. In M. Carrier and G. Wolters, eds., *Homo Sapiens und Homo Faber: Festschrift für Jürgen Mittelstraß*, 157–169. Berlin: De Gruyter.

Moebus, Joachim. 1977. Bemerkungen zu Kants Anthropologie und physischer Geographie. In K.-H. Braun and K. Holzkamp, eds., *Kritische Psychologie: Bericht über den ersten Internationalen Kongreß Kritischer Psychologie, 13.–15. Mai 1977 in Marburg*, vol. 2, 365–380. Cologne: Pahl-Rugenstein.

van Hoorn, Tanja. 2004. *Dem Leibe abgelesen: Georg Forster im Kontext der physischen Anthropologie des 18. Jahrhunderts*. Hallesche Beiträge zur Europäischen Aufklärung 23. Tübingen: Niemeyer.

Zedler, Johann Heinrich. 1732–1750. *Großes vollständiges Universal-Lexicon aller Wissenschaften und Künste*. Halle and Leipzig.

13 The Delayed Linkage of Heredity with the Cell Theory

François Duchesneau

The cell theory as the foundation of a systematic conception of organic units came to influence scientific views on heredity in a stepwise fashion. Initially, when Theodor Schwann (1810–1882) established in the late 1830s that the formation of complex organisms might be reduced to combinations of cells and transformed cells, cell reproduction, starting from the fertilized ovum, would bear the material influence of the organism's "cytoblastemas" and imply global processes responsible for transmission of hereditary types and production of individual variants. Then, in the revised sections of his *Handbuch der Physiologie des Menschen*, Johannes Müller (1801–1858) blended Schwann's theory with a more holistic approach to organic reproduction, development, and heredity. In a second phase culminating in the late 1850s, Robert Remak (1815–1865) and Rudolf Virchow (1821–1902) abandoned the hypothesis that cells might form from "blastemic" crystallization. Though they retained a holistic approach to hereditary transmission of organic dispositions, their principle "*Omnis cellula e cellula*" segmented that approach and opened the way to the attribution of reproductive and differentiating dispositions to inner structures of cell nuclei. Ultimately, in cell theory, the holistic or holistic-analytic interpretation of genetic factors was replaced by a combinatorial conception of postulated nuclear microstructures: these were presumed to possess a special power of expressing similarities and differences in cell replication and of structuring complex organisms through functional operations at the cellular and nuclear levels, later identified and analyzed as mitosis and meiosis.

Evidently, neither of the original phases of the cell theory produced a "scientific" explanation for the hereditary transmission of individual traits and for the ensuing differentiation of organisms comprising a given type. These topics became objects of scientific investigation when the cellular mechanisms in mitosis and meiosis were discovered and genetic models were proposed to account for replication, differentiation, and variation of elementary organic structures and operations. But it is worth inquiring about the thematic space occupied in earlier versions of the cell theory by what would be identified as hereditary processes in later versions. Since these initial versions involved considerable tensions between holistic and reductionistic strategies, it may prove instructive to analyze and compare the models present in both explanatory styles, because they conditioned the later insertion of concepts of heredity into the cell theory.

13.1 Schwann's Cell Theory as a Reductionistic Theory of Type Replication with Variation

In his *Mikroskopische Untersuchungen über die Übereinstimmung in der Struktur und dem Wachstum der Thiere und Pflanzen* (1839), Schwann compares two conceptions of generation, or more precisely, two conceptions of the powers involved in the production of complex organisms from elementary structures consisting of cells.[1] These conceptions can be respectively described as teleological and physical. The teleological conception seems to account adequately for the production of complex organisms building up from germ cells, a process requiring the production and integration of a manifold of organic structures in accordance with the specific type represented by the parent organisms. For Schwann, this view is ridden with insuperable methodological and theoretical shortcomings. The fact of invoking the ontogenetic power of organisms as the unique and self-sufficient cause of so many different chemical and physiological processes does not provide any satisfactory explanation for the emergent complex structure and its variations relative to the integrative pattern of the parent organisms. Methodologically, it is worth trying to derive from the observation of elementary, and more particularly cellular, morphological processes an analogical account of the complex living entities that are reproduced with architectonic and functional similarities and variations. From a theoretical point of view, also, the teleological postulate presents significant deficiencies. By supposing that the germinal cell or cells possess a partial simile of the architectonic power owned by the adult organism, how can we hope to account at once for the segmentation of this integral power in multiple sets of specialized cells when organic parts develop successively from the germinal originals, and for the production of an integrative whole that supersedes the morphogenetic capacity of its compounding elements? At the same time, empirical investigations have shown that organic architectures, even those of the more complex type, rely on elementary parts endowed with specific properties: organisms are either cells or combinations of more or less transformed cells. Appealing, as Johann Friedrich Blumenbach (1752–1840) had done, to a *Bildungstrieb* for framing up complex organisms as exemplars of a similar specific archetype[2] appears as a simplifying strategy that discards the complex molecular processes taking place in cells during the various ontogenetic phases.

The physical conception, on the other hand, besides avoiding the shortcomings of the teleological conception, develops analogies based on experimental evidence. If we suppose that the functional properties of organisms depend on material dispositions, such as specific combinations of organic molecules occurring in the egg, it is relatively easy to find morphogenetic connections linking cellular constructs to this primordial structure; and from there, higher-level cellular constructs, through various interactions, would bring forth the functional dispositions of the integral organism. This model is congruent with the morphogenetic processes as far as they have been and can be observed. Cells frame up in a liquid cytoblastema contained inside, or surrounding, previously existing cells. Then, a typical stratification occurs from nucleolus to nucleus and from nucleus to membrane, according to constant determinations, which may nevertheless entail factors of differentiation to be determined. The point is not here to define the fundamental forces operating in organization, but to account for the phenomena of cell development and differentiation underlying all histological and physiological processes. In this perspective, it is conceivable that anomalies will be progressively accounted for, that problematic phenomena will be solved, that the law of development and its corollaries will be adjusted to match the mechanism of phenomena more closely. Thus, the synthetic order of the new general anatomy provides a comprehensive view of the formation of elementary organic parts. First, the properties attributed to the cytoblastema indicate how new cells are presumed to generate. In turn, these modes of generation suggest principles that would regulate the processes of cell evolution and account for such metamorphoses as produce the various elementary and complex structures of organisms.

In a Schwannian approach, the predisposing and predetermining factors of cell formation reside in the cytoblastema, due to its special physical and chemical composition. In a way, as historians of science have shown, the cytoblastema assumes the functions of what Claude Bernard (1813–1878) will later call "milieu intérieur."[3] The structures that will emerge from a given cytoblastema will tend to reproduce cells of the type or types that had originally determined the formation of this precise cytoblastema. The special constitution of a cytoblastema will thus be reflected in the sedimentation process Schwann identifies as capable of producing nucleoli, nuclei, and membranes in succession. Inside the triple sedimentary structure, physical and chemical processes take place that will in turn affect—that is, determine and diversify—the potential of both interior and surrounding cytoblastema for further cell generations.

Considered the ruling structure in cytogenesis, the nucleus is always dependent on the typical features of the cytoblastema it originates from, and it redetermines in due time what may be viewed as the architectonic properties of subsequent cytoblastemas. Since the composition of a given cytoblastema, which is structurally determined by its cellular surroundings, is also contingent on external factors—for instance, nutriments—this cytoblastema will support a line of replication of original organic forms fostering sequences of specific cell types, and a line of diversification of organic productions reflecting the action of contingent factors on cytoblastemic compounds.

Plastic phenomena depend on the special arrangements of molecules forming cell structures; metabolic phenomena express changes in chemical composition and processes occurring inside the cell or in the surrounding cytoblastema as a consequence of the action of existing cell structures.[4] The analysis of cell processes will imply the constant conjunction of both types of phenomena: on the one hand, the mode of apposition of molecules to form cellular strata determines activities that pertain to cellular metabolism; on the other hand, this metabolism interferes in turn with the morphogenetic potential inherent in the cytoblastema. Triggered by stratification, metabolic activity increases structural differentiation in the cell and fosters the emergence of specific physiological properties. In reference to such a process, Schwann introduces the assimilation of nuclei to first-order cells, and of nucleated cells to second-order cells. This theoretical view would be supported by the homology of stratification processes in both cases; and it would provide explanation for the fact that intussusception and metabolic activity are nucleus-dependent before they expand more intensely and more fully inside the subsequently formed cell membrane.

The synthetic presentation of Schwann's theory needs to be completed by offering empirically derived and analogically conceived views on the way cells are modified into derivative organic forms. Explanation here conjoins chemical combinations, mechanical processes, and the end effects of a developmental principle whose action varies according to surrounding conditions. Chemical combinations pertain to the membrane and cell contents and determine stratifying effects through deposits on the membrane. Mechanical dispositions can provoke important changes in form: this is, for instance, the case when osmosis determines the flattened shape of blood globules. But, in the end, physical and chemical conditions fall short of explaining, as it seems, the phenomena of differential proliferation and of the more significant metamorphoses: "We must admit modifications in the principle of development of the cells, of such a nature, as that the force, which affects the general growth of the cells, is enabled to occasion an equable disposition of new molecules in one cell, and an unequal one in another."[5] The force mentioned can be similar to those intervening in inorganic nature. Instead of appealing to a *Bildungstrieb*, it may suffice to consider that the phenomena of differential growth and morphogenesis result from the interaction of structural devices specific to organized matter and of forces analogous to those governing inorganic formations. But, for these various types of phenomena representing the modes of derivation of apparently noncellular composite structures, the analysis of processes involved still remained for the most part to be achieved at the time Schwann was formulating his theory.

As noted by several writers,[6] Schwann's conception of cells was structured on two founding principles: (1) the universal derivation of organic structures from cells—a principle which will remain; and (2) the production of cells from cytoblastemic "crystallization"—a principle that will be challenged and replaced by the principle of the formation of cells from cells. But Schwann's theory is also original insofar as it anticipates, through its insistence on the structuring power of the cytoblastema and the stratification processes involving the cell's inner molecular dispositions, such protoplasmic hypotheses as Hugo von Mohl (1805–1872) and Ferdinand Cohn (1828–1898) will develop during the 1850s.[7] The force of cellular attraction would act the more intensely as the concentration of fluids is weaker, and this would account for the production of strata nested in one another. The same relation would hold for the differential development of organic tissues, once a constant degree of structuring force is ascertained by the available empirical data.

Metabolic processes are subject to the same theoretical representation as plastic processes. The different classes of cells modify chemically the organic fluids they absorb in various ways and they are reciprocally and similarly modified in their own specific structures. Insofar as the plastic and metabolic powers involved are connected with imperceptible microstructures and microprocesses, which exceed our present analytic capacity, the methodological procedure will consist in identifying in the process of cell building such determining conditions as underlie the transmission and differentiation of higher-level organic traits. To sum up, the expansion of research on the phenomena of cell life, whether plastic or metabolic, must fit analogical schemas that remain within bounds of what can be deduced from observed morphogenetic phenomena. In this context, the elements of a cytological theory of heredity are only sketched in very abstract and provisional fashion, insofar as all morphogenetic sequences, from primary cell formations in embryonic development to the production of whole organic structures through combinations of specialized cell layers, point to complex chemical reactions generating jointly type replication of cellular structures and individual variations of the transmitted organizational pattern.

13.2 Müller's Version of Schwann's Cell Theory as a Holistic Theory of Type Replication with Variation—Remak's and Virchow's Revised Theory

Schwann framed statements of laws concerning the phenomena of cell life and presumed that the whole architectonics of vital operations in the most complex of organism would result from the interrelation of dynamic devices in individual cells: these devices themselves would emerge from the potential of typical replication and variation inherent in the cytoblastema. Opposing a physiology governed by a synthetic approach to organic processes, even when these are analyzed through experimental means, Schwann rejected the subordination of organic effects to an integrative design pertaining to the construction plan of the whole organism. Such a position was doomed to raise criticisms and provoke adjustments of the theoretical premises involved, at the very moment when Schwann's experimental findings were fostering experimental investigations on the inner structures and processes of cells and cell-type elements. This assimilation and remodeling of Schwann's ideas took place initially among Johannes Müller's research team at the University of Berlin. The chief investigator of the group, Müller himself, proposed a recasting of Schwann's views to make them fit into his own research program. By so doing, he most probably influenced transformations of the cell theory that will be achieved later by Remak and Virchow. Müller's physiological "system" and the theoretical principles underlying his empirical investigations were already set when Schwann published his influential papers in 1838 and his *Mikroskopische Untersuchungen* in 1839. Müller took Schwann's discoveries into account in his small treatise on cancerous tumors, *Über den feinern Bau der krankhaften Geschwülste* (1838), in some sections of volume II (1840) of his *Handbuch der Physiologie des Menschen*, as well as in the revised *Prolegomena* for the third edition of volume I (1844) of that handbook. Some commentators, principally Brigitte Lohff and Timothy Lenoir,[8] have rightly stressed the strong coherence of the cell theory adopted by Müller with the physiological doctrines of the *Handbuch*, including the *Lebensprincip* theory. The main questions relate to what Müller decided to integrate and how he proceeded in so doing: selection and modification were rather effective in this instance and went along with the production of new frame principles to contain Schwann's empirical generalizations and analogies.

Müller's position on cell life is governed by the idea that the whole organism exerts a regulative and architectonic role over the fragmentary processes that occur in it.

Indeed, a single cell may be considered a miniature organism, but then it should be conceived as programmed according to a global functional order expressing itself in, and by, plastic and metabolic activity. Once a cell exerts its activity in a complex organic system, it is determined by an encompassing vital force matching the organism's morphological complexity. This hegemonic vital power of the organism is reflected by, or expresses itself through, determining factors in the composition and formative dispositions of cytoblastemas: hence the production of subordinate cellular structures characterizing the inherited, as well as diversified, formative potential of archetypal organisms. If Müller adopts Schwann's model for the morphogenetic sequence of cancer cells, he integrates it under a concept of organism that is more unifying than aggregative. To use Müller's own phrase, the "monadic" life of cells refers to a system of harmony ruling over the whole composite organism.

This global thesis on the relation between cells and the architectonic power embodied in the whole organism is to be found in the sections of the *Handbuch* in which Müller annexes Schwann's findings. Schwannian conceptions are involved in phrasing a "theory of vegetation and organization,"[9] in providing an explanation for nutrition and growth, and, finally, in accounting for generation and development by reference to the cell considered as a reproductive organ. For Müller, Schwann's theses possess a capacity of theoretical synthesis that can unify the various aspects of experimental physiology. First, Müller finds in Schwann's *Mikroskopische Untersuchungen* a developmental law that applies universally to all vegetal and animal organisms. According to this law, the elementary parts of all organisms are cells or structures deriving from transformed cells that emerge in definite ways through specific cytoblastemic organization. The Schwannian cell is not only a morphological, but a physiological, unit. This definitional feature pertains to the capacity of the cell to produce similar cells, either inside or outside itself, by such modifications of the cytoblastema as result from its metabolic activity. This is a function of the proper life of cells in the global organism. From that starting point, analysis can detect the phenomena of assimilation-disassimilation, growth, and metamorphosis that characterize the organism's more complex and architectonic functions presupposing an organization plan relating the parts to the whole. Since the cell possesses implicitly as its essential disposition the dynamic capacity of the whole organism, cellular morphogenetic sequences will unfold the correlative organizational plan as a law of type replication and differentiation for all organic structures and processes.

As concerns the use of the cell theory in explaining phenomena of generation and ontogenesis, the basic facts are as follows: one must admit the cellular nature of the germs of animals and plants; on the other hand, the embryo is initially reducible to a cell or a cluster of cells; that embryo grows by replication of cell formations similar to the original ones; metamorphoses provide the primary derivative elements. At least for lower organisms, a cell detached from the whole organism owns the power to multiply in such a way as to reconstitute an analogous whole. These facts can justify two different sorts of hypothesis on the generative mechanism.

The first hypothesis supposes, in each cell of an organism, a reproductive potential relative to the combination of cells characterizing that organism. It is based on a regressive inference. The various types of cells combined in the actualized organism originate from "analogous" cells in the embryo's elementary parts. Ultimately, one or a few germinal cells are responsible for the embryonic architecture, which is later subject to diverse replications and morphological differentiations. In certain rudimentary organisms, such as trychomycetes and hydras, every cell can serve as an "embryological model" and reconstitute the whole organism. Indeed, this formula appears less likely to be generalized to more complex and differentiated modes of organization. If we suppose, however, that each cell is an autonomous whole comprising full organogenetic power for a specific type, how can the various cells constitute an organic complex by reciprocal limitation of this distributed potential? The hypothetical solution consists in admitting a reciprocal action of the diverse cells determining such a regulation, with the possibility also that certain cells can play a more or less marked hegemonic role toward others (by reference to the Republic, hive, or polyp metaphors). Such a model can only match the facts if the reproductive function of most cells can be restricted to the sole replication of their specific and differential type. Müller suggests that a particular metabolism occurring in that type of cells might suspend their vital activity insofar as ordained for reproducing the integral organism: a gradual functional death of specialized cells would therefore occur. The cells of the hydra's arms would not be able to reproduce the whole organism they were generated from, while the cells of this rudimentary animal's body would have preserved more of such a power. Indeed, this first hypothesis appears more congruent with Schwann's theses and the analytic principle of the individual interaction of cells with one another. Müller's criticisms are here quite meaningful: "In this hypothesis, too much importance is attributed to cells. It is so difficult to apply it to the higher animals that it becomes improbable as a general theory, while no one will deny its relevance insofar as the lower animals are concerned."[10]

The second hypothesis is that which Müller retains. It supposes that the germinal cell or cells are the only ones to possess full organogenetic force. This force can undergo some expansion in the initial phases of embryonic development, but once the typical functional disposition of the organism is determined, the cells composing that structure are only called on to reproduce multiple copies of themselves, but not the whole itself. Müller distinguishes between the "implicit whole" that can be attributed to the germinal cell or the reproductive cells of the bud, and the "explicit whole" consisting in a complex set of cells of distinct types and comprising for each type an indefinite number of copies. The explicit whole can thus be taken at its minimal stage, in the embryo displaying the archetypal set of specialized cells initially gathered. The implicit whole subsists nevertheless within the explicit whole. On the one hand, it affords a sufficient reason for the processes of regeneration of amputated parts; on the other hand, it affords an even more general sufficient reason for the reconstitution of germinal cells and reproductive buds in the adult organism. In the epigenetic tradition inherited from Blumenbach, Müller underlines that the reproductive power, in short the "*Bildungstrieb*" or "*Lebensprincip*," manifests the notable property of not undergoing invalidating alterations when there occurs a multiplication of rudimentary organisms through natural or artificial division. This phenomenon is therefore relatively autonomous from the interrelation of the multiple cellular parts in the organism. Hence a negative analogy against the former hypothesis that counts at the same time as a positive analogy for the latter. In these conditions, Müller concludes in favor of the second hypothesis in a passage in which he compares the generative power in the multicellular organism with the architectonic force expressing itself in the formation of the embryonic archetype:

> Not only do all organized beings, from the first moment of their development, produce cells, but, while in this way they constantly increase the sum of their constituent parts, they form cells or clusters of cells, which are virtually a whole, that is which possess the power to produce all cells destined to particular ends. The growth of all organized beings comprises therefore two very different things: first the implication of the individual form by the multiplication of its constituent particles; then the multiplication of the specific form in an undeveloped state in which everything that must be separated later remains fused together, in a word, either under the shape of a bud containing itself everything it is to develop, or under the shape of a germ that can do the same only after having undergone fertilization.[11]

Müller sets the necessary correlation between the two modes of reproduction in every multicellular organism: on the one hand, the replication of the individual organic form (the "mechanism of individual organization") by indefinite multiplication of the cells of a specialized type; on the other hand, the reproduction of the virtual whole corresponding to the specific organic form (the "potential or not yet developed manifold"). This latter form is embodied in "primordial cells" or in given combinations of such cells, for instance in the generation by sexual interaction, or in the initial phases of embryonic development, between the fertilization of the ovule and the framing up of a combination of archetypal cells (which corresponds to von Baer's "*Urtypus*"). But when trying to conceive the reproduction of these potential manifolds, there is no need to fall back on Schwann's model recast in the first hypothesis. For, by itself no cell or set of cells seems to suffice to embody, by the sole intervention of determining cytoblastemic conditions and the interaction of individual cells, the power of regular and differentiated organogenesis that will yield the effective multicellular organism. Neither the germinal cell nor the primary structures deriving from it could express by their inner disposition the integral "*Bildungstrieb*": the latter surpasses the mere correlation of structures, for it anticipates potentially the emergence of a complex structure not yet present in the dispositions of germinal cells and immediately derived formations. The cell appears then as a mere vector for the manifold to be developed, a point of articulation for the *Bildungstrieb*: it triggers on by its inherent activity a morphogenetic process whose sufficient reason is of a higher and more global order. Hence the need to admit an archetypal role for the implicit form of the individual organism, which presupposes the preexistence and influence of previous complex structures, those of the parent organisms, in determining the contingent outcome of the generative process.

While annexing the sequences of cellular developments revealed by Schwann, Müller did not discard the synthetic approach to morphogenesis. By criticizing the first hypothesis more congruent with Schwann's analyses, he would underline the inadequacy of a model that grants too much to the functional features of individual cells in accounting for the morphogenesis of higher-order organized beings. Only the correlation of a "*Bildungstrieb*" with the inner dispositions of germinal cells might afford the sufficient reason for this type of process.

For Müller, the formation of cells, in the embryo itself or in the cytoblastemic milieu in which the embryo develops, is an instrumental mode of operation for the "*Lebensprincip*": the fundamental architectonics of the achieved organism could not be reduced to this sole instrumental mode. If certain cases might support a different interpretation, they would probably be found in the manner in which relatively less complex organized beings are formed and reproduce their type with variations: in such instances, each cell of the actualized organism seems to possess the entire embryogenic power. Those cases would be the only ones providing some positive analogy in support of the Schwannian hypothesis. Let us consider the regeneration of the polyp, truncated and parceled out, reduced at the limit to a single of its constituent cells and still able to recompose the actual whole. Müller illustrates the mechanism involved in this transformation as follows:

If we regard the entire polyp as a system of molecules, of cells, all similar in virtue of the determinate force they are animated with, remaining subordinate to the individual organizing principle only as long as they have a certain affinity with it, and if, on the other hand, we regard the individual organizing force as resulting from the convergent interaction of molecules, we conceive that the cut portions still contain systems of similar molecules. Here also the organizing principle acts in such a way that, because of the affinity of molecules with one another, the detached portion reconstitutes the organization of a new polyp.[12]

The nature of the individual organizing principle or force needs to be clarified. The relation it holds to the activity of individual cells is presented as an affinity relation, expressing itself through the collaborative action of cells. In this type of organism, the organizing principle operates on more or less large sets of homologous cells, provided these sets comprise the instrumental dispositions required for recombining the organic structure. Let us suppose that the organizing principle can exert itself in a single cell homologous in its dynamic disposition to some multicellular sets that comprise multiples of similar cells. Even then, the organizing principle differs from this individual dynamic disposition that serves as its vector, for it represents the embryonic whole of a multicellular organism to be actualized. The cellular model affords a system of determining conditions for organic formation and differentiation, but Müller maintains the specificity of a vital functional reason for such processes. Thus, he opposes the reductionistic approach set forth by Schwann.

From a general viewpoint, Müller's position consists in reinserting Schwann's models in a theory dominated by a synthetic notion of living organism. As a consequence, the analytic procedure that Schwann had followed takes on a derivative meaning and plays an instrumental role relative to the regulative concepts of Müller's theory. Normal and pathological cytogenetic phenomena support the representation of a determining architectonic design for complex individual organizations congruent to specific types and inherited with variations because of the cytoblastemic circumstances affecting their reproduction. Since he distinguishes between cells that possess the power of reproducing the virtual whole organism and derived cells that have ceased being able to enact that power, but can only replicate individual units of their own type, Müller acknowledges in a way the need to identify the conditions of cell reproduction that underlie this functional distinction and explain the type modeling, as well as the particularizing individuation of complex organisms.

Among Müller's immediate disciples at the University of Berlin, the research agenda will bear directly on the inner metabolism of cells on the one hand, and on the sequence of cell development in the embryo on the other. Concerning the mechanism of cell reproduction, the trend is toward favoring endogenesis as evidenced in division processes affecting previously formed cells. Robert Remak will be the first to officially reject Schwann's principle of cytoblastemic formation. This talented histologist's interest in the cell theory had developed as he was undertaking ambitious embryological investigations. As a result, he proposed an inductive synthesis on the modes of cell reproduction in his *Untersuchungen über die Entwickelung der Wirbelthiere* (1855). Since 1839, the observations of Albert Kölliker (1817–1905) on the embryo's blastomeres had established that they consist in cellular structures. Also, a certain number of observations, especially those of Karl Nägeli (1817–1891), were suggesting modes of division of the cell's inner structures—protoplasm and nucleus. For instance, in 1846, Karl Ernst von Baer (1792–1876) had provided a remarkable empirical analysis of nuclear division in the embryo of *Paracentrotus lividus*. Along the same lines, Remak argues that cell replication supposes that the unitary organic disposition of the parent cell is maintained through the process that generates the new cell, and this process entails a regular division of the cell from nucleus on. Based on the architectonic role played by the nucleus and its integrative elements, the new theory focuses on a developmental process from center to periphery, but, reciprocally, this process would be determined from periphery to center by the metabolic alterations occurring in the protoplasm. As a new formation principle, cell division conjoins a duly specified morphogenetic sequence that conforms to the organic type, and an integrative conception of the modifications taking place in the cell. Remak undertakes to discover in the fertilized egg how blastomeres are formed and multiply: he thus hopes to trace back the progressive differentiation of the various sets of cells in the embryo. In cell derivation, a uniform division mechanism would prevail, but, correlatively, under the uniform mode of cell organization, structural differentiations would condition the emergence of the more complex organisms and their diversified functions.

When he states his axiom "*Omnis cellula e cellula,*" Rudolph Virchow resumes the principle of cell reproduction Remak had proposed, but he uncovers a special territory scarcely explored by Müller and his disciples—cell pathology—thus inflecting significantly the common research program on elementary structures of the living. Already sketched in 1855, Virchow's version of the cell theory is principally developed in his masterwork, *Die Cellularpathologie in ihrer Begründung auf physiologische und pathologische Gewebelehre* (1858). Thus the notion of cell condenses in itself a manifold of meanings. It is at once a morphological and morphogenetic category that makes for the integration of physiological activities. In particular, the cell is viewed as the sole integrative form enabling vital nutrition processes. As nucleated protoplasm, the cell, whatever its diverse modalities of inner functioning, represents the basic developmental form for all vital organizations. Even the highly differentiated cells that yield specific functions are only characterized by their original modes of nutritive metabolism. On the other hand, the cell, as the ultimate living element, is a structure to which the principle of totality applies: the organism is only a derived whole resulting from the association of structurally and functionally diversified sets of cells. The nucleus assumes the principal architectonic role in cell reproduction, since the division of primordial cells starts with that of their nuclei. Virchow opens up for observation and analysis the vast area of those metabolic dispositions that underlie at once type replication and specific transformation of cells and derivative organic structures. Above all, the functional micromorphology of living units underpins theoretical interpretations. Thus, Virchow invents the concept of "cellular territory," meaning thereby the zones of metabolic intervention of types of cells exerting their action on organic fluids and on cell-secreted interstitial substances. Through their respective territories, interacting sets of cells may influence the way each develops and diverges from its original type.

Working out a research program like Virchow's implies a considerable range of possible adjustments. For instance, Virchow's contemporaries and successors proposed alternative interpretive models that were more or less mechanistic and reductionist, or vitalist and holistic. But all such analytic developments had to prove compatible with the notion that each elementary living unit possesses potentially the power of type replication and that of particularized histogenesis. A materialist like Ernst Wilhelm von Brücke (1819–1892), promoted a protoplasmic theory and professed that the most complex morphogeneses emerge from chemical interactions involving interrelations between the inner components of cell protoplasm and impinging extracellular agents, but he would yet maintain the principle that cells are elementary organisms endowed with special functionalities, even though this specificity depends on complex molecular dispositions and should ultimately reduce to these.[13]

13.3 A Contingent Epistemological Link with Later Cytological Theories of Heredity

Not surprisingly, this founding stage of the cell theory represented a latent speculative and problematic view relative to what later biological science will identify as hypotheses about heredity, empirically derived and capable of experimental confirmation. But, at the same time, followers of Müller, Remak, and Virchow would attempt to link their cytological principles to a concept of transmission of specific life forms (*Lebensformen*) through successive generations of animals and plants conceived as alterable varieties of cell-composed organisms. Müller's preferred option involved transmission of individuals' organismic potential from one generation of germ cells to the next. The expression of that potential in the next generation implied the subsequent differentiation of structures and functions according to diverse cell types combining to form the organism. Such a theoretical approach connects easily with the model analyzed by Ohad Parnes in chapter 14 of the present book, which stressed the inheritance and variation of organic features between generations of individuals. Nägeli—who adhered to that model—also adhered to a concept of cell replication and differentiation resembling Virchow's. His version of the cell theory integrated Müller's principle of transmission of morphogenetic potential from a generation of germ cells to the next with subsequent causal effects on the differentiation of the various types of organic cells. Nägeli mentioned transmission of predispositions (*Anlagen*) that combined specific features and individually gained variations along sequences of generations.[14] The very last section of Müller's handbook was devoted to "*Schlussbemerkungen über die Entwickelungsvariationen der thierischen and menschlichen Lebensformen auf der Erde.*"[15] While ascertaining that there are constant specific types through generations, Müller argued that each such type constituted a circle of variation ("*Variationkreis*") from which the various races would form. In the succession of time, a manifold of internal and external conditions modified individual organisms in various ways. In some cases, similarly affected organisms interbred persistently so as to transmit a significant and sufficiently stable racial pattern through subsequent generations. But such patterns would remain capable of further variations, which might even imply reversion to ancestral features or production of alternate racial traits. In Müller's words, "The more unions recur among similar individuals, without outside blends, longer remains the type to which parents belong. In this way and apart from any external influence, a persistent race can be produced, which fits inside the circle of possible variations for that species, while resulting from variations that internal causes generate."[16] In post-Müllerian physiology, the presumption behind this statement will be that internal causes of that type must be accounted for by transmission of the full morphogenetic potential involved in germ cells, a potential that will be differentially expressed into specialized breeds of organic cells within the developed individual. The model also implies that transmission should be conceived in such a way that a potential for further variation may characterize subsequent generations of organisms.

It is not without significance that Émile Littré (1801–1881), editor of the 1851 second French edition of Müller's *Handbuch*, felt the need to add a "Note additionnelle sur l'hérédité" to the last chapter about varieties among animals and the human species.[17] In this note, Littré exposed speculative views somehow connected with Müller's version of the cell theory, and these illustrated the provisional character of such notions about heredity as were available at the time. Littré simply summarized the main tenets of Prosper Lucas's (1805–1885) *Traité philosophique et physiologique de l'hérédité naturelle dans les états de santé et de maladie du système nerveux* (1847–1850). This work was replete with speculative schemes borrowed from the *Naturphilosophie* of Goethe, Oken, Schelling, and others, the main physiological reference being to Karl Friedrich Burdach's (1776–1847) treatise *Die Physiologie als Erfahrungswissenschaft* (1826–1840).[18] On the other hand, it contained an immense collection of more or less verified empirical observations about hereditary transmission and variation of traits in the most diverse species, a reason why Charles Darwin (1809–1882) among others referred to it. Natural heredity being properly inaccessible in its anatomical and physiological substrate and causal determination, it was scarcely possible to reach beyond plain empirical generalizations, for instance on artificially induced breed variations, or on transmission of so-called hereditary diseases. But these inductions on pedigrees were themselves rendered confused by the hypothesis Lucas had framed up: the intangible fixity of the type constituted the law of heredity ("*loi d'hérédité*"); this law was opposed by all forms of idiosyncratic tendency embodied in the individual or the race, which were found to be in a constant state of change according to what Lucas called the law of innateness ("*loi d'innéité*"). For instance, a given state of disease would be transmitted from one generation to the next as a "tendency to disturb and destroy the functional order and vital state" deriving from type inheritance.[19] Innate dispositions to variation would thus counterbalance, alter, or even defeat the morphological and functional norm embodied in hereditary type transmission.

However, the cell theory in its early versions (those of Schwann, Müller, Remak, and Virchow), notwithstanding the differences involved, pointed to a conception of heredity at once more analytic and more unified, though essentially programmatic and speculative. To justify this statement, it is worth considering very briefly a later stage of the cell theory. This stage developed in the last quarter of the nineteenth century when the processes of mitosis and meiosis were disclosed.[20] As a result of these later discoveries that pointed to a new and more profound principle—"*Omnis nucleus e nucleo*"—it seemed natural for cytologists to include as an integral part of their doctrine some hypothesis about the anatomical and physiological basis of heredity in cell reproduction and differentiation. That hypothesis was presumed to be experimentally confirmable, at least in its main conceptual features, if not in its ultimate details, due to actual limitation in analytic and experimental means. A remarkable illustration of this outcome is the very title of Edmund B. Wilson's (1856–1939) treatise *The Cell in Development and Inheritance*, whose first edition was published in 1896.

According to O. Hertwig,[21] the advances in cell theory from the mid-nineteenth century on provided a solid basis for the theories of generation and heredity. Thus developed and perfected, the cell theory would combine three principles. First, the ovum and spermatozoon are simple cells that separate from the organism for the sake of reproduction; the developed organism is an orderly association of multiple cells, adapted to diverse roles, but all of them deriving from the divisions of blastomeres in the fertilized egg. Second, a cell is an extremely complex entity, constituting an elementary organism, whose complexity is being progressively disclosed. Third, the mitotic and meiotic processes afford knowledge on nuclear transformation that conditions the theoretical interpretation of generation and heredity.

In the initial period we have been considering, which preceded the investigations on fertilization and nuclear replication and division, only the first two principles came to be accepted, except for the schematic representation of the nucleus as the primary formative and regulative structure in all cell reproduction—a representation already present in Schwann's and Müller's notion of cytoblastemic formation, but significantly modified by Remak and Virchow in conformity with the axiom of morphogenetic derivation of cells from cells. Not surprisingly, to connect heredity with cell theory, one needs to suppose that sexual cells possess numerous latent properties and characters that render possible the formation of a final product: the complex organism. These latent properties and characters reveal themselves progressively in development: they act as immanent architectonic tendencies or dispositions ("*Anlagen*"). The developed organism is at least schematically preformed or potentially contained in the complete set of "*Anlagen*" that defines the archetype of such and such organisms. Clearly, this model is the Blumenbachian model of the "*nisus formativus*" or "*Bildungstrieb*," but restricted to a context of integrative microstructures and metabolic processes. It fits better with Müller's reappraisal of the cell theory than with Schwann's reductionistic approach. On the basis of the knowledge of mitotic and meiotic processes progressively gained in the period 1860–1890, O. Hertwig points to the common features of the more representative heredity theories of the time, which he attributes to Darwin, Herbert Spencer (1820–1903), Nägeli, Eduard Strasburger (1844–1912), August Weismann (1834–1914), Hugo De Vries (1848–1935), and himself. The "*Anlagen*" form the hereditary substance, which relates to the adult organism as a microcosm to its corresponding macrocosm. This microcosm is composed of numerous different particles, organized in orderly fashion, which would possess specific energetic dispositions and bear the hereditary characters. Evidently, these particules, differently named by the various authors ("gemmules," "physiological units," "idioplasmic particles," "groups of micelles," "pangenes," "idioblasts," and so on), afford a hypothetical representation of the structural and dispositional micro-substrate for the contingent, but determined, unfolding of organs and for processes involving further cell divisions and specializations, up to the final integrative organic structure.[22]

One essential mechanism borrowed from the cell theory is presumed to operate on, and among, the multiple particles forming the hereditary substance or idioplasm: it consists in the property of multiplying by division, like cells themselves viewed as higher-order units reproducing by a process that starts with nucleus division. A certain number of lawlike statements aim to translate the rules that are presumed to govern this mechanism: (1) the male and female hereditary substances are equivalent; (2) the hereditary substance, when it multiplies, distributes itself uniformly in all cells deriving from the fertilized egg; (3) the hereditary substance is prevented from augmenting in the next generation; (4) the protoplasm is isotropic. Concerning proposition 2, it is worth noting that, in this instance, O. Hertwig sides against Weismann, who restricted the persistence of hereditary substance to germinal cells, as isolated from somatic cells.[23] His argument is that every cell, or at least almost every cell, possesses in latent state the full set of hereditary characters defining the type of organism to which it belongs. This principle can be fully generalized to lower animals and plants, while in the case of the more complex organisms most cells will not develop those latent properties, because of ambient circumstances, such as the highly specialized and differentiated composition of their protoplasm. In this instance, Müller's famous model serves as a reference for Hertwig.[24] Proposition 4 is significant in that, based on certain experimental evidence drawn from Oscar and Richard Hertwig (1850–1937), Hans Driesch (1867–1941), Laurent Chabry (1855–1894), and Theodor Boveri (1862–1915), it disqualifies protoplasm as a potential bearer of hereditary "*Anlagen*," restricting this role to the complex set of hereditary characters forming the nuclear idioplasm.

The process by which the hereditary "*Anlagen*" are developed into complex molecular structures and emergent functional properties could only be speculatively conceived for the time being. Indeed, one could presume that the idioplasm is differently distributed in the various cells issuing from the division of the fertilized egg, because of the specific composition of the various protoplasmic milieus in which nuclei operate—a quasi-Schwannian position. But Hertwig professed that every cell in the organism receives the full set of hereditary tendencies, since its nucleus reproduces the nucleus of its parent cell and ultimately the combined halved contents of the nuclei of the sexual cells from which all cells of the fertilized egg derive. The mechanism involved in the differential expression of these characters in the various cells might have to do with the selective interaction of some idioblasts with the specific particles forming their protoplasm. And it might happen, as a result, that certain complex protoplasmic structures would emerge and develop a propensity for further selective reproduction in the next generation of cells. Hence the two modes of multiplication for the particles of hereditary substance: one would involve the full set of particles and be supported by nuclear division and uniform distribution of hereditary characters in the daughter cells; the other would consist in a functional multiplication of certain nuclear particles interacting with the particles of a given protoplasm and triggering specific metabolic actions in partial and selective implementation of the whole hereditary potential.

13.4 Conclusion

The connection of the theory of heredity with the cell theory happened with considerable historical delay. The link became manifest only after the mechanisms of cell replication and of morphological and functional differentiation had been investigated, as well as the processes of nuclear transformation involved in fertilization. The first cell theory that Schwann had developed at the end of the 1830s comprised two principles: (1) the universal presence of cells as the basic morphological and physiological units of all living organisms; (2) the formation of cells by successive crystallization of their structural components: nucleoli, nuclei, and membranes, in a more or less fluid cytoblastema. The unoccupied place for a concept of heredity in that system lay in the presumption that germinal cells inherit specific as well as individual physicochemical dispositions from the cytoblastemas they emerge from. Hence the belief that one might derive analytically the characters of a complex organism from the transformations undergone by its multiple constituent cells. These transformations would be occasioned by the composition of the cytoblastemic milieus in which the germinal cells of each species of organisms are generated, and ambient physicochemical conditions would eventually provoke additional individual changes in morphological processes. In Schwann's theory, no set of elements could be found in cell microstructures that would determine hereditary characters to be transmissible from one generation to the next. Subverting Schwann's reductionistic approach, Müller subordinated the emergence of cell structures in the various cytoblastemas to the intervention of a "*Lebensprincip*" that embodied, as Blumenbach's "*Bildungstrieb*" did, the architectonic plan of the complex organism to be achieved and the progressive unfolding of a corresponding system of differentiated cells. In the germinal cells and the fertilized egg that will subsequently divide and produce the diverse types of cells, a set of morphological and physiological dispositions—hereditary "*Anlagen*"—would account for the progressive emergence of the cell-composed complex organism. Any cell would contain the reproductive "*Anlagen*" of the whole organism, but in most cases, notably among higher animals, the determinations proper to the more specialized cells would block any expression of the full immanent power needed to reproduce the whole organic structure. At the same time, the programmed differential production of cell types might undergo alterations because of the ambient conditions acting on cytoblastemas. In this Müllerian holist framework the second principle of Schwann's cell theory came to be drastically revised. Remak and Virchow replaced the presumed cytoblastemic epigenesis of cells by a mode of reproduction that implied the division of the inner structures of preexisting cells and the splitting of nuclei as a universal condition for cytogenesis. This critical revision when it occurred in the 1850s brought further strength to Müller's hypothesis that morphological and physiological "*Anlagen*" wrapped in the microstructures—especially nuclei—of the germinal cells would be transmitted to the embryo's cells and the adult organism's derivative structures. Indeed, a parallel reductionistic trend would continue to consider the whole cell structure, including its protoplasm, as embodying the entire specific and individual determinations that underlie the cellular framework of complex organisms. But the tendency to particularize hereditary determinations, to envision them as latent intranuclear dispositions, to locate them along a germinal sequence conceived in a holist fashion, and to isolate them from the metabolic processes occurring in nongerminal organic cells would be reinforced with the experimental discovery of mitosis and meiosis. This is why, once these processes were unveiled, the main trend in cytology would be toward adding to the two founding principles of the cell theory a third principle for which Müller's speculative hypothesis afforded a preexperimental expression. That hypothesis presumed that the "implicit organic whole" would be immanent in the primordial embryonic cells; it would generate by successive differentiations such sets of specialized cells as were required for composing the "explicit whole": these sets of specialized cells would only form partial expressions of the "implicit whole" since they could not transform themselves if needed into full expressions of the preordained ontogenetic power. This archetypal power located in germinal cells and fertilized eggs would be required for producing cellular architectures representing species of organisms, even if the resulting architectures entailed sequences of operations that ambient conditions might slightly modify. In the versions of the cell theory that will emerge toward the end of the nineteenth century, the developed organism will appear as preformed or potentially contained in the entire set of tendencies

("*Anlagen*") identified as latent characters present in the transmittable microstructures of cell nuclei. Only then will a theory of heredity seem to emerge from the cell theory, while earlier versions of the latter, like Müller's, had envisioned heredity in a more speculative and holist fashion and inferred that successive generations of variant life forms might result from processes of cell replication and morphogenetic differentiation.

Notes

1. Schwann 1969, 186–187.
2. Blumenbach 1789.
3. Holmes 1963; Duchesneau 1997.
4. Schwann 1969, 193.
5. Schwann 1969, 183.
6. Among them Albarracín Teulón 1983 and Duchesneau 1987.
7. Geison 1969.
8. Lohff 1978; Lenoir 1982.
9. Müller 1851, vol. 2, 772.
10. Müller 1851, vol. 2, 613.
11. Müller 1851, vol. 2, 603.
12. Müller 1851, vol. 1, 323.
13. Brücke 1861.
14. Nägeli 1866, cited by Parnes in chapter 14, this volume.
15. Müller 1834–1840, vol. 2, 768–778.
16. Müller 1834–1840, vol. 2, 770.
17. Müller 1851, vol. 2, 799–807.
18. Balan 1989.
19. Lucas 1847–1850, vol. 2, 509.
20. Baker 1988.
21. Hertwig 1903, vol. 1, 317.
22. Hertwig 1903, vol. 1, 319.

23. Weismann 1892.

24. Hertwig 1903, 327.

References

Albarracín Teulón, Agustin. 1983. *La teoría celular.* Madrid: Alianza.

Baker, John R. 1988. *The Cell Theory: A Restatement, History, and Critique.* New York: Garland.

Balan, Bernard. 1989. Prosper Lucas. In C. Bénichou, ed., *L'Ordre des caractères: Aspects de l'hérédité dans l'histoire des sciences de l'homme.* 49–71. Paris: Vrin.

Blumenbach, Johann Friedrich. 1789. *Über den Bildungstrieb.* Göttingen: Dieterich.

Brücke, Ernst Wilhelm von. 1861. Die Elementarorganismen. *Sitzungsberichte der Kais. Akademie der Wissenschaften in Wien, Mathematisch-naturwiss. Klasse* 44 (2): 381–406.

Burdach, Karl Friedrich, ed. 1826–1840. *Die Physiologie als Erfahrungswissenschaft.* Leipzig: Voss.

Duchesneau, François. 1987. *Genèse de la théorie cellulaire.* Paris: Vrin; Montréal: Bellarmin.

———. 1997. Claude Bernard et le programme de la physiologie générale. In C. Blanckaert, C. Cohen, P. Corsi and J.-L. Fischer, eds., *Le Muséum au premier siècle de son histoire.* 425–445. Paris: Éditions du Muséum d'Histoire naturelle.

Geison, Gerald L. 1969. The protoplasmic theory of life and the vitalist-mechanist debate. *Isis.* 55:273–292.

Gley, Émile. 1910. *Traité élémentaire de physiologie.* Paris: J.-B. Baillière.

Hertwig, Oscar. 1903. *Éléments d'anatomie et de physiologie générale: La Cellule.* Paris: C. Naud.

Holmes, Frederic Lawrence. 1963. The "milieu intérieur" and the cell theory. *Bulletin of the History of Medicine.* 37:315–335.

Lenoir, Timothy. 1982. *The Strategy of Life: Teleology and Mechanics in Nineteenth Century German Biology.* Dordrecht: Reidel.

Lohff, Brigitte. 1978. Johannes Müllers Rezeption der Zellenlehre in seinem *"Handbuch der Physiologie des Menschen." Medizinhistorisches Journal* 13:247–258.

Lucas, Prosper. 1847–1850. *Traité philosophique et physiologique de l'hérédité naturelle dans les états de santé et de maladie du système nerveux.* Paris: J.-B. Baillière.

Müller, Johannes. 1834–1840. *Handbuch der Physiologie des Menschen für Vorlesungen.* Coblenz: J. Hölscher.

———. 1838. *Ueber den feinern Bau und die Formen der krankhaften Geschwülste.* Berlin: G. Reimer.

———. 1851. *Manuel de physiologie.* 2nd ed., revised and annotated by É. Littré. Paris: J.-B. Baillière.

Nägeli, Carl. 1866. Die Individualität in der Natur. Mit vorzüglicher Berücksichtigung des Pflanzenreiches. *Monatsschrift des wissenschaftlichen Vereins in Zürich*, 1:171–212.

Rather, Lelland J. 1978. *The Genesis of Cancer: A Study in the History of Ideas*. Baltimore: Johns Hopkins University Press.

Remak, Robert. 1855. *Untersuchungen über die Entwickelung der Wirbelthiere*. Berlin: Reimer.

Schwann, Theodor. 1839. *Mikroskopische Untersuchungen über die Übereinstimmung in der Struktur und dem Wachstum der Thiere und Pflanzen*. Berlin: Sander'schen Buchhandlung.

———. 1969. *Microscopical Researches into the Accordance in the Structure and Growth of Animals and Plants*. New York: Kraus Reprint Co.

Virchow, Rudolph. 1858. *Die Cellularpathologie in ihrer Begründung auf physiologische und pathologische Gewebelehre*. Berlin: Hirschwald.

Weismann, August. 1892. *Das Keimplasma. Eine Theorie der Vererbung*. Jena: Fischer.

Wilson, Edmund B. 1896. *The Cell in Development and Inheritance*. New York: Macmillan.

14 On the Shoulders of Generations: The New Epistemology of Heredity in the Nineteenth Century

Ohad S. Parnes

14.1 Introduction

The notion of heredity prevalent throughout most of the nineteenth century was essentially different from that underlying modern genetics. The most important difference concerned a hypothesized dichotomy between two kinds of organismal traits: stable and nonstable. The stable traits were conceived as hereditary, and constituted the "type" of the organism—notably its species character. The nonstable characters were transitory and transient, appearing merely in the individual organism, and only rarely transmitted to its offspring. Heredity was thus considered in terms of two opposite dynamics—the one conserving the type, the other generating variations, as Karl Ernst von Baer described it in 1827:

> A species consists of a large number of individuals—these are usually not identical, yet still sufficiently similar to be recognized as belonging to the same kind. Some individuals within a species, however, exhibit such remarkable variations that one could seriously doubt their relation to that specific form. However, such variations [*Abweichungen*] are quite rare, and are sustained by nature only very weakly. Thus they have the tendency to disappear and return to the norm again—either already during the individual's life or in the next generation.[1]

As late as 1863 Thomas Huxley still defined hereditary "atavism" as the tendency in the offspring to take the characters of the parental organism, in contrast to "variation," namely, "a tendency to vary in certain directions." Both tendencies are always in action, "as if there were two opposing powers working upon the organic being, one tending to take it in a straight line, and the other tending to make it diverge from that straight line, first to one side, and then to the other."[2]

In contrast, modern genetics, as we know it since the beginning of the twentieth century, is based on a very different concept of heredity. No significant distinction is made between individual variability and species stability, and both the species-specific aspects and the individual peculiarities of the organism are assumed to be brought about by the same mechanism, namely, "heredity." Heredity, in turn, is essentially a mechanism of intergenerational transmission: discrete entities, corresponding to discrete traits, are transferred from one generation to the next in an ordered manner. For modern genetics, understanding heredity amounts to the study of the regularities of these transmissions.[3]

In this chapter, I argue that the establishment of heredity as a universal scientific category was the direct result of a wider epistemological shift, which took place during the first half of the nineteenth century and reaching far beyond the borders of biology, namely, the conceptualization of populations in terms of generations.[4] I will begin with a brief discussion of the origin of the modern notion of generations, which I trace back to the period around 1800. I will continue by describing the way generational reasoning constituted a new order of heredity. Here, too, the establishment of the modern notion of heredity cannot be discussed solely in its biological context, but has to be seen as a more general manner of reasoning characterizing not only the natural, but also the human and social sciences in the middle of the nineteenth century. Finally, I will argue that this new mode of reasoning laid the epistemological foundations for Mendelian genetics—and this regardless of the work of Mendel himself.

14.2 Generations and Progress in Late Enlightenment Thought: The Transmission Problem

The idea that society can be stratified according to a generational order is relatively new, and did not prevail in Western science and culture before about 1800. Strictly seen, there is no logical necessity in assuming that people (or, for that matter, other organisms), born in about the same *time*, would have anything in common apart from their age. There is even less logical or semantic necessity in using the same word for the act of organismal creation and for a group of individuals sharing nothing but a vaguely defined age. And yet, since about 200 years ago the word *generation* has been used, in most European languages, for both these phenomena.[5] This has not always been the case. Until the end of the eighteenth century, the word *generation(s)* meant either the actual act of procreation (*Zeugung*), or a typical interval of time, usually between twenty and thirty years.[6] This second meaning, however, was very different from the one associated today with the social notion of generations, and referred merely to the postulated distance between two succeeding acts of procreation (usually assumed to be around thirty years, sometimes even exactly 33.3 years, to enable a convenient accommodation of three generations per century).[7]

Toward the end of the eighteenth century, a new notion of "generations" emerged, referring to well-defined social and cultural collectives of individuals. New was the assumption that generations are identifiable units within society, and that the members of generations share significantly more than merely the simultaneity of existence. The reasons for the emergence of this new way of understanding society cannot be discussed here in detail. Obviously, the use of generational arguments was especially attractive as an alternative to the old model of royal genealogies, because generations were egalitarian units, including all citizens living within a specific time period. At the same time, the basic principle of genealogy was also kept, ensuring a regular and orderly transfer of political power over time—among generations.[8] "Every generation," Thomas Paine wrote 1791 in his *The Rights of Man*, "should be considered as an independent unit—it has its own character, its own existence, and thus naturally also its own right to act for itself." Power is transmitted from one generation to the next, but with each transmission a new, wholly independent population arises, one that is and must be "competent to all the purposes which its occasions require," thus "it is the living, not the dead that are to be accommodated."[9] The American president and political thinker Thomas Jefferson, apparently influenced by Paine, suggested distinguishing between political rights and forms of government, the former being noninheritable, the latter passed down over generations. Jefferson likened generations to the political electorate. An analysis of European tables of mortality, he claimed in 1813, showed that "of the adults living at any one moment of time, a majority will be dead in about nineteen years. At the end of the period, then, a *new majority* is come into place; or, in other words, a *new generation*."[10] Note how important it is for Jefferson, writing at the beginning of the nineteenth century, to justify his use of the generational terminology. Generations, he insists, constitute stable epistemic units, and "may be considered as bodies or corporations."[11] Accordingly, they offer a new manner by which the economy of transmission, conservation, and variation can be calculated:

> Each generation has the usufruct of the earth during the period of its continuance. When it ceases to exist, the usufruct passes on to the succeeding generation, free and unencumbered, and so on, successively, from one generation to another forever. We may consider each generation as a distinct nation, with a right, by the will of its majority, to bind themselves, but not to bind the succeeding generation, more than the inhabitants of an other country.[12]

For Jefferson, generations are thus much more than merely a formal way of counting within history. Every generation, as he puts it, must be considered as if it is a distinct nation, notably a *sovereign* nation, comprised of a collective of individuals sharing their national-generational identity. The important difference between real nations and generations, however, is that generations are temporal units; they succeed and inherit each other. This point is of special importance for Jefferson, namely, that generations are the carriers of political sovereignty, which (in a democracy as he envisions it) is transmitted from one generation to the next. Generations, therefore, are the agents of political heredity: they ensure continuity, while also being the units of change.

Generations, thus, played a central role in late Enlightenment thought, notably in discussions of progress and the ways it can be enhanced. Accordingly, generations became highly relevant to the pedagogical discourse of the time. In 1791, Jean-Antoine-Nicolas de Caritat, known also as Marquis de Condorcet, submitted to the Legislative Assembly a short report, aimed at the restructuring of the educational system in France. The *First Memorandum on Public Instruction* quickly became one of the most influential programmatic writings of the French Revolution. The goal of instruction, Condorcet maintains, is "not to make men admire legislation that is fully completed, but to render them capable of evaluating and correcting it." Education is not concerned with "subjecting each generation to the opinions and the will of that preceding it, but to enlighten successive generations more and more, so that each becomes more and more worthy to govern itself by its own reason."[13] However, in order to guarantee such progress one has to identify the objects of transfer, the carriers of knowledge, and the dynamics of its transmission over time. For this, Condorcet invokes generations. Culture, he insists, "can improve whole generations and [this] improvement of individual faculties can be transmitted to their descendants."[14] Society is thus considered as composed of generations, and the appearance or disappearance of collective traits is explained in terms of *causal* relations among generations. Indeed Condorcet even explicitly argues for hereditary processes as corroborating his view of progress: "Observation of domesticated animals seems to offer an analogy consonant with this opinion. The way these breeds have been raised not only changes their size, shape, and purely physical characteristics; it also seems to affect their natural traits and their character." Moreover: "If several generations of men receive an education directed toward a constant goal, if each of the individuals comprising them cultivates his mind by study, succeeding generations will be born with a greater propensity for acquiring knowledge and a greater aptitude to profit from it."[15]

Incidentally, a comparable argument can be found in the pedagogical writings of Immanuel Kant, who, in the early 1780s, similarly considers "generations" as the basic subjects and objects of education. Kant, too, uses "generations" for biological units of reproduction as well as for social and cultural dynamics. What both have in common is their role as mediators of transmission. Thus Kant is comparing the way human generations educate their successors to the "very wonderful" fact that "each species of bird has its own peculiar song, which is preserved unchanged through all its generations; and the tradition of the song is probably the most faithful in the world."[16] An argument along similar lines was put forward by another influential pedagogical theoretician of the time, Friedrich Schleiermacher. The whole point of scientific pedagogy, he contended in 1826, is to offer a systematic method for the transmission of knowledge from the "older generation" to the "younger."[17] Generations, therefore, do not simply follow each other; they relate and interact, each generation inheriting its predecessors. A good education amounts to the proper "transmission" of knowledge and manners among generations and the role of pedagogy is to understand the mechanism of this transmission.[18] Schleiermacher termed this dynamics "generational alternation" (*Wechsel der Generationen*).[19]

14.3 The Alternation of Generations and the Problem of Individuality

Schleiermacher first used the phrase "generational alternation" (*Wechsel der Generationen*) in a lecture on the "Foundations of Pedagogy" held in Berlin in 1826. Just a couple of years earlier, the author und naturalist Adalbert von Chamisso had invented similar terminology (*Generationswechsel*)—albeit for the representation of a very different kind of phenomenon. In 1819, Chamisso published a short treatise describing a discovery he made on board the brig *Rurik*, during an expedition to the South Seas and the Bering Strait between 1815 and 1818.[20] Among others, he performed a series of investigations of the minute marine animal *Salpa*. This species was known to appear in two different forms: either as independently living individuals (solitary), or in form of long aggregate chains, comprising a multitude of individuals (zooids). The relation between these two forms was not clear. Chamisso's solution was quite unconventional: in fact, he argued, these two forms are genetically related, and exhibit a regularity of forms in consequent generations, so "that the same species appears in two totally distinct forms in alternating generations."[21] Chamisso was quite explicit about his semantic move—that is, about his decision to use the category "generation" for the depiction of a zoological phenomenon: "Salps appear in two distinct forms: an offspring remains throughout its life dissimilar to its parents, but is producing offspring that again resemble them. Therefore every salp does not look like its mother or like its daughters, and at the same time bears a resemblance to its grandmother, to its granddaughters and to its sisters."[22]

Chamisso's terminology has been in use for the last two centuries and is still valid today. Simply put, the alternation of generations describes a situation in which the offspring (second generation) differs in a radical way from their parents, whereas the original (parental) form appears again in one of the following (usually third or fourth) generations.[23] It took several years before Chamisso's generational reasoning met with general approval. By the 1840s, the alternation of generations was one of the most hotly debated ideas in biology.[24] The most obvious use of the new notion was in botany. It offered an alternative to viewing a plant as a singular organism, or as an arbitrary aggregate of minute constituents (e.g., cells). Instead, the generational scheme suggested deconstructing the plant into layers of generations, analogous to human society. "The plant or tree," wrote the zoologist Johannes Jupetus Steenstrup in 1842, is a "colony of individuals arranged in accordance with a simple vegetative principle, or fundamental law," namely, that the colony "unfolds itself through a frequently long succession of generations into individuals, becoming constantly more and more perfect."[25] Considered as a collective, individuals seemed to exhibit a regularity not detectable when viewed in isolation: "It is the generation, not the individual, which is undergoing metamorphosis," as the Norwegian naturalist Michael Sars postulated in 1841.[26]

Considering living phenomena in terms of generations may seem almost trivial to us today, but this has not always been the case. Ascribing generational identity to cells, germs, or buds was far from obvious in about 1850 and was often disputed. William Carpenter, arguably one of the most influential zoologists in nineteenth-century Britain, vehemently attacked the "so-called Alternation of Generations" in 1848. "We are not in the habit of speaking of the leaf-buds and the flowerbuds of a plant as of two distinct generations; nor, if our comparison be correct, have we any ground for giving such a designation to the polypoid larva and the medusa-imago, which are continuous developments from the same germ," he argued.[27] Carpenter maintained that these phenomena are simply examples of metamorphosis or change of form, attending the evolution of successive products from the original germ. "Every botanist knows," contended the British naturalist Edward Forbes in 1848 in reply to Carpenter, "that [a plant] is a combination of individuals, and if so, each successive series of buds must certainly be strictly regarded as generations." To argue that because "we are not in the habit of speaking of the leaf-buds and the flower-buds of a plant as of two distinct generations," we are, therefore, not to regard the alternations of polyps and medusae as such, is "to bring the common, popular, unscientific, and untrue notion of the nature of a plant into a scientific discussion of the nature of animals."[28]

Generational reasoning offered a new way of coping with individuality in a biological realm that, especially after 1840, became increasingly chaotic. One of the more radical tenets of the cell theory, already in its original formulation by Theodor Schwann, was the ascription of biological agency to the single cell. “We must attribute to all cells an independent life,” Schwann wrote in 1838. And: “the cause of nutrition and growth resides not in the organism as a whole, but in the separate elementary parts—the cells.”[29] This highly influential doctrine, combined with rapid improvement in the resolution and magnification power of microscopes, brought about an upheaval in the understanding of organismal individuality. Suddenly a whole new spectrum of potential individuals opened up. Was every cell an individual? Carl Nägeli, like most of his contemporaries, argued in 1844 that, indeed, cells are “individual organisms in form of a globule.”[30] This, in turn, gave rise to the notion of “unicellular organisms” being full-fledged individuals: “Each such cell conducts its own individual animal life; they move, sense, grow and even have their own peculiar mode of reproduction,” as the German zoologist Albert Kölliker maintained in 1845.[31]

But if that was the case, how were structures consisting of a multitude of cells to be judged? Microscopic entities seldom appear in isolation, and often tend to form “colonies”—that is, bundles and conglomerates, often very difficult to distinguish from more complex, multicellular organisms. “We are facing a real problem,” wrote the anatomist and physiologist Karl B. Reichert in 1852: “In the past, scientists were content to ascribe individuality to each and every isolated and independently existing entity.” Recent discoveries, however, have falsified this view, and have shown that many of these alleged individuals are in fact composed of “systematically organized bundles” of smaller organisms.[32] In other words, just by observing an entity scientists may not be able to decide whether it is a full-fledged “individual” or merely one of its components. The best solution to this problem, according to Reichert, would be to shift the perspective from the living unit to the more general developmental scheme. Instead of beginning with a single individual, and then reconstructing its genealogy, the scientists must first understand the generational regularity of the species, its modes of reproduction and development.[33] “Through the generational succession,” explained the botanist (and Gregor Mendel’s teacher) Franz Unger in 1852, “we are able to give a unified sense to a series of individuals that are spatially disparate, to understand them as a unit.”[34] The orderly phenomena of the alternation of generations resemble “a family of very close relatives, all mutually supporting each other.”[35] “The individual,” stated the botanist Alexander Braun in 1851, must be conceived as part of an “inner unity, which runs through the series of phenomena.”[36] When we view the totality of the organism’s life in the manifold of its generations, we realize that it is only through the unfolding of generations that the "essence" [*Wesen*] of the organism is expressed. An individual—be it an individual animal or even an individual cell—can never express the ideal type of its species, because it is only one "link" [*Glied*], a member of one generation in a long succession. In particular, the case of the alternation of generations was for Braun a demonstration of the fact that the "biological individual [is] breaking up into a limited or unlimited series of subordinate (morphological) individuals."[37] Accordingly, the aim of biology should be the "comprehension of the individual phenomenon as a member of a series of essential correlative representations."[38] Further, "Such a generational succession may be homogenous [*gleichartig*] or differentiated, simple or complex (including also divisions). The process may have a constant or a graduated and cyclic pattern—which explains the phenomena of the division and alternation of generations."[39]

Jan Evangelista Purkynje, one of the most influential physiologists of the time, compared the phenomena of the alternation of generations to the "problem" of sleep; in both cases, he argued, individuality remains intact, even though the integrity of the living entity seems disrupted. Individuality should thus be considered continuous across generations, whereas every "generation" (awakening) is but an "individualization" of one and the same "principle."[40] It would be wrong to acknowledge the individual as really existing, argued Alexander Braun along similar lines in 1851, and "deny the natural activity of the more comprehensive complexes of the organism of nature." We tend to make this mistake, he claimed, because we are "deceived" into believing that the individual is immediately given in the phenomenon. In fact, it is easy to see that the individual as such—that is, as a single being—"can only be conceived indirectly, in the recognition of the inner unity which runs through the series of phenomena, in which it displays itself."[41]

This new mode of reasoning had far-reaching consequences for a theory of heredity. It made a claim about a regular, lawful alternation of forms within a population of organisms, while the reappearance of specific traits (in this case bodily form) was explained in terms of a lawful expression of these traits over generations. Hence, the alternation of generations showed the way to a departure from the old dichotomy of "hereditary type" versus variations. Instead, traits that were traditionally considered as variations were now incorporated into the hereditary scheme, albeit not because they were part of the essence of the species, but on the basis of a structure of appearance, a genealogical regularity, a pedigree of the expressions of these traits.[42] This was a hitherto unrecognized aspect of the living world: a fascinating lawfulness, a seemingly mathematical regularity by which forms come and go, appear and disappear.

And it was out of this new understanding of generational dynamics that the modern notion of heredity emerged.

14.4 Generations, Heredity, and the Science of Society around the Middle of the Nineteenth Century

Sociological work in the first decades of the nineteenth century was motivated by a kind of intellectual desperation.[43] The world was rapidly changing, faster than ever before. Everything was in constant flux—knowledge, wealth, and human conduct—possibly even the living world. Historians, philosophers, and political thinkers were challenged to account for this new apprehension, to try to explain the unremitting transformation of society, the constant modification of commodities, the seemingly infinite growth of knowledge. There was a general need to understand this permanent change, to contain it within a framework, to constrain it. However, traditional philosophy and political economy did not seem to provide the proper theoretical tools for the understanding of such rapid historical change, or of the role of individuals within this process. Who were the agents of this change? Neither the individual nor the nuclear family appeared to be a plausible candidate for this role. Generations, on the other hand, offered such a framework: they could explain both change and tradition, both revolution and stability. As constituents of generations, individuals could be considered both in their role as links in a genetic chain, and in terms of their ahistorical, contemporary social unit. At the same time, the generational scheme imposed two main constraints on the process of historical change. First, the extent to which a new generation could be transformed in comparison to its predecessors was limited by the quantity and quality of the inherited assets, the "seed capital" it had inherited from the previous generation. The second constraint on change was imposed by the further transmission of the modified commodity to the next generation, which guaranteed a minimal level of continuity to the process. These order-governing abilities of generational explanations made them so attractive both to social and biological scientists.[44]

A paradigmatic example of an attempt to ground a science of society in biological metaphors was the work of the French social thinker Auguste Comte, in many senses the founder of the modern science of sociology. Society, Comte said around 1850, develops in accordance with biological laws. One of the more fundamental of these is the "law of reproduction," which states that "each living body emanates from another like itself."[45] Consequently, the social sciences must begin their work from the fundamental recognition that "the dead govern the living"—that is, that every contemporary state of society is contingent on the society that preceded and gave rise to it.[46] Any analysis of social states, therefore, is essentially a genetic analysis, based on the dynamics of transmission across generations. Comte's fellow positivist, the philosopher John Stuart Mill, elaborated this position in 1843 giving it a sharper analytic formulation. Mill insisted that generations follow each other in a lawful manner, and that the regularity of transmission among generations must constitute the heart of a modern science of society.[47] States of society, Mill contended, are "like different constitutions or different ages in the physical frame; they are conditions not of one or a few organs or functions, but of the whole organism." A simultaneous state of affairs in society, therefore, is not arbitrary; "there exist uniformities of co-existence between the states of the various social phenomena."[48] These "uniformities" are the expression of structural regularities, whose analysis requires an understanding of both their synchronic and diachronic interrelations. Therefore, such uniformities of structure can only be analyzed using the frame of reference of "generations."[49]

A very similar notion of generations—albeit embedded in a totally different ideology—can be found in the early writings of Karl Marx. "History," Marx wrote in 1845, is nothing but "the succession of distinct generations, each of which exploits the materials, the capital funds, the productive forces banqueted to it by all preceding generations, and thereby, on the one hand, continues the traditional activity in completely changed circumstances and, on the other, modifies the old conditions through a new kind of activity."[50] This totality of inherited productive forces, capital and means of social communication, which "each individual and each generation finds as given," constitutes, according to Marx, "what philosophers have always considered as the substance and essence of human beings."[51] Thus human existence cannot be understood outside its historical context, and this context must be analyzed in terms of generational units. Goods, knowledge, and cultural and national characteristics are all transmitted across succeeding generations. Each generation constitutes a well-defined historical unit, able to modify its inherited load. Society, therefore, is by definition the manifestation of these intergenerational relations, "the result of the activity of a whole succession of generations, each standing on the shoulders of the preceding one."[52]

The power of generations lay in their mediating role, in their function as a prism through which society, or culture, could be observed both in its temporal evolution and as a contemporary unit. At the same time, this approach also attached every individual to its very specific generational context, a particular link within a complex web of hereditary transmissions.[53] Generations, argued the German philosopher Wilhelm Dilthey in 1865, express not merely a given historical situation, but also the relation of this situation to past situations (generations), as well as signaling the constraints on the development of future generations. Dilthey, the founder of the scientific study of methodology of the human science, stresses first and foremost the methodological utility of generations. We must use generations as analytic tools, Dilthey maintains, because they provide the necessary perspective for the study of the synchronic relations between individuals and their environments, as well as the diachronic effects among successors and predecessors.[54] It is precisely this mediating role, this ability to explain both change and stability, that is making generations such a "highly useful concept."[55] The most "fortunate" circumstances, he concludes, are where "a generation can be recognized and isolated very clearly, so that we could study it as such."[56] Only then are we in a position to recognize the genetic processes underlying the development of culture:

> The process begins with a given acquired state of intellectual culture, as it exits before the generation under consideration begins its development. At a further stage, the rising generation is taking possession of the accumulated intellectual content and attempts to proceed and progress out of it. Here we meet the second aspect influencing the development of culture, namely the environmental factors: the *surrounding* life, actual circumstances, social, political and numerous other kinds of conditions. Therefore, there are clear limits to the extent to which that which has been transmitted from earlier generations can be further modified and developed.[57]

Marx, as we saw before, considered human society in terms of transmission dynamics. Transmitted was the sum total of capital, which "each individual and each generation finds as given [*etwas Gegebenes*]."[58] Dilthey's imagery is not much different: instead of "capital" and "productive forces" he envisions "a given acquired state [*Besitzstand*] of intellectual culture" as the inheritance, which is transmitted among and modified by generations. In both cases the explanation of social and cultural development is reduced to a regular transmission across generations. Each generation begins its life on the basis of what it received from the parent generation, and later transmits its own goods/capital/culture/traits to the next generation. In 1856, the Swiss-German biologist Carl Nägeli summarized his own understanding of (biological!) hereditary processes as follows:[59] "It must be the case that under favorable conditions the hereditary predisposition [*Anlage*] will keep growing and developing, through a series of generations, like a capital, which grows by gaining a yearly interest. For every generation inherits from its predecessors not only the possibility to realize the seed capital, but also the possibility to add the gained interest to it."[60]

Biology as well as the social sciences thus took recourse to a novel imagery, a new metaphorology aimed at endowing the generational scheme with a mechanistic explanation. Traits were likened to commodities, bodily constitution compared to a capital investment. Essentially, heredity was becoming a process of acquisition and delivery (with the possibility of modification—if and when the inheritance of acquired characters was allowed[61]). Traits thus attained an independent existence; instead of being seen as part of the essence of the organism they were now considered as its possession, or, more accurately, its lease. Physiological characters were in the possession of each generation for a limited time, later to be bequeathed to its offspring.

14.5 Generations and Heredity (1850–1860): A Notion in Flux

Out of the ongoing generational discourse, heredity gradually emerged as an independent entity, an epistemic structure that can be studied and investigated on its own terms. Understanding heredity in terms of cross-generational relations was not an obvious move, and required a strained argumentation. How was one to give a formal account of the complex set of appearances and reappearances of traits over dozens of generations? No simple way of describing such processes existed in biology before the middle of the nineteenth century. During the 1850s and the 1860s, investigators were still struggling to find the proper conceptual language for these phenomena. The German biologist (and later translator of Darwin[62]) Victor Carus, for instance, propagated a mathematical solution. Carus's point of departure was a terminology elaborated by his teacher and mentor, the Dorpat physiologist Carl Bogislaus Reichert. Reichert proposed, as early as 1840, that developmental processes be understood as "series of differentiation" [*Differenzierungsreihen*]. Every difference between two succeeding developmental steps could, according to Reichert, be experimentally quantified and then defined as a single "differentiation." Organismal development could thus be formalized according to its series of differentiations so that each kind of organism is inscribed with its own specific pattern of differentiation—its "specific plan" or the "imprint" [*Gepräge*] of the organism.[63] Originally, Reichert drafted this scheme for the explanation of individual (ontogenetic) development. Carus, working with Reichert in Dorpat between 1844 and 1846, tried to adapt this scheme for the analysis of the alternation of generations. Not only the development of a single organism, he contended, but also the transitions across generations could be considered as a series of calculable differentiations. Interestingly, Carus in 1848 did not even consider using pedigrees as a tool for depicting generational regularities. Instead, he was advocating the employment of mathematical series for calculating generational series and was even trying to work out such an expression for the specific case of the alternation of generations: every step from one generation to the next could be considered a measurable "difference," and the complete "series of differentiations" will attain the form of a typical mathematical series (arithmetic, geometric, or a combination of some sort).

Each individual within a series will then attain a mathematical expression, and the hereditary state of a single organism would amount to the sum total of all those preceding it (see figure 14.1: Carus's attempt to produce a mathematical expression for a simple case of the alternation of generations).[64]

Carus maintained that the fact that the alternation of generations could be expressed in terms of a mathematical series is proof of the hidden order or "specific plan" behind these phenomena. But what is a "specific plan"? What is it that governs the regular alternation of traits and endows generations with a "specific imprint"? Obviously, argued Carus, generations and generational transmutations constitute some kind of series. But then again: "What is a series?"[65] And what is the nature of the agency responsible for the regularity and stability of this series? Writing in 1849, Carus was desperately searching for the appropriate terminology. The "type [*Typus*]" of an organism, he stated, must be understood as "an *idea* specifying the structure and distribution of the organic elements and the disposition of the organs for the embryonic development." Thus, it is this "idea" that is responsible for "generating [*bilden*] an independent and self-reliant organism out of a pile of grain."[66] The "idea," therefore, is the agent governing the transgenerational regularities. However, Carus explicitly acknowledged that the word *idea* is misleading, and that he was using it only for lack of a better expression: "I would have been happy to replace the word 'idea' with 'force', or with 'law'," he notes, "as it is quite obvious that the evolution of forms follows some kind of law." Alas: "This law is still totally unknown to us."[67]

It should be noted that speculations such as Carus's about a generational "plan" were essentially different from the older notions of formative life forces—for example,

$$\frac{Ax^{n} =}{1.2.3\ldots n-1.n}$$

$$\frac{(Ax + d)^{n} - \left(\frac{n}{1} Ax^{n-1} d + \frac{n.n-1}{1.\ 2.} Ax^{n-2} d^{2} + \frac{n.n-1.n-2}{1.\ 2\ 3.}\right.}{1\ \ 2\ \ 3.\ 4.\ \ldots\ldots}$$

$$\frac{\left. Ax^{n-3} d^{3} \ldots\ldots \frac{n.\ n-1.\ n-2.\ \ldots\ n-(n-1)}{1.\ 2.\ 3.\ \ldots\ n-2.\ n-1} Ax^{n-(n-1)} d^{n-1} + d^{n}\right)}{\ldots.\ n-2.\ n-1.\ n.}$$

Figure 14.1

An attempt to give a mathematical formulation to the regularity behind the alternation of generations. From Victor Carus, *Zur naeheren Kenntnis des Generationswechsels. Beobachtungen und Schlüsse* (1849).

in the tradition of Johann Blumenbach's *Bildungstrieb*.[68] The main difference was in the emphasis on the duality of generation—that is, the stability of the type on the one hand, and a regular variation among the individuals belonging to this type on the other. Alexander Braun described this duality as the "rhythmic-spiral" nature of living beings, namely, their tendency to unremittingly repeat the same pattern, while also alternating between different patterns and even giving rise to completely new forms. But how was one to explain this phenomenon in biological terms? What "property of life" is it, Braun asked, that declares itself in these phenomena, and that is responsible for the "orderly succession of phenomena," yes for the "adherence to plan" [*Planmäßigkeit*]? Surely, Braun reasoned, this must have its grounds, its "internal cause," within the physiology of the organism.[69] It must be through the activity of this determining impulse that "under like external conditions of existence, the organism shapes itself in a peculiar, specific, nay even individual way."[70] Writing in 1851, Braun defined the ability to transmit and repeat patterns of organismic life over several generations as a "gift" [*Gabe*], specifically as "memory" [*Erinnerung*]. "Memory" is here a central physiological concept, the "cornerstone" for the "structure of all of Nature."[71] All living phenomena, Braun contended, should be considered to be determined by cross-generational memory, a mental-like process of recalling the characters of the individual's ancestors and reviving these traits within the present context of the organism's life: "Must we not call it memory when, in the course of generation, the old specific nature returns to life with each new individual? Is it not still more strikingly a memory when, in the course of the Rejuvenescences, a long-relinquished parent-form suddenly returns to existence? as we have seen in the striking back of varieties and the recurrence of hybrids into the mother species."[72]

The memory of the "inherent design of life" thus "presents itself to us in the organism" through the "repeated representation of the same vital form." Moreover, "memory" can also be invoked for the explanation of the appearance and reappearance of traits after a long series of generations: "It presents itself to us, retrogressively, in the reproduction of long overpassed, or in advancing development, in the production of new vital forms."[73]

Heredity about 1860 was a notion in flux.[74] On the one hand, it was frequently used in its traditional sense—that is, as the stabilizing, type-conserving force. At the same time, it was increasingly employed to account for the more complex regularities of generational phenomena. Thus Carl Nägeli, in 1856, defined *Vererbung* as the "tendency [*Bestreben*] of the individual to become similar to its parents." However, Nägeli also acknowledged the fact that there is more to generational relations than a simple dichotomy of type constancy versus individual variations. Very often, he remarked, offspring exhibit traits that were not present in their parents but that did exist in their grandparents or great-grandparents, or even in "very distant ancestors." This, contended Nägeli, is also part of "heredity," which is apparently able to "operate [*wirken*]" not only from parents to offspring, but also throughout several generations.[75] Similarly, Ernst Haeckel in 1866 was still trying to accommodate "conservative heredity," corresponding to the old notion of a stabilizing force, with "disrupted or hidden or alternating heredity." The latter was supposed to account for all those cases where the recurrence of similar traits was not continuous, yet followed a regularity of disappearance and reappearance across generations.[76]

14.6 Picturing Generations

The shift toward generational reasoning in biology about 1860 can be traced not only in texts and arguments, but also in attempts to visualize hereditary processes. The growing awareness of the need to understand individual organisms as part of a generational series (*Generationenfolge, Generationenreihe*) encouraged several botanists and zoologists to also try to give this stratification a pictorial form. But how does one visualize a generation? Finding the appropriate imagery was not a simple task. The generational regularity had both a temporal and a spatial aspect: generations were temporal units, while the simultaneous existence of different kinds within a generation was a synchronic phenomenon. How could the regularities of the appearance and reappearance of traits over generations be expressed visually? No convention for such a depiction existed. Thus, for example, of the numerous plates accompanying his influential 1865 work on hybridization, Charles Naudin dedicated only one plate to the depiction of several (four) "generations" of a hybridization succession (figure 14.2 and plate 1). Clearly, Naudin did not try to depict the actual hereditary relations between the generations, or the position of each individual within its particular generation. At the same time, the plate is quite remarkable for its representation of each generation as an (admittedly small) population of hybrids. By placing four generations of hybrids horizontally, each manifesting an exemplary ensemble of members of the corresponding generation, Naudin managed to invoke a sense of hereditary order despite the fact that the transformation of traits did not follow a simple pattern, as he also elaborated in the accompanying text.[77]

A very different attempt to visualize generational relations can be found in the work of Alexander Braun. As mentioned above, Braun insisted on the importance of generational stratification for biological analysis. In every plant, he argued, the

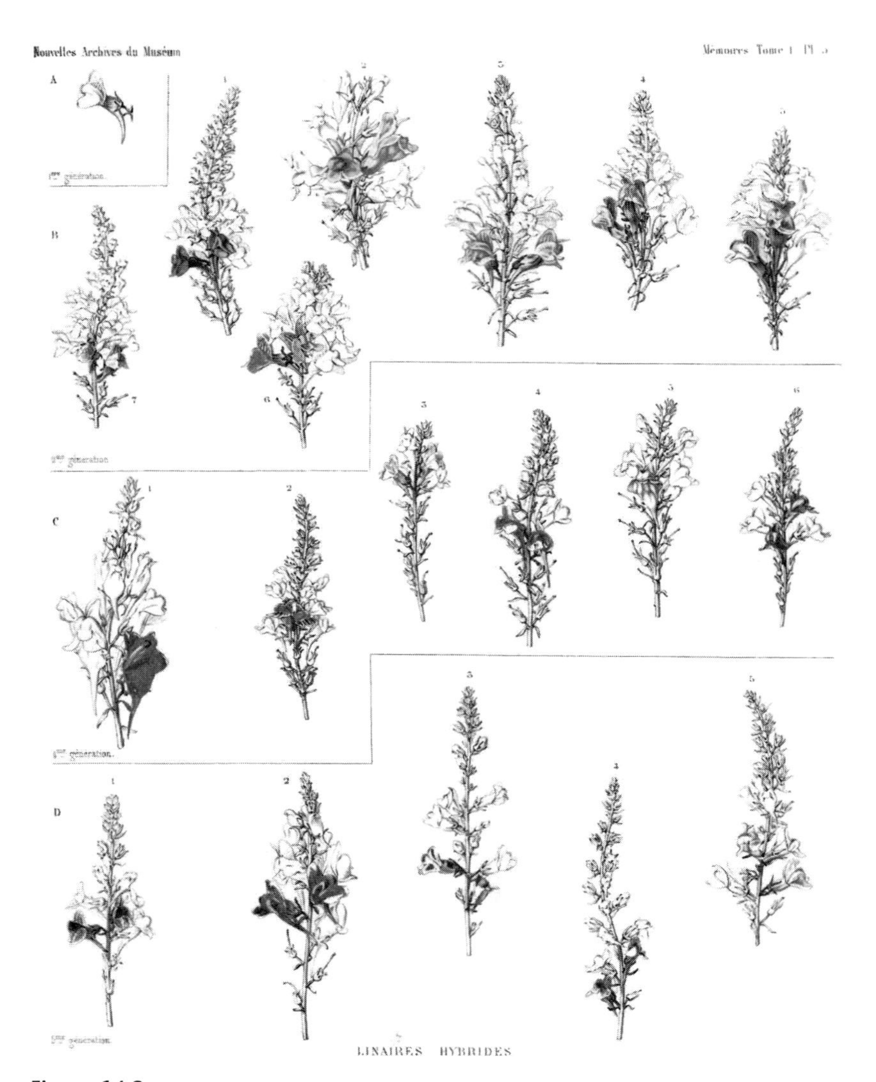

Figure 14.2
Four generations of *Linaria purpurea-vulgaris*. From Charles Naudin, "Nouvelles recherches sur l'hybridité dans les végétaux," *Nouvelles archives du muséum d'histoire naturelle de Paris* 1 (1865).

generational stratification is expressed by various parts and tissues of the plant—for example, the layers of leaves. Braun's idea of using pictures of plants as abstract models for the unfolding of generations was quite remarkable, notably the way he superimposed the old iconographic convention of the tree as a representation of genealogy (*Stammbaum*) on a new theory about the generational stratification of plants. Indeed Braun did not claim to have based his images on actual observations; the figures, he emphasized, should not be understood as actual representations of real plants, not even as idealizations. Instead, they should be considered a "pictorial language," a "hieroglyphic explanation of the accompanying text," serving as schematic representations of the generational order in nature.[78] In Braun's plates, the generational order was signified through Roman numerals, representing the inner developmental hierarchy of the plant: "I" for the first generation, "II" for the second, and so forth (see figure 14.3). At the same time, Braun argued that these figures could also be considered representations of a universal regularity characterizing all living entities, including the human race and its history: "each generation building upon what earlier generations had begun, similar to the way shoots or branches spring from the main stock of a tree and derive their nutrition from it . . . generations upon generations, generation cycles upon generation cycles."[79]

The genealogical tree and its various uses have a long history, which has been studied in some detail and cannot be discussed here.[80] However, there is clearly a close relation between the introduction of pedigree-like iconography into biology and the shift to generational analysis as discussed in this chapter. In 1865 Max Wichura, a relatively unknown lawyer from Breslau, published a book compiling his long series of tedious experiments on the hybridization of the willow.[81] The *Bastardbefruchtung im Pflanzenreich* is remarkable for its extensive use of a graphic method for the representation of experimental generations. Wichura contended that visual representations are indispensable for calculating the relative composition of hybrids, especially those generated out of more than two species. Accordingly, he opted for the pedigree (*Stammbaum*) as an especially useful and "lucid" tool, "enabling us to grasp, at a single glance, the relative ratios of the various species which contributed to the generation of the hybrid."[82] (See figure 14.4.) Note that Wichura is using the pedigree in a "reverse" way: the first and highest level of his diagrams represent the original wild-type generation, while the subsequent downward levels stand for a series of hybridizations ending with one single individual. Each pedigree, therefore, is finite and represents one specific individual (in contrast to the branching tree, which is infinite). However, at least in one case Wichura is hinting at the later use of the pedigree for genetic analysis. Referring to one of his illustrations (figure 14.4), he notes that one

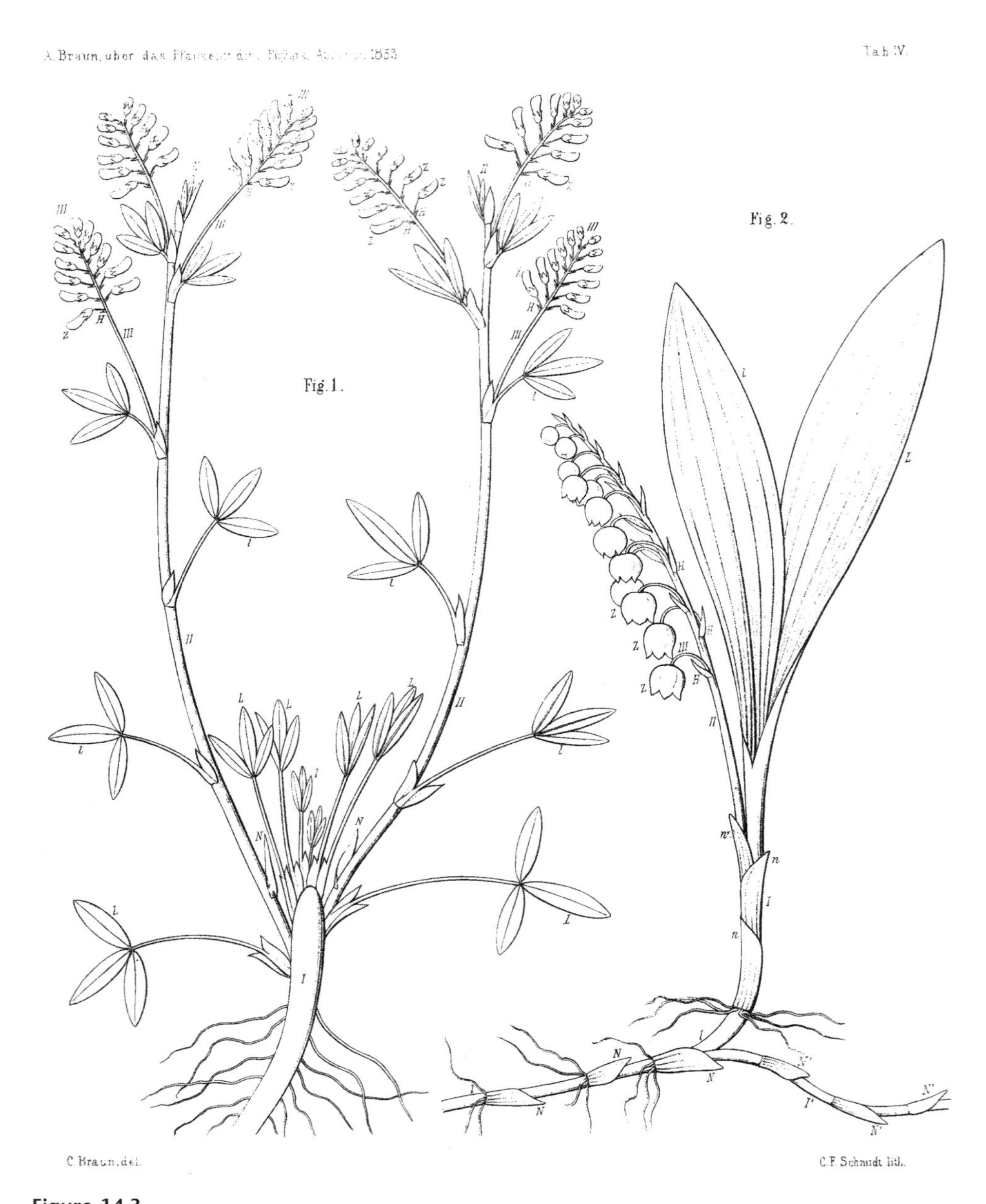

Figure 14.3
Representation of generations in *Trifolium montanum* (Mountain Clover) and in *Convallaria majalis L.* (Lily of the Valley). Alexander Braun, "Das Individuum der Pflanze in seinem Verhältniß zur Species. Generationsfolge, Generationswechsel und Generationstheilung der Pflanze," *Abhandlungen der königlichen Akademie der Wissenschaften zu Berlin* 1853 (1854).

a

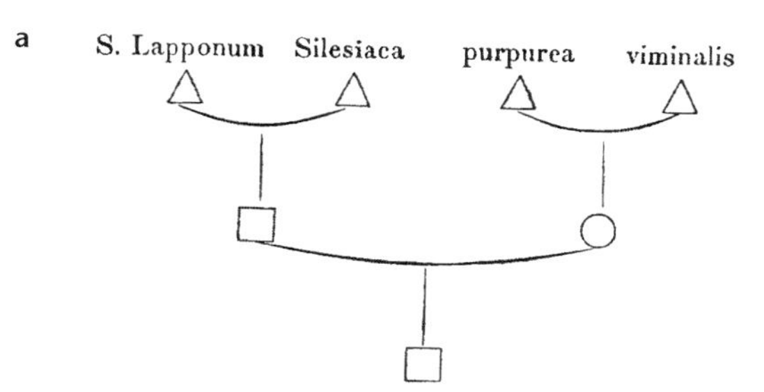

27. 1854. ♀ *S. (aurita L. + repens L.) spont.*
+ ♂ *S. (caprea L. + viminalis L.)*
spont.

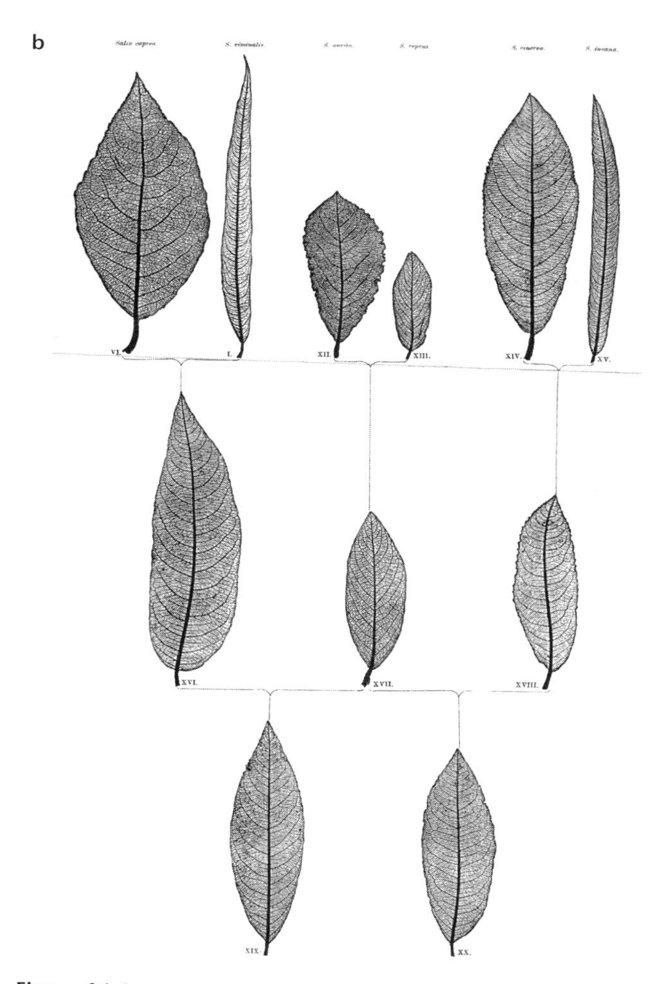

Figure 14.4

Leaves of two quaternary hybrids, depicted both in a schematic way (above) and as a shaded figure (below). From Max Wichura, *Die Bastardefruchtung im Pflanzenreich erläutert an den Bastarden der Weiden* (1865).

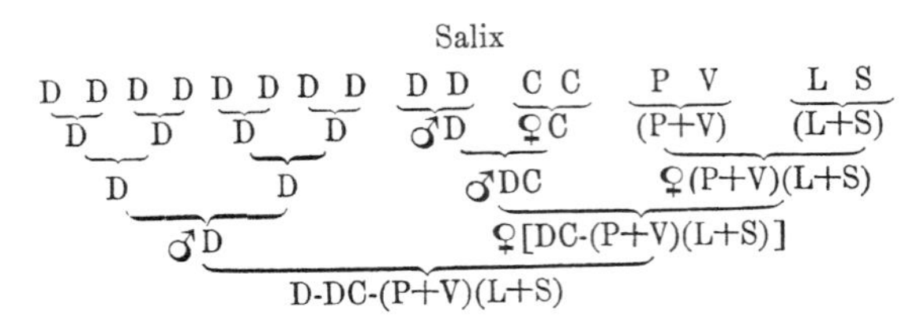

Figure 14.5
A pedigree of a hybrid. From Carl Nägeli, "Ueber die abgeleiteten Pflanzenbastarde," *Sitzungsberichte der königlichen bayerischen Akademie der Wissenschaften* 1 (1866).

can learn from the figure that the same individual (e.g., leaf no. XVII) yields very different offspring when it is crossed with two different types of leaves (XVI and XVIII, respectively).

One of the earliest uses of the pedigree in its modern form in the context of hereditary analysis appeared in 1866 in a paper on hybrids by Carl Nägeli (figure 14.5).[83] Nägeli may have been the first to suggest the systematic use of genealogical trees similar to those still in use today: the generational stratification rolling vertically downward, while the distribution of traits within a single generation is represented horizontally, each generation distributed along each axis. In Nägeli's illustrations, the pedigree is not merely a representation of family history, but an actual tool—to use Gayon's terminology[84]—for the understanding of heredity. The visual model is constructed according to very specific assumptions about the biological process of heredity, and the iconography is explicitly elaborated in terms of those assumptions. For example, Nägeli emphasized the need to represent all individuals belonging to a single generation in one horizontal line, arguing that this structure is critical, because it is only on the basis of such an analysis that the underlying hereditary processes can be understood. Heredity, according to Nägeli, is not a symmetrical process—"the material shares of the father and of the mother are not equal." This must "inevitably" lead to the conclusion that "the transmission of traits must be unequal as well."[85] Accordingly, Nägeli suggested the following procedure for the reconstruction of hereditary processes: first, we have to reconstruct the pedigree of a hybrid [*Abstammungsformel*], in which each generation is clearly identifiable as a horizontal line within the pedigree (figure 14.5). However, this pedigree is not complete without knowledge about the proportion of heredity—that is, the percentage of influence of each parent. For this, Nägeli put forward what he called the "formula of inheritance" [*Erbschaftsformel*], providing the information about the proportions according to which the traits are transmitted (figure 14.6).

Generation	Abstammungsformel		Erbschafts-formel [3])
I.	AB		a + b
II.	A-AB		3a + b
III.	A-A-AB	oder 2A-AB	7a + b
IV.	A-A-A-AB	„ 3A-AB	15a + b
V.	A-A-A-A-AB	„ 4A-AB	31a + b
VI.	A-A-A-A-A-AB	„ 5A-AB	63a + b
VII.	A-A-A-A-A-A-AB	„ 6A-AB	127a + b

Figure 14.6
"Inheritance formula" of a hybrid; *a* and *b* stand for the respective proportions of maternal and paternal transmission. From Carl Nägeli, "Ueber die abgeleiteten Pflanzenbastarde," *Sitzungsberichte der königlichen bayerischen Akademie der Wissenschaften* 1 (1866).

14.7 Conclusion—and a Note on Mendel

In modern genetics, the specific physiological identity of every organism is assumed to be directly determined by characters transmitted from its parents. The distribution of these characters follows a mathematical regularity, which cannot be observed in a single organism, but only as it is expressed in large populations—generations. Genetics cannot tell a meaningful story of heredity within a population before classifying this population according to generations—that is, the population must undergo a generational stratification. The "traveling" of discrete characters over hereditary paths makes sense only if we assume that this travel takes place between clearly identifiable realms, namely, two (or more) succeeding generations. Similarly, a segregation of characters would be meaningless without a generational context. We *must* know to what generation individuals belong before postulating a hereditary regularity.

My aim in this chapter has been to show the central role of the concept of generations in the establishment of a new vision of heredity in European science and culture in the nineteenth century. The chapter has concentrated on the first half of the nineteenth century. A discussion of the later part of the story will have to await another opportunity. However, I could scarcely conclude a chapter on heredity in the nineteenth century without at least mentioning Gregor Mendel and his 1866 *Versuche über Pflanzenhybriden*.[86]

It has often been noted that in order to pursue Mendelian genetics one has to employ a statistical method. Several historians, like Ernst Mayr, emphasized this as the most innovative aspect of Gregor Mendel's work, notably Mendel's decision to study whole, large populations instead of single individuals. Indeed Mayr even claims that Mendel's method anticipated modern population genetics.[87] But what was the nature of Mendel's population analysis? What were his "populations"? Mendel himself, in 1866, emphasized that "in order to discover the relations in which the hybrid forms stand towards each other and also their progenitors it appears to be necessary that all members of the developmental series in each successive generation should be, without exception, subjected to observations."[88] Crucial for Mendel's work was, therefore, not merely a diligent population analysis, as Mayr contends, but also his assumption that "a generation" can indeed be considered a closed population, an object for experimentation and analysis. Needless to say, Mendel's work was original in various respects, notably in his formulation of a law of segregation and the realization of the 3:1 relation. However, these discoveries, too, were only made possible on the basis of Mendel's strict classification of individuals according to their generational identity, and the corresponding assumption that each of these generations carries a finite load of traits transmitted to it an orderly fashion from its predecessor "generation." In this sense, Mendel's work is not as far "ahead of its time" as has often been argued, and is in full agreement with my argument about the centrality of generational reasoning in nineteenth-century biology.[89]

Notes

1. Baer 1827, 742. All translations are mine except where otherwise indicated.

2. Huxley [1863] 1943, 212. Around this time, Henry Thomas Buckle wrote: "Such . . . is our ignorance of physical laws, and so completely are we in the dark as to the circumstances which regulate the hereditary transmission of character, temperament and other personal peculiarities, that we must consider [any] alleged progress as a very doubtful point." Moreover: "We often hear of hereditary talents, hereditary vices, and hereditary virtues; but whoever will critically examine the evidence will find that we have no proof of their existence" (Buckle [1857] 1904, 134–135).

3. Historians have been aware of this problematique for quite a while. In his classic work *The Mendelian Revolution*, Peter Bowler (1989, 6) argued that Mendelism became meaningful only "within a theoretical model of heredity in which *(a)* the transmission of characters from one generation to the next constitutes a distinct and worthwhile field of study and *(b)* it is assumed that the characters can be treated as distinct units." In a landmark paper, Frederick B. Churchill (1987, 161) discussed what he called the "transmission problem" in the second part of the nineteenth century, claiming that the Mendelian revolution was only possible on the basis of a shift from "heredity" to "Vererbung," the latter manifesting the

modern notion of hereditary processes ("new heredity"): "Heredity as a process became restricted to the function of transmission between generations of organisms." More recently, López-Beltrán (2003, 11) similarly argued that in nineteenth-century French science one could observe a gradual process through which "the general term *physiological heredity* became accepted for referring to the *normal* mechanism by which bodily resemblances are transmitted through the generations." An analytically sharper description of this historical development was given by the French historian of biology Jean Gayon, who contended that the main characterization of the "Mendelian break" was in its new employment of pedigrees. For Mendelian genetics a pedigree was a tool, and no longer the "fundamental concept" it had been for nineteenth-century biology. Instead of being considered the "sum total of influences received from the ancestors," heredity—in its "Mendelian-chromosomal phase"—became "a question of structure in a given generation." Thus "the parental generation was, by definition, the generation that by its genetic structure was to be expressed in the context of specific crosses." Heredity was thus practically "synonymous with descent or lineage" (Gayon 2000, 75–77). See also Jacob [1970] 1974; Rheinberger 1983.

4. Note that all the definitions quoted above (by Bowler, Churchill, López-Beltrán, and Gayon) uncritically presuppose the existence of well-defined "generations."

5. A discussion of the prehistory of the concept of generations before 1800 would take us beyond the scope of this chapter. Its origin may be traced to the Hebrew word *Dor,* which appears in the Old Testament more than 200 times with various meanings. Usually, *Dor* stands for a unit or collective of people, and very often it is used as a reference to a genealogically defined group. See Ackroyd 1968; Orelli 1871. See also Biberger 2003.

6. Johnson 1755; Newton [1728] 1964.

7. For example, Süßmilch 1775–1776, vol. 2, 374.

8. On the newly established relations between the "old" and the "young" around 1800 see Nora 1996; Spitzer 1987. See also Weigel 2001.

9. Paine [1791] 1989, 55.

10. Jefferson 1984, 1281.

11. Jefferson 1984, 1280.

12. Jefferson 1984, 1280.

13. Condorcet [1792] 1976, 56.

14. Condorcet [1792] 1976, 47.

15. Condorcet 1976, 47.

16. Kant [1803] 1899. For a detailed discussion of Kant's views on heredity see McLaughlin, chapter 12, this volume.

17. Schleiermacher 2000, vol. 2, 13.

18. Schleiermacher 2000, vol. 2, 114.

19. Schleiermacher 2000, vol. 2, 13.

20. Chamisso 1819. Chamisso (1978) also published a long account of the whole expedition. See also Geus 1972; Parnes 2004.

21. Chamisso 1978, 35.

22. Chamisso 1819, 2–4; German translation in Chamisso 1983, 49–50.

23. In modern genetic terms, alternation of generation occurs when the haploid phase is able (at least sometimes, at least under certain circumstances) to undergo mitosis. The resulting haploids may undergo several more successive divisions, before fertilization and a diploid stage appear again.

24. For example, Sars 1841; Steenstrup [1842] 1845; Purkynje 1846. The theory was also received with much approval in Britain. See Huxley 1851; Owen 1849; Thomson 1859. See also Churchill 1979; Farley 1982; Farley 1989–1990; Geus 1972.

25. Steenstrup [1842] 1845, 115.

26. Sars 1841, 11.

27. Carpenter 1848, 205.

28. Forbes 1848, 87.

29. Schwann 1839, 192. The translation follows Schwann 1847, 228–229. See also Parnes 2000. On the cell theory and heredity in the nineteenth century, see Duchesneau, chapter 13, this volume.

30. Nägeli 1844, 27.

31. Kölliker 1845, 98.

32. Reichert 1852, 3.

33. Reichert 1852, 3.

34. Unger 1852, 120.

35. Unger 1852, 36.

36. Braun [1851] 1853, 322.

37. Braun [1851] 1853, 322–323.

38. Braun [1851] 1853, 11.

39. Braun 1854, 25.

40. Purkynje 1846, 230–232.

41. Braun [1851] 1853, 322. Some thirty years later, Carl Nägeli (1884, 275) will argue in very similar terms when drafting his theory of the Idioplasm: "The whole pedigree is essentially one single continuous individual made of Idioplasm. . . . With each generation this individual is wearing another dress—that is, it forms a new individual body." On Nägeli's later work on heredity see Rheinberger 1983.

42. Recall the discussion in the first part of this chapter.

43. Needless to say, the term *sociology* is used here anachronistically, referring to theoretical elaborations of social issues within contemporary disciplines like philosophy or political economy.

44. Did this line of reasoning evolve in the social sciences, to be adopted later by biology, or was it the biological metaphors that influenced social thinking in the nineteenth century? I do not think there is much point in trying to give a simple answer to this question. While the original metaphor of generations surely came from the biological realm, its elaboration and the formulation of an explicit theory of intergenerational relations were worked out within a social discourse and largely independent of biological knowledge.

45. Comte 1875–1877, vol. 4, 479.

46. Comte 1875–1877, vol. 1, 381. See also Comte 1839–1841.

47. Mill 1843, 595.

48. Mill 1843, 595.

49. Mill 1843, 596.

50. Marx [1845–1846] 1969, 45. The translation is mine, albeit partially relying on a translation available on the web: http://www.marxists.org/archive/marx/works/1845/german-ideology/ch01b.htm#b1.

51. Marx [1845–1846] 1969, 38.

52. Marx [1845–1846] 1969, 43. Obviously, Marx is paraphrasing here an old idiom about "standing on the shoulders of giants." See Merton 1985.

53. Elsewhere, I have tried to trace the role of the figures "generations" and "alternation of generations" through a comparative reading of literary and biological texts of this period. See Parnes 2004.

54. Dilthey 1913, 272.

55. Dilthey 1913, 271. Originally published as Dilthey 1865.

56. Dilthey 1913, 271.

57. Dilthey 1913, 270, italics added.

58. Marx [1845–1846] 1969, 38.

59. Carl Nägeli's work is often discussed in relation to that of Gregor Mendel, who probably read Nägeli's work and wrote him about his own experiments in the 1860s. As has been mentioned numerous times, Nägeli seemed to have overlooked the importance of Mendel's work, and in any case did not refer to it in his own attempts to formulate a theory of heredity. See Mendel 1924; Orel 1996. In this context, historians have usually consulted only Nägeli's later work, notably his *Mechanisch-physiologische Theorie der Abstammungslehre* (1884). See, for examples, Rheinberger 1983; Bowler 1989.

60. Nägeli 1856, 203. Note that in German the word *Anlage* could mean either a financial investment or a hereditary predisposition.

61. The new epistemology of heredity did not exclude the possibility of the inheritance of acquired (or at least modified) characters. This is another important point that I have to refrain from discussing in this chapter. See Jablonka and Lamb 1995.

62. For example, Darwin 1868.

63. Reichert 1840; also 1852, 129–133. See also Carus 1849, 40.

64. Nägeli 1884, 41–46.

65. "Was ist eine Reihe?" (Carus 1849, 55).

66. Carus 1849, 51–52, my italics.

67. Carus 1849, 51.

68. Blumenbach 1791.

69. Braun 1851 347; [1851] 1853, 324–325.

70. Braun 1851, 17; [1851] 1853, 15. Note the departure, in Braun's theory, from the traditional assumption that the type (or *Art*) remains the same as long as the external conditions remain unaltered. On this see Müller-Wille, chapter 8, this volume.

71. Braun 1851, 348; [1851] 1853, 326.

72. Braun 1851, 348; [1851] 1853, 325. I have chosen to translate *Erinnerung* as "memory" instead of "recollection" or "reminding," which the 1853 translator preferred. Alexander Braun's discussion of generations and heredity in 1851 was part of a more general attempt to understand what he called the phenomenon of "Rejuvenescence" [*Verjüngungserscheinungen*] in nature.

73. Braun 1851, 347–348; [1851] 1853, 325.

74. Elkana 1970.

75. Nägeli 1856, 204.

76. Haeckel 1866, 181.

77. Naudin 1865, notably 96–105.

78. Braun 1854, 107.

79. Braun 1854, 105–106.

80. On the history of the iconography of pedigrees see Resta 1993 and Bouquet 1996, as well as Heck 2002, Castaneda 2002, and Weigel 2002. Earlier than 1860, pedigrees were rarely used for the depiction of hereditary traits. Occasionally, these were superimposed on specific and concrete family genealogies. For example, Pliny Earle, an American physician, published a pedigree in 1845 in which circles and squares were used to depict the inheritance of color blindness in one family. See Resta 1993, 238.

81. On Wichura see Olby 1986.

82. Wichura 1865, 10–11.

83. Note the lack of any comparable iconography in Carl F. Gärtner's influential *Versuche und Beobachtungen über die Bastarderzeugung* (1849). Interestingly, three of the works discussed here, all incorporating visual representations of hybridization experiments, appeared almost at the same time—Naudin's and Wichura's in 1865, and Nägeli's in 1866 (but reporting his work from 1865).

84. See note 3.

85. Nägeli 1866, 74.

86. Mendel [1866] 1965.

87. Mayr 1982.

88. Mendel [1866] 1965, 58. The translation follows a modified version of Bateson's 1901 version, available at http://www.esp.org/foundations/genetics/classical/gm-65.pdf, p. 3.

89. Orel 1996. For a contrary view see Olby 1974, 1979.

References

Ackroyd, P. R. 1968. The meaning of Hebrew *dor* considered. *Journal of Semitic Studies* 13 (1): 3–10.

Baer, Karl Ernst v. 1827. Beiträge zur Kenntnis der niedern Thiere. *Verhandlungen der kaiserlichen-leopoldinisch-carolinischen Akademie der Naturforscher* 13:523–762.

Biberger, Bernd. 2003. *Unsere Väter und wir: Unterteilung von Geschichtsdarstellungen in Generationen und das Verhältnis der Generationen im Alten Testament*. Berlin: Philo.

Blumenbach, Johann Friedrich. 1791. *Über den Bildungstrieb*. Göttingen: Johann Christian Dieterich.

Bouquet, Mary. 1996. Family trees and their affinities: The visual imperative of the genealogical diagram. *Journal of the Royal Anthropological Institute* 2 (1): 46–66.

Bowler, Peter J. 1989. *The Mendelian Revolution: The Emergence of Hereditarian Concepts in Modern Science and Society*. London: Athlone Press.

Braun, Alexander. 1851. *Betrachtungen über die Erscheinung der Verjüngung in der Natur, insbesonders in der Lebens- und Bildungsgeschichte der Pflanzen*. Leipzig: Wilhelm Engelmann.

———. [1851] 1853. Reflections on the phenomenon of rejuvenescence in nature, especially in the life and developments of plants. In A. Henfrey, ed., *Botanical and Physiological Memoirs*. London: Ray Society.

———. 1854. Das Individuum der Pflanze in seinem Verhältniß zur Species: Generationsfolge, Generationswechsel und Generationstheilung der Pflanze. *Abhandlungen der königlichen Akademie der Wissenschaften zu Berlin* 1853:19–122.

Buckle, Henry Thomas. [1857] 1904. *History of Civilization in England*. London: Henry Frowde.

Carpenter, William. 1848. On the development and metamorphoses of the zoophytes. *British and Foreign Medico-Chirurgical Review* 1:183–214.

Carus, Victor. 1849. *Zur naeheren Kenntnis des Generationswechsels: Beobachtungen und Schlüsse*. Leipzig: Wilhelm Engelmann.

Castaneda, Claudia. 2002. Der Stammbaum: Zeit, Raum und Alltagstechnologie in den Vererbungswissenschaften. In S. Weigel, ed., *Genealogie und Genetik: Schnittstellen zwischen Biologie und Kulturgeschichte*, 57–69. Berlin: Akademie Verlag.

Chamisso, Adelbert von. 1819. *De animalibus quibusdam e classe vermium Linnaeana in circumnavigatione terrae auspicante vomite N. Romanzoff duce Ottone De Kutzbue annis 1815, 1816, 1817, 1818*. Berlin: Ferdinand Dümmler.

———. 1978. *Reise um die Welt*. Berlin: Rütten & Loening.

———. 1983. . . . *Und lassen gelten was ich beobachtet habe: Naturwissenschaftliche Schriften mit Zeichnungen des Autors*. Ed. R. Scheeli-Graf. Berlin: Dietrich Reimer.

Churchill, Frederick B. 1979. Sex and the single organism: Biological theories of sexuality in mid-nineteenth century. In vol. 3 of *Studies in History of Biology*, ed. W. Coleman and C. Limoges, 139–178. Baltimore: Johns Hopkins University Press.

———. 1987. From heredity theory to Vererbung: The transmission problem, 1850–1915. *Isis* 78:337–364.

Comte, Auguste. 1839–1841. *Cours de philosophie positive*. Paris: Bachelier.

———. 1875–1877. *System of Positive Philosophy*. London: Longman.

Condorcet, Jean-Antoine-Nicolas de Caritat. [1792] 1976. The nature and purpose of public instruction. In K. M. Baker, ed., *Condorcet: Selected Writings*. Indianapolis, IN: Library of Liberal Arts.

Darwin, Charles. 1868. *Das Variiren der Thiere und Pflanzen im Zustande der Domestication*. Trans. J. V. Carus. Stuttgart: Schweizerbart.

Dilthey, Wilhelm. 1865. Novalis. *Preußische Jahrbücher* 15:596–650.

———. 1913. *Das Erlebnis und die Dichtung: Lessing, Goethe, Novalis, Hölderlin*. Leipzig: B. G. Teubner.

Elkana, Yehuda. 1970. Helmholtz' *Kraft*: An illustration of concepts in flux. *Historical Studies in the Physical Sciences* 2:263–299.

Farley, John. 1982. *Gametes and Spores: Ideas about Sexual Reproduction 1750–1914*. Baltimore: Johns Hopkins University Press.

———. 1989–1990. Cells as individuals: The life-cycles of plants. *Folia Mendeliana* 24–25:65–73.

Forbes, Edward. 1848. *A Monograph of the British Naked-Eyed Medusæ: With Figures of All the Species*. London: Ray Society.

Gärtner, Carl Firedrich. 1849. *Versuche und Beobachtungen über die Bastarderzeugung im Pflanzenreich*. Stuttgart: K. F. Hering.

Gayon, Jean. 2000. From measurement to organization: A philosophical scheme for the history of the concept of heredity. In P. J. Beurton, R. Falk, and H.-J. Rheinberger, eds, *The Concept of the Gene in Development and Evolution: Historical and Epistemological Perspectives*, 69–90. Cambridge: Cambridge University Press.

Geus, Armin. 1972. Der Generationswechsel: Die Geschichte eines biologischen Problems. *Medizinhistorisches Journal* 7:159–173.

Haeckel, Ernst. 1866. *Generelle Morphologie der Organismen*. Vol. 2 of *Allgemeine Entwickelungsgeschichte der Organismen*. 2 vols. Berlin: Georg Reimer.

Heck, Kilian. 2002. Das Fundament der Machtbehauptung: Die Ahnentafel als genealogische Grundsttruktur der Neuzeit. In S. Weigel, ed., *Genealogie und Genetik: Schnittstellen zwischen Biologie und Kulturgeschichte*. Berlin: Akademie Verlag.

Huxley, Thomas Henry. 1851. Observations upon the anatomy and physiology of Salpa and Pyrosoma. *Philosophical Transactions of the Royal Society of London* 141:567–593.

———. [1863] 1943. The perpetuation of living beings, Hereditary Transmission and Variation. In *Man's Place in Nature and Other Essays*. London: J. M. Dent.

Jablonka, Eva, and Marion J. Lamb. 1995. *Epigenetic Inheritance and Evolution: The Lamarckian Dimension*. Oxford: Oxford University Press.

Jacob, Francois. [1970] 1974. *The Logic of Life: A History of Heredity*. Trans. B. E. Spillman. London: Allen Lane.

Jefferson, Thomas. 1984. *Writings*. New York: Library of America.

Johnson, Samuel. 1755. *A Dictionary of the English Language*. London.

Kant, Immanuel. [1803] 1899. *On Education*. Trans. A. Churton. London: Kegan Paul.

López-Beltrán, Carlos. 2003. Heredity old and new: French physicians and l'hérédite naturelle in early 19th century. In *Conference: A Cultural History of Heredity II: 18th and 19th Centuries*, 7–19. Vol. 247, preprints. Berlin: Max-Planck-Institute for the History of Science.

Marx, Karl. [1845–1846] 1969. *Die deutsche Ideologie: Kritik der neuesten deutschen Philosophie in ihren Repräsentanten Feuerbach, B. Bauer und Stirner und des deutschen Sozialismus in seinen verschiedenen Propheten*. Vol. 3 of *Karl Marx—Friedrich Engels—Werke*. Berlin: Dietz Verlag.

Mayr, Ernst. 1982. *The Growth of Biological Thought*. Cambridge, MA: Harvard University Press.

Mendel, Gregor. 1924. Gregor Mendels Briefe an Carl Nägeli, 1866–1873. In F. V. Wettstein, ed., *Carl Correns: Gesammelte Abhandlungen zur Vererbungswissenschaft aus periodischen Schriften*, 1233–1297. Berlin: Julius Springer.

———. [1866] 1965. Versuche über Pflanzen-Hybriden. In J. Krizenecky and B. Nemec, eds., *Fundamenta genetica: The Revised Edition of Mendel's Classic Paper with a Collection of 27 original Papers Published during the Rediscovery Era*. Prague: J. A. Barth.

Merton, Robert K. 1985. *On the Shoulders of Giants: A Shandean Postscript*. San Diego: Harcourt Brace Johvanovich.

Mill, John Stuart. 1843. *A System of Logic, Ratiocinative and Inductive*. London.

Nägeli, Carl. 1844. Ueber die gegenwärtige Aufgabe der Naturgeschichte, insbesondere der Botanik (I). *Zeitschrift für wissenschaftliche Botanik* 1 (1): 1–33.

———. 1856. Die Individualität in der Natur: Mit vorzüglicher Berücksichtigung des Pflanzenreiches. *Monatsschrift des wissenschaftliches Vereins in Zürich* 1:171–212.

———. 1866. Ueber die abgeleiteten Pflanzenbastarde. *Sitzungsberichte der königlichen bayerischen Akademie der Wissenschaften* 1:71–93.

———. 1884. *Mechanisch-physiologische Theorie der Abstammungslehre*. Munich, Leipzig: R. Oldenburg.

Naudin, Charles. 1865. Nouvelles recherches sur l'hybridité dans les végétaux. *Nouvelles archives du muséum d'histoire naturelle de Paris* 1:25–176.

Newton, Isaac. [1728] 1964. The chronology of antient kingdoms amended. In Vol. 5 of *Opera quae exstant omnia*, reprint ed., 1–291. Stuttgart–Bad Cannstatt: Frommann (Holzboog).

Nora, Pierre. 1996. Generation. In *Realms of Memory*. New York: Columbia University Press.

Olby, Robert. 1974. The origins of genetics. *Journal of the History of Biology* 7 (1): 93–100.

———. 1979. Mendel no Mendelian? *History of Science* 17:53–72.

———. 1986. Mendels Vorläufer: Kölreuter, Wichura und Gärtner. *Folia Mendeliana Musei Moravia* 21:49–65.

V Anthropology

15 *Las Castas:* Interracial Crossing and Social Structure, 1770–1835

Renato G. Mazzolini

15.1 Introduction

What happens when individuals belonging to different populations mix and have offspring? And how do the offspring look when this mixture goes on for generations? Such questions were first raised during the last decades of the eighteenth century by a small group of naturalists working on the "natural history of man," a subject that we would now call physical anthropology. In Europe, these questions could not be answered empirically because marriages between Europeans and non-Europeans were extremely rare. Only mulattoes could be observed, and then mainly residing in the Iberian Peninsula, England, France, and Holland, where they were known to be the offspring of blacks and whites. The physical characteristic of these offspring that drew attention was their skin color, which was neither black nor white, but more often than not seemed to be a mixture of the two, thus suggesting that both parents contributed materially to the makeup of the fertilized egg. This was not an insignificant suggestion, given that, according to contemporary theories of generation such as those held by preformationists, the contribution of the male to procreation consisted only in stimulating the female egg to unfold.

The questions posed above had considerable implications. In a period in which the very concept of race was being introduced to the study of the natural history of man to denote a stable biological entity, to ask how races could mix and provide fertile offspring had paramount importance for the more general question of the unity of the human species. Given, however, that Europe did not present sufficient examples to be studied, naturalists were forced to turn to information provided by missionaries, travelers, and physicians who had spent some time in Central and South America. What they reported about the populations living in these faraway territories were mainly the social classifications that the Spaniards and the Portuguese had developed and implemented as early as the late sixteenth and early seventeenth centuries. The purpose of such classifications had been to control the inhabitants of their vast empires, which had increasingly become peopled by individuals of mixed ancestry.

15.2 Naturalists and Interracial Crossing

In 1777, Georges-Louis LeClerc Comte de Buffon (1707–1788) published an important essay titled “Addition à l'article qui a pour titre, Variétés dans l'espèce humaine,” in which he included two sections describing, on the one hand, four mixed generations of whites and blacks and, on the other, four mixed generations of whites and Indians.[1] Buffon's simplified scheme consisted of the following categories:

1. White × black = *mulatto*, half white, half black
2. White × *mulatto* = *quarteron*, tanned
3. White × *quarteron* = *octavon*, less tanned than the quarteron
4. White × *octavon* = a perfectly white boy or girl[2]

The inverse mixture gave the following results:

1. Black × white = *mulatto* with long hair
2. Black × *mulatto* = *quarteron*, three-fourths black and one-fourth white
3. Black × *quarteron* = *octavon*, seven-eighths black and one-eighth white
4. Black × *octavon* = a perfectly black boy or girl[3]

It is interesting to note that the main physical trait to be considered was skin color, and that the first scheme illustrated “four generations mixed in order to make the color of Negroes disappear,” while the second was the inverse and showed how “to blacken Whites.”[4] Buffon then went on to describe the crosses between whites and Indians.

1. European woman × American = *mestizo*, two-fourths from each parent and tanned
2. European woman × *mestizo* = *quarteron*, one-fourth of an American, less tanned
3. European woman × *quarteron* = *octavon*, one-eighth of an American
4. European woman × *octavon* = *puchuella*, a perfectly white boy or girl[5]

Buffon derived the above schemes from the two-volume work *Recherches philosophiques sur les Américains*, which Cornélius de Pauw (1739–1799) published anonymously in 1768–1769, and lamented that the author made no reference to his authorities.[6] Buffon also added that he himself had sought similar information, but could not find any as precise as the information provided by de Pauw. But he urged caution on the part of his readers because, he said, de Pauw had often advanced many statements that had been shown to be either misleading or false.

It seems most probable that de Pauw gathered his information from two Spanish sources that had also been translated into French. The first is *El Orinoco ilustrado*, a book authored by the Jesuit José Gumilla (1686–1750), which appeared in its second Spanish edition in 1745 and was translated into French in 1758.[7] The second is *Relación histórica del viage a la América meridional* by Jorge Juan y Santacilia (1713–1773) and Antonio de Ulloa (1716–1795), which appeared in 1748 and was translated into French in 1752.[8] Gumilla, for instance, provided the following scheme for the white-black generations:

1. European × *negra* = *mulata*, two-fourths from each parent
2. European × *mulata* = *quarterona*, one-fourth of *mulata* [sic]
3. European × *quarterona* = *octavona*, one-eighth of *mulatta* [sic]
4. European × *octavona* = *puchuela*, completely white[9]

For crosses between Europeans and Indians Gumilla provided the following scheme:

1. European × *India* = *mestiza*, two-fourths from each parent
2. European × *mestiza* = *quarterona*, one-fourth of *India*
3. European × *quarterona* = *octavona*, one-eighth of *India*
4. European × *octavona* = *puchuella*, completely white[10]

In both sequences of generations the male was always a European and the female either a black or an Indian woman. Gumilla knew that there were much more complex crosses to describe, but did not add many more details because what he intended to maintain was that, according to the Catholic Church, the *quarterones* and *octavones* had to be considered as *blancos*, while only the *Indios* and *mestizos* could be considered newly converted.

On the basis of a substantial number of authorities, Johan Friedrich Blumenbach (1752–1840) described four generations of mixed human varieties when discussing the question of pigmentation in his epoch-making doctoral dissertation *De generis humani varietate nativa*, published in Göttingen in 1775.[11] An indefatigable supporter of monogenesis, he drew the conclusion that whatever might be the cause of human skin color, intermixture of human varieties demonstrated that color was changeable and could never be considered a character constituting a diversity of species. He also reported the case of a black woman who gave birth to twins of different color, the male being a mulatto and the female as black as her mother.[12] In the third edition of *De generis humani varietate nativa*, which appeared in 1795 and should be considered a new book altogether, Blumenbach went into great detail with respect to the various names he had discovered in the literature concerning human mixed varieties found in the

Americas. He restricted himself to three generations but on this subject added no new observations to those made in 1775.[13]

In 1799 the old and distinguished anatomist and physiologist of Padua University, Leopoldo Marc'Antonio Caldani (1725–1813), published an essay consisting of "conjectures" on the cause of the various skin colors displayed by humans.[14] He stated that he had discussed the question in his private correspondence with his friend, the Genevan naturalist Charles Bonnet (1720–1793). Caldani had furthermore asked one of his patients in Padua, the ex-Jesuit Francesco Berengher, who had spent a number of years in the Americas, to put on paper what Berengher had told Caldani in conversations on the intermixture of Indians, blacks, and whites. Caldani published Berengher's report in his essay and had an engraving made of the drawings the latter had added to his report. The engraving represented the faces of individuals generated by parents belonging to different populations, and emphasis was placed on rendering their skin color and head shape accurately. In examining the classification of the different crosses, Caldani noticed two things. On the one hand, the report substantially confirmed the classification provided by Buffon, drawing on de Pauw's work. On the other, some new terms designating crosses appeared, such as the "*Zamba di Nero figlio d'un Mulatto e di una Nera*" and the "*Parda, figlio d'una Nera e di un Quarteron.*" In his report, Berengher wrote that there were "some remarkable peculiarities." For instance, he claimed to have noticed that at times the offspring of a mulatto and a white woman had a complexion that was fairer than that of his mother and was therefore called *salta-adelante,* meaning someone who jumps ahead "because he has made such a surprising and gigantic step toward whiteness."[15]

Caldani used the data provided by Berengher to confirm the unity of the human species, and to maintain that skin color was not so much a result of the action of climate, as held by Buffon, but of the generation process. He argued that the color of the Malpighian layer of the skin resulted from the male's fecundating fluid, which not only stimulated and nourished the germ, but also modified it in such a way as to "alter the distribution of the tiny vessels that made up the skin of the germ."[16] Notwithstanding his support for Bonnet's preformation theory of generation, Caldani was led by his investigations into procreation by individuals belonging to populations with different skin color to speculate on the action of sperm on the germ, which he now regarded not only as a stimulus but as a modifying agent of the germ's structure.

It was only in 1811, when Alexander von Humboldt (1769–1859) published his *Essai politique sur le royaume de la Nouvelle-Espagne,* that non-Iberian scholars learned about the social system in which the interracial mixtures they were studying were embedded. Humboldt had been greatly interested in the natural history of man since his youth.[17] However, uncertainties concerning the common origin of all humankind and awareness of the political and social implications of the topic required him to be very cautious when expounding his ideas. During a journey through the Americas he paid close attention to the pigmentation of the indigenous peoples, which he believed was a clue to their possible origin. In 1803–1804 Humboldt stayed for a year with his friend Aimé Bonpland (1773–1858) in Mexico. There, as well as in Cuba, he was able to study at first hand the effects of the *las castas* system, which he considered deleterious because of the social inequality to which it gave rise, *las castas* being the terminology employed by the Spanish in the seventeenth century for establishing the ancestry of each individual and his or her social standing. In two chapters of his 1811 book—which contain an extraordinary wealth of data, historical-social analysis, and political prophecy—Humboldt described the Mexican population according to the classification used by the local governors. There were seven main racial groups in Mexico: three pure ones (the *casta de los blancos*, comprising *Españoles* and *Créoles*; Indios; and blacks) and three mixed (*mestizos, mulattos*, and *zambos*):

1. Spanish, or whites, born in Europe (termed *gachupines*)
2. Spanish, or whites, born in America (*Créoles*)
3. Descendants of a white and an Indian (*mestizos*)
4. Descendants of a white and a black (*mulattos*)
5. Descendants of a black and an Indian (*zambos*)
6. *Indios*
7. African blacks

In descriptions that still today exert a powerful impact, Humboldt reconstructed the consequences for the indigenous population of oppression first by the Aztec rulers and then by the Spanish. He described the physical constitution, the intellectual faculties, and the social conditions of the *Indios*. Mexico, he argued, was a country where inequality reigned: "Le Mexique est le pays de l'inégalité."[18] All economic and political power was in the hands of the whites, and only the *Indios* and the colored castas did the hard manual labor. Humboldt quoted long passages from a manuscript by a bishop that asserted that the *Indios* would never progress given the conditions of absolute inequality in which they were forced to live. Black descendants who were no longer slaves had to nonetheless pay tribute, which constituted a kind of prolonged slavery.

Skin color signaled membership in a particular social stratum: "In America more or less white skin determines the rank occupied by man in society."[19] According to Humboldt, a society structured this way inevitably generated conflicts of interest, prejudice, and resentment. Even in a counterfactual scenario in which an enlightened government had gradually eliminated the monstrous inequalities of rights and opportunities, it would nevertheless have encountered the obstacles created by centuries-long barriers of blood and color. It would have been extremely difficult to construct a net of social cooperation and to ensure that all inhabitants of Mexico could consider themselves "*concitoyens*."[20] It was the racial rather than the social implications of miscegenation, however, that captured the attention of many with an interest in natural history.

In his *Lectures on Physiology, Zoology, and the Natural History of Man*, a book dedicated to Blumenbach and published in 1819, the English physician William Lawrence (1783–1867) showed great interest in the problems of heredity and selection. He believed that a "superior breed of human beings could only be produced by selections and exclusions similar to those so successfully employed in rearing our more valuable animals" and insisted that sexual selection in "savage nations" prevented deformed individuals from propagating their deformities, while intermarriage among the nobility caused a "degradation of race," as could be noticed "in the present state of many royal houses in Europe."[21]

In a chapter dedicated to the color of the human species, Lawrence wrote about the complexion of mixed races such as mulattoes, tercerons, quarterons, quinterons, mestizos, and zambos, drawing on the works of Buffon, Blumenbach, Humboldt, and a few other authors. On the one hand, he stressed that the offspring of two varieties of the same species could not be termed "hybrid," because this name properly applied only to the offspring of different species such as the horse and donkey; on the other hand, he stated that in the case of humans one could speak of "hybrid races" when referring to the crosses between Europeans, black Africans, and Indians from America.[22] The term *hybrid*, which since antiquity had also denoted the offspring of crosses between savage and civilized peoples (Pliny, VIII 213) and which had crept into scholarly literature from the sixteenth century onward, was to become from the 1820s until the end of the Second World War one of the main scientific terms to denote people generated by interracial mixing.[23]

Describing the "genealogy" of the white-black crosses, as already illustrated in the works of Gumilla, de Pauw, and Buffon, Lawrence stated that "we have the power of changing one species [!] into another by repeated intermixture" and that "in this manner the colour of the race may be completely changed in three or four generations; while it never has been changed by climate, even in the longest series of ages."[24] It should be noted that, according to Lawrence, color was not an essential character of race and that having the same skin color did not imply descent from a common stock. Lawrence's ideas about intermixture were indeed paradoxical. In fact, while he stated that crosses between varieties of the same species of domestic animals were wonderfully advantageous, he emphasized that whites could never be improved by any mixture, as the following passage clearly demonstrates:

The dark races, and all who are contaminated by any visible mixture of dark blood, are comprised under the general denomination of people of colour. It is not, however, merely by this superficial character that they are distinguished; all other physical and moral qualities are equally influenced by those of the parents. The intellectual and moral character of the Europeans is deteriorated by the mixture of black or red blood; while, on the other hand, an infusion of white blood tends in an equal degree to improve and ennoble the qualities of the dark varieties.[25]

It is well known that in British and French colonies half-breeds were excluded from military and government offices. Mulattoes, in particular, were described as human hybrids, and the English doctor Edward Long (1734–1813) had cast doubt on their fertility by suggesting that crosses between mulattoes were infertile just like those of mules. This idea gave rise in the middle of the nineteenth century to much scientific literature on human hybrids by anthropologists such as Samuel G. Morton (1799–1851) and Pierre-Paul Broca (1824–1880). In popular fiction they were often represented as fakes, troublemakers, biological frauds, and ambivalent creatures torn between different cultures and loyalties.[26] Work by eighteenth- and nineteenth-century European naturalists of the Americas thus developed and popularized quite extensive classification systems of human hybrids and dealt with both the social and racial implications of such systems.

15.3 Las Castas

Before the publication of Humboldt's work, naturalists discussing interracial crossing were not completely aware that what they were dealing with was an administrative system, known as *las castas*, set up by the Spanish colonial government in South and Central America.[27] During the second half of the eighteenth century the meaning of the term *casta* was little known to non-Iberian scholars. For instance, in providing the French version of the *Relación histórica del viage a la América meridional* published in 1752, the translator felt it necessary to explain the word *casta* by adding the word *race* as a synonym: thus the Spanish expression "distinction de Castas" was rendered as "différence des *Castes* ou *Races*."[28]

Many verbal and nonverbal sources clarify the workings of this system. These include books written by Spanish officials, parish registers, censuses, and tribute records, as well as paintings such as those reproduced in figures 15.1 to 15.5. (See also plates 2 and 3) The system was intended to establish the ancestry of each individual in such a way that he or she could be included in a specific caste with a specific name.

Figure 15.1
De Español y Negra Mulata (*Spaniard and Black woman, Mulatta*), oil painting, eighteenth century. Courtesy Museo de América, Madrid.

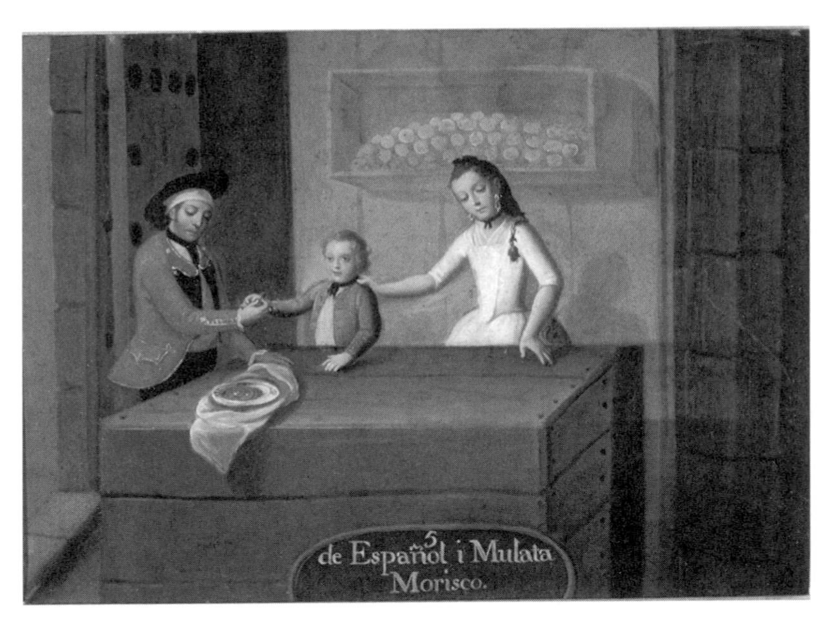

Figure 15.2
De Español i Mulata Morisco (*Spaniard and Mulatto woman, Morisco*), oil painting, eighteenth century. Courtesy Museo de América, Madrid.

Figure 15.3
De Español y Morisca Alvino (*Spaniard and Morisco woman, Albino*), oil painting, eighteenth century. Courtesy Museo de América, Madrid.

Figure 15.4
De Español y Alvina Negro Torna atras (*Spaniard and Albino woman, Negro thrown-back*), oil painting, eighteenth century. Courtesy Museo de América, Madrid.

Figure 15.5
De Español è Yndia Mestizo (*Spaniard and Indian woman, Mestizo*), oil painting, eighteenth century. Courtesy Museo de América, Madrid.

Given the extraordinary expansion of the Spanish Empire in the Americas, the system presented local variations from Peru to Mexico and its use was subject to both changes in time and imprecise and inconsistent application.[29] Its aim, however, was quite clear: to create a hierarchical social structure based on place of birth and, especially, ancestry. The most important clues to establishing the caste into which newborn children fell were the castes of their parents and their skin color.

This system had great consequences because the different rights and obligations of individuals depended on affiliation with particular castes, which enabled or precluded access to certain careers and religious orders. Separate fiscal statuses and legal systems also existed. In his impressive investigation into the state of the Spanish colonies in America expounded in his *Politica Indiana*, published in 1647, the jurist Juan de Solórzano Pereira (1575–1655) discussed the tensions existing within the caste system. He held, for instance, that the *Criollos*—individuals born in the Americas but with Spanish ancestry—had the same rights as the *Españoles* residing in the Americas but born in Spain. He added that *Criollos* should also have the possibility of undertaking an ecclesiastical career, something that had been contested around the time Solórzano

Pereira's study was published.[30] He also argued that *mestizos* and *mulatos*, known in Latin as *varios* and *hybridas*, had the same right of access to offices and religious orders as long as they were the offspring of a legitimate marriage.[31] But this was the problem since, he wrote, very few Spaniards of honor ("*Españoles de honra*") would marry an Indian or black woman, and only a legitimate marriage could eliminate the stain of color ("*la mancha del color vario*") in a child born of a mixed relationship.[32] He added that the number of illegitimate *mestizos* and *mulatos* was steadily increasing and that they were potentially very dangerous since they refused to work in the mines like the poor *Indios* ("*los pobres Indios*"), thus aggravating the conditions of the latter.[33] The illegitimate among the castes of *mestizos* and *mulatos* were not to be privileged in any way, because they were the offspring of lust and were themselves vicious.

Some further aspects of the hierarchical social structure established in Spanish America may be perceived by looking at the extraordinary cycles of paintings illustrating the caste system, which were produced in Mexico from the late seventeenth century.[34] Usually a cycle was formed by a series of twelve to sixteen paintings showing four generations of crosses beginning respectively with that of a Spaniard and an Indian woman, a Spaniard and a black woman, and an Indian and a black woman. Other paintings showed more complex intermarriages, such as that between an Indian and a mulatta, whose offspring were called *coyotes* and had Indian, African, and European blood running through their veins. Because of their relevance for the problem of heredity, we will consider four items only, representing crosses commencing with that of a Spaniard and a black woman (figure 15.1).[35] The caption to this painting explains that the girl born from this relation is called a *Mulata*. The scene that is represented clearly shows both parents quarreling furiously in a peasant kitchen. The most striking thing about this painting is its didactic message: when a white man and a black woman marry, they inevitably quarrel. This theme appears in a good number of other cycles. It should be noted, in passing, that the child of a Spaniard and an Indian woman is often portrayed weeping (figure 15.5).

Figure 15.2 shows a Spaniard, a mulatto woman, and their *Morisco* son (the *quarteron* of Gumilla and Buffon) placed in the friendly atmosphere of a small shop. Figure 15.3 depicts a Spaniard of some distinction writing at his desk with his wife, a *Morisca*, and their son named *Albino* (the *octavon* of Gumilla and Buffon). The ambience is socially more elevated than in previous scenes and the mother wears a better dress than her mother and grandmother. The most significant element in this scene is that the *Morisca* has very white skin with a black freckle at the level of her left temple. The clear message in this case is for white men to beware of white women with dark freckles, because they might have African ancestry. Finally, figure 15.4 illustrates the fourth generation of crosses, that between a Spaniard and an albino woman, who wear very fine clothes and are located in a patrician environment with tapestries hanging on a wall. Their son is not a perfectly white boy, as one would expect from the schemes presented by Gumilla and Buffon, but is completely black and is named *Negro Torna atrás*, meaning something like a "black throwback." The mother of the boy has a very white complexion but also an evident black freckle at the level of her left temple. Here again the message is very clear: beware of white women with dark freckles, because they might have African ancestry and you might get a son with black skin.

These paintings reveal a parallelism between social structure and skin color. The higher a Spaniard rises in the social hierarchy, the whiter his wife's complexion should be. The paintings also reveal their didactic and disciplining functions—to remind males that in their choice of a partner the appearance of a white complexion was not enough, because it did not guarantee that the person in question had no black ancestry. Thus the message was for white men to check the ancestry of their sweetheart before getting married. Unfortunately, these Mexican cycles of paintings were unknown to non-Iberian scholars of the late eighteenth and early nineteenth centuries. It should be noted, however, that the system of las castas was not exclusive to Spanish America, but was also adopted, if in a less formalized way, in the British and French West Indies.[36]

15.4 Historians and Caste Systems

It is well known that, in the early decades of the nineteenth century, historians, naturalists, and philosophers made increasingly frequent use of the term *race*, to which, despite its indeterminacy, they gave an explanatory rather than a simply descriptive meaning. Thus Barthold Georg Niebuhr (1776–1831) interpreted the conflicts between patricians and plebeians in ancient Rome as racial conflicts. French historians in the 1820s claimed that they could explain the French Revolution as the inevitable consequence of the centuries-long struggle between two races: that of the victorious Franks and that of the subjugated Gauls. The notion of a war between races thus assumed the status of historical explanation.[37] It is less well known that in those same years the term *caste* came into use among historians to denote social systems deemed similar to those existing in India. To my knowledge the origin, spread, and application of this term in European culture has not been studied, in contrast to the notion of race. So the following discussion should be considered provisional.

The word *casta* is a European invention and is first recorded in the Iberian Peninsula as denoting progeny, stock, lineage, breed, and race as early as the fifteenth century.[38] In 1516, the Portuguese used it in the plural to designate the hereditary social classes in India, which were subsequently referred to by this term in the other European languages as well. In the seventeenth century *casta* was used by the Spanish in the Americas to denote the descent of a single people, but also the progeny of interbreeding between races. At the end of the eighteenth century, as a result of studies of ancient Indian literature, a number of historians used the term *caste* as a "translation of two quite different indigenous concepts, *varna* and *jati*," and applied it to other systems of social differentiation in other ancient civilizations, like Egypt.[39] As we have seen, during the 1810s, the las castas system applied by the Spanish in the Americas was made known to non-Iberian European culture mainly by Alexander von Humboldt.

In the following section of this chapter I will argue that the racial theory of caste—usually attributed to Herbert Hope Risley's (1851–1911) works published in the 1890s—was in fact developed by European scholars between the 1790s and the 1830s, that is, during the period in which British rule over India was firmly established.[40] According to the theory developed in this period, the determining principle of the Indian caste system was the community of race and not that of occupation, and its hierarchical stratification was a consequence of the struggles and antagonism of the races peopling the country. According to this theory, *caste* was a synonym of *race* when used to designate social groups subjected to barriers against some forms of social contact and with respect to specific marriage and occupational customs.

Whatever the origins of the Indian caste system may be, recent historical research has shown that it should not be thought of as a monolithic structure from its inception, but as a social configuration subject to historical change.[41] Contemporary scholars of Indian history claim, in fact, that it was the British colonial government of the early nineteenth century that reinforced the Indian caste system and shaped it in the way that we have come to know it. It has also been claimed that the dialectics of black-white group identities were introduced to India by Europeans.[42] Given these reservations concerning the application of notions of caste to Indian history, I now go on to consider how such notions were developed by Europeans in the first place.

At the University of Göttingen, Christoph Meiners (1747–1810), professor of philosophy from 1775 until his death, but also an active historian, placed emphasis on the human body as an object of historical inquiry.[43] Humans, he argued, were of historical interest not only for their actions, customs, the products of their minds and personalities, but also for their bodies. There was a history to be written, which had never been narrated, whose protagonist would be the human body. In *Grundriß der Geschichte der Menschheit*—his best-known work, published in 1785—Meiners maintained that peoples were substantially unequal because their dispositions and talents, manifested in historical events, sprang from intrinsic physical differences.

In his examination of bodily features, Meiners considered stature, strength, adiposity, beauty, hirsuteness, and the shape of the body and of the eyes, in addition to skin color. He obtained his data from travel literature and on this basis believed that he could classify humans into two stocks, Tartaric-Caucasian and Mongoloid, with each category being further subdivided into races and varieties. Thus, for example, the Tartaric-Caucasian stock consisted of two races—Celtic and Slav—which in turn were divided into numerous subgroups. In addressing these issues, Meiners took part in a debate that for years animated scholars in the small university city of Göttingen. Indeed, in this period great interest was shown in non-European peoples not only by historians but, as we have seen, also by doctors and naturalists like Blumenbach, whose contributions to natural history brought him great fame. Yet Meiners's attitude toward non-European peoples differed greatly from that of his colleagues, Blumenbach or Georg Christoph Lichtenberg (1742–1799), for instance, who had declared their opposition to the trade in Africans and slavery. Meiners, by contrast, favored the slave trade not only because he considered it an economic necessity, but also because he believed that Africans were physically inferior to Europeans and that their natural inequality entailed their inequality of rights.

In a series of essays published between 1785 and 1793, the year in which a second and extended edition of *Grundriß* came out, Meiners developed his distinction between the two racial stocks further.[44] He used for this purpose the notion of beauty and the color of the skin. He maintained that humanity consisted of two breeds. One was handsome, white, and hirsute (Caucasians), the other was ugly, colored, and with little hair (Mongols). Moreover, the moral perfectibility of a people depended on its physical constitution. The dark-skinned peoples were stupid and could not hope to improve themselves, while the white peoples were ingenious and could progress. Although Meiners was aware of the infinite gradations of skin color, he classified humankind chromatically into white, yellow, red, and black, believing that the original colors had been yellow and white.

For Meiners, skin color was a physical feature that indicated an individual's social status, his or her rank in society. Significantly, he was one of the first scholars, if not the first, to argue that there was a link between membership in a particular caste and skin color. He extended this theory from India to ancient Egypt, hoping to provide an explanation of the social structure of Egyptian society.[45] He believed that he could rightly claim that ancient Egypt possessed a caste system based on skin color. Meiners's purpose was to prove that in relationships between whites and people of color, whites had always predominated, whereas people of color had always been subordinate because they possessed fewer rights. For Meiners, therefore, the white domination of the colored peoples was not a recent phenomenon, but a constant historical feature that sprang from an original physical inequality.

Meiners's historical-ethnographic studies provoked much criticism, and after 1793 his disputes with Georg Forster (1754–1794) and Blumenbach induced him to be more cautious, or to remain silent, at least as regards ethnography and the natural history of mankind. While the diffusion of Meiners's ideas among German scholars, and particularly historians, in the early nineteenth century has yet to be studied, the subtle but substantial influence of Meiners's racial history on a handful of significant French scholars of the late 1700s and early 1800s has been noted. Many years later, Arthur de Gobineau (1816–1882), who thematized Meiners's notion of inequality, recalled that he had "invented the most simple classification, which was composed of only two categories: the beautiful, that is to say the white race; and the ugly, which comprised all the others."[46]

In Germany a major historian like Arnold Heeren (1760–1842), whose handbooks were also translated into French and English, took up Meiners's theory of caste and applied it both to India and ancient Egypt, influencing more than one generation of historians.[47] The idea, however, that the Indian caste system resulted from a race war and that the higher castes were white, while the lower were black and belonged to a subjugated race, met with some criticism. For instance, the Catholic Bishop Nicholas Wiseman (1802–1865) wrote in 1836 that Heeren's thesis had been contradicted by the Anglican Bishop of Calcutta, Reginald Heber (1783–1826), who had written that color did not depend on caste "since very high caste Brahmins are sometimes black, while Pariahs are comparatively white."[48] The American anthropologist Charles Pickering (1805–1878), who inquired about castes during his visit to Bombay, wrote in *The Races of Man* (1848) that "from repeated inquiries it appeared that the rules of 'caste' are independent of colour or physical difference . . . and that, unlike what takes place in other countries and in respect to other races, no such distinction is recognised by the people themselves."[49]

During the 1820s and 1830s, French historians, doctors, and naturalists shared an interest in the notion of "race." The text titled *Des caractères physiologiques des races humaine considérés dans leur rapports avec l'histoire*, which the naturalist William-Frédéric Edwards (1777–1842) addressed in 1829 to the historian Amédée Thierry (1797–1873), is symptomatic.[50] In this work, which for decades was the manifesto of French physical anthropology, Edwards argued that when different races settle in the same territory, the blacks obey the yellows and both obey the whites, while people of intermediate skin color form the intermediate ranks of society. Edwards was convinced of the fixedness of race and of the permanence of hereditary features like color. Similar notions were put forward in 1827 by the jurist François-Charles-Louis Comte (1782–1837) in his *Traité de législation*, which devoted hundreds of pages to describing the supposed superiority of the white race over peoples of color.[51] Comte criticized Buffon's climatic theory, maintaining that human pigmentation was entirely unrelated to environmental factors or to the action of light and heat. It was instead an original property of race, and any modification of it in a particular individual could not be transmitted to his or her descendants.

It is evident that these discussions were tied, although rarely explicitly, to political and social issues like the abolition of slavery and France's colonial expansion in North Africa. This becomes immediately apparent when one considers the papers on race, language, and ancient and modern slavery given at the *Congrès historique européen* held in Paris between November 15 and December 15, 1835—an event of considerable cultural importance. The congress, which was attended by historians like Jules Michelet (1798–1874) and anthropologists like Jean-Baptiste Bory de Saint-Vincent (1778–1840), was divided into six "classes" that assembled for several sessions, each on a particular theme. The first class—the most important—opened with a session that was titled "Quel est le but de l'histoire?" It was immediately followed by a session addressed by Victor Courtet de l'Isle (1813–1867) on a theme of great topicality: "Déterminer par l'histoire si les diversités physiologiques des peuples sont entre elles comme les diversités des systèmes sociaux auxquels ces peuples appartiennent."[52]

Courtet pointed out that another session of the congress would discuss definitions of the terms *genus*, *species*, and *race* applied to humans. Nevertheless, he said, those definitions must also be addressed in light of the theme for his session. He defined as members of the same "species" those similar individuals who by reciprocal fertilization generated fertile individuals. A "race" was instead constituted by individuals characterized by a mutation that persisted through subsequent generations, while a "variety" comprised individuals in whom mutations did not persist. After clarifying the terms further and criticizing their inaccurate use by historians, Courtet de l'Isle went on to distinguish two schools of scholars: those like Buffon, Blumenbach, and Cuvier who considered humans to be a unique species with several varieties; and those like Virey, Duméril, Desmoulins, and Bory de Saint-Vincent who maintained instead that there were several human species. Whatever the origin of humankind, Courtet declared, there was one incontestable fact: "The original and constant diversity of the human types."[53] Camper's facial angle, an anatomical measure, and the development of the nervous system in the various human types proved that there was an undeniable gradation of human beauty and intelligence that started from crude forms and ascended, as Virey put it, "to the white man, to the most industrious and enlightened European."[54] Just as humans and animals differed in their moral inequality, so the human races differed in their physique, morality, and intellect. In short, Courtet argued that the twofold inequality of physical and moral features in the human races determined their different social positions.

Courtet gave great prominence to recent historical studies on the caste system and slavery. Then, summarizing Edwards's theory of the existence of a social scale parallel to that of the gradations of skin color, he claimed first that the caste systems, whether Indian, Egyptian, or American, reflected a division among races, and second that social inequalities were due to the original inequality among the races. Like others before him, he advanced the opinion that "the civilisation of the Europeans was the effect and not the cause of their superiority."[55] Courtet's analysis then moved from inequality to the principle of equality. This, he maintained, prevailed in the French population because it no longer comprised castes, classes, orders, ranks, or estates. Yet this principle, too, was a social manifestation of a physiological principle, namely, fusion of the races. This process, Courtet argued, had been brought about in France by invasions, wars, and revolutions, which had from time to time overthrown the ruling class and mingled it with the rest of the population, which had consequently become increasingly homogeneous. Courtet developed this theme further in a book published in 1838, which anticipated numerous ideas generally attributed to Gobineau.[56]

15.5 Concluding Remarks

The first concluding remark I wish to offer is that the studies carried out by naturalists on interracial crossing of humans in the period 1770–1835 were based exclusively on written accounts of administrative practices in the Americas, where individuals were classed according to their ancestry. From the point of view of the history of biology, the system of las castas may be considered one of the earliest models of heredity and a vast field of pre-Mendelian investigation concentrated mainly on only one characteristic: that of skin color.

Second, both the notion of "caste" and the notion of "race" may be considered European inventions as biopolitical concepts—that is, concepts that had, at the time they were used, a biological as well as a political meaning.

Third, by the 1820s, the sons and daughters from marriages of European and non-European individuals were usually named hybrids by scientists, and viewed as creatures threatening the boundary between white and nonwhite humanity. Drawing from a well-established Christian moral discourse on illegitimacy, they were more often than not considered to be living signs of shame, ambivalent creatures torn between different cultures and loyalties, and therefore a threat to empire and race.

Fourth, the symbolic significance attributed to the inheritance of complexion by numerous early-nineteenth-century authors betrays the European obsession with genealogy. It also reveals the valorization by Europeans of their own complexion, as well as excessive self-love and exaltation of self over others, who were instead denigrated and stigmatized. In short, this obsession was a form of narcissism. From the late Enlightenment until the full-blown positivism of the latter part of the nineteenth and early twentieth centuries, this ethnocentrism was manifested by reactionaries, conservatives, liberals, and socialists alike. Only occasionally was it criticized.[57] Moreover, whereas in the early campaigns to abolish slavery numerous intellectuals argued that humankind could be civilized and perfected, in the early nineteenth century the idea gained increasing ground that civilization was almost exclusively the work of the white race. This association between skin color and the civilizing process was the distinctive contribution made by historians to the construction of a European collective somatic identity, which I have elsewhere called "leucocracy" (supremacy of whites).[58]

Finally, for the first time in history, during the last decades of the eighteenth century and the first decades of the nineteenth, various historians attributed to the body alone, and to skin pigmentation in particular, the prime role in explaining both great historical events and such phenomena as social stratification by using the notion of racial struggle. Having adopted a chromatic classification of humanity—which was inaccurate and arbitrary—those historians began to speculate on the invariability of allegedly original qualities of the human races, on the impossibility of peaceful coexistence among them, and on their inevitable conflict. They then elevated this chromatic classification into a heuristic model, which they used to determine how social inequalities and caste systems had arisen in other cultures. But in doing so they concealed the only certain historical fact—that it was Europeans who established a worldwide caste system based on a rigid distinction between whites and people of color.

Notes

1. Buffon 1777, 502–505, 526–527.

2. Buffon 1777, 504.

3. Buffon 1777, 504.

4. Buffon 1777, 504.

5. Buffon 1777, 526–527.

6. Pauw 1768–1769.

7. Gumilla 1745, 1758.

8. Juan y Santacilia and Ulloa 1748; Juan y Santacilia and Ulloa 1752.

9. Gumilla 1745, 86.

10. Gumilla 1745, 83. Gumilla's definition of the *quarterona* being "one-quarter of *mulata*" and the *octavona* being "one-eighth of *mulatta*" is an obvious a mistake; *mulata* and *mulatta* should read *negra*.

11. Blumenbach 1775, 55–57. Compare Blumenbach 1865, 111–113. Blumenbach and Buffon were the likely sources for Immanuel Kant's anthropological treatises; on these treatises see McLaughlin, chapter 12, this volume.

12. Blumenbach 1775, 57. Compare Blumenbach 1865, 113.

13. Blumenbach 1795, 139–149. Compare Blumenbach 1865, 216–218.

14. Caldani 1799.

15. Caldani 1799, 449.

16. Caldani 1799, 451. For eighteenth-century investigations of the Malpighian layer of the skin, see Mazzolini 1990.

17. On Humboldt's anthropological ideas, see Mazzolini 2003.

18. Humboldt 1811, 103.

19. Humboldt 1811, 136.

20. Humboldt 1811, 143.

21. Lawrence 1819, 459–460.

22. Lawrence 1819, 293–296.

23. On the Latin origin of the term *hybrid*, see Warren 1884.

24. Lawrence 1819, 297.

25. Lawrence 1819, 300.

26. See, for instance, Vedder, chapter 4, this volume.

27. A still useful, but rare, book on las castas from the point of view of hereditary studies is León 1924. A good introduction to the historical problems concerning the study of American populations is provided by the collective volume edited by Mörner (1970), while an extensive study on the Spanish terminology of castes is provided by Forbes (1988). For special reference to *mestizos* and *mulatos*, see Ramos 1999.

28. Compare Juan y Santacilia and Ulloa 1748, vol. 1, 40, to Juan y Santacilia and Ulloa 1752, vol. 1, 27.

29. As demonstrated by Jackson 1999.

30. Solórzano Pereira 1647, 244.

31. Solórzano 1647, 246.

32. Solórzano Pereira 1647.

33. Solórzano Pereira 1647, 248.

34. By far the most complete collection of reproductions of paintings representing las castas is provided by García Sáiz 1989. Also see Katzew 2004.

35. The presence of blacks in Mexico from the sixteenth century has been studied by Palmer 1976.

36. Edwards 1793, vol. 2, 15–17; Boniol 1990.

37. Seliger 1958; Sommer 1990; Piguet 2000.

38. Dalgado 1919–1921, vol. 1, 225–229; Coromines 1992, 622–624.

39. For British scholarship of Indian literature, see Trautmann 1997. The quotation is from Quigley 1993, 4. It is interesting to note that already in 1796 the Italian missionary Paolino da S. Bartolomeo (1796, 230) expressed dissatisfaction in translating *varna* as "castes" ("dagli Europei malamente chiamate *Caste*") and preferred to call them "*classi e tribù*." In his very influential work, Dubois (1817, 1) wrote: "The word *Cast* is a Portuguese term, which has been adopted by Europeans in general, to denote the different classes or tribes into which the people of India are divided." In the French edition of his work, Dubois (1825, vol. 1, 1) stated: "Nous désignons en Europe par la dénomination de *castes*, mot emprunté du portugais, les différentes tribus qui composent les peuples de l'Inde." It should be noted that Dubois left out the reference to "class." On the meanings of the word *varna*, see Sharma 1975.

40. Risley 1891, vol. 1, i–ii, xxvii–xxviii, xxxii–xl. See also Ghurye 1924; 1932, 101–123. On Risley, see Trautmann 1997, 198–204.

41. Cohn 1987, 136–171; Quigley 1993; Smith 1994; Klass 1998, 35–63; Bayly 1999, 25–143; Dirks 2001.

42. Chaudhuri 1993.

43. For references to various works by Meiners and secondary literature on him, see Mazzolini 2002, 51–55.

44. Meiners 1793, 91–97.

45. Meiners 1791.

46. My translation, Gobineau 1853–1855, vol. 1, 179: "Il eut imaginé une classification des plus simple; elle n'était composée que de deux catégories: la Belle, c'est-à-dire la race blanche, et la laide, qui renfermait toutes les autres."

47. See Heeren's note on Meiners's essay of 1791 in Heeren 1821–1826, vol. 9, 361. On the caste system in ancient Egypt and India, see Heeren 1821–1826, vol. 9, 361–400, and vol. 12, 282–307. See also Bohlen 1830, vol. 1, 42–50, and vol. 2, 11–41.

48. Wiseman 1836, vol. 1, 221. Compare Heber 1829, vol. 1, 9.

49. Pickering 1848, 181.

50. Edwards 1829. On Edwards's conception of race and history, see Sommer 1990.

51. Comte 1826–1827, vol. 2, 69–93; vol. 3, 421–465.

52. Courtet de l'Isle 1836, 74.

53. Courtet de l'Isle 1836, 77.

54. Courtet de l'Isle 1836, 78.

55. Courtet de l'Isle 1836, 81.

56. Courtet de l'Isle 1838; Boissel 1972.

57. Salles 1849, 179: "La science européenne, qui accepte l'inégalité intellectuelle des races, se fait solidarie d'une sorte d'orgueil national, puisque les races blanches sont à la fois juge et partie dans la question. Par ce trait elles ressemblent déjà à d'autres races qui se sont, elles aussi, faites centre du monde et dernier mot de la perfection physique et morale."

58. Mazzolini 2002.

References

Bayly, Susan. 1999. *Caste, Society and Politics in India from the Eighteenth Century to the Modern Age.* The New Cambridge History of India, IV/3. Cambridge: Cambridge University Press.

Blumenbach, Johann Friedrich. 1775. *De generis humani varietate nativa.* Goettingae: typis Frid. Andr. Rosenbuschii.

———. 1795. *De generis humani varietate nativa.* 3rd ed. Gottingae: apud Vandenhoek et Ruprecht.

———. 1865. *The Anthropological Treatises.* Transl. and ed. Thomas Bendyshe. London: Published for the Anthropological Society by Longman and Green.

Bohlen, Peter von. 1830. *Das alte Indien mit besonderer Rücksicht auf Aegypten.* 2 vols. Königsberg: im Verlage der Gebrüder Bornträger.

Boissel, Jean. 1972. *Victor Courtet (1813–1867) premier théoricien de la hiérarchie des races.* Paris: Presses Universitaires de France.

Boniol, Jean-Luc. 1990. La couleur des hommes, principe d'organisation sociale: Le cas antillais. *Ethnologie Française* 20:410–417.

Buffon, Georges-Louis LeClerc Comte de. 1777. Addition à l'article qui a pour titre, Variétés dans l'espèce humaine. *Histoire naturelle, générale et particulière, avec la description du Cabinet du roi*, supplèment 4, 454–582. Paris: Imprimerie Royale.

Caldani, Leopoldo Marc'Antonio. 1799. Congetture intorno alle cagioni del vario colore degli Africani, e di altri popoli; e sulla prima origine di questi. *Memorie di Matematica e Fisica della Società Italiana* 8 (1): 445–457.

Chaudhuri, K. N. 1993. From the barbarian and the civilised to the dialectics of colour: An archaeology of self-identities. In Peter Robb, ed., *Society and Ideology: Essays in South Asian History Presented to Professor K. A. Ballhatchet*, 22–48. Delhi: Oxford University Press.

Cohn, Bernard S. 1987. *An Anthropologist among the Historians and Other Essays*. Delhi: Oxford University Press.

Comte, François-Charles-Louis. 1826–1827. *Traité de législation*. 4 vols. Paris: A. Sautelet.

Coromines, Joan. 1992. *Diccionari etimològic i complementari de la llengua catalana*. vol. 2. Barcelona: Curial Edicions Catalanes.

Courtet de l'Isle, Victor. 1836. Déterminer par l'histoire si les diversités physiologiques des peuples sont entre elles comme les diversités des systèmes sociaux auxquels ces peuples appartiennent. In *Congrès historique européen, réuni à Paris, au nom de l'Institut historique*, vol. 1, November-December 1835, 74–95. Paris: P. H. Krabbe.

———. 1838. *La science politique fondée sur la science de l'homme, ou étude des races humaines sous le rapport philosophique, historique et social*. Paris: chez Arthus Bertrand.

Dalgado, Sebastião Rodolfo. 1919–1921. *Glossário luso-asiático*. 2 vols. Coimbra.

Dirks, Nicholas B. 2001. *Castes of Mind: Colonialism and the Making of Modern India*. Princeton, NJ: Princeton University Press.

Dubois, Jean-Antoine. 1817. *Description of the Character, Manners and Customs of the People of India; and of Their Institutions*. Translated from the French manuscript. London: Printed for Longman, Hurst, Rees, etc.

———. 1825. *Moeurs, institutions et ceremonies des peoples de l'Inde*. 2 vols. Paris: chez J.-S. Merlin.

Edwards, Bryan. 1793. *The History, Civil and Commercial, of the British Colonies in the West Indies*. 2 vols. London: Printed for John Stockdale.

Edwards, William-Frédéric. 1829. *Des caractères physiologiques des races humaines considérés dans leurs rapports avec l'histoire*. Paris: Compère jeune.

Forbes, Jack D. 1988. *Black Africans and Native Americans: Color, Race and Caste in the Evolution of Red-Black Peoples*. Oxford: Blackwell.

García Sáiz, María Concepción. 1989. *Las castas mexicanas: Un género pictórico americano.* Segrate (Milan): Olivetti.

Ghurye, G. S. 1924. The ethnic theory of caste. *Man in India* 4:209–272.

———. 1932. *Caste and Race in India.* London: Kegan Paul, Trench, Trubner.

Gobineau, Arthur. 1853–1855. *Essai sur l'inégalité des races humaines.* 4 vols. Paris: Librairie de Firmin Didot.

Gumilla, José. 1745. *El Orinoco ilustrado, y defendido, historia natural, civil, y geographica de este gran rio.* Madrid: per Manuel Fernandez, impressor de el Supremo Consejo de la Inquisition, y de la Reverenda Camera Apostolica.

———. 1758. *Histoire naturelle, civile et geographique de l'Orenoque.* Avignon: chez la Veuve de F. Girard and Marseille: chez D. Sibié et Jean Mossi.

Heber, Reginald. 1829. *Narrative of a Journey through the Upper Provinces of India from Calcutta to Bombay, 1824–1825.* 4th ed. 3 vols. London: John Murray.

Heeren, Arnold Hermann Ludwig. 1821–1826. *Historische Werke.* 15 vols. Göttingen: bei Johann Friedrich Röwer [from vol. 10] Vandehoek und Ruprecht.

Humboldt, Alexander von. 1811. *Essai politique sur le royaume de la Nouvelle-Espagne.* Voyage de Humboldt et Bonpland. Troisième partie. Paris: chez F. Schoell.

Jackson, Robert H. 1999. *Race, Caste, and Status: Indians in Colonial Spanish America.* Albuquerque: University of New Mexico Press.

Juan y Santacilia, Jorge, and Antonio de Ulloa. 1748. *Relación histórica del viage a la América meridional.* 2 vols. Madrid: Impressa de orden del rey nuestro señor por Antonio Marin.

———. 1752. *Voyage historique de l'Amérique meridionale.* 2 vols. Amsterdam and Leipzig: chez Arkstée and Merkus.

Katzew, Ilona. 2004. *Casta Painting: Images of Race in Eighteenth-Century Mexico.* New Haven, CT: Yale University Press.

Ketkar, S. V. 1979. *History of Caste in India.* Jaipur: Rawat Publications.

Klass, Morton. 1998. *Caste: The Emergence of the South Asian Social System.* New Delhi: Manohar.

Lawrence, William. 1819. *Lectures on Physiology, Zoology, and the Natural History of Man, Delivered at the Royal College of Surgeons.* London: Printed for J. Callow.

León, Nicolas. 1924. *Las castas del Mexico colonial o Nueva España: Nóticias etno-antropologicas.* Mexico City: Museo Nacional de Arqueologia, Historia y Etnografia.

[Long, Edward]. 1774. *The History of Jamaica.* 3 vols. London: Printed for T. Lowndes.

Mazzolini, Renato G. 1990. Anatomische Untersuchungen über die Haut der Schwarzen (1700–1800). In Gunter Mann and Franz Dumont, eds, *Die Natur des Menschen: Probleme der Physischen Anthropologie und Rassenkunde (1750–1850)*, 169–187. Soemmerring Forschungen, 6. Stuttgart: Gustav Fischer Verlag.

———. 2002. Leucocrazia o dell'identità somatica degli Europei. In Paolo Prodi and Wolfgang Reinhard, eds, *Identità collettive tra Medioevo ed Età Moderna*, 43–64. Bologna: Clueb.

———. 2003. Physische Anthropologie bei Goethe und Alexander von Humboldt. In Ilse Jahn and Andreas Kleinert, eds, *Das Allgemeine und das Einzelne—Johann Wolfgang von Goethe und Alexander von Humboldt im Gespräch*, 63–79. Acta Historica Leopoldina, 38. Halle: Deutsche Akademie der Naturforscher Leopoldina.

Meiners, Christoph. 1785. *Grundriß der Geschichte der Menschheit.* Lemgo: im Velage der Meyerschen Buchhandlung.

———. 1791. Ueber den Unterschied der Casten in alten Aegypten, und im heutigen Hindostan. *Neues Göttingisches Historisches Magazin* 1 (3): 509–531.

———. 1793. *Grundriß der Geschichte der Menschheit.* Zweyte sehr verbesserte Ausgabe. Lemgo: im Verlage der Meyerschen Buchhandlung.

Mörner, Magnus, ed. 1970. *Race and Class in Latin America.* New York: Columbia University Press.

Palmer, Colin A. 1976. *Slaves of the White God: Blacks in Mexico, 1570–1650.* Cambridge, MA: Harvard University Press.

Paolino da S. Bartolomeo. 1796. *Viaggio alle Indie orientali umiliato alla Santità di N. S. Papa Pio sesto pontefice massimo.* Roma: Presso Antonio Fulgoni.

Pauw, Cornélius de. 1768–1769. *Recherches philosophiques sur les Américains, ou Mémoires intéressants pour servir à l'histoire de l'espèce humaine.* 2 vols. Berlin: chez George Jacque Decker.

Pickering, Charles. 1848. *The Races of Man and Their Geographical Distribution.* Boston: Charles C. Little and James Brown.

Piguet, Marie-France. 2000. Observation et histoire: Race chez Amédée Thierry et William F. Edwards. *L'Homme* 153:93–106.

Quigley, Declan. 1993. *The Interpretation of Caste.* Oxford: Clarendon Press.

Ramos, Demetrio. 1999. El incremento y la función de los mestizos y mulatos. In Demetrio Ramos, ed., *Historia de España*, vol. 27, La formación de las sociedades iberoamericanas (1568–1700), founded by Ramón Menéndez Pidal. Madrid: Espasa-Calpe.

Risley, Herbert Hope. 1891. *The Tribes and Castes of Bengal.* 2 vols. Calcutta: Printed at the Bengal Secretariat Press.

Salles, Eusèbe-François de Salles. 1849. *Histoire générale des races humaines ou philosophie ethnographique.* Paris: Benjamin Duprat libraire et Pagnerre éditeur.

Seliger, M. 1958. Race-thinking during the Restoration. *Journal of the History of Ideas* 19:273–282.

Sharma, K. N. 1975. On the Word "Varna." *Contributions to Indian Sociology* (NS) 9:293–297.

Smith, Brian K. 1994. *Classifying the Universe: The Ancient Indian Varna System and the Origins of Caste.* New York: Oxford University Press.

Solórzano Pereira, Juan de. 1647. *Politica Indiana.* Madrid: Officina de Diego Diaz de la Carrera.

Sommer, Antje. 1990. William Frédéric Edwards: "Rasse" als Grundlage europäischer Geschichtsdeutung. In Gunter Mann and Franz Dumont, eds, *Die Natur des Menschen: Probleme der Physischen Anthropologie und Rassenkunde (1750–1850),* 365–409. Soemmerring Forschungen, 6. Stuttgart: Gustav Fischer Verlag.

Trautmann, Thomas R. 1997. *Aryans and British India.* Berkeley: University of California Press.

Warren, Thomas. 1884. On the Etymology of Hybrid (Lat. Hybrida). *American Journal of Philology* 5:501–502.

Wiseman, Nicholas. 1836. *Twelve Lectures on the Connexion between Science and Revealed Religion.* 2 vols. London: Joseph Booker.

16 Acquired Character: The Hereditary Material of the "Self-Made Man"

Paul White

It is Character which builds an existence out of circumstance. Our strength is measured by our plastic power . . . thus it is that in the same family, in the same circumstances, one man rears a stately edifice, while his brother . . . lives for ever amid ruins.
—G. H. Lewes, *The Life and Works of Goethe*, 1855

The inheritance of acquired characteristics has often been treated as a timeless theory, inherently plausible prior to the isolation of chromosomes as the material of hereditary transmission, and therefore requiring little historical explanation.[1] And yet the natural historical concept of character as developed in nineteenth-century studies of breeding, hybridization, and inherited disease, was closely linked to prevalent concerns about inborn or acquired character in the wider, moral sense. The term *character*, together with *characteristic*, had been used in natural historical contexts to signify a distinctive feature of a species or variety since the mid-eighteenth century, although its wider meaning as an "essential quality" and its use with reference to moral and mental attributes began much earlier.[2] The Latin term became widespread in natural history through Linnaeus's *Genera plantarum* (1737), while translations of Theophrastus's famous character sketches appear from the beginning of the eighteenth century. An 1824 English edition of Theophrastus nicely conjoins moral and natural historical character by way of the physiognomic tradition, with a series of portraits emphasizing facial shape and expression added to the text.[3] The continued importance of character studies as a literary genre is evident, for example, in the last work of George Eliot, *The Impressions of Theophrastus Such* (1874). An accretion of habit, history, and custom, of social influence and environment, and of will and "natural constitution," character was central to psychological and medical accounts of agency and affliction in individuals. As Stefan Collini has argued, the concept also held special importance for political writers of the period who, in attributions of "national character," assigned it a great causal role in history.[4]

Historians have located a widespread shift in Victorian Britain from models of hierarchical order based on innate capacity—the blue blood of nobility—to one in which character of the highest sort could be acquired. Debates of this type were much older, but the story has often been told as part of the rise of the middle classes and the professions and the consequent decline of the hereditary aristocracy.[5] This social transformation was far from straightforward, and models of inborn and acquired character frequently overlapped or were employed in different ways according to setting.[6] Ongoing debates about the role of ancestry, of landed property, of money, and of moral conduct in the constitution of social status often focused on the figure of the gentleman. Once largely reserved to those of noble birth, "gentlemanliness" gained a much wider currency over the course of the eighteenth century. By the mid-Victorian period, it came to designate a form of moral or social approval in some degree severed from traditional rank and riches.[7] Research on heredity in Victorian Britain may thus be situated within a wide-ranging set of debates about self-fashioning, individual improvement, and the formation of gentlemanly character. The hereditary transmission of acquired characters, so crucial to progressive theories of evolution and to theories of degeneration in the period, must itself be understood in relation to the broader controversy over the acquisition of moral character.

16.1 Breeding Character

In Victorian discourses of breeding, natural historical and moral character were considerably intertwined. What was "character" for the breeders? At one level, it was simply a physical or behavioral trait, taken as distinctive of a given variety or species: a particular shape of beak or horn, the sheer size and mass of an animal's body, a manner of masculine display, such as the inflating of the chest. Such traits were not only variable, but mutable, and breeders often presented themselves as able to shape a plant's or animal's character, almost at will. As Roger Wood shows in chapter 10, the selective care of the breeder, through the isolation of favorable traits, progeny testing, and other techniques, was elevated above the influences of ancestry and environment by eighteenth-century agricultural improvers. Thus William Youatt, veterinary surgeon to the Zoological Society of London, spoke of selection as "that which enables the agriculturalist, not only to modify the character of his flock, but to change it altogether. It is the magician's wand, by means of which he may summon into life whatever form and mould he pleases."[8] This power to mold character through the manipulation of conditions and the exercise of will, was in fact the operative principle of another Victorian creature: the self-made man.

Elaborated in numerous didactic works and novels of the early Victorian period, the culture of self-help and self-fashioning was epitomized in the writings of Samuel Smiles, first in his biography of the manufacturer George Stephenson, in many subsequent lives of workingmen-made-good, and in a series of companion volumes on particular virtues such as thrift, duty, and, in 1871, on *Character* itself.[9] According to Smiles, character was "formed by a variety of minute circumstances, more or less under the regulation and control of the individual." Every action, thought, and feeling contributed to its formation: "Man is not the creature, so much as he is the creator, of circumstances. . . . Energy of will—self-originating force—is the soul of every great character." By locating the virtues of moral character in individual deeds and habits, rather than in ancestry, Smiles legitimized, indeed lionized, the rapid social ascent of Britain's industrial and commercial leaders, while holding out the promise of improvement to many. Revealed in the "transactions and commonplaces of daily duties," a respectable character could be acquired by anyone.[10]

How then was this culture of individual improvement and character formation operative in the discourse and practice of breeders? One way of getting at this connection is to look at the various forms by which character, both in its natural historical and moral senses, was displayed. Figure 16.1 shows the prize shorthorn bull, the VIIIth Duke of York, as he was depicted in Thomas Bates's history of the improved Durham cattle.[11] Harriet Ritvo's work on animal husbandry in Britain has argued that breeders often affixed their own social status to that of their animals: through breed characteristics, no less than prize competitions, the criteria of social hierarchy were defined, and the differences established between a person (or animal) of noble pedigree and a parvenu.[12] The nobility of an animal was always demonstrated by its progeny, as in figure 16.2, a pedigree of Canterbury Pilgrim from the *Suffolk Stud Book*.[13] Such registers typically added a further layer of representation, in the form of biographical entries, noting the prizes that the animal had won, the prices it had fetched, and also its leading characteristics. Another noble animal, Cup-Bearer, was depicted as "a large horse with a grand fore end, great depth of girth, and splendid muscular shoulders . . . although not an elegant walker," its sons were legion, numbering many prizewinners.[14]

This sort of iconography could assume the more elaborate and costly form of oil on canvas, drawing on the tradition of gentry portraiture in which lords of the manor were foregrounded against their estates. Such animals—for example the prize bull of Sir Charles Morgan, baronet—were not just part of a gentleman's large holdings. Their distinguishing properties, their formidable size, beauty, and elegance, were those of nobility, and were also distinctive of their owners.[15] Corpulence, in conjunction with

Figure 16.1
Shorthorn bull from Thomas Bell, *The History of Improved Short-Horn or Durham Cattle, and of the Kirklevington Herd, from the Notes of . . . Thomas Bates* (1871). By permission of the syndics of Cambridge University Library.

elegance, was a virtue among the English gentry, signifying power, wealth, and generosity. The nobility of size is evident in the medical literature on gout, which physicians widely associated with the great and their glamour, men of stong natural constitution, large appetites, and long life.[16] In portraiture, such qualities of man and beast were typically displayed by a rounded, decorous body on slender legs. In figure 16.3 the "fathers" of the shorthorn cattle, the Yorkshire farmers Charles and Robert Colling, show the same prodigious girth and delicacy of limb (or calf) as their noble breed. The Collings had been highly successful in producing a line of beef cattle through intense inbreeding from stock of exceptional merit, but unknown or indifferent ancestry, according to the method popularized by Robert Bakewell. The son of a small dairy farmer, Robert was apprenticed to a grocer before striking out as an

Figure 16.2
Pedigree from Herman Biddell, *The Suffolk Stud-Book: A History and Register of the Country Breed of Cart Horses* (1880). By permission of the syndics of Cambridge University Library.

PEDIGREE OF MR. BARTHROPP'S CANTERBURY PILGRIM 85.

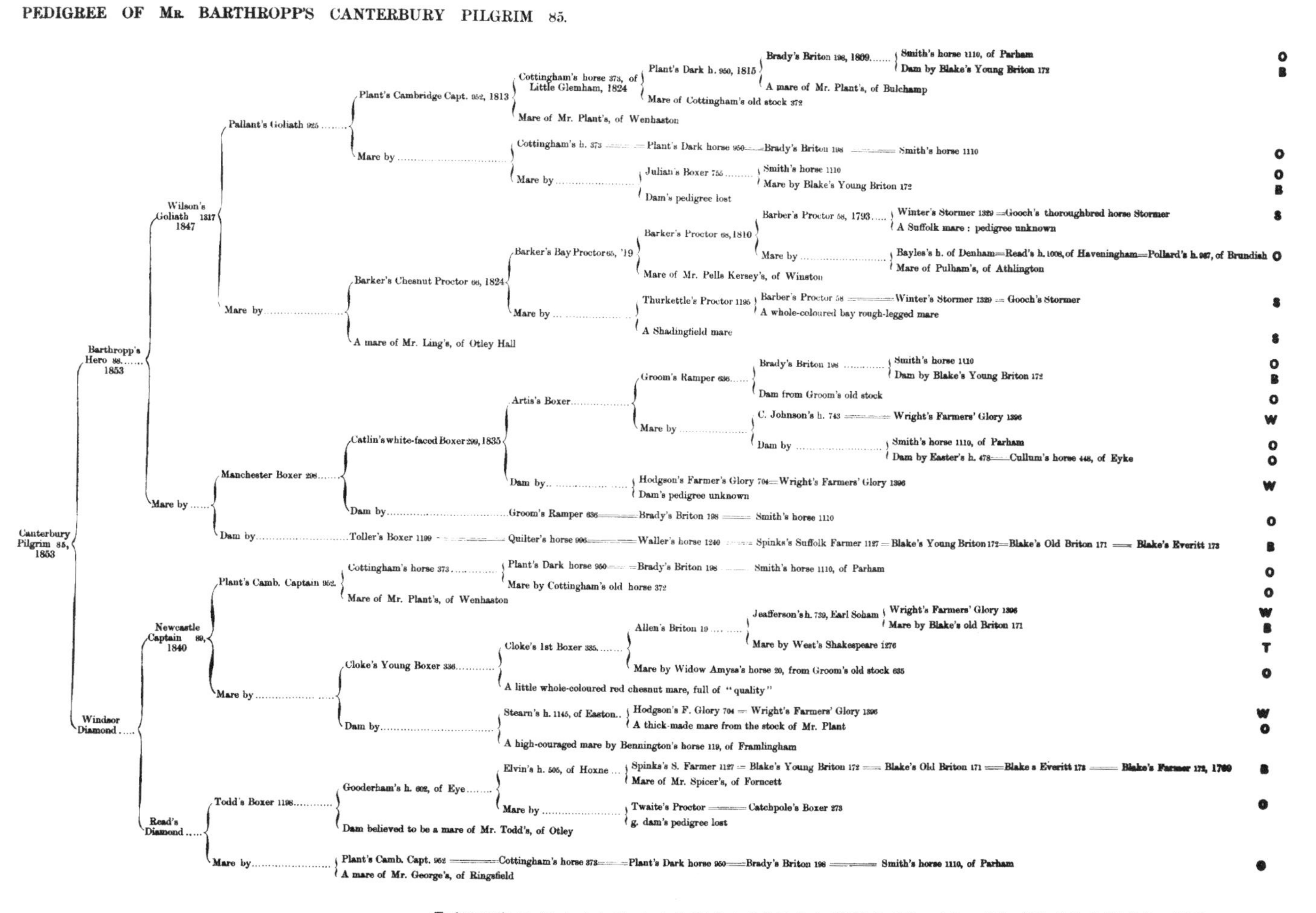

The letters at the end of the terminal origins denote: O *Old Breed*—B *Blake's Strain*—W *Wright's Attleborough Farmers' Glory Tribe*—S *Shadingfield Stock*—and T *Trotting Origin.*
The double line = denotes "son of," giving the direct descent in the male line.

Figure 16.3
Charles and Robert Colling from *Shorthorn Transactions* (1868–1869), frontispiece. By permission of the syndics of Cambridge University Library.

agriculturalist, founding his pedigree herd with a yellow-red and white bull, Hubback, whom he had seen along the roadside and purchased from a local bricklayer for eight guineas.[17] In the same volume of *Shorthorn Transactions* that carried their portrait, a moral pedigree of the self-made breeders is given on the page opposite, declaring their "enlarged views, generous disposition, love of order, nobility of bearing, and extensive means."[18]

The role of breeders in modifying structure through habit and conditions thus bore equally on their own status as gentlemen and conservative improvers. These forms of representation, the family tree, portraiture, and biography, together enforced the view that gentlemanly character was natural historical—composed of heritable traits, rooted in good breeding—and yet highly malleable through the breeders' art. The great estate owners who commanded this culture of improvement, some of whom were new members of England's ruling classes, were, as much as their prize animals, living examples of the power of selective breeding to modify character, and thereby enrich the stock of the nation.

16.2 Powers of Discrimination

In response to criticism that in *Origin of Species* he had described natural selection as an agent, productive of species, Darwin replied that he had merely adopted the language of breeders.[19] James Secord has shown the attention to the work of animal and plant hybridists in the framing of Darwin's theory.[20] Indeed, the skill and art of breeders in fashioning their stock was repeatedly emphasized in Darwin's acknowledgments of individual fanciers and horticulturalists in the *Variation of Animals and Plants under Domestication*. As Darwin wrote in *Origin*, they "habitually speak of an animal's organisation as something quite plastic, which they can model almost as they please."[21] In drawing the analogy between natural and artificial selection, Darwin conceded a power to breeders that was a source of consternation among some of his gentlemanly correspondents, such as Charles Lyell. On reading the first chapter of *Origin*, Lyell remarked:

> Your comparison of Selection to the Architect . . . I . . . cannot accept. The architect who plans beforehand & executes his thoughts & invents . . . must not be confounded in his functions with the humble office of the most sagacious of breeders. . . . It is the deification of Natural Selection.
>
> The idea of [selection] being ever allowed to play such pranks as the breeder has played & to sport with God's creatures & the laws of reproduction so as to perpetuate pouter pigeons, & other monstrosities, would have been scouted by a philosopher of the Wealden Period.[22]

The deificiation of natural selection, and by implication, the elevation of breeders to creative status, remained an ongoing dispute. Darwin actively defended his analogy as appropriate. Yet Darwin's views on the relative power of external agency to shape the character of species and varieties differed substantially from those of the leading breeders he consulted. While agreeing that the conditions of life, especially those of domestication, had profound effects on the variability of character, Darwin urged that "we must never forget that the nature of the organisation which is acted on, is by far the most important factor in the result."[23]

Unlike the great horse and cattle breeders whose land and wealth enabled them to enter England's ruling class to some degree, the majority of the poultry men with whom Darwin associated were of considerably humbler status. Many were or had been artisans, a point of considerable importance to a community that prided itself on gentlemanly character. Though a tailor and thus an artisan by trade, John Matthews Eaton, one of Darwin's leading authorities in *Variation*, clearly regarded himself and his fellow fanciers as gentlemen—not by birth, or wealth, but by virtue of their pastime. Some gentlemen, even noblemen, were indeed breeders—a fact reiterated in countless treatises on domestic animals.[24] At the same time, fancy-pigeon breeding for artisans was closely linked to ideals of craft skill within the fashion trade.[25] Yet the crucial point of Eaton's claim was that the practice of breeding was itself gentlemanly in character. It was a "study and science" especially "adapted to the professional gentlemen of law, physic, and divinity, or any other person engaged in long continued and excessive exertion of the intellectual faculties."[26]

Breeders were renowned for their unsurpassed powers of observation and discrimination, able to detect differences of character between individuals that were invisible to anyone not a master of the art. Their way of seeing, their watchfulness, vigilance, and discernment, were as important to Darwin as their knowledge of variation and inheritance, for they embodied nature's selective power. Darwin's own account of the changes wrought by breeding practice in *Origin of Species* culminated in a character sketch of the breeders as self-made men, succeeding according to the Smilesian virtues of perseverance, hard work, and the resolute pursuit of a single goal:

> Not one man in a thousand has accuracy of eye and judgment sufficient to become an eminent breeder. If gifted with these qualities, and he studies his subject for years, and devotes his lifetime to it with indominatable perseverance, he will succeed, and may make great improvements; if he wants any of these qualities, he will surely fail. Few would readily believe in the natural capacity and years of practice requisite to become even a skilful pigeon-fancier.[27]

And yet Darwin's own relations with breeders such as Eaton clearly indicate that he regarded them as social inferiors, joking privately about their diminutive stature and their bad grammar. To his eldest son, he wrote: "NB all Pigeon Fanciers are little men."[28] When the zoologist Thomas Huxley asked Darwin for information on breeding for a talk at the genteel Royal Institution, Darwin included some extracts from Eaton's treatise on the Almond Tumbler that he thought "would make the audience laugh." After citing passages that displayed the gentlemanly pretenses of the fancy, together with the "little tailor's grammar" to comic effect, Darwin cautioned Huxley not to mention Eaton's name, "as he is my friend."[29]

As a man of wealth and property, and the recipient of a traditional classical education, Darwin could have rested his superiority to the breeders on these grounds, and yet he made common cause with Huxley, of relatively humble birth, modest means, and largely self-educated. That Darwin and Huxley could laugh at the breeders' expense, and yet regard them as friends, and look to them as authorities on matters of science, raises questions about the social status of their own experimental practice, and the grounds on which they distinguished themselves from such associates. The particular gifts that Darwin praised in breeders—their accuracy of eye and powers of concentration and discrimination—were also widely regarded as characteristic of the Victorian "man of science."

16.3 Hereditary Genius

How then did the "man of science" stand in relation to models of innate and acquired character? In fact, as with the breeders, the status of men of science as gentlemanly, and as self-made, was also uncertain. A romantic tradition was still operative in Victorian scientific biography and self-presentation, in which the scientific practitioner was distinguished by specially endowed powers of mind. In his essay "Characteristics," Thomas Carlyle described genius as a heroic intellectual force: innate, like nobility, but often residing in those of humble birth. Its characteristics—intuition, mental suppleness, refined discrimination—marked a few individuals as destined to be leaders of men and spirits of the age.[30]

Such romantic models were appropriated by those scientific practitioners, such as Huxley and John Tyndall, whose backgrounds did not confer gentlemanly status. And yet in the Victorian period, genius was transformed by the culture of industry and work.[31] What had been conceived as a largely effortless and innate capacity by eighteenth-century writers became an endowment of the self-made man, who had to labor and struggle after truth. Even Darwin, a gentleman of secluded leisure, continually characterized his scientific activity as "hard work" and "hard labour," tallying up the months and years, right up to the day, that he spent in the production of each of his books.[32] Francis Galton's 1874 survey of men of science compiled over a hundred comparable accounts: men who walked fifty miles a day without fatigue in search of specimens, men who worked habitually until two or three in the morning. Using autobiographical testimony, Galton documented their leading characteristics as perseverance, steadiness, determination, and "the secretion of nervous force"—that is, energy: "Many have laboured as earnest amateurs in extra professional hours long into the night . . . they have climbed the long and steep ascent from the lower to the upper ranks of life."[33]

Despite this emphasis on self-fashioning through hard work, however, certain traits of men of science remained inborn. In his 1869 work, *Hereditary Genius,* Galton used his own family history, and the pedigrees of others, to argue that intellectual aptitude was hereditary. Galton drew heavily on the tradition of family trees to map out the inheritance of illustriousness over generations, producing pedigrees, for example, of the natural philosopher, Robert Boyle.[34] Like Carlyle, Galton viewed men of genius as great forces of the age, but recast their influence in terms of hereditary transmission, as the sports of nature by which new higher types of mankind were established. In his theory of sports, a new character suddenly made an appearance in one individual and was transmitted to subsequent generations, causing a change in the center of the type and a step forward in the course of evolution.[35]

Much has been written about Galton's critique of Darwinian evolution, which emphasized the gradual accumulation of minute variations, and of his theory of the germplasm or "stirp," which located hereditary material in the reproductive organs, continuously inherited from generation to generation, and isolated from the effects of the environment, habit, and disposition.[36] Yet despite this abandoment of acquired characteristics as a theory of inheritance, Galton nonetheless called on his countrymen to further their own evolution "deliberately and systematically," as a matter of "religious duty," having done so in the past only "half-consciously" and out of "personal interest."[37] By shifting the focus of evolution to populations, Galton adapted the tradition of self-fashioning from the formation of individual character, to selective breeding for the improvement of the nation or race. Such selection could not bring about sports, but it could foster their inheritance. The family trees of Victorian cultural elites were indeed considerably interwoven, showing, in effect, the application of principles of inbreeding among those of proven mental character, so as to establish an "intellectual aristrocacy."[38]

16.4 Diseases of Character

In the preface to the 1892 edition of *Hereditary Genius*, Galton noted recent work by Lombroso and others that traced the proximity of genius to insanity, commenting that he did see some evidence in families for a "painfully close relation between the two" conditions.[39] The very characteristics of mind possessed by leaders of humanity, the geniuses and self-made men, brought them close to less favorable qualities, commonly identified with degeneration. One form, or stage, of degeneration became in the Victorian period a particular malady of the self-made man: intemperance, or as it came to be identified in the 1860s, alcoholism.[40] This equation between drink and self-fashioning is especially prominent in the medical literature and novels of the mid-Victorian period, decades of great prosperity for the middle classes.

Once widely regarded as a problem of the lower orders, of the poor and destitute, medical theory now suggested that poverty was a result of habitual drinking, not a cause.[41] In mid-Victorian medical writing, the onset of the condition was often linked to the pressure of work, and commentators noted its high concentration among the merchant, manufacturing, and professional classes.[42] Regarded as a hereditary or congenital condition from the early Victorian period onward, intemperance remained the most controversial of "diseases" because of its relationship to the will. Medical treatments and physiological research that identified alcohol with a material toxin nevertheless continued to regard patients as morally responsible for their illness or cure. In an 1850 essay, the physiologist and medical author William Carpenter characterized the disease as a premature exhaustion of nervous power, a weakening of the controlling power of the will, and an augmenting of the automatic or impulsive part of nature, giving a constitutional predominance to the lower feelings and passions. Habitual intemperance thus produced a progressive degradation of moral character by repeated excitement of the lower propensities and diminution of the will, leaving the person in "a state of complete [self-induced] slavery."[43]

The moral and physical effects of alcohol, and the perils of drink for a self-made man, are explored in Anthony Trollope's *Doctor Thorne* (1858). The novel also problematizes the Smilesian version of character as a product of self-determination, as well as the traditional equation of gentlemanly character with noble descent. It contains two self-made men. One is a paragon of Smilesian heroism, a lowly stonemason by the name of Roger Scatcherd, who rises to be a captain of industry, building bridges, canals, and railroads across the British Empire, and is made a baronet—a hereditary title, though not an aristocratic one. Though conquering the world, Scatcherd succumbs to the habit of drink and dies, leaving a sickly son, who inherits the craving for alcohol, suffers delirium tremens, and dies at the age of twenty, extinguishing the family line. The other self-made man, the eponymous Doctor Thorne, is of ancient family and prides himself on being a poor man of high birth. Yet he rises to eminence as a physician through displays of great generosity, honesty, and duty. It is Thorne's belief in his superior birth, rather than any necessary inherited tendency, that motivates his gentlemanly conduct. Thus Trollope implies that inheritance can operate as a cushion against degeneration in a period of rapid social change, but it does so as a form of culture—as a structure of beliefs, values, and behaviors.

The problem of intemperance is medicalized to some degree in the novel. Dr. Thorne is Scatcherd's physician. The physical effects of the toxin are delineated and inescapable. "I feel the alcohol working within your veins," says the doctor, having taken Scatcherd's hand on the pretext of feeling his pulse.[44] And yet Thorne is ultimately powerless to effect a cure without the patient's cooperation. His methods, consisting largely in severity and admonishment, are ultimately unable to effect a cure either in the father or the son. Nor is the condition, following the pattern of a congenital disease, clearly separated from the social world it inhabits. As Trollope remarks, had Scatcherd or his son been born a duke or an earl, provision would have been made for such behavior: it would have been redefined and accommodated within the culture of aristocracy. Great bouts of drinking were appropriate at the parties and banquets of great houses, and within gentlemen's clubs. Within the highest circles, occasions for such social drinking could be multiplied almost indefinitely. But Scatcherd has no place in this level of society. He is a drunkard, at least in part, because he is *not*, despite his social climb, a gentleman. Alcoholism is in fact a disease of the self-made man. The condition is not attributed to waywardness, or improvidence, a common critique of working-class modes of celebration in the temperance literature. Though a propensity in Scatcherd when still an apprentice mason, the condition is worsened during the course of his rise to eminence, with each bout of heavy drinking coinciding with the anxieties of competing for a new government contract.

Significantly in *Doctor Thorne*, the nature of the son's condition is more advanced. Scatcherd junior suffers from a weakness of character, while the father is of exemplary strength of character save for this one weakness. The hereditary nature of alcoholism is such that in the second generation, it seems to erode character and will, as well as physical health. Alcoholism was a preeminent case of how habits acquired by individuals could progress over generations, through transmission to their offspring, into more degenerate conditions, such as criminality and madness.[45] In the second half of the century, a medical specialty grew up around the condition, closely allied with alienists and asylum culture, and more elaborate medical treatises were produced over the course of the second half-century. Toward the end of the nineteenth century, in what was perhaps the most exhaustive study, the physician Norman Kerr pronounced that no "natural law was more patent than [that] of alcoholic heredity." Men and women of the highest culture, the most irreproachable morals, and the strongest will, had come to know that they could "never dally with strong drink," and that doing so would in fact alter their character, and turn "perfect gentlemen" into criminals or lunatics. And yet, despite the inexorable material and hereditary basis of the disease, Kerr cast the alcoholic within the terms of the Smilesian self-made man, his cure an epic struggle of will against an oppressive disease: "The continuous and victorious struggle of such heroic souls with their hereditary enemy—an enemy the more powerful because ever leading its treacherous life within their breasts, presents to my mind such a glorious conflict, such an august spectacle, as should evoke the highest efforts of the painter and the sculptor."[46] Other treatises on alcoholism typically included chapters on the treatment of the disease through a strict moral regime, involving the modification of habits and conditions of life in order to reinvigorate the will.[47]

Such literary and medical accounts of alcoholism bring to the surface the fundamental circularity, or indeed Calvinism, of the Smilesian account of character: it was both a property (a "noble estate") and a product, a cause and an effect. The self-made man was formed through a process of struggle and the overcoming of hardships; and yet one could only succeed amidst life's difficulties if one's heart was strong and upright, and one's spirit lively—in other words, unless one was already endowed with a noble character. Both the biographical tradition in which Smiles worked, and the medical temperance writing of Carpenter and others, drew heavily on Christian evangelical models of the virtuous life, in which salvation was won through a battle of will against corrupt influences and lower impulses.[48]

The equation between drink and self-fashioning did not typically extend to science as well, although there were associations between drink and poetry. Rather, science was often seen as sobering and, particularly in the literature on rational education and useful knowledge, as a potential remedy for the dissolute behavior of urban workers.[49] At the same time, scientific activity and sociable drinking were frequently conjoined in artisan communities, from botanists to breeders. Anne Secord's work on Elizabeth Gaskell and on Smiles has shown that models of character among different social groups were often discontinuous. The fashioning of self-made heroes in didactic or consciousness-raising literature entailed the abstraction of such men from the social settings and communities in which their character as scientific practitioners had actually been acquired, especially that central institution of artisanal and working-class life, the public house.[50]

16.5 Conclusion: Eugenic Self-Fashioning

Tying together theories of heredity, acquired character, and degeneration, the medical literature on alcoholism would figure prominently in early eugenic writings. Like the pedigrees of prize animals, noble families, and hereditary genius, eugenics-inspired studies of the hereditary effects of alcoholism were displayed as family trees. The families themselves were not sired by self-made men, merchants, or professionals, however, but by paupers. Like the social surveys and industrial novels of the 1840s, the problem of alcoholism at the end of the century—a period that saw the rise of socialism and the formation of a Labour Party—was again located in the lowest ranks of society.[51] Figure 16.4 shows a chart of alcoholic degeneration drawn at the beginning of the twentieth century by the research committee of the Eugenics Society. Displaying the profound influence of one characteristic on a population after only two generations, such evidence was used to argue that, Weismann's experiments notwithstanding, the

CHART I.

SHOWING GOOD HEREDITY CONTAMINATED BY SLIGHT ALCOHOLIC HEREDITY AND TOWN LIFE.

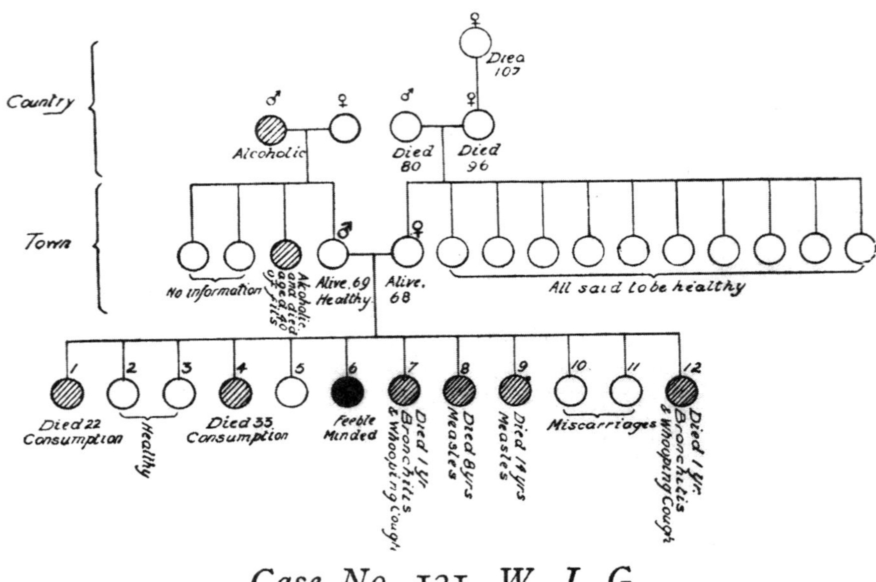

Figure 16.4
Pedigree from Alfred F. Tredgold, *Mental Deficiency* (1908). By permission of the syndics of Cambridge University Library.

inheritance of acquired characteristics still prevailed with some traits, especially diseases such as alcoholism that affected the entire organism, including the germplasm.[52] Such pedigrees, which had once served as a mode of self-fashioning for members of the gentry, professionals, and some artisans, were now generated in biometrics laboratories, asylums, and workhouses, with the aim of enlisting the state to direct the improvement of the race or nation. Significantly, the other modes of representation through which the characters of the great and good had been displayed—biography and portraiture—were now absent. "The effect upon any community of the continued propagation of the unfit," the author noted, "is simply a question of mathematics."[53]

The eugenic chart of alcoholic heredity formed part of a series indicating that character was also transmitted by females. Although contributions of the female sex to bloodline had long been acknowledged by animal breeders, Bakewell's system followed established traditions of property and title inheritance in fixing breed type through the male line, a convention still recognized. The official recording of female contributions to pedigree emerged in the last quarter of the nineteenth century, partly through the practice of Wilfrid and Lady Anne Blunt, and their daughter Judith Blunt Lytton, who bred Arabian horses according to Middle Eastern conventions that attached "purity" to the female line.[54] While Galton summoned men of science to direct eugenic policy, the closing decades of the century also saw the rise of the "new woman," who sought emancipation from patriarchal customs and legislation that had long governed reproduction, inheritance, and property. Civic motherhood was indeed embraced by some leading feminists, who claimed independence from established scientific and medical authorities by assuming eugenic responsibility in the rational selection of their reproductive partners.[55]

The possibilities and ironies of eugenic self-fashioning are further indicated in the life and work of Galton's biographer and intellectual heir, Karl Pearson. As Ted Porter has shown, Pearson was both a feminist and a socialist, dedicated to the study and improvement of humankind through biometric methods, and yet remained passionately committed to a model of individual development and achievement, conducting his own life as a literary character and composing a series of autobiographcial writings, including an extravagantly Goethian epistolary novel, *The New Werther*.[56] Whereas George Lewes's *Life of Goethe*, written some three decades earlier in the heyday of liberal individualism, celebrated the poet-playwright-philosopher as one able through sheer force of character to be the "architect of circumstance," what Pearson found most heroic in Goethe's life was the renunciation of self, and the dissolving of personality into a universal humanity.[57] Pearson would later cultivate his own impersonality through the science of statistics and the pursuit of iron laws of heredity, with a view to the better fashioning of human collectives.

As the "father" of the new biometrics had urged, in order to determine the laws of heredity, it was essential both to deal statistically with populations, and to regard human characters like the traits of a pea.[58] Such enthusiasm for statistics and deterministic social laws had been criticized on moral grounds, as eliminating personality and the force of individual character from human history. Galton defended the "charm of statistics" against its detractors, and called for a "scientific priesthood" that could use statistics to minister to the "health and well-being of the nation."[59] The paring down of "data" in favor of symbols denoting the presence or absence of a single characteristic, is of course a style of representation that would become fundamental in modern genetics and population biology. The hyperbolic claims of Victorian breeders to shape an animal's nature would perhaps make few genetic engineers of today blush, while physicians, alienists, and social scientists like Galton and Pearson, the new custodians of the poor, as well as the "new women" of the eugenic age, seemed to appreciate that good breeding depends very much on who descriminates between desirable and undesirable characters, and on what grounds.

Notes

1. On the inheritance of acquired characteristics, see Bowler 1989; Churchill 1976, 1987; Lomax 1979; Robinson 1979; Windholz 1991; Zirkle 1946.

2. See Fristrup 2001.

3. [Taylor] 1824. On the editions of *Characters*, see Smeed 1985.

4. Collini 1991, 91–118.

5. For example, Corfield 1995; Perkin 1989; Reader 1966. Revisionist approaches to British social history and the traditional Marxist framework of class structure and consciousness are discussed in Cannadine 1998, Kidd and Nicholls 1999, and Pilbeam 1990.

6. On the enduring structures of aristocratic power and deference, see Beckett 1986. On the relative breadth and permeability of the English ruling class, see Gash 1979.

7. On the "debasement of gentility" in the eighteenth century, see Langford 1989, 63–68. Its Victorian usage is discussed in Crossick 1991, 164–165.

8. Youatt 1837, 60.

9. For a general discussion of Smiles's work, see Jarvis 1997. On the literary construction of the self-made man by Smiles and others, see Hilkey 1997; Meckier 2001; Pettitt 1999; Rodrick 2001; Wyke 1999.

10. Smiles 1871, 1–6. In fact, Smiles encountered great difficulties in applying his model of character across class boundaries; see A. Secord 2003.

11. Bell 1871, 354 facing.

12. Ritvo 1987, 45–63. On general breeding practices in Britain, see also Derry 2003 and Russell 1986.

13. Biddell 1880, 654 facing.

14. Biddell 1880, 166 and facing.

15. See Ritvo 1987, 49. The original painting of Morgan by J. H. Carter is in a private collection.

16. On the virtue of corpulence, see Porter and Rousseau 1998.

17. On the Collings, see Bates 1899.

18. *Shorthorn Transactions*, 1868–1869, frontispiece.

19. Darwin 1859, 30, and letter to Charles Lyell, January 22 [1865], in Burkhardt et al. 1985–2002, vol. 13, 35.

20. Secord 1981, 1985.

21. Darwin 1859, 31.

22. Letters from Charles Lyell, June 15 and June 19, 1860, in Burkhardt et al. 1985–2002, vol. 8, 255, 260.

23. Darwin 1868, vol. 2, 502. See also Winther 2000.

24. See, for example, Dixon 1851, 6, and Eaton 1858, iii–vi; see also Ritvo 1987, 90, on dog fanciers.

25. On pigeon breeding as craft knowledge and its significance in nineteenth-century theories of descent, see Feeley-Harnik 2004.

26. Eaton 1851, iii–vi.

27. Darwin 1859, 32.

28. Letter to William Erasmus Darwin, [November 29, 1855], in Burkhardt et al. 1985–2002, vol. 5, 509.

29. Letters to Thomas Huxley, November 27 [1859], and December 13 [1859], in Burkhardt et al. 1985–2002, vol. 7, 404–405, 428.

30. Carlyle 1831.

31. On the cult of genius and work among Victorian men of science, see White 2003, chapter 1. See also Tyndall 1868.

32. Darwin 1958.

33. Galton 1874, 75.

34. Galton 1869, 197. On the tradition of family trees, see Klapisch-Zuber 1991.

35. Galton 1892, xvii–xix.

36. See for example Cowan 1985; Norton 1973; Provine 1971.

37. Galton 1883. As Roger Wood suggests in chapter 10, Galton's "population thinking" had long been practiced by animal breeders.

38. Annan 1955.

39. Galton 1892, ix.

40. On the social history of temperance in Britain, see especially Harrison 1971; the first usage of the term *alcoholism* is discussed on pp. 21–22.

41. Bynum 1984; Huertas 1993; Levine 1978; Valverde 1997; Warner 1994.

42. This contrasts with certain eighteenth-century representations of drink as a malady of the noble, or affluent. See the discussion of gout as rooted in the habits of consumption, as well as the blood, of the aristocracy, in Philip Wilson's chapter on Erasmus Darwin in this volume (chapter 6).

43. Carpenter 1850, 47.

44. Trollope 1858, 9.

45. Huertas 1993; Lubinsky 1993.

46. Kerr 1894, 17, 55.

47. Carpenter 1853; Clouston 1884; Maudsley 1874; Sankey 1884; Talbot 1898.

48. See, for example, Hyde 1875. On the importance of evangelicalism in Victorian writing on character, see Collini 1991, 105.

49. Harrison 1971, 160–162, 322–331, 387–388. See also Vincent 1981, 169–172.

50. On science and pub culture among artisan botanists, see A. Secord 1994; on pigeon breeders, see Feeley-Harnik 2004. On Smiles and Gaskell, see A. Secord 2003, 2005.

51. Hawley 1989; Mazumar 1992.

52. Tredgold 1908, 35–48. See also Mazumdar 1992, 81.

53. Tredgold 1908, 360.

54. See Wood, chapter 10, this volume; Derry 2003, 110–113. Although the "breed line" remained attached to the stallion, mare lines were recognized through a naming procedure by which offspring took a name beginning with same initial as the mother.

55. Richardson 2003.

56. [Pearson] 1880. On Pearson's literary self-fashioning, see especially Porter 2004, 43–68, 297–314.

57. Lewes 1855, vol. 1, 5–16, 29–30.

58. See the corresponding tables in Galton 1892, 206, 226.

59. Galton 1889, 62; 1874, 195. See especially Porter 1986, 128–146, 165.

References

Annan, Noel. 1955. The intellectual aristocracy. In J. H. Plumb, ed., *Studies in Social History*, 241–267. London: Longmans and Green.

Bates, C. J. 1899. The brothers Colling. *Journal of the Royal Agricultural Society of England*, 3rd ser. 10:1–30.

Beckett, J. V. 1986. *The Aristocracy in England 1660–1914*. Oxford: Blackwell.

Bell, T. 1871. *The History of Improved Short-Horn or Durham Cattle, and of the Kirklevington Herd from the Notes of the Late Thomas Bates*. Newcastle: Robert Redpath.

Biddell, Herman. 1880. *The Suffolk Stud-Book; a History and Register of the County Breed of Cart Horses*. Diss: Francis Cupiss.

Bowler, Peter J. 1989. *The Mendelian Revolution: The Emergence of Hereditarian Concepts in Modern Science and Society*. Baltimore: Johns Hopkins University Press.

Burkhardt, Frederick, et al., eds, 1985–2002. *The Correspondence of Charles Darwin*. 13 vols. Cambridge: Cambridge University Press.

Bynum, W. F. 1984. Alcoholism and degeneration in nineteenth-century European medicine and psychiatry. *British Journal of Addiction* 79:59–70.

Campbell, Margaret. 1983. The concepts of dormancy, latency, and dominance in 19th-century biology. *Journal of the History of Biology* 16:409–431.

Cannadine, David. 1998. *History in Our Time*. New Haven, CT: Yale University Press.

Carlyle, Thomas. 1831. Characteristics. *Edinburgh Review* 51:351–383.

Carpenter, William. 1850. *On the Use and Abuse of Alcoholic Liquors, in Health and Disease*. London: Charles Gilpin.

———. 1853. *The Physiology of Temperance and Total Abstinence*. London: Henry Bohn.

Churchill, Frederick B. 1976. Rudolf Virchow and the pathologist's criteria for the inheritance of acquired characteristics. *Journal of the History of Medicine and Allied Sciences* 31:117–148.

———. 1987. From heredity theory to "Vererbung": The transmission problem, 1850–1915. *Isis* 78:337–364.

Clouston, T. S. 1884. *The Effects of the Excessive Use of Alcohol on the Mental Functions of the Brain*. Edinburgh: Andrew Elliot.

Collini, Stefan. 1991. *Public Moralists: Political Thought and Intellectual Life in Britain, 1850–1930*. Oxford: Clarendon Press.

Corfield, P. J. 1995. *Power and the Professions in Britain 1700–1850*. London: Routledge.

Cowan, Ruth Schwartz. 1985. *Sir Francis Galton and the Study of Heredity in the 19th Century*. New York: Garland.

Crossick, G. 1991. From gentlemen to the residuum: Languages of social description in Victorian Britan. In P. Corfield, ed., *Language, Class and History*, 150–178. Cambridge: Cambridge University Press.

Darwin, Charles. 1859. *On the Origin of Species by Mean of Natural Selection*. London: John Murray.

———. 1868. *The Variation of Animals and Plants under Domestication.* 2 vols. London: John Murray.

———. 1958. *The Autobiography of Charles Darwin.* Ed. Nora Barlow. London: Collins.

Derry, Margaret. 2003. *Bred for Perfection: Shorthorn Cattle, Collies, and Arabian Horses Since 1800.* Baltimore: Johns Hopkins University Press.

Dixon, E. S. 1851. *The Dovecot and the Aviary.* London: John Murray.

Eaton, John Matthews. 1851. *A Treatise on the Art of Breeding and Managing the Almond Tumbler.* London: J. M. Eaton.

———. 1858. *A Treatise on the Art of Breeding, Managing, and Taming Domesticated, Foreign, and Fancy Pigeons.* London: J. M. Eaton.

Feeley-Harnik, Gillian. 2004. The geography of descent. *Proceedings of the British Academy* 125:311–364.

Fristrup, Kurt. 2001. A history of character concepts in evolutionary biology. In Günther Wagner, ed., *The Character Concept in Evolutionary Biology*, 13–35. San Diego: Academic Press.

Galton, Francis. 1869. *Hereditary Genius, an Enquiry into its Laws and Consequences.* London: Macmillan.

———. 1874. *English Men of Science: Their Nature and Nurture.* London: Macmillan.

———. 1883. *Inquiries into Human Faculty and Its Development.* London: Macmillan.

———. 1889. *Natural Inheritance.* London: Macmillan.

———. 1892. *Hereditary Genius, an Enquiry into Its Laws and Consequences.* 2nd ed. London: Macmillan.

Gash, Norman. 1979. *Aristocracy and People: Britain 1815–1865.* London: Edward Arnold.

Harrison, Brian. 1971. *Drink and the Victorians: The Temperance Question in England, 1815–1872.* London: Faber and Faber.

Hawley, John C. 1989. Mary Barton: The inside view from without. *Nineteenth Century Studies* 3:23–30.

Hilkey, Judy. 1997. *Character Is Capital: Success Manuals and Manhood in Gilded Age America.* Chapel Hill: University of North Carolina Press.

Huertas, Rafael. 1993. Madness and degeneration, part II: Alcoholism and degeneration. *History of Psychiatry* 4:1–21.

Hyde, John. 1875. *Character, Its Elements and Development, by a Bible Student.* London: James Speirs.

Jarvis, Adrian. 1997. *Samuel Smiles and the Construction of Victorian Values.* Stroud: Sutton.

Kerr, Norman. 1894. *Inebriety or Narcomania, Its Etiology, Pathology, Treatment and Jurisprudence.* 3rd ed. London: H. K. Lewis.

Kidd, Alan, and David Nicholls, eds. 1999. *Gender, Civic Culture and Consumerism: Middle-Class Identity in Britain, 1800–1940.* Manchester: Manchester University Press.

Klapisch-Zuber, Christiane. 1991. The genesis of the family tree. *I Tatti Studies* 4:105–129.

Langford, Paul. 1989. *A Polite and Commercial People: England, 1727–1783.* Oxford: Oxford University Press.

Levine, Harry. 1978. The discovery of addiction. *Journal of Studies of Alcohol* 39:143–174.

Lewes, George. 1855. *The Life and Works of Goethe.* 2 vols. London: David Nutt.

Lomax, Elizabeth. 1979. Infantile syphilis as an example of 19th-century belief in the inheritance of acquired characteristics. *Journal of the History of Medicine and Allied Sciences* 34:23–39.

López-Beltrán, Carlos. 1994. Forging heredity: From metaphor to cause, a reification story. *Studies in History and Philosophy of Science* 25:211–235.

Lubinsky, Mark S. 1993. Degenerate heredity: The history of a doctrine in medicine and biology. *Perspectives in Biology and Medicine* 37:74–90.

Maudsley, Henry. 1874. *Responsibility in Mental Disease.* London: Henry King & Co.

Mazumdar, Pauline. 1992. *Eugenics, Human Genetics and Human Failings: The Eugenics Society, Its Sources and Its Critics in Britain.* London: Routledge.

Meckier, Jerome. 2001. *Great Expectations* and self-help: Dickens frowns on Smiles. *Journal of English and Germanic Philology* 100:537–554.

Norton, B. J. 1973. The biometric defense of Darwinism. *Journal of the History of Biology* 6:283–316.

[Pearson, Karl]. 1880. *The New Werther.* By Loki. London: Kegan Paul & Co.

Perkin, Harold. 1989. *The Rise of Professional Society.* London: Routledge.

Pettitt, Clare. 1999. "Every man for himself, and God for us all!" Mrs Oliphant, self-help, and industrial success literature in *John Drayton* and *The Melvilles. Women's Writing* 6:163–179.

Pilbeam, Pamela M. 1990. *The Middle Classes in Europe, 1789–1914: France, Germany, Italy and Russia.* Basingstoke: Macmillan Education.

Porter, Roy, and Rousseau, G. S. 1998. *Gout: The Patrician Malady.* New Haven, CT: Yale University Press.

Porter, Theodore. 1986. *The Rise of Statistical Thinking, 1820–1900.* Princeton, NJ: Princeton University Press.

———. 2004. *Karl Pearson: The Scientific Life in a Statistical Age.* Princeton, NJ: Princeton University Press.

Provine, William. 1971. *The Origins of Theoretical Population Genetics.* Chicago: University of Chicago Press.

Reader, W. J. 1966. *Professional Men: The Rise of the Professional Classes in Nineteenth-Century England.* New York: Basic Books.

Richardson, Angelique. 2003. *Love and Eugenics in the Late Nineteenth Century: Rational Reproduction and the New Woman.* Oxford: Oxford University Press.

Ritvo, Harriet. 1987. *The Animal Estate: The English and Other Creatures in the Victorian Age.* Cambridge, MA: Harvard University Press.

Robinson, Gloria. 1979. *A Prelude to Genetics: Theories of a Material Substance of Heredity, Darwin to Weismann.* Lawrence, KS: Coronado Press.

Rodrick, Anne Baltz. 2001. The importance of being an earnest improver: Class, caste, and self-help in mid-Victorian England. *Victorian Literature and Culture* 29:39–50.

Russell, Nicholas. 1986. *Like Engend'ring Like: Heredity and Animal Breeding in Early Modern England.* Cambridge: Cambridge University Press.

Sankey, W. H. O. 1884. *Lectures on Mental Disease.* London: H. K. Lewis.

Secord, Anne. 1994. Science in the pub: Artisan botany in early nineteenth-century Lancashire. *History of Science* 32:269–315.

———. 2003. "Be what you would seem to be": Samuel Smiles, Thomas Edward, and the making of a working-class scientific hero. *Science in Context* 16:147–173.

———. 2005. Elizabeth Gaskell and the artisan naturalists of Manchester. *Gaskell Society Journal* 19:34–51.

Secord, James. 1981. Nature's fancy: Charles Darwin and the breeding of pigeons. *Isis* 72:163–186.

———. 1985. Darwin and the breeders: A social history. In D. Kohn, ed., *The Darwinian Heritage,* 519–542. Princeton, NJ: Princeton University Press.

Shorthorn Transactions. 1868–1869. London: John Thornton.

Smeed, J. W. 1985. *The Theophrastan "Character": The History of a Literary Genre.* Oxford: Clarendon Press.

Smiles, Samuel. 1871. *Character.* London: Murray.

Talbot, Eugene. 1898. *Degeneracy, Its Causes, Signs, and Results.* London: Walter Scott Ltd.

[Taylor, Isaac]. 1824. *The Characters of Theophrastus; Translated from the Greek, and Illustrated by Physiognomical Sketches. To which Are Subjoined the Greek Text, with Notes, and Hints on the Individual Varieties of Human Nature.* By Francis Howell [pseud.]. London: Josiah Taylor.

Tredgold, Alfred. 1908. *Mental Deficiency (Amentia).* London: Baillière, Tindall and Cox.

Trollope, Anthony. 1858. *Doctor Thorne*. London.

Tyndall, John. 1868. *Faraday as a Discoverer*. London: Longmans.

Valverde, Mariana. 1997. "Slavery from within": The invention of alcoholism and the question of free will. *Social History* 22:251–268.

Vincent, David. 1981. *Bread, Knowledge and Freedom: A Study of Nineteenth-Century Working Class Autobiography*. London: Europa Publications.

Wagner, Günter P., ed. 2001. *The Character Concept in Evolutionary Biology*. San Diego: Academic Press.

Waller, John C. 2001. Ideas of heredity, reproduction and eugenics in Britain, 1800–1875. *Studies in History and Philosophy of Biological and Biomedical Sciences* 32:457–489.

Warner, Jessica. 1994. "Resolv'd to drink no more": Addiction as a pre-industrial construct. *Journal of Studies on Alcohol* 54:685–691.

White, Paul. 2003. *Thomas Huxley: Making the "Man of Science."* Cambridge: Cambridge University Press.

Windholz, George. 1991. Pavlov's view of the inheritance of acquired characteristics as it relates to theses concerning scientific change. *Synthese* 88:97–111.

Winther, Rasmus G. 2000. Darwin on variation and heredity. *Journal of the History of Biology* 33:425–455.

Woiak, Joanne. 1994. "A medical Cromwell to depose king alcohol": Medical scientists, temperance reformers, and the alcohol problem in Britain. *Social History* 27:337–365.

Wyke, Terry. 1999. The culture of self-improvement: Real people in Mary Barton. *Gaskell Society Journal* 13:85–103.

Youatt, William. 1837. *Sheep*. London: Baldwin & Cradock.

Zirkle, Conway. 1946. The early history of the idea of the inheritance of acquired characters and of pangenesis. *Transactions of the American Philosophical Society* N. S. 35:91–151.

17 "Victor, l'enfant de la forêt": Experiments on Heredity in Savage Children

Nicolas Pethes

17.1 Observing the Innate: Cases of Wolf Children

Although natural historians such as Linnaeus, Buffon, and Maupertuis had already suggested theories on the generation of human beings and the inheritance of their traits, it was a philosophical debate that dominated the discussion surrounding the innateness of human faculties at the end of the eighteenth century. This discussion transferred the issue of heredity into a discursive context structured by the distinction between "nature" and "culture." Is an education in society responsible for every positive trait within otherwise "savage" humans, as Claude Adrien Helvétius (1715–1771) claimed? Or do we have to strip away everything society has taught us in order to find our "essential nature," as Jean-Jacques Rousseau (1712–1778) suggested? Rousseau was well aware of the methodological problems associated with the latter choice, given the fact that all of humanity seems to live within larger or smaller social units. Consequently, he asked at the beginning of his *Discours sur l'origine et les fondements de l'inégalité parmi les hommes* from 1755: "What experiments would be necessary to achieve knowledge of natural man? And what are the means for making these experiments in the midst of society?"[1] Due to the time necessary to observe the transmission of traits in humans, such "experiments" (*expériences*) were hard to imagine, and one seemed left with speculation or, at best, analogous reasoning.[2] There seemed to be one exception, however, and Rousseau mentions it in a footnote of his *Discours*.[3] Empirical data could be provided by the cases of so-called wolf or savage children that had been documented numerous times in the eighteenth and nineteenth centuries.[4] As the natural historian Johann Friedrich Blumenbach wrote about one particular case, that of Wild Peter of Hanover, in 1811: "Peter's celebrity met a time in which the question whether innate ideas existed was vitally and heavily discussed. Peter appeared as an ideal subject to decide this issue."[5]

Earlier reports about wolf children had not yet addressed the problem of heredity. Instead, they were interested in the affinities and differences between humans and animals, and the savage child represented a hybrid "in-between state," or "monster."[6] Thus, Linnaeus subsumed *homo ferus* as a subspecies in his classification and mentioned nine examples of children who were found without knowledge of language, covered with fur, and walking on all fours, in the twelfth edition of his *Systemae naturae* (1766). From this description, wolf children seem to be more closely related to the animals who had been their companions than to the human beings who were their parents. Transferred to a model of inheritance, this implies that it was their way of life rather than their genealogical descent that determined their place in the natural order.

It was only at the turn of the nineteenth century that this focus on the natural history of humans was replaced with a focus on human beings as the objects of experimental science. This shift also redefined the image of savage children as well as the significance attributed to them with regard to the question of human heredity. In their *Practical Education* (1798), Maria and Richard Edgeworth were the first to report an "experiment" they had conducted with Peter, the wild child—an experiment that supposedly proved his inability to develop abstract ideas.[7] Subsequently, savage children who grew up isolated from any social influence were regarded as "'experiments with nature'"[8] that were supposed to answer the question as to which faculties were essentially "human." Did these faculties include reason, language, memory, self-consciousness, morality, and social skills? If any of these attributes were found in children that had never experienced any kind of education, they could rightfully be called part of the basic and general heredity of humanity. The traits that the children lacked, however, had to be attributed to society, culture, and education alone. The experimental analysis of savage children contributed to models of either social or biological determinism.[9]

But the appearance of a savage child is not only a spectacular event for scientists, it is also "spectacular" in the core meaning of the word: it is a sensation, an attraction, or public entertainment. Most reports on savage children—be it Peter at King George's court, Victor in Rodez, or Kaspar Hauser in Nuremberg—pointed out that these children spent their first days in society like animals in a zoo or, for that matter, like the native inhabitants of European colonies, who were put on display at fairs, in vaudeville acts, and at world exhibitions later in the nineteenth-century.[10] But it was not just the wild child who was a spectacular phenomenon. So too were the new methods of their examination. As Barbara Maria Stafford has shown, the early nineteenth-century demonstrations of experimental science were received like magic shows in front of stunned audiences.[11] The exotic and spectacular dimensions of experiments on savage children were readily taken up by popular culture. Theater performances, cabaret shows, and novels regularly responded to the observations made in anthropological case studies, and sometimes even preceded their scientific analogues. In his analysis of the various documents, texts, and discourses that paved the way and accompanied the discovery of Kaspar Hauser, Jochen Hörisch states: "In fact, Kaspar's appearance at Nürnberg resembles and doubles the discursive events that preceded it."[12] Wild children, who are not able to speak, enter a discourse that, in this sense, always precedes them.

The constellation of scientific experiments on heredity and the popular discourse on human nature is precisely what interests me here. Around 1800, the question of heredity was an anthropological, physiological, linguistic, psychiatric, and pedagogical one and therefore could not be answered by any of these disciplines alone. What I offer in the following discussion is therefore an analysis of three different discourses on the isolation of humans and experiments that attempt to undo the effects of this isolation: first, hypothetical projects in science and literature to deliberately isolate children that measure the relation of nature and nurture in humans; second, actual case studies conducted on isolated children that mainly point to the significance of education; third, observations of the same cases that come to a completely different conclusion, namely, the dominance of inherited faculties.

17.2 Isolation and Education: The Experimental Paradigm of Science and Literature around 1800

Among the main features that connect the experimental approach and the discovery of savage children is the element of "isolation" that structures the laboratory setup as well as Rousseau's vision of the "natural man" living in isolation.[13] Both notions were combined when in 1799 the Société des Observateurs de l'homme was founded in Paris. In his opening address, one of the founding members, Louis-François Jauffret (1770–1850), suggested:

> One day the society will certainly need to consider if, in order to observe the intellectual and moral abilities of humans in a manner not only new but also comprehensive, it would not be profitable, with the approval of the government, to conduct an *experiment on the natural human*, consisting in the *careful observation of twelve children, half of them male, half of them female, who would be left, at the same peaceful location, to the development of their own ideas and their own language through their natural instincts, far removed from every societal construction*. Doubtlessly, one would gain countless exceptionally useful observations that would enlighten us with certainty on the development of our abilities, in that philosophy observed a few individuals separated

> from birth on from our customs, our constructions, our prejudices and even our language, who would act and express themselves only on the basis of the natural instinct and the natural condition which is common to all humans. [14]

However, because the Société lacked both a government grant and suitable experimental subjects, there were only two ways to realize Jauffret's proposal. The first alternative was fiction. In 1844, Marivaux published his drama *La dispute,* which anticipated the Société's project by staging the encounter of two men and two women brought up in isolation from each other and everybody else during the eighteen years of their life—in order to resolve the question as to which of the sexes is responsible for infidelity. In the course of the eighteenth century, literature reinvestigated the topic of isolation on lonely islands that Defoe's *Robinson Crusoe* had introduced, and there were numerous novels dealing with children growing up without any education.[15] In 1797, François-Guillaume Ducray-Duminil (1761–1819) published *Fanfan et Lolotte,* the story of two English children who are washed up on a desert island in the Carribbean at the age of three and who manage to survive the fierce conditions and the proximity of native inhabitants until they are discovered four years later by the person who will become their enlightened teacher of civilization.

The "scientific" question addressed by Ducray-Duminil's fictive case study is obvious. Will the children turn out to be savages, reduced to their natural faculties? Or will they instead be able to develop some of the basic principles of the culture from which they stem? Ducray-Duminil's novel demonstrates that both alternatives are able to coexist. The shipwrecked Colonel Carlton finds that his "pupils of Nature"[16] are normally developed as far as their "heart[s]" and their "intelligence" is concerned.[17] At the same time, the Colonel decides that the children nevertheless need further instruction. He teaches them to bury the dead (they had kept the corpse of their sole travel companion in their hut for over four years), to build a house in which separate bedrooms are allocated to each sex, and to tell their story, which the Colonel takes as evidence of the providence of God, who provided the children with the natural instincts that helped them survive. After a few days, the acculturation of the savage children is perfect. The Colonel takes the boy hunting, while the girl is "employed in baking the bread or other little offices of domestic economy."[18] The goal of education seems to be the separation of the sexes. The report about the children's progress also makes a clear statement on the relationship between heredity and culture in humans. Human beings are born with innate faculties sufficient to keep them alive and to develop the basic ideas that distinguish them from animals. But they need further education in order to develop higher cultural qualities that distinguish them from the savages who visit the island occasionally.

Thus, Ducray-Duminil's novel is not merely an escapist fairy tale; it contributes to the contemporary debate on human heredity and its possible observation in isolation. Another of his novels is worth mentioning here. It tells the story of a boy who grows up among a band of robbers in the woods where he is found by a nobleman, who grants him higher education. The novel—titled *Victor, ou l'enfant de la forêt*—came out in 1796,[19] four years before Jauffret proposed his experiment to the Société. By 1800, almost everybody in Paris knew what Ducray-Duminil's earlier novel was about. This is because it anticipated the manner in which Jauffret's wish would be fulfilled just a few months after his opening address, this time not by fictional means, but in reality.

On January 8, 1800, a boy of about eleven years of age, who had been seen running about naked in the woods like a wild animal on several prior occasions, entered a farmer's workshop and started his career as the only actual experimental subject of the Société des Observateurs de l'homme. From all evidence, it seemed that the boy—who could not speak, wore no clothes, preferred smelling to his other senses, and appeared to be unable and unwilling to engage in any kind of contact with other human beings—had spent most of his life alone in the forest. He was the perfect *enfant de la forêt*,[20] and although his teacher Jean Itard would later report different reasons for the name he gave his pupil, it seems only too obvious why the name he chose had to be "Victor."[21]

The various reports of the initial examination of the savage child from Aveyron indicate the shift of interest in human nature at the turn of the eighteenth century. The Abbé Pierre-Joseph Bonnaterre (1747–1804), who took care of the child in Rodez, reported to the Société that Victor showed "purely animal functions." He had no knowledge, no wants, no memory, no imagination, and no moral sentiment whatsoever. "You might say his mind is in his stomach,"[22] wrote Bonnaterre. The latter's interest was mainly in natural history, and he used the common comparison to animals in order to classify the savage child. The two experiments he conducted on the boy emphasize this interest. Bonnaterre stripped his charge naked, exposed him to temperatures below zero degrees Celsius, and noted with amazement that the boy "appeared glad indeed to be rid of these garments."[23] On another occasion, Bonnaterre tested the ability of the "child of nature" to share food with others and observed: "He has no idea of property, wants everything for himself, because he thinks only of himself."[24] The Abbé concluded his report with a prescient remark: "Such an astonishing phenomenon will furnish philosophy and natural history with important ideas about the essential nature of man and the development of his intellectual faculties, provided that the state of imbecility we have noticed in this child places no obstacle in the way of his instruction."[25]

The two options presented here—a possible education or the diagnosis of idiocy—are precisely the ones the wild boy from Aveyron faced when he was brought to Paris by request of the Société des Observateurs de l'homme. Two of the leading *Observateurs* represented the two different approaches with which the savage child was confronted. On the one hand, the Abbé Roche-Ambroise Sicard (1742–1822), principal of the first school for deaf and dumb children in Paris, was one of the most progressive teachers of his time. On the other hand, Philippe Pinel (1745–1826), the renowned reformer of mental hospitals, had been systematizing diagnostic methods for the retarded.[26] The wild boy's trip from Bonnaterre's Rodez to Pinel's Paris was also the journey from natural history to natural science. Whereas Bonnaterre had likened the boy to animals, Pinel, in his report to the Société in 1800, added a different comparison: "He is much inferior to many individuals who are locked away in our hospitals. I should not be afraid to say that in this respect even elephants have a marked advantage over him."[27] The imbeciles Pinel knows so well showed "clear parallels with that of the child of Aveyron." Like him, they were mute, emotionally unstable, and without memory. But more important than the parallels is the verdict that follows immediately from the chosen comparison. Since there is no hope of ever educating or integrating the wild boy into society, he ought to be locked away with his fellow sufferers at Bicêtre.[28]

Obviously, both Bonnaterre's and Pinel's diagnoses made implicit statements on the hereditary makeup of the boy from the woods. Bonnaterre's approach was still part of the eighteenth century's tradition of dealing with *homines feri*: his experiments attempted to discern the relation between the boy's behavior and animal behavior. Pinel, on the contrary, replaces this approach with clinical diagnosis and an interest in the boy's relation to society rather than nature.[29] The question Pinel raised with regard to heredity now read: "finding out the nexus of ideas and moral sentiments which are independent of socialization."[30] Since he was unable to establish this nexus, Pinel concluded that Victor was not a suitable subject for further experimentation.[31]

At this point Jean Itard arrived on the scene, protesting vehemently against Pinel's view without siding with Bonnaterre. Itard, who shortly after Pinel's report took the boy into his household and kept him there for six years, conducted one of the first and longest systematic educational experiments, and simultaneously introduced a new approach to the study of human nature. He agreed with Pinel that Rousseau's belief in the "nobility" of savage humans had to be rejected on the basis that the intimidated, spastic, and autistic child did not provide much evidence that human beings develop best in a natural state. What Itard refused, however, this time contrary to Pinel, was the belief in the innateness of the boy's defects. To Itard, the fact that Victor was a child of nature proved how much the development of our faculties depends on training. Drawing both from Étienne Bonnot Condillac's (1715–1780) theory of sensory education and from Sicard's success with the deaf and dumb, Itard planned to prove this theory by turning Victor into a well-educated, conversational, social being. Should he succeed, he would have evidence for his basic hypothesis:

> Cast on this globe, without physical powers, and without innate ideas; unable to obey the constitutional laws of his organization, which call him to the first rank in the system of being; MAN can find only in the bosom of society the eminent station that was destined for him in nature, and would be, without the aid of civilization, one of the most feeble and least intelligent of animals—a truth which, although it has often been insisted upon, has not as yet been rigorously demonstrated.[32]

17.3 Laboratory Nature: Itard's "Sincere" Observations

Itard's two case studies on Victor from 1801 and 1806 will serve as my second example of the discourse on isolation, education, and experiment surrounding the concept of heredity at the turn of the nineteenth century. It is important to note that Itard, despite his refusal of the Société's recommendation to leave Victor in the asylum, still shared the basic methodological approach sketched out by Jauffret. After the news about the wild boy had spread from Aveyron to Paris, Jauffret wrote a letter to the orphanage of Saint-Afrique where the spectacular discovery, the boy from the woods, was initially kept:

> If it is true that you have currently in your orphanage a young wild boy, twelve years old, who was found in the woods, it would indeed be important for the progress of human knowledge that a zealous and sincere observer take him in charge and, postponing his socialization for a little while, examine the totality of his acquired ideas, study his manner of expressing them, and determine if the state of man in isolation is incompatible with the development of intelligence.[33]

This "zealous and sincere observer" turned out to be Itard. Whether he was in fact "postponing" Victor's progress cannot be decided retrospectively. In any case, he shared the experimental attitude implied in Jauffret's proposal: to design a realm in which a savage child could be fruitfully observed.

Remarks like these reveal that Itard's incentive for taking care of the child was motivated not primarily by philanthropy. His enthusiasm was directed more toward the laws of human nature and less toward compassion for individual representatives of the species. Whereas Pinel's devastating diagnosis had at least been directed at the boy as an individual, for Itard the boy was but a means to gain general knowledge.[34] As Itard notes at the beginning of his first report, published in 1801, this was a rare opportunity that had remained unexplored on all earlier occasions. People were interested only in "seeing" the spectacular discoveries, whereas "actual observation was reckoned of no value; and these interesting facts tended little towards improving the natural history of man."[35]

To contradict Pinel's verdict effectively, Itard had to revise the diagnosis of the wild boy's idiocy and instead trace Victor's developmental backwardness to his isolation: "The Savage of Aveyron is much less a simple youth, than an infant of ten or twelve months old."[36] In his attempt to avoid Pinel's medical rhetoric, however, Itard invoked colonial stereotypes similar to those already encountered in Ducray-Duminil: "Like some savages in the warmer climates, he was acquainted with four circumstances only; to sleep, to eat, to do nothing, and to run about in the fields."[37] Itard's principle of education entailed first tending to these basic—obviously innate—needs and thus mitigating the hostility the boy felt toward the society that ended, in such an aggressive and irretrievable way, the life to which he was accustomed. A further method of education Itard applied to his increasingly accepting charge, though, took no regard of any need the boy might feel. Rather, Itard designed a series of experiments to verify his basic theoretical conviction, according to which human senses and understanding are not innate but—as Locke and Condillac had claimed—subject to experience and training. When he first met the boy, he had no sense of heat and cold, no fear of anything, and no sense of justice. Itard, consequently, exposed him to warm baths, extreme height, calculatedly unfair treatment, and many more stimuli intended to awaken his dormant senses.

If we summarize rather than enumerate these various experiments, two general conclusions can be drawn from Itard's approach. The five steps of his educational program—socialization, sensualization, development of ideas and concepts, language acquisition, and general education and instruction—mirror the order that Condillac had introduced in his theory of human development. This theory consisted of first the establishment of sensation and perception, then consciousness and attention, and finally imagination and memory—the latter requiring the ability to use signs.[38] In this model, Victor's progress was a complete repetition of the process of becoming human. Just as in *Fanfan et Lolotte* the process of education followed the steps of humanity's cultural evolution, Victor's education mirrors the order of each individual human being's intellectual evolution as outlined by Condillac.

The second anthropological implication of Itard's experimental setup was closely related to the first. The experiments resulted in Victor's growing ability to distinguish his impressions and reactions: hot and cold, pleasant and unpleasant, comfort and fear, love and anger. These distinctions are the ones that "civilized" and "socialized" the wild boy in the first place. Itard's experiments were a means not only of observing Victor, but also of training and shaping him. Scholarly criticism of Itard's work so far has focused on the therapeutic nihilism expressed in Victor's treatment, which was more likely to produce an "experimental neurosis"[39] in the boy than to help him fight his previous fears. But from a theoretical perspective what is even more interesting is that Itard's experiments in fact produced what they were supposed to analyze: "human nature."

Itard's own account of his first results does not go that far. After one year of treatment, he notes the "mental equality between the boy and the brute"[40]—to which Victor had been inferior on his arrival in Paris. The crucial distinction applied to Victor in order to measure his humanness is the distinction between "speaking" and "nonspeaking." One question is at the core of Itard's experiments: "'Does the savage speak'?"[41] This is related to Condillac's and Rousseau's theoretical treatises on the development of language in the "natural state," well known to Itard. Victor could hear and was able to distinguish sounds, but he did not utter words. Still, Itard rejected the assumption of hereditary defects and attempted to prove that Victor *did* have the language competence necessary to express his needs and wishes:

> I satisfied myself of this one day by an experiment of the most conclusive nature; I chose, from amongst a multitude of others, a thing for which I was previously assured that there did not exist between him and his gouvernante any indicating sign; such was, for example, the comb, which was kept for his purpose, and which I wished him to bring to me. I should have been much mistaken, if, by disordering my hair with my hand, and showing him my hand in this state, I had not been understood. Many persons see, in all these proceedings, only the common instinctive actions of an animal; as for myself, I confess, that I recognize in them the *language of action*, in all its simplicity.[42]

Again, it is Condillac's description of a "language of action" that accounts for Itard's judgment that Victor possessed the capacity for language, and it is a judgment derived from experimental observation.

Condillac, in his *Essai sur l'origine des conaissaces humaines*, proposed a thought experiment: "But I am assuming that two children, one of either sex, sometime after their deluge, had gotten lost in the desert before they would have known the use of any sign."[43] He concluded that the two children would habitually develop signs for situations that repeated themselves, such as the search for food: "These details show how the cries for passion contributed to the development of the operations of the mind by naturally originating the language of action, a language which in its early stages, conforming to the level of this couple's limited intelligence, consisted of mere contortions and agitated bodily movements."[44]

In Itard's closely related observation of an actual experiment, it is again the experiment itself that produces the anthropological feature. Humans are humans as long as they engage in language, and if Victor demonstrated sign usage, he has entered the realm of human nature. For the time being, he is at the stage of primitive humans—a stage identical to the stage of the isolated children in Condillac—but Itard planned to demonstrate by further educational experiments that the inherited faculties of the species could be acquired by the single specimen under proper circumstances. In a sense, this was the converse view to that of Jean Baptiste Lamarck, according to which *acquired* faculties can be *handed down* to descendants.

That Itard himself called his ensuing attempts to teach Victor how to talk a failure is mainly due to the inflexible theoretical concepts that guided his observations. Victor did indeed learn to use written letters in order to denote certain objects, but Itard did not acknowledge this achievement because his concept of language required spontaneous voiced articulations. The "language of action," on the contrary, is supposed to be but a deficient supplement for true human language, which not only serves as a means of communication, but also enables the development of ideas.[45] While conceding his failure, Itard concludes his report with the remark that there are "most important inferences relative to the philosophical and natural history of man, that may be already deduced from this first series of observations!" Itard claimed that Victor was already "endowed with the free exercise of all his senses; that he gives continual proofs of attention, reflection, and memory; that he is able to compare, discern and judge, and apply in short all the faculties of his understanding to the objects which are connected with his instruction."[46]

Thus, both Pinel's rejection of the possibility of Victor's education and Ducray-Duminil's Rousseaueian belief in innate human qualities are repudiated: "The moral superiority which has been said to be *natural* to man, is merely the result of civilization."[47] Itard called this statement on human heredity the destruction of "prejudices . . . which . . . constitute the most amiable, as well as the most consoling illusions of social life."[48] Yet it was through Itard's own prejudices that the experimental production of human qualities within Victor at last succeeded to a certain degree.

The second report—which Itard submitted five years later, in 1806, at the behest of the French Secretary of State (the Société had ceased to exist two years earlier)—provides even more examples of the productive use of experiments. This time, Itard immediately begins with the acknowledgment that his report is "less a story of the pupil's progress than an account of the teacher's failure."[49] Victor's language training failed whenever he was supposed to perform something beyond mere imitation. Itard consequently brought this instruction to an end and turned to another essential realm of human nature, emotions. Again, in order to be successful, the "emotional faculties"[50] must be considered as the result of education rather than heredity, since Victor was emotionally completely indifferent when first found. Therefore, Itard engaged Victor in a training process involving his emotions—for example, the feeling of remorse after his final attempt to escape. He is brought back and encounters his benefactor: "But when he saw that instead of going to him I stood where I was with a cold demeanour and an angry face, he . . . began to cry."[51] Itard deliberately provoked Victor's emotional reaction by staging an appropriate stimulus. The same holds true for the second example, this time aimed at Victor's sense of justice. After Victor successfully accomplished something, Itard refused to congratulate him but instead criticized him. As a result of this "test," Victor bit Itard:

> It could only delight me, for the bite was a legitimate act of a vengeance; it was an incontestable proof that the idea of justice and injustice, the permanent basis of the social order, was no longer foreign to my pupil's mind. By giving this feeling to him, or rather by stimulating its development, I had raised savage man to the full stature of moral man through the most striking of his characteristics and the most noble of his powers.[52]

As in the case of his sensual and intellectual training, Itard's experiments provoked emotional reactions in Victor claimed to be absent in the first place. This power of the mode of observation to produce its object, however, makes it impossible to distinguish between preexisting faculties and those generated by the observation. The omnipresence of the experimental gaze in Itard's pedagogical laboratory prevented the visibility of any kind of "natural" processes whatsoever. Victor, the savage child, could only be analyzed if he was, at the same time, deprived of his genuine savagery through education.[53] This paradox is not an avoidable failure of experimental design, but the result of the experimental approach itself: the need to replace the wild boy's isolation from society with Victor's isolation from nature. Itard had to draw "Victor back into a cocoon in which life alternated between the household and the classroom, nearly abolishing all Victor's opportunities for unstructured contact with the natural and social environment; instruction became the predominant means of learning."[54]

Thus, the question as to whether innate human faculties can be experimentally refuted cannot be answered precisely because laboratory observation fundamentally changes the situation in which such faculties might be observed. Contrary to Ducray-Duminil's example, the experiment takes place in "culture," not in "nature," and consequently results in the assumption that education is more significant than heredity. Indeed, the only possible experiment that could reveal Victor's nature and hereditary makeup would bring about the deconstruction of its intrinsic observational hierarchy. Had Itard followed Victor into the woods in order to see him in his "natural state," it would soon have become clear that the deficiency in coping with the environment changed sides. In the woods of Aveyron, Itard would have learned from Victor that it is less important whether one's faculties are inherited or acquired, but imperative that they adjust to the requirements of the immediate environment, be it cultural or natural.[55]

17.4 Idiot, Not Student: Gall and Spurzheim Visit the "Pretended Savage"

I will now turn to my final example of the various discourses on experiments in heredity with savage children, and this time it will be neither a fictive nor an actual case study, but a systematic scientific treatise. From the perspective of contemporary science, there was another problem with Itard's hypotheses, aside from the methodological concerns just mentioned: they were all wrong. Parallel to Itard's efforts, the early nineteenth century witnessed amazing progress in physiological and neurological research that arose from revolutionary models of physical and cognitive functions of humans. The result of these models was a renewed optimism in the possibility of explaining human nature according to its own principles and without the need to refer to external factors such as society and education. It was the beginning of the positivist age that would in fact develop theories that were able, for the first time, to suggest a scientific model of heredity. Returning to the concept of innate human faculties, but altogether abandoning the philosophical realm in which authors such as Rousseau had been elaborating this concept, observations of savage children headed in a new direction.

In 1809, Johann Christoph Spurzheim (1776–1832) published a volume that summarized his and Franz Joseph Gall's physiological theory of the nervous system. There is no doubt that this field, according to its theorists, was directly connected to the topic we have dealt with thus far. "The first question perhaps in Anthropology," wrote Spurzheim in his treatise, "is, Whence has man his faculties? Is man born indifferent; or does he come into the world endowed with determinate faculties?"[56] There are two possible answers. One is the answer given by Itard, which Spurzheim represented in the following words: "According to this opinion, not only man but also animals are born without determinate faculties—indifferent—as tabulae rasae or blank paper. All the instincts and aptitudes of animals, from the insect to the dog and elephant, are the effects of instruction. . . . It must be answered that neither in animals nor in man does education produce any faculty whatever."[57]

The other answer is contained in Gall's and Spurzheim's attempts to "demonstrate that external influences are not the cause of the internal faculties of the mind."[58] Spurzheim mentions the physical condition of humans and their unconscious "automatic life" as an example of innate qualities. With regard to aspects of conscious "animal life," this evidence also applieds to faculties such as motion and the senses. They are obviously "determined by creation," insofar as they can also be found in animals.[59] The authors then went on to consider qualities specific to human beings, such as affect and understanding. These could be explained "either by external impressions, or by internal causes." Spurzheim argued that the former, as random events, could not account for the emergence of internal structures: "It is certain that external circumstances must be presented, otherwise internal faculties cannot act; but opportunities do not produce faculties. Without food I cannot eat; but I am not hungry because there is food. . . . How many children are exposed to the same influences without manifesting the same energy of faculties?"[60]

But at the same time, Gall and Spurzheim were strongly opposed to the assumption of a "natural state," drawing on their rejection of "cultural" influences on innate abilities. The postulation of a natural human state lacks empirical evidence and is therefore absurd: "According to this hypothesis, man is made for solitude."[61] That humans are never—or only seldom—found in solitude is not proof of the deprivation of their natural instincts through having the constraints of society foisted on them, but the very result of these instincts. Spurzheim claimed that "society itself is a natural institution" following the "social instinct" of human nature.[62]

This was nothing less than a complete reversal of Rousseau: the natural inheritance of human beings is their striving to surpass the merely natural state. Society is part of human nature and the faculties we need to be able to live in it are not acquired but inherited. Spurzheim presented two empirical proofs for this claim. One is the example of geniuses: "Children sometimes show particular disposition and faculties before they have received any kind of instruction. Almost every great man shows in his infancy the character of future greatness." Moreover, if faculties could be taught and acquired, "why are we not all men of genius?"[63] We encountered the other empirical proof of the inheritance of human nature earlier, albeit in a different sense:

> The pretended savage of Aveyron, who is kept in the Institution of Deaf and Dumb at Paris, is an idiot in a high degree. His forehead is very small and much compressed in the superior region. . . . We could not convince ourselves that he hears; for it was impossible to make him attentive to our calling him, or to the sound of a glass struck behind him! His attitude and manner of sitting are decent, but his head and body are incessantly in motion from side to side. He knows several written signs and words, and points out the objects noted by them.

The most remarkable instinct in him is the love of order. As soon as any object is displaced, he puts it in order. Such unfortunate creatures therefore are idiots, not because they have not received any education, but because they cannot be educated on account of their imbecility.[64]

It is stunning to see to what extent the interpretation of the observations of the same experimental subject differ from one another: what for Itard proved the influence of education on human behavior was for Gall and Spurzheim an example of the determination of the same behavior by innate faculties. Education is replaced by physiology, and physiognomic details make up the description of Victor, the idiot. For Spurzheim, there was no doubt that the "first principle of anthropology," according to which "every special faculty is innate,"[65] holds true even when confronted with the case of wild children. Individual deviations such as Victor's prove the influence of education no more than cultural differences. Only the use of faculties may be dependent on external circumstances, whereas the existence of these faculties is granted by heredity. "In one word, all that man does he did at first through nature alone."[66]

17.5 Conclusion

Both the strict opposition between Itard's and Spurzheim's interpretations of Victor's behavior and the fictive combination of both viewpoints in *Fanfan et Lolotte* hint at one important observation. As soon as scientific observation takes the stage, there is nothing left in isolated children that can be called their "natural inheritance." All that remains are theories, and Victor is judged far more according to the principles of these theories than according to the principles guiding his behavior. Although wolf children seem to provide the empirical material for establishing experimental evidence, the relation between innate and acquired faculties remains open to interpretation.

Instead of trying to add yet another interpretation to the endless and irresolvable debate on whether Victor was genetically or socially deprived, it seems more rewarding to take a step back and examine the circumstances under which all these interpretations arose. As I have tried to show, these interpretations either combined heredity and education (Rousseau/Ducray-Duminil), claimed the sole influence of education (Condillac/Itard), or emphasized the exclusive significance of heredity (Gall/Spurzheim). The three discourses on human heredity lead to the development of three models for a philosophy of innateness. Human faculties are (1) the result of *noble nature plus education*, (2) the result of *savage nature and education only*, or (3) the result of *only nature without education.* None of these models represents a "master discourse" on heredity in the first half of the nineteenth century.[67]

As for the debates arising from further discoveries of savage children, the contribution of such cases to a theory of heredity was limited. When Kaspar Hauser appeared in 1828, he too became subject to experiments by his foster father, Georg Friedrich Daumer (1800–1875). The goal of these experiments, however, was no longer the identification of the innate faculties of human nature. Instead, Daumer tested Kaspar's magnetic abilities to detect metal and to react to gestures that had not actually touched him.[68] But Daumer did not speculate as to whether these abilities are innate or result from the circumstances of Kaspar's imprisonment, during which he had no contact with humans for over ten years. Rather, the evidence he reported was designed to reject assumptions that Hauser's story may have been a hoax.[69]

Thus, a new discourse arose that addressed the issue of savage children: the legal discourse. Kaspar's exposure is considered not an experiment in human nature, but a *Crime against the Human Soul* ("*Verbrechen am Seelenleben des Menschen*"), as Anselm Feuerbach called his case study in 1832. Instead of questioning Kaspar's natural faculties, this discourse focused on the biographical childhood that was stolen from him. The question of heredity enters a different realm here, in which the main issue surrounding the discovery of Kaspar Hauser was his possible descent from the Duke of Baden.[70] Where experiments failed to prove that savages are noble, they might at least demonstrate that nobles can be savages.

Notes

1. Rousseau [1755] 1992, 13.

2. See Terrall, chapter 11, this volume.

3. Rousseau [1755] 1992, 68–70.

4. For a survey on all reported cases see Singh and Zingg 1942 as well as Malson [1964] 1972, 62–82.

5. Blumenbach 1811, 18; my translation.

6. See Douthwaite 1997.

7. Edgeworth and Edgeworth [1798] 1815, 61–62.

8. Malson [1964] 1972, 49.

9. See Singh and Zingg 1942, xvi, versus Malson [1964] 1972, 35. According to Shattuck [1980] 1994, 41, it is this interest in experiments with savage children that makes them "forbidden" experiments.

10. See Barker 1991.

11. See Stafford 1994. Daston (1998) argues that the description of "strange facts"—among others monstrous appearances—marks the beginning of modern natural philosophy in the seventeenth century.

12. Hörisch 1979, 270–271; my translation. See Gineste 1993, 29–32. As Malson [1964] 1972, 37–38, put it: "Every so often the press reports the discovery of yet another Mowgli waiting for his Kipling."

13. See Foucault [1976] 1977, 204. Even for Rousseau [1762] 1911, 5, however, the perfect isolation of a human being would lead to the development of a monster.

14. Quoted from Moravia [1970] 1973, 215; translation by Anne Beryl Wallen.

15. See Pethes 2007.

16. Ducray-Duminil [1797] 1807, 10.

17. Ducray-Duminil [1797] 1807, 31.

18. Ducray-Duminil [1797] 1807, 51.

19. See Gumbrecht [1981] 1992, 206–211.

20. There was, in fact, a novel that dealt with the case of Victor de l'Aveyron. Lane [1976] 1977, 352, refers to a book by J. A. Neyer—*Rodolph ou le sauvage de l'Aveyron*—that I could not track down in any library. Gineste (1993, 482) records that its publication was announced in the *Journal général de la littérature française* on October 2, 1800, but has also been unable to retrieve it.

21. The assumption that Itard's choice was influenced by Ducray-Duminil's novel and its adaptation for theater is expressed in Shattuck [1980] 1994, 92.

22. Quoted from the first English translation in Lane [1976] 1977, 35–54, here 41–42. Bonnatere's report was published in 1800 as *Notice historique sur le sauvage de L'Aveyron et sur quelques autres indivus qu'on a trouvés dans les forêts à différentes époques* with Pancoucke in Paris.

23. Lane [1976] 1977, 49.

24. Lane [1976] 1977, 47.

25. Lane [1976] 1977, 53.

26. On Pinel see Cartron, chapter 7, this volume.

27. Quoted from the English translation in Lane [1976] 1977, 64–79, here 66. The original text is now available in Gineste 1993, 249–260, 271–278.

28. Lane [1976] 1977, 69.

29. See the title of Gineste 1993, *Dernier enfant sauvage, premier enfant fou*, as well as Douthwaite 1997, 191: "Earlier, the wild child had been embraced as an example of nature, but by the nineteenth century his antisocial habits and insensitivity gave rise to claims of mental pathology."

30. Lane [1976] 1977, 64.

31. See Moravia [1970] 1973, 116.

32. Quoted from the English translation in Malson [1964] 1972, 91. Itard's first report was published in 1801 as *De l'éducation d'un homme sauvage ou des premiers développements physiques et moreaux du jeune suivage de l'Aveyron* with Goujon in Paris.

33. Quoted from Lane [1976] 1977, 15. Lane also quotes various contemporary newspaper articles that express hopes for "pure" laboratory conditions for the study of human nature in Victor (p. 20).

34. See Moravia [1970] 1973, 102–103.; Lane [1976] 1977, 104.

35. Malson [1964] 1972, 92.

36. Malson [1964] 1972, 101.

37. Malson [1964] 1972, 103.

38. Condillac [1746] 2001, 15–40.

39. Lane [1976] 1977, 136.

40. Malson [1964] 1972, 104.

41. Malson [1964] 1972, 106.

42. Malson [1964] 1972, 126.

43. Condillac [1746] 2001, 113.

44. Condillac [1746] 2001, 115. See Lane [1976] 1977, 92–93.

45. This is why Edgeworth and Edgeworth [1798] 1815, 60, called Wild Peter an idiot—precisely the diagnosis Itard wanted to avoid.

46. Malson [1964] 1972, 137.

47. Malson [1964] 1972, 138.

48. Malson [1964] 1972, 140.

49. Quoted from the English translation in Malson [1964] 1972, 141–179, here 141.

50. Malson [1964] 1972, 168.

51. Malson [1964] 1972, 170.

52. Malson [1964] 1972, 174.

53. See Moravia [1970] 1973, 103–104.

54. Lane [1976] 1977, 191f.

55. See Lane [1976] 1977, 198.

56. Spurzheim [1809] 1815, 53.

57. Spurzheim [1809] 1815, 68.

58. Spurzheim [1809] 1815, 60.

59. Spurzheim [1809] 1815, 59. Spurzheim emphasizes: "It results from these considerations, that the comparison of man with other beings (not only with animals, but also with plants and minerals) must be admitted, and cannot be repugnant to our feelings."

60. Spurzheim [1809] 1815, 62.

61. Spurzheim [1809] 1815, 63.

62. Spurzheim [1809] 1815, 64. In the same way, Blumenbach (1811, 44), in his reconstruction of the story of Wild Peter of Hanover, rejects any notion that children who were exposed at an early age would develop back toward an "original savage race" of humans.

63. Spurzheim [1809] 1815, 70. On the discourse of genius in the early nineteenth century see Willer, chapter 18, this volume.

64. Spurzheim [1809] 1815, 71–72.

65. Spurzheim [1809] 1815, 94, 74.

66. Spurzheim [1809] 1815, 92.

67. George Sand's novel about a foundling—*François le champi*—presents a nice example of the entangling of all three models. See Sand [1852] 1894, 5, 18, 31.

68. Daumer [1832] 1983, 28.

69. See the documentation in Hörisch 1979, 214–222.

70. For the switch from inherited to acquired characteristics within the discourse on English nobility, see White, chapter 16, this volume.

References

Barker, Andrew. 1991. "Unforgettable people from paradise!" Peter Altenberg and the Ashanti-visit to Vienna of 1896–1897. *Research in African Literatures* 22 (2): 57–71.

Blumenbach, Johann Friedrich. 1811. *Beyträge zur Naturgeschichte*. Vol. 2. Göttingen: Heinrich Dieterich.

Condillac, Étienne Bonnot. [1746] 2001. *An Essay on the Origin of Human Knowledge*. Translated from the French and edited by Hans Aarslef. Cambridge: Cambridge University Press.

Daston, Lorraine. 1998. The language of strange facts in early modern science. In Timothy Lenoir, ed., *Inscribing Science: Scientific Texts and the Materiality of Communication*, 20–38. Stanford, CA: Stanford University Press.

Daumer, Georg Friedrich. [1832] 1983. *Mitteilungen über Kaspar Hauser*. Ed. Peter Tarkowsky. Dornach: Rudolf Geering.

Douthwaite, Julia. 1997. Homo ferus: Between monster and model. *Eighteenth Century Life* 20:203–221.

[Ducray-Duminil, François-Guillaume. [1797] 1807. *Fanfan et Lolotte*.] *Ambrose and Eleanor*. Translated from the French with alterations, adapting it to the perusal of youth by Lucy Peacock. 3rd ed. London: Johnson & Harris.

Edgeworth, Maria, and Richard Lovell Edgeworth. [1798] 1815. *Practical Education*. 2nd U.S. ed. Vol. 1. Providence: J. Francis Lipitt/Boston: T. B. Wiat & Sons.

Foucault, Michel. [1976] 1977. *Discipline and Punish: The Birth of the Prison*. New York: Vintage.

Gineste, Thierry. 1993. *Victor de l'Aveyron: Dernier enfant sauvage, premier enfant fou. Édition revue et augmentée*. Paris: Hachette.

Gumbrecht, Hans Ulrich. [1981] 1992. Outline of a literary history of the French Revolution. In Hans Ulrich Gumbrecht, ed., *Making Sense in Life and Literature*, 178–225. Minneapolis: University of Minnesota Press.

Hörisch, Jochen. 1979. *Ich möchte ein solcher werden wie . . . Materialien zur Sprachlosigkeit des Kaspar Hauser*. Frankfurt am Main: Suhrkamp.

Lane, Harlan. [1976] 1977. *The Wild Boy of Aveyron*. New York: Bantam.

Malson, Lucien. [1964] 1972. *Wolf Children and the Problem of Human Nature*. Translated from the French by Edmund Fawcett and Peter Ayrton. New York: Monthly Review Press.

Moravia, Sergio. [1970] 1973. *Beobachtende Vernunft: Philosophie und Anthropologie in der Aufklärung*. Translated from the Italian by Elisabeth Piras. Munich: Hanser.

Pethes, Nicolas. (2007). *Zöglinge der Natur. Der literarische Menschenversuch des 18. Jahrhunderts*. Göttingen: Wallstein.

Rousseau, Jean-Jacques. [1755] 1992. Discourse on the origins of inequality (Second Discourse). Translated from the French by Judith R. Bush. In Roger D. Masters and Christopher Kelly, eds., *The Collected Writings of Rousseau*. Hanover/London: University Press of New England.

———. [1762] 1911. *Émile*. Translated from the French by Barbara Foxley. London: Dent/New York: Dutton.

Sand, Georges. [1852] 1894. *François the Waif.* Translated from the French by Jane Minot Sedgwick. Boston: Little, Brown.

Shattuck, Roger. [1980] 1994. *The Forbidden Experiment: The Story of the Wild Boy of Aveyron.* New York: Kodansha International.

Singh, J. A. L., and Robert M. Zingg. 1942. *Wolf-Children and Feral Man.* New York: Harper and Row.

Spurzheim, Johann Christoph. [1809] 1815. *The Physiognomical System of Drs. Gall and Spurzheim; Founded on Anatomical and Physiological Examination of the Nervous System in General, and of the Brain in Particular; and Indicating the Dispositions and Manifestations of the Mind.* London: Baldwin, Cradock, and Joy.

Stafford, Barbara Maria. 1994. *Artful Science: Enlightenment, Entertainment, and the Eclipse of Visual Education.* Cambridge, MA: MIT Press.

18 Sui Generis: Heredity and Heritage of Genius at the Turn of the Eighteenth Century

Stefan Willer

In 1802 Johann Nikolaus Forkel (1749–1818), the founder of musicology as a modern academic discipline, published a biography of Johann Sebastian Bach (1685–1750). It not only marked the rediscovery of Bach by bourgeois music culture, but initiated his elevation to the status of one of the greatest artistic saints of the nineteenth century. Among other things, Forkel accounted for Bach's stature as a "true great artistic genius" with the following statements: "The works with which Johann Sebastian Bach bequeathed posterity are part of an invaluable national heritage, which can comparably be claimed by no other peoples," wrote Forkel. He continued: "If there has ever been a family in which a distinguished predisposition for one and the same art was, so to speak, inheritable, it was most certainly the Bach family."[1] These statements linked genius with cultural inheritance in a twofold way. First, genius was seen as a decisive factor in the creation and perpetuation of cultural legacy. Second, it could be described as an effect of familial hereditary transmission.

Of the two statements, the second one might seem most surprising on account of its expression at so early a date, given our understanding of the history of the idea of hereditary genius. Only in the late nineteenth century did questions of whether or not genius is hereditary, whether it can be passed on to one's descendants, and whether there are familial predispositions that make the birth of a genius probable, become the object of systematic and coherent discussion. The commencement of this discourse, which in demonstrating a precarious affinity between hereditary and racial biology betrayed a close connection with eugenics, can be ascribed, with relative accuracy, to Francis Galton's (1822–1911) writings on *Hereditary Talent* and *Hereditary Genius*.[2] Consequently, genius persisted well into the twentieth century as an exemplar of hereditary theory. Here, exemplar connotes a case that confirms a norm by effectively transgressing it, such that the biology of genius remained allied to its pathology.[3] Musical talent was often chosen as the object of inquiry within this context.[4] And Bach and the Bach family in particular awakened exceptional interest. The monumental biography written by Galton's contemporary Phillip Spitta (1801–1859) dedicated nearly 200 pages to the ancestors of Johann Sebastian. In the 1950s, Karl Geiringer gave his monograph the programmatic subtitle *Seven Generations of Creative Genius*.[5]

While a copious amount of literature treating genius and inheritance was published in the decades on either side of 1900, this literature clearly maintains a distance from an aesthetic or poetic determination of genius. Already under Galton, the actual creative activity of the genius was merely postulated and remained unanalyzed, while the concept of genius itself was understood as a complex of predispositions or talents that could be verified individually by genealogical analysis. In later attempts to situate these predispositions, as in Lewis M. Terman's (1877–1956) *Genetic Studies of Genius* (1925), the retreat from aesthetic reflection even intensified.[6] Nevertheless, the persistence of talk of genius in the context of hereditary theories is neither mere mystification nor simply a persuasive technique. On the contrary, even prior to Galton, genius, generation, and inheritance entered into a close theoretical and argumentative relation that was grounded in the very attempt to give a physical account of creative innovation.

Forkel's 1802 biography of Bach is testament to the fact that the connection of genius and inheritance had long been of significance. Although Forkel's work did not occupy the center of discourse on hereditary genius, it proves to be a symptom of unarticulated or forgotten anthropological presuppositions. The text is particularly amenable to a symptomatic reading, precisely because Forkel wrote neither as a theoretician of genius, nor as a prominent aesthetician. He remained largely indifferent in decisive respects toward the programmatic tenets of contemporary philosophical aesthetics. What mattered for him was the singular and concrete case of Johann Sebastian Bach. Forkel treated Bach as a "mentor" and "man," probed his "ancestry," "life history," and "aesthetic development," to quote some of the chapter titles. In Forkel's biography the "recognition of genius," as the concluding chapter is titled, was only derivable from concrete reality. From this anthropological perspective, on the one hand, inheritance designates a set of cultural impressions, which in principle permitted the transfer of culture from familial to national heritage. On the other hand, inheritance simply signifies the quotidian experiential fact that "like engenders like." Indeed, a most telling reservation can be discerned in Forkel's assessments of Bach's genius quoted above, when the musical predisposition of the Bach family is characterized as being "so to speak, inheritable' (*gleichsam erblich*)."[7]

To better understand this metaphorical reservation—"gleichsam"—some of the general implications of the genius debate around 1800 must initially be explained, which I will do in the first section of this chapter. I then intend to probe, in the second section of the chapter, the role played by the genre of artistic biographies within this debate, and the value specifically ascribed to music in cultural inheritance. My intention is to read the title of Forkel's book, *On Johann Sebastian Bach's Life, Art, and Works of Art* (*Über Johann Sebastian Bachs Leben, Kunst und Kunstwerke*), as an anthropological formula. By emphasizing "life" as the center of the idea of genius, the concept of inheritance becomes crucial in transforming "life" into "works of art." In the concepts of generation, birth, and self-reproduction, as they were used in the discourse on genius at Forkel's time, an idea of "nature" and of "life" emerges that articulates a desire for originality equivalent to contemporary fascination with the copy.

18.1 The Physiology of Genius: Generation, Gender, and Genus

Rudolf Hildebrand's (1824–1894) article "Genius," which appeared in 1897 in the *German Dictionary*, provides a truly encyclopedic survey of the literary and philosophical use of the word in the eighteenth and beginning of the nineteenth century.[8] No less than thirteen semasiological aspects, together with dozens of further nuances of meaning, are differentiated, creating a continuous spectrum from the abstract-transcendent "genius within" to the human spirit, or "menschengeist."[9] Moreover, Hildebrand provides an etymology that begins with influences stemming from the Latin *ingenium* and French *génie*. His discucssion encompasses philosophical research on the concept since the 1760s, the "authentic period of genius" of *Sturm und Drang* (storm and strife), and addresses its further conceptual consequences "in the great times" of Goethe and Schiller.[10] This teleological survey leading up to the Weimar Classic period implies an increasingly idealized reformulation of the concept of genius, which Hildebrand endorses in accordance with the intentions of his article to clear up the "curious confusion"[11] surrounding the semantics of the word. Yet this confusion inevitably manifests itself in Hildebrand's attempt to render countless subsidiary meanings and acts of translation under one concept, despite his desire for clarification.

Already one of the first meanings of the term *genius* put forward by Hildebrand raises an objection to the ideality and spiritualtiy of this concept. It is concerned with the equivalence between *genialis* and *naturalis* and with the early High German definition of genius as "god of nature." Both derive from the word *genius* itself, which can be traced back to begetting and generating: "*genius* from *gignere* as the creative force of all which is living."[12] In spite of this etymological observation, Hildebrand does not assign an independent place within the systematic context of his dictionary article to this physiological derivation of the term *genius*. This can be accounted for by the fact that the majority of the texts evaluated in the rest of Hildebrand's article, whether they stem from Gotthold Ephraim Lessing (1729–1781), Johann Gottfried Herder (1744–1803), Johann Wolfgang von Goethe (1749–1832), Friedrich Schiller (1759–1805), or Jean Paul Friedrich Richter (1763–1825), do not systematically explicate the analogy between ingenious artistic creation and natural generation. Despite this, the analogy occupies center stage in the entire discourse on genius. Indeed, it is hardly an overstatement to altogether attribute the persuasive power of the category of genius to the idea that generative force evolves through genius. At the turn of the eighteenth century, the intersection of physiological and aesthetic aspects of creative acts gave rise to what can be called a *poetics of generation.*[13]

This concept manifested itself in one of the first philosophical explorations of the new "epigenetic" paradigm of generation. In Herder's *Reflections on the Philosophy of the History of Mankind* (*Ideen zur Philosophie der Geschichte der Menschheit,* 1784), especially in his chapter on the "genetic force" as "mother of all culture on earth," the term *genetic* is neither a presentiment of modern genetics nor an opposition to *epigenesis,* but instead signifies the natural process of coming-into-being in general. Herder here explicitly referred to work in contemporary embryology. Caspar Friedrich Wolff (1733–1794) in his *Theoria generationis* conceptualized a "certain" and "essential" force ("gewisse Kraft,"[14] "vis essentialis"[15]) that distributed the fluid generative matter before its solidification and formation.[16] Despite the empirical knowledge cited in the footnotes, Herder's discourse is thoroughly marked by the rhetoric of potential. His enthusiastic, almost rhapsodic prose persistently alternates between the conjunctive and the indicative. He simultaneously accentuates the miraculous character of procreation as well as its reality, thus stressing its unobservable and observable nature at the same time:

> How must the man have been astonished, who first saw the wonders of the creation of a living being! Globules, with fluids shooting between them, become a living point; and from this point an animal forms itself. . . . What would he who saw this wonder for the first time call it? There, he would say, is a living organic power: I know not whence it came, or what it intrinsically is: but that it is there, that it lives, that it has acquired itself organic parts out of the chaos of homogeneal matter, I see: this is incontestable.[17]

In spite of this "undeniable" evidence, Herder uses the word *gleichsam* (*quasi* or "as it were," as in Forkel) twice when referring to the relation between the "force" on the one hand and the effects it actually produces on the other. He states that each of the organic parts is formed "quasi in an act, as an effect on its own" ("gleichsam actu, in eigner Wirkung"). Furthermore, he understands the "invisible force" of epigenesis as a "quasi-manifestation of its inner nature" ("daß die unsichtbare Kraft . . . sich ihrer innern Natur nach gleichsam nur *offenbare*").

Thus the same structure of simile emerges as the one presented in Forkel's description of the hereditary character of genius. Indeed, given that a more or less numinous force, bearing divergent, if not contradictory, names ("genetic," "epigenetic," "organic," "procreative," "formative," "essential," "vital"), produces analogical resemblances, one could attribute metaphorical potential to this very force itself. That genius plays a decisive role in these transformations is articulated by Herder in the chapter mentioned above through an explicit identification of genetic force and internal inborn genius: "This faculty [of living itself] is inborn, organic, genetic: it is the ground of my natural forces, the inner genius of my being." ("Angeboren, organisch, genetisch ist dies Vermögen: es ist der Grund meiner Natur-Kräfte, der innere Genius meines Daseins.")[18]

Innateness is a topos in the aesthetics of genius throughout the second half of the eighteenth century. Birth was no longer regarded as occasioning the encounter with the genius standing guard over it—akin to an older meaning in the vein of "guardian spirit"—but became the instance of the originality of human genius itself.[19] The classical formulation is found in Immanuel Kant's (1724–1804) *Critique of Judgment* (1790): "Since talent, as an innate productive faculty of the artist, belongs itself to nature, we may put it this way: *Genius* is the innate mental aptitude (*ingenium*) *through which* nature gives the rule to art."[20] In the identification of genius with the original, the "inborn" functions as a cipher of innovation.[21] The embryological context of the Herder quotation nonetheless illustrates that birth does not so much mark the emergent unconditional genesis of the new. It is much more the completion of a development, itself an effect of a genetic and, according to Herder, geniuslike force. In this respect, the coupling of natural birth and aesthetic originality deserves further examination.

In Edward Young's (1683–1765) *Conjectures on Original Composition* (1759), one of the initial texts treating the aesthetic of genius in the second half of the eighteenth century, originality is defined as the quintessence of naturalness. Nature "brings us into the world all *Originals*: No two faces, no two minds, are just alike; but all bear Nature's evident mark of Separation on them. Born *Originals*, how comes it to pass that we die *Copies*?"[22] Yet, a more careful perusal of Young's writing reveals that "Nature" proves to be a dubious witness in regard to the process of copying.[23] Right at the beginning of the *Conjectures*, naturalness and originality are brought into close connection with imitation: "*Imitations* are of two kinds; one of Nature, one of Authors: The first we call *Originals*, and confine the term *Imitation* to the second."[24] From this linguistic convention, it becomes evident that in speaking of originals, Young is also generally concerned with imitation. This is directed toward nature, not as an object, but rather as a means of reproduction. This genuinely modern expression of the concept of nature marks an updating of the traditional opposition between *natura naturata* and *natura naturans*. Seen in this light, the emphasis on nature, which can be found in numerous texts treating genius and originality, far from being tautological, serves as a precise argument for the original self-reproducing force of nature. This argument can be observed in the distichs entitled "The genius" from the *Tabulae votivae* (1797) jointly published by Schiller and Goethe:

Wiederholen zwar kann der Verstand, was da schon gewesen,
Was die Natur gebaut, bauet er wählend ihr nach.
Über Natur hinaus baut die Vernunft, doch nur in das Leere,
Du nur, Genius, mehrst *in* der Natur die Natur.[25]

The multiplication (*mehren*) of nature through genius can be read in this context as an indication that the process of copying need not have anything to do with a deficit in relation to the original. The etymology refers copy to its Latin origin *copia*, abundance, or quantity. In English *copy* is evocative of this meaning, whereas in German *Kopie* is but the unequivocal conceptual opposite of *Original*.[26] Yet, in works written around 1800, the concept of the unconditional new overlaps with the concept of a similar preceding entity. This overlap pertains not only to creation through genius, but also to the question of how genius, in turn, is generated. It is hardly possible to point out clear argumentative oppositions here. Indeed, lack of clear delineations is the most prominent characteristic of this metaphorical constellation. Such is the case with Young, where it is difficult throughout to determine whether the originals, to which he refers as "unprecedented births,"[27] are the ingenious authors or their works.

That the generation of genius is intimated through such unclear relations is depicted in Jean Paul's *School for Aesthetics* (*Vorschule der Ästhetik*, 1804). There, "family resemblances" that mediate between the exceptional genius and the "heterogeneous masses" are discussed.[28] The question of the genealogy of genius can be traced through a succession of paragraphs of the *Vorschule*, which deals with various "grades of phantasy" from those of common people to true genius. Jean Paul's humorous, elastic imagery constantly alludes to the genealogical subtext, though never making it explicit. Thus, whereas mere talent could die out with "posterity," "genius as a species cannot."[29] To these true geniuses, Jean Paul ascribes "ingenious originality" as well as "ingenious identity," whereby this identity itself is said to be "generated."[30] The concluding formulation claims that genius depicts "the entirety of life."[31]

The importance ascribed to the generative within this context is further illustrated through Jean Paul having devised a typology of the genius on the basis of gender. Following the predominant prototype of gender that prevailed in the late eighteenth century, he characterizes the so-called passive geniuses as impressionable and feminine, designating the active as generative and masculine. What is interesting about this distinction is that it goes far beyond a mere stereotype. Jean Paul devotes considerable space to an exploration of "bordering geniuses (*Grenzgenies*)," where the division along gender lines defies strict delineation. Such is the case with the "ingenious viragoes, who believe in generation through conception," and to which Jean Paul notably affiliates his contemporary Friedrich von Hardenberg, "Novalis."[32] Indeed when one thinks of Novalis's idea that the work of genius "consists in procreation, birth and rearing of one's equals" and manifests itself as an "act of self embrace,"[33] Jean Paul's characterization seems to be quite appropriate. In his latent bisexuality and plausible self-impregnation, the genius appears to verge on the hermaphrodite who may even be regarded as the aim of ingenious hypersexuality. As opposed to intimating infertility, an existence straddling the sexes unites the generative forces of both sexes.[34]

Reflections concerning gender difference in turn imply those treating the reproduction of genus—whereby in Germany around 1800 both could still be referred to with one and the same word, *Geschlecht*. As such, the connection between gender and genus is likewise central to the debate on genius. In 1795, in his work *On Gender Difference and Its Influence on Organic Nature* (*Über den Geschlechtsunterschied und dessen Einfluss auf die organische Natur*), Wilhelm von Humboldt (1767–1835) derives from the organic life of genius a force that generates genus: "For every work of genius is in turn inspiring for the genius and further propagates his own genus."[35] A few years later in his late work *Kalligone* (1800), Herder summarized this complex of terms and their relations by referring to the etymological relation of *genius*, *genus*, and *gignere*. Here once again, it becomes clear that the concern is both with the generation and proliferation of genius. First: "Genius ist *angeboren* (genius est, quod una genitur nobiscum, in cuius tutela vivimus nati; ingenium ingenitum est)." Second: "Der Genius *schaffet, erzeuget, stellt sich selbst dar* (genius gignit, sui simile procreat, condit genus)." If the genuine of the genius "sustains and propagates itself (*sich erhält und fortpflanzt*)," Herder continues, it becomes "*Geschlecht* (genus) oder *Gattung*."[36] Genius is not a sui generis appearance in the sense of a solitary original beyond all resemblance, but it is constituted in the act of generating a *genus suum*.

This emphasis on the logic of gender and species in the later work of Herder can be read as an objection against the sublimated concept of genius in Kant's *Critique of Judgment*. Despite the previously cited quotation from Kant, claiming that through genius "nature endows art with rules,"[37] physiological traces are effaced in this iteration. For Kant, the question of a species grounded in genius is predominantly understood as one of both an exemplary model and conscious emulation. In contrast, Herder's wording is a reminder of the fact that the problem of emulation and imitation within the aesthetics of genius is always discussed in relation to the concept of generation. Since Young's *Conjectures*, a fundamental distinction has been made between a disdainful "Dingnachahmung," that is, the imitation of canonical artwork, and a productive "Kraftnachahmung," or imitation of nature.[38] Yet, what remains striking is that the undesirable form of "thing-imitation" is also discussed in a natural context, namely, that of genealogy. Thus according to Young, the "similitude" with the "great Predecessors" is dangerous precisely because it evokes an undesirable lineal descent, which should be transformed into another form of kinship: "The farther from them in *Similitude*, the nearer are you to them in *Excellence*; you rise by it into an *Original*; become a noble Collateral, not a humble Descendant from them."[39] The question of posterity remains the decisive touchstone of the aesthetic of genius well into the late nineteenth century with the problem of epigons or *epi-gonai*, literally meaning "descendants."[40]

18.2 Forkel's Bach as Hereditary Genius: Life, Works, and Bequest

In the previous section I have given an overview of the problematic circumstances to which Forkel's characterization of Bach's genius as "so to speak inheritable" refers. The question remains as to how the general program of genius—the reproduction of originality or vice versa—can be reconciled with more pragmatic interests directed toward describing and understanding the unique case of the genius and the concrete form of his creation. For Forkel, it is the narration of an artist's biography that allows this pragmatic realization of genius. As mentioned in the introduction, the mere title of Forkel's book, *On Johann Sebastian Bach's Life, Art, and Works of Art*, suggests that it can be conceived of as an abridged version of an anthropology of genius. The emphasis on "life" in the contemporary discourse on genius intimates, at the very least, that it should not be taken as a self-evident and value-free category in its coupling with "art" and "works of art." Life is not a mere background to the art and works of art but rather their analogue. If one understands life with a view to its generative and genealogical implications, Forkel's book can be read as a performance of this subtext of organic integration among the three thematic categories.

Whereas in the preface of his book, Forkel attests above all to Bach's singularity as a sublime and exceptional artistic figure, in his first chapter, titled "Descent," he maintains exactly the opposite thesis by invoking the quasi-hereditary origin of Bach's art. It involves tracing "six generations" of the Bach family, marked throughout by denominations rooted in the vocabulary of a genealogical tree with extensively outspread "branches" and "limbs." Originating with "forefather" Veit Bach in the sixteenth century, many male family members began pursuing musical vocations from generation to generation, until finally Johann Sebastian Bach appeared as the "crowning ornament" of the family. This was not a matter of an abrupt emergence of genius, but rather one of a genealogical derivation. Bach's relation to his forefathers is a necessary one, insofar as he "arose from them"—a process that seems to be conceived of by Forkel as an accumulation of hereditary musical assets. Moreover, Bach's forefathers appear to be simultaneously indebted to him, in that without this exceptional figure they would have been consigned to the "oblivion of posterity."[41]

This transgenerational family connection is mirrored in the traditionally positive solidarity of the Bachs. The "members of this family" always had a "great devotion for one another." This familial regard was most obviously expressed in their making music together, particularly in the singing of quodlibets—meaning the humorous combination of songs with different texts or tunes—at the Bach family reunions.[42] It is not insignificant that Forkel emphasizes the hybrid genre of quodlibets at such an early place in his book. Familial music-making marks an opposition to the concept of a great integral musical work of art ascribed to a single author. Family "devotion" was also directed toward the compilation of family histories with which Forkel himself worked. The Bachs were assiduous genealogists in their own right. Bach's son, Carl Philipp Emanuel (1714–1788), authored his father's obituary together with a handful of biographical supplements with which he furnished Forkel as early as the 1770s—complete with a wealth of genealogical material. In addition, C. P. E. Bach commented on and supplemented a genealogy for Forkel titled "Origins of the Musical Bach Family" ("Ursprung der musicalisch-Bachischen Familie"), completed in 1735 by Johann Sebastian Bach.[43]

Thus, the author of the Bach biography inherited genealogical representations, with which members of the family identified and interpreted themselves. Much discussion within Johann Sebastian's "Ursprung" text centers on wives and daughters, deaths and remarriages, as well as many cases in which no male descendant is heir to patriarchal talent or pursues a musical vocation. Notwithstanding Forkel's attempt at a reduction of Johann Sebastian's genius to its patrilineal derivation, he could not ignore such genealogical accidents. Musical talents within the extended family were by no means unique. Noteworthy is the mention of an uncle, the twin brother of Johann Sebastian's father, "who so resembled him, that even the two men's respective wives could not distinguish between them except through their attire." The twins are described as "perhaps the only of their kind," where "everything resembled each other," particularly "the style of their music."[44] C. P. E. Bach's commentary on this twin ancestry stresses that the existence of two exact musical copies is testament to a "miracle."[45] Although the story of the twins figures merely as one of numerous anecdotes in the layout of Forkel's book, above all intended to enrich the following chapter, the protagonist's "life history," it nonetheless signifies a moment of duplication, namely nonoriginality to be found within the radius of Bach's birth as a genius. The Bach family resemblances and devotions are, so to speak, concentrated in these twins.

It is this affirmation of the family that differentiates Forkel's biography from many of the artistic portraits of his time. If one compares this genre with romantic novels about artists, it becomes evident that here familial ties are, in contrast, undesired. In no case, however, is the genealogical embedment of the genius completely abandoned. It figures as an inalienable component of the underlying narrative, having been codified as belonging to the genre of the *Bildungsroman* since Karl Morgenstern.[46] The entelechy of the developing individual thus stands in complex reciprocal relations with sociocultural as well as biological conditioning and inheritance. The context of a historically young nuclear family presented an intimate and all the more problematic representation of genealogy.[47] Already Goethe's Wilhelm Meister, in the novel bearing the same name (1795), has to defend his proclivity for the theater against the commerce oriented reasoning of his father. In his story titled *The Peculiar Musical Life of Musician Joseph Berglinger* (*Das merkwürdige musikalische Leben des Tonkünstlers Joseph Berglinger*, 1797), Wilhelm Heinrich Wackenroder (1773–1798) characterizes the birth of his hero by noting that "nobody could be more out of place in this family than Joseph."[48] Similarly, the painter Franz Sternbald, the protagonist in Ludwig Tieck's (1773–1853) novel *Franz Sternbald's Journey* (*Franz Sternbalds Wanderungen*, 1798–1799), receives explicit confirmation of his feelings of being out of place, when his father, an aging farmer, one day reveals to him: "You are not my son, dear child."[49]

Both in the motif of the undesired family connection as well as in the notion of a mysterious origin of unprecedented talent, novels about artists and the education of the young present themselves as family romances in the Freudian sense. In these novels, what is at stake is to ascertain or devise the true genealogy appropriate to one's own artistic genius.[50] The texts mentioned are marked not least by the fact that they problematize this search for origins, in depicting their heroes' course of life as being in jeopardy of failure. In the lives of increasingly stylized characters—as with E. T. A. Hoffman's artistic figures—these searches often end in aporias.[51]

Thus, by around 1800, the end of the family line belonged in certain respects to the topology of genius, crystallizing in the course of the nineteenth century through the close linkage between genius and degeneration as a central element of the pathology of genius. In contrast with this, Forkel portrays Bach's life as a happy and successful one. His unmistakable character is not distinguished by a break with his origins and Forkel does not portray him as the terminating point of a genealogy. For Forkel, Bach embodies the kind of genius that propagates itself. Thus, as mentioned at the end of the chapter titled "Life History," "I merely add that he was married twice, with seven children resulting from the first marriage and thirteen from the second, namely eleven sons and nine daughters. The sons collectively displayed excellent musical predispositions; nonetheless only a few of the elder ones underwent complete musical training."[52]

Of the sons only six and of the daughters but three survived beyond childhood.[53] The latter were in any event precluded from undergoing thorough musical training. Of the former, in contrast, four would become renowned musicians. This in turn raises Bach's quota of procreation to four musical geniuses out of his six surviving sons. As an explanation for this transmission, in addition to the familial "predisposition" argument, Forkel draws foremost on the fact that the musical practice and culture within the Bach household artistically habitualized the sons: "They were accustomed to first-class art from early on, prior even to their receiving lessons."[54] This form of cultural appropriation was institutionalized, in that Johann Sebastian Bach musically instructed his own sons. Notwithstanding the fact that even for the eighteenth century, such a form of familial instruction was not uncommon, Forkel's explicit reference to this practice invokes once again the motif of an overlap of natural and cultural inheritance. For this reason, it is important for Forkel to emphasize that Bach only gave lessons to those students in which "what one calls musical genius" was discerned.[55] Furthermore, he assesses the respective sons according to how clearly the "originality" or the "original spirit (*Originalgeist*)" of their father is manifest in their works.[56] This concern for originality among the progeny suggests that although the genius of Johann Sebastian himself is not conceived as the effect of degeneration, his offspring might be threatened by degeneration. C. P. E. Bach articulated this idea in one of his genealogical and biographical communications to Forkel: "In the question of music, the present generation degenerates."[57]

The distance the Romantics affected toward Forkel is in no small part due to the latent critique addressed to this new generation in his Bach biography. Wackenroder and Tieck heard Forkel in musicology lectures at Göttingen, developing their paradoxical concept of a musical language of the ineffable in opposition to Forkel's *General History of Music* (*Allgemeine Geschichte der Musik*), the first volume of which appeared in 1788.[58] Forkel's codification of musical grammar and rhetoric is rightly characterized as a "pedantic attempt to normatively lay down a musical system of expression with the help of 'stipulations or artistic rules.'"[59] On the path toward autonomy of expression within the changing paradigm of musical aesthetics, Forkel undoubtedly played a conservative role. Nevertheless, the question remains whether this conservatism is sufficiently described by claiming that Forkel "put his trust in the artistic concept of *techné*" and consequently fell behind all attempts to promote the metaphysics of music.[60] Forkel's very preoccupation with the genius of Bach illustrates that the opposition of technique on the one hand, and the expression of "ineffable" inspiration and metaphysics on the other, fails to lead very far, at least in the matter of musical praxis. For, above all else, the pragmatic technical aspects of music account for its relative importance within the confines of the discourse on genius.

This pertains not only to Forkel's biography of Bach. Even an emphatic authority on musical genius and "heartfeltness" such as Christian Friedrich Daniel Schubart's (1739–1791) *Ideas on the Aesthetic of Tonal Art* (*Ideen zu einer Ästhetik der Tonkunst*), dating from the 1780s, concedes that "without culture and practice," musical genius is destined to remain imperfect: "art must perfect and complete what nature cast down raw." What emerges is the ideal figure of the "cultivated genius."[61] The concern here is not only about bridging the supposed opposition between nature and art but also about the specifics of a musical medium. The predominance of instrumental and operative capabilities in realizing musical genius is illustrated through the progression of Schubart's text, which treats the details of musical composition, style, and acoustics. This technical aspect is decisive for the question concerning inheritance.

This explains why Bach's education received a great deal of Forkel's attention. According to Forkel, Bach's genius had first to be cultivated through "some instruction."[62] On the one hand, the ease with which Bach learned, as manifest through his autodidactic grasp, is illustrated. On the other hand, and in unresolved opposition to this ease, the young Bach is depicted as toiling and being led astray in matters of style.[63] For Forkel, in every case, studies are an integral constituent of genius rather than a mere ingredient. What results from this is a peculiar combination of erudition and titanic greatness: "Having put to use his time until this period with much study, playing and composing, Johann Sebastian Bach had thus turned 32 years old. Through this persistent diligence and zeal he obtained such power over the entire art, that he stood there as a giant, capable of trampling everything underfoot into dust."[64]

A similar picture applies to Bach as a mentor. Even though, according to Forkel, the presence of "genius" was a fundamental prerequisite for his pupils, and his composition lessons as a whole were aimed at a "natural coherence,"[65] Forkel mentions explicitly that aside from Bach's own compositions, his pupils "were not permitted to study or become acquainted with anything other than classical works."[66] This canonization of model form is similarly evidenced in his instrumental instruction of which it is said "he first played them the piece which they were supposed to practice once through, then exclaiming: this is how it must sound."[67] Indeed, Bach himself was trained in a similar fashion during his youth, listening and playing back what he heard. According to Forkel, Bach's autodidactic composition lessons consisted of arranging "superb musical pieces" for other instruments through a process of rewriting and reworking.[68]

Here, it is not a matter of imitation set in opposition to an artistic creation of one's own right, but rather of a mimetic method, which due to the technical character of music as an art form proves necessary. This coincides with the fact that Forkel interprets true production not as a sudden emergence but instead as a work process. It is in relation to this very process that the concept of genius is once again taken into account. For, as explicated in Forkel's concluding chapter, the "mark of a true genius" entails not only imagination and inventiveness, but also "ingenuity in the expedient use of the finest and most penetrating artistic means," together with "dexterity of execution."[69] By contrast, in Wackenroder's already mentioned *Berglinger* narrative, one is struck by an emergence of the musical, which bypasses both the study of composition and playing technique as well as the process of creative production. Berglinger's musical training is not part of the story, but it falls within a void between two chapters. When he composes, the music almost seems to burst forth automatically from him. In Forkel's account on the other hand, the paragraphs about Bach as a composer fall exactly in the middle of the book and contain an entire array of technical details of counterpoint, modulation and rhythm. It thus becomes evident that composing is a polymorphic and even syncretic process—a bringing together, which is already implied in the word *componere*. The genius, as invoked by Salomon Maimon around this time against the notion of an original genius, is accordingly a genius of assembling, a "methodical inventor."[70] Once again, this illustrates the problem implied in the emphasis on unconditional self-expression in the aesthetics of genius. Perceived in this light, *Original Composition* as called for by Young is an explicit conceptual contradiction, an oxymoron.

To some extent, Forkel's biography and its invocation of the genius of Bach can be seen as a legitimation of imitation. Nonetheless, this is neither simply a compositional-technical objection against the concept of originality, nor a mere rationalization in order to repudiate the idea of a numinous genius. Rather, the idea of inheritance supplies an argument for the dialectical nature of the opposition between originality and derivation. The declaration, previously cited in the opening, with which Forkel begins his preface and claims Bach as a "national heritage," addresses exactly this point, namely the question of how one should inherit from Bach. Within the "musical Bach family" this inheritance is brought about through a direct transmission of aptitude and skills. The Bach house and lineage portray one and the same context of tradition; the household cultural economy of inheritance runs in synchrony with biology. Outside of the confines of this family, another economy must take hold, which proves at once valid "for patriotic admirers of true musical art" (Forkel's subtitle) and remains suitable for the cultural market.

This is why the pragmatic context of Forkel's publication is by no means insignificant. Originally planned as the last volume of his *Allgemeine Geschichte der Musik*, its appearance in 1802 served as an introduction to the "praiseworthy undertaking of the Hoffmeister und Kühnelschen Musikhandlung in Leipzig in organizing a complete and critically acclaimed edition of Sebastian Bach's works."[71] Critical editions of musical and literary lifeworks were to establish themselves in the course of the nineteenth century—much like monuments of cultural legacy. As in the particular case of Bach, this was often a matter of publishing works that had as of then remained unprinted, thus for the first time making them known as an inheritance. Already in Young's *Conjectures*, the question of publication was of significance. According to Young, in cases where the original works were lost, the successors, "on their Father's Decease, enter, as lawful Heirs, on their Estates in Fame." Nonetheless, where the "perpetuating power of the Press" presides, one may have to wait exceedingly long before the "Decease."[72] Forkel turns the argument around, whereby he preserves the centrality of the concept of "inheritance" so as to legitimate the formation of aesthetic tradition.

Indeed, only by means of documentation and canonization of a critical edition does the genius Bach become the rightful composer of his own works—the "spiritual father" of a national legacy who transmitted his works to descendants transcending his family. Bach could thereby be studied and even imitated, without being plagiarized. Consequently, the legacy of genius becomes an "economimetic"[73] complex combining original and copy, as well as economic and cultural value creation. This goes back to strategies of naturalization of cultural inheritance without being reducible to them. It is especially in music that the problem of original and copy need not only be argued in terms of elaborate philosophical dialectics, but can be solved, more importantly, on a technical basis, and with a pragmatic understanding of artistic production as *composition*. Even the musical genius Bach himself, being presented as an effect of genealogy and study, may thus be interpreted as a *composed genius*.

18.3 Conclusion

The anthropology of genius as displayed in Forkel's biography of Bach demonstrates that around 1800 heredity was discussed in epistemic and discursive fields where the "natural" and the "cultural" were in close connection with one another. In this respect, Forkel himself seems to be heir to ancient knowledge on hereditary transformations in the broadest sense, rather than a precursor to hereditary biology of genius of the later nineteenth century. For instance, by continuously conflating familial disposition and the labors of study—in both the making and self-making of the musical genius Bach, as well as in his self-reproductions as father and teacher—Forkel perpetuated the early modern nondistinction of heredity and environment ("nature and nurture"). In the same vein, the structure of the family history as unfolded in the Bach biography can be said to represent an ancient narrative model, which had testified to the relative constancy of traits from generation to generation long before that constancy was a scientific issue of heredity.

Nevertheless, it was especially this narrative model that came to a crisis around 1800. This is why the family history told by Forkel means more than just the perseverance of commonsense knowledge. The very concept of genius arouses a protest against this commonsense understanding of family inheritance, because genius claims a kind of absolutism that always goes along with a rupture in familial, genealogical continuity. If it is true that the modern concept of heredity comes into being where the interests for the life of the species and for the life of the individual intertwine,[74] then the meaning of genius for this concept can hardly be overestimated. The trouble area of difference and similarity between individual and species is addressed in the aesthetic debate on genius with its physiological derivation. The emphasis on the "life" of genius and its generative forces, which I have laid open in the first section of this chapter, is likely to establish a logic of genus at the very point where individuality comes into its own: in the highly idiosyncratic and absolutist figure of the *Originalgeist*.

The guiding role of genius for a cultural history of heredity shows up, moreover, in the juridical and economic aspects of the genius discourse. In its pragmatic context, Forkel's narrative of genius as family business was intended to serve the capitalist economy of a publishing company—which, among other things, points to a transfer of copyright from the Bach family to the publishers Hoffmeister and Kühnel. There is no hint in Forkel's book as to how this transfer was actually negotiated, but it was definitely at issue around 1800 whether an author's copyright could or could not be inherited by his descendants, or by the descendants of his "original" publisher.[75] This is where the dialectics of original and copy, which constitutes the aesthetics of genius, plays its crucial part in modern inheritance law. The tendency of nineteenth- and twentieth-century cultural economics to socialize the legacy of geniuses in order to form a "national heritage" was never uncontested by the inherited proprietary rights of the authors.[76] All of these legacy contests converge in the genius/genus debate around 1800 and in its key question, who may claim to be "sui generis," relative to the original genius.

Notes

1. Forkel [1802] 2000, 131, 21–22, 28. Translations are my own in the following, if not indicated otherwise.

2. Galton 1865; 1869.

3. See Flourens 1861; Lombroso 1864; Lange-Eichbaum 1928.

4. Feis 1910; Haecker and Ziehen 1923; Mjöen 1934; Rittershaus 1935.

5. See Spitta 1873–1880; Geiringer 1954.

6. See Terman 1925–1959.

7. Forkel [1802] 2000, 28.

8. J. Grimm and W. Grimm 1897, cols. 3396–3450.

9. J. Grimm and W. Grimm 1897, col. 3411.

10. J. Grimm and W. Grimm 1897, col. 3422–3440.

11. J. Grimm and W. Grimm 1897, col. 3398.

12. J. Grimm and W. Grimm 1897, col. 3397.

13. Willer 2005a; see also Müller-Sievers, chapter 19, this volume; Begemann and Wellbery 2002; Derrida 2003.

14. Wolff [1764] 1966, 37.

15. Wolff [1759] 1966, 13.

16. See Terrall, chapter 11, this volume.

17. Herder [1784–1791] 1989, 270–271. Translation follows Herder [1789–1791] 1968, 20–21.

18. Herder [1784–1791] 1989, 273. For further reflections on the connection between epigenesis and the concept of genius, see Müller-Sievers, chapter 19, this volume.

19. See Schmidt-Dengler 1978.

20. Kant [1790] 1983, 405–406, §46. Translation follows Kant [1790] 1952, 168.

21. See Greiner 2000.

22. Young [1759] 1970, 24.

23. See De Renzi, chapter 3, this volume.

24. Young [1759] 1970, 7.

25. Schiller 1958, 311: "Understanding can repeat what has already been,/ What nature builds, eclectic understanding builds accordingly./ Beyond the bounds of nature, reason builds in emptiness,/ Only you, genius, multiply in nature nature itself." My translation.

26. See Willer 2004, 56.

27. Young [1759] 1970, 11.

28. Jean Paul 1963, 49.

29. Jean Paul 1963, 51.

30. Jean Paul 1963, 63.

31. Jean Paul 1963, 64.

32. Jean Paul 1963, 53–54.

33. Novalis 1978, 329.

34. See Neymeyr 1995.

35. Humboldt 2002, 275. See Leroux 1948; Müller-Sievers 1993, 25–28.

36. Herder [1800] 1998, 836–837.

37. Kant 1983, 405–406, §46.

38. See Weimar 1989, 60.

39. Young [1759] 1970, 13–14. This coincides with David W. Sabean's thesis of a shift from verticality to horizontality in kinship concepts during the eighteenth century. See Sabean, chapter 2, this volume.

40. See Schmidt 1985, 83–128; Meyer-Sickendiek 2001.

41. Forkel [1802] 2000, 28–32.

42. Forkel [1802] 2000, 31–32.

43. These documents are reprinted in the edition of Forkel quoted (pp. 138–166).

44. Forkel [1802] 2000, 33.

45. Forkel [1802] 2000, 161.

46. See Morgenstern [1817] 1988, 45–54.

47. See Minden 1997.

48. Wackenroder 1994, 70.

49. Tieck 1979, 49.

50. See Freud [1909] 1999, 227–231.

51. See Liebrand 1996.

52. Forkel [1802] 2000, 42–43.

53. See the editor's note, Forkel [1802] 2000, 171.

54. Forkel [1802] 2000, 88–89.

55. Forkel [1802] 2000, 86.

56. Forkel [1802] 2000, 91–92.

57. Forkel [1802] 2000, 157: "Die jetzige Generation, qvoad Musicam, artet aus."

58. See Wessely 1967, xxiii; Dahlhaus 1978, 105–108; Naumann 1990, 24–25, 29.

59. Lubkoll 1995, 55.

60. Naumann 1990, 29. At least in his discussion of Bach's compositions, Forkel is quite up to date, because he gives much more space to the instrumental works than to the vocal ones. It was in instrumental music that the metaphysical character of music was presumed. See Lubkoll 1995, 64–69.

61. Schubart 1977, 278–280.

62. Forkel [1802] 2000, 61.

63. Forkel [1802] 2000, 75, 81–82.

64. Forkel [1802] 2000, 36.

65. Forkel [1802] 2000, 85.

66. Forkel [1802] 2000, 88.

67. Forkel [1802] 2000, 83.

68. Forkel [1802] 2000, 35, 63.

69. Forkel [1802] 2000, 132.

70. See Maimon [1795] 1971.

71. Forkel [1802] 2000, 21.

72. Young [1759] 1970, 10.

73. See Jacques Derrida's notion of "économimésis," Derrida 1975.

74. See chapter 1 of this book.

75. See Bosse 1981, 126–129.

76. See Willer 2005b.

References

Begemann, Christian, and David E. Wellbery, eds. 2002. *Kunst—Zeugung—Geburt: Theorien und Metaphern ästhetischer Produktion in der Neuzeit*. Freiburg: Rombach.

Bosse, Heinrich. 1981. *Autorschaft ist Werkherrschaft: Über die Entstehung des Urheberrechts aus dem Geist der Goethezeit*. Paderborn: Schöningh.

Dahlhaus, Carl. 1978. *Die Idee der absoluten Musik*. Kassel/Munich: Bärenreiter/Deutscher Taschenbuch Verlag.

Derrida, Jacques. 1975. Économimésis. In Sylviane Agacinski, Jacques Derrida, Sarah Kofman, Philippe Lacoue-Labarthe, Jean-Luc Nancy, Bernard Pautrat, *Mimésis des articulations*, 55–93. Paris: Flammarion.

———. 2003. *Genèses, généalogies, genres et le génie: Les secrets de l'archive*. Paris: Galilée.

Feis, Oswald. 1910. *Studien über die Genealogie und Psychologie der Musiker*. Wiesbaden: Bergmann.

Flourens, Marie-Jean-Pierre. 1861. *De la raison, du génie et de la folie*. Paris: Garnier.

Forkel, Johann Nikolaus. [1802] 2000. *Über Johann Sebastian Bachs Leben, Kunst und Kunstwerke: Für patriotische Verehrer echter musikalischer Kunst*. Leipzig: Hoffmeister und Kühnel. (New edition by Claudia Maria Knispel. Berlin: Henschel.)

Freud, Sigmund. [1909] 1999. Der Familienroman der Neurotiker. In *Gesammelte Werke*, vol. 7, ed. Anna Freud, 227–231. Frankfurt am Main: Fischer.

Friedrich, Christian. 1977. *Daniel Schubart, Ideen zu einer Ästhetik der Tonkunst*. Leipzig: Reclam.

Galton, Francis. 1865. Hereditary talent and character. *Macmillan's Magazine* 12:157–166, 318–327.

———. 1869. *Hereditary Genius: An Inquiry into Its Laws and Consequences*. London: Macmillan.

Geiringer, Karl. 1954. *The Bach Family: Seven Generations of Creative Genius*. London: George Allen & Unwin.

Greiner, Bernhard. 2000. Genie-Ästhetik und Neue Mythologie: Versuche um 1800, das Neue als Neues zu denken. In Maria Moog-Grünewald, ed., *Das Neue: Eine Denkfigur der Moderne*, 39–52. Heidelberg: Winter.

Grimm, Jacob, and Wilhelm Grimm. 1897. *Deutsches Wörterbuch: Vierten Bandes erste Abtheilung. Zweiter Theil: Gefoppe-Getreibs*, by Rudolf Hildebrand und Hermann Wunderlich. Leipzig: Hirzel.

Haecker, Valentin, and Theodor Ziehen. 1923. *Zur Vererbung und Entwicklung der musikalischen Begabung*. Leipzig: Barth.

Herder, Johann Gottfried. [1784–1791] 1968. *Reflections on the Philosophy of the History of Mankind*. Transl. E. O. Churchill. Chicago: University of Chicago Press.

———. [1784–1791] 1989. Ideen zu einer Philosophie der Geschichte der Menschheit. In *Werke*, vol. 6, ed. Martin Bollacher. Frankfurt am Main: Deutscher Klassiker Verlag.

———. [1800] 1998. Kalligone. In *Werke*, vol. 8, ed. Martin Bollacher. Frankfurt am Main: Deutscher Klassiker Verlag.

Humboldt, Wilhelm von. 2002. Über den Geschlechtsunterschied und dessen Einfluss auf die organische Natur. In *Werke in fünf Bänden*, vol. 1, ed. Andreas Flitner and Klaus Giel, 268–295. Darmstadt: Wissenschaftliche Buchgesellschaft.

Kant, Immanuel. [1790] 1952. *The Critique of Judgement*. Transl. J. C. Meredith. Oxford: Oxford University Press.

———. 1983. Kritik der Urteilskraft. In *Werke*, vol. 8, ed. [1790] Wilhelm von Weischedel. Darmstadt: Wissenschaftliche Buchgesellschaft.

Lange-Eichbaum, Wilhelm. 1928. *Genie, Irrsinn und Ruhm*. Munich: Reinhardt.

Leroux, Robert. 1948. L'esthétique sexuée de Guillaume d'Humboldt. *Études Germaniques* 3:261–280.

Liebrand, Claudia. 1996. *Aporie des Kunstmythos: Die Texte E.T.A. Hoffmanns*. Freiburg: Rombach.

Lombroso, Cesare. 1864. *Genio e follia*. Milan: Chiusi.

Lubkoll, Christine. 1995. *Mythos Musik: Poetische Entwürfe des Musikalischen in der Literatur um 1800*. Freiburg: Rombach.

Maimon, Salomon. [1795] 1971. Das Genie und der methodische Erfinder. In *Gesammelte Werke*, vol. 6, ed. Valerio Verra, 397–420. Hildesheim: Olms.

Meyer-Sickendiek, Burkhard. 2001. *Die Ästhetik der Epigonalität: Theorie und Praxis wiederholenden Schreibens im 19. Jahrhundert*. Tübingen: Francke.

Minden, Michael. 1997. *The German Bildungsroman: Incest and Inheritance*. Cambridge: Cambridge University Press.

Mjöen, Jon Alfred Hansen. 1934. *Die Vererbung der musikalischen Begabung*. Berlin: Metzner.

Morgenstern, Karl. [1817] 1988. Über den Geist und Zusammenhang einer Reihe philosophischer Romane. In Rolf Selbmann, ed., *Zur Geschichte des deutschen Bildungsromans*, 45–54. Darmstadt: Wissenschaftliche Buchgesellschaft.

Müller-Sievers, Helmut. 1993. *Epigenesis: Naturphilosophie im Sprachdenken Wilhelm von Humboldts*. Paderborn: Schöningh.

Naumann, Barbara. 1990. *Musikalisches Ideen-Instrument: Das Musikalische in Poetik und Sprachtheorie der Frühromantik*. Stuttgart: Metzler.

Neymeyr, Barbara. 1995. Das Genie als Hermaphrodit? Schopenhauers ästhetische Fertilitätsmetaphorik und ihr Verhältnis zu Nietzsche. *Zeitschrift für Ästhetik und allgemeine Kunstwissenschaft* 40:199–217.

Novalis. 1978. *Werke, Tagebücher und Briefe*. Vol. 2. Ed. Hans-Joachim Mähl and Richard Samuel. Munich: Hanser.

Paul, Jean. 1963. Vorschule der Ästhetik. In *Sämtliche Werke*, sec. 1, vol. 5, ed. Norbert Miller. Munich: Hanser.

Rittershaus, Ernst Ludwig Johann. 1935. Die Vererbung musikalischer Eigenschaften. *Archiv für Rassenbiologie* 29:132–152.

Schiller, Friedrich. 1958. *Sämtliche Werke*. Ed. Gerhard Fricke and Herbert G. Göpfert. Munich: Hanser.

Schmidt, Jochen. 1985. *Die Geschichte des Genie-Gedankens in der deutschen Literatur, Philosophie und Politik 1750–1945*. Vol. 2. Darmstadt: Wissenschaftliche Buchgesellschaft.

Schmidt-Dengler, Wendelin. 1978. *Genius: Zur Wirkungsgeschichte antiker Mythologeme in der Goethezeit*. Munich: Beck.

Schubart, Christian Friedrich Daniel. 1977. *Ideen zu einer Ästhetik der Tonkunst*. Leipzig: Reclam.

Spitta, Philipp. 1873–1880. *Johann Sebastian Bach*. Leipzig: Breitkopf & Härtel.

Terman, Lewis M., et al., eds. 1925–1959. *Genetic Studies of Genius*. Vols. 1–5. Stanford, CA: Stanford University Press.

Tieck, Ludwig. 1979. *Franz Sternbalds Wanderungen: Eine altdeutsche Geschichte*. Ed. Alfred Anger. Stuttgart: Reclam.

Wackenroder, Wilhelm Heinrich. 1994. Das merkwürdige musikalische Leben des Tonkünstlers Joseph Berglinger. In Barbara Naumann, ed., *Die Sehnsucht der Sprache nach der Musik: Texte zur musikalischen Poetik um 1800*. Stuttgart: Metzler.

Weimar, Klaus. 1989. *Geschichte der deutschen Literaturwissenschaft bis zum Ende des 19. Jahrhunderts*. Munich: Fink.

Wessely, Othmar. 1967. Einleitung. In Johann Nikolaus Forkel, *Allgemeine Geschichte der Musik*. Leipzig: Hoffmeister & Kühnel, reprint Graz: Akademische Druck- und Verlagsanstalt.

Willer, Stefan. 2004. Was ist ein Beispiel? Versuch über das Exemplarische. In Gisela Fehrmann and Eckhard Schumacher, eds., *OriginalKopie: Praktiken des Sekundären*, 51–65. Cologne: DuMont.

———. 2005a. "Eine sonderbare Generation": Zur Poetik der Zeugung um 1800. In Sigrid Weigel, Ohad Parnes, Ulrike Vedder, and Stefan Willer, eds., *Generation: Zur Genealogie des Konzepts-Konzepte von Genealogie*, 125–156. Munich: Fink.

———. 2005b. Heritage-appropriation-interpretation: The debate on the Schiller legacy in 1905. In *A Cultural History of Heredity III: 19th and Early 20th Centuries*, 167–177. Max Planck Institute for the History of Science, preprint 294.

Wolff, Caspar Friedrich. [1759] 1966. *Theoria generationis*. Halle: Litteris Hendelianis, reprint. Hildesheim: Olms.

———. [1764] 1966. *Theorie von der Generation*. Berlin: Birnstiel, reprint. Hildesheim: Olms.

Young, Edward. [1759] 1970. *Conjectures on Original Composition: In a Letter to the Author of Sir Charles Grandison*. Dublin: Wilson, reprint. New York: Garland.

Epilogue

19 The Heredity of Poetics

Helmut Müller-Sievers

At the same time as the life sciences developed a new concept of heredity, poetics, too, confronted the question of how to account for the fixity of its genres, for the variety of its individual expressions, and for the regularity of their tradition. The question first appeared as a pedagogical concern over the transmission of poetological knowledge that had accumulated in the universities, schools, journals, and handbooks over the past three centuries. This concern became ever more pronounced from the beginning to the last quarter of the eighteenth century, when the teaching of literature gradually became an integral part of the university curriculum, and general literacy and public interest had generated an unprecedented number of readers.[1] In the last quarter of the eighteenth and the first decades of the nineteenth century, however, the pedagogy of poetry was eclipsed by a metaphysical inquiry into the essence of poetic genres and the nature of poetic creativity. One example of this shift can be found in pedagogical treatises. The old poetological handbooks sought to advise future authors, often in minute detail, on how to construct a tragedy. The new metaphysical aesthetics instead attempted to define the nature of tragedy and its place in human history and thought.

Poetics, then, followed the same path that many of the contributors to the present book have retraced in scientific discourses. Concentration on the transitive bequeathing and inheritance of definable traits was replaced by an exploration of the intransitive transmission of forms. In the complex semantics of the German *Vererbung*, to which Peter McLaughlin has drawn attention,[2] we could characterize this path as leading from the active meaning involving subject and direct object of *etwas vererben* (to hand something down) to the reflexive voice of *sich vererben* (to engage itself or oneself in the process of inheritance). It is as if the hereditary process had acquired a substance and an agency all its own.

The following pages seek to illustrate this transition, and the reader will find ample occasion, beyond the cited instances, to see connections between developments in the life sciences and the arguments made in poetics and the metaphysics of art. However, the goal of this chapter is not only to document a remarkable contemporaneity of developments but also to put forth a stronger hypothesis: Poetics, and aesthetics more generally, actively facilitated and supported the transition in the sciences from a model that located the cause for the fixity of forms in the will and design of the creator, to a model that presumed the principle of regularity in the hereditary process. One way to conceive of this aesthetic support of scientific developments is to invoke the patron philosophers associated with either pole of the transition: Plato and Aristotle. Plato's conception of a reservoir of immutable forms is clearly behind the deistic idea of created and constant species, as Linnaeus, for example, employed it in his classificatory system.[3] Aristotle, on the other hand, insisted on immanent causes as principles of change and identity in nature. Emerging theories of reproduction and heredity in the late eighteenth and early nineteenth century—for example, the theory of epigenesis—freely employed his metaphysical vocabulary. Indeed, the period can be understood as a return to Aristotelian principles of natural philosophy.[4] However, this shift in the philosophical underpinnings of concepts of inheritance has a curious twist, since in the field of poetics and the philosophy of art we can observe a transition in the opposite direction. Aesthetic debates moved away from the Aristotelian view of poetics as a set of teachable rules and serviceable examples toward a Platonic enthusiasm for the transcendence of poetic forms and the adventitious nature of poetic inspiration. This tendency can be followed in the aesthetic theories and idealist systems of Hegel and, as we will see in this chapter, Schelling, in the projects for a philosophical history of literature in the lectures of the brothers Schlegel, as well as in the reflections of practicing poets like Goethe and Schiller.

What we witness, then, is a crossing over of the argumentative strategies with which scientists and aestheticians sought to explain the regularity and the comprehensibility of forms over time, be they natural species or poetic genres. The life sciences rejected transcendent reasons for the coincidence of regularity and variety that will later be named heredity at the same time that poetics and aesthetics adopted them. Furthermore, I go on to suggest that this chiasm of argumentations served a specific discursive purpose. The Platonism of aesthetics helped the natural sciences to survive outside of the waters of theological transcendence. If it was possible that the forms of poetry and art are governed by principles that, although not theologically motivated, are nonetheless stable, then natural forms might participate in the same kind of source of identity. The discourse of formative drives and plastic forces that emerges in the life sciences in the 1780s with the publication of Johann Friedrich Blumenbach's *Über den Bildungstrieb und das Zeugungsgeschäfte*[5] attempts to participate in the Platonic model of explanation. In fact, the peculiar amalgam of science and aesthetics known as Romantic *Naturphilosophie* enacted that moment of crossover. It compensated for the loss of transcendence in the sciences until separate protocols could gain acceptance in the narrower realms of each of the discourses.[6] The natural sciences moved into the laboratories where forces could be quantified and speculation replaced by experimentation. In contrast, aesthetics and poetics became part of the large historicist projects of literary history and the history of style that were undertaken in the universities.

In the following discussion I present two instances in the development of eighteenth-century poetics in which some of the investments and costs in this chiastic exchange of attributes and approaches come to light. The first is from the last systematic attempt to adapt Aristotelian poetics to the demands of German poetry, Johann Christoph Gottsched's *Versuch einer Kritischen Dichtkunst,* first published in 1730, but continuously enlarged and improved in the following years.[7] The second example is taken from Friedrich Joseph Wilhelm Schelling's lectures on his *Philosophie der Kunst* (Philosophy of Art), first given at Jena in 1802–1803 but not published until in 1859.[8] While Gottsched represents what one might call "preformationist poetics" in their autumnal splendor, Schelling, borrowing freely from his own natural philosophy, seeks to systematize the consequences flowing from his "epigenetic" conception of art.[9] At issue in both texts is the question of how form (genre, species) is bequeathed and inherited.

19.1 Aetas Horatiana

The connection between natural species and the genres of poetry is made right at the beginning of what proved to be the most influential poetological treatise until the end of the eighteenth century, Horace's *ars poetica*:

> If a painter were willing to join to a human head the neck of a horse, and to spread multi-colored feather on limbs collected from just anywhere, so that what began above as a pretty woman ends up horribly in a black fish—would you, dear Pisos, be able to hold back your laughter if you were admitted to such a spectacle? . . . "Painters and poets always had an equal right to dare to do whatever they wanted." We know that, and we demand such license and grant it in return. But not to such a degree that the pleasant mate with the savage, that snakes are coupled with birds, lambs with tigers.[10]

Horace's poem, in typical Roman fashion, comes along as a mixture of explication, legislation, and exhortation: here is what odes, what tragedies are, here is how you should go about writing one, and these are the reasons why. Historical form, active manipulation, and pedagogical intent are all contained in the old (but not Horatian) title *ars*, the felicitous Latin translation of the Greek *techne*. Mundane considerations, like the attention span of spectators under the influence of alcohol, are as important as are reflections on art and mortality. Common sense, the observance of the golden mean (*aurea mediocritas,* Od. II, 10, 5), and a certain rustic simplicity, such is Horace's conviction, will help produce wholesome literary works—just as they help farmers keep their fields fertile and their livestock healthy. This intertwining of agricultural and poetological advice is characteristic of Horace's writing and has contributed a great deal to the naturalization of poetic forms during the many centuries of his influence. As Aristotle had shown in an equally influential treatise, the generation of animals and their special aspect can be explained entirely without recourse to transcendent forms or agents, just by the interplay of four, and more particularly of two immanent causes.[11] Mutatis mutandis—extracting the formal and final causes from the animal body and placing them in the soul of the poet—poetry can be reduced to rules that are not discontinuous with the rules nature herself observes in the generation of animals. Poetry, Aristotle and Horace uphold against certain tendencies in Platonic doctrine and polemic, is neither dependent on mystic visions nor on divine influence, it is neither carried by temporary insanity nor by enthusiasm.[12] Horace, although making due allowances for inspiration and inventiveness, defines his task as that of explicating (although not fully justifying) rules. His is a *Regelpoetik,* as German aestheticians still call it with no small amount of disdain.

While the *ars poetica*—itself a poem composed in the hexameter of the epic—has been ceaselessly read, translated, adapted, commented on, and interpreted, the problems of its internal structure have only recently been clarified. It was no longer evident to the commentators of the nineteenth and early twentieth century what lines of argument Horace might be following, and how the various topics of the poem were related. The desire for exactitude on the part of nineteenth-century philologists—a legacy of their attempt to approximate the standards of the natural sciences—was partially satisfied when it became apparent that Horace followed schemata that were known from traditional handbooks of rhetoric.[13] This collusion of rhetoric and poetics, which was so obvious to preromantic readers as not to merit much attention, would become a crucial factor in the adoption and in the rejection of Horace's poetics. Rhetoric and its rules, it was argued, are defined by the specificity of purpose. Orators want to persuade and convince, be this in the courtroom, in public forums, or in private debate. Poets have no such immediate purpose. Their art seeks to persuade by more vaguely definable means, which ideally contribute to the object that contains a purpose within itself: beauty.

And yet, for Horace, and later for Gottsched, beauty is not born from unfettered originality, to the contrary—that way monsters and horrors lie. Nor is beauty recognized as something formerly espied by the immortal soul during its heavenly sojourn. It is rather the result of imitation and variation: a play with what is given, known, and revisable. Imitation does not only extend to the means of expression, but also to the topics that may become objects of expression. The idea of a tragedy, for example, that would not be based on one of the stories known from the Trojan War is an extreme improbability for Horace, even though he does not exclude it as a possibility (vv. 125–127). Finding the right topic to suit one's linguistic means, one's purpose, and one's intended audience is the labor of *inventio*, which must not be mistaken for the German *erfinden* but rather should be equated with *auffinden*, because the topics are already "there" and only need to be discovered.[14] Variation then "works" with these topics, both by searching for the right perspective from which to present them (Horace's famous advice: "in medias res," v. 148) and by deciding on the proper way they should be dressed and addressed. This latter step is the domain of *elocutio*, with its main categories, figures and tropes, and the endless subcategories in each field. Along with this mating of topic and terms (of *res* and *verba*) goes the convenient and reasonable ordering of the poem, its *dispositio*, which guarantees the regularity of succession and transition, the domain in which logic plays the largest, and variation the smallest role. The remaining two parts of oratory—memorizing and pronunciation (*memoria* and *actio*)—are less important in this context, though they are of course relevant in the training of actors.

The indebtedness of poetics to rhetoric in general, and of Horace's *ars poetica* to Roman adaptations of Greek rhetoric in particular, is buoyed by the conviction that poetry is composition, not creation. Its practitioners are bound to deal with natural, given forms, the rules of producing these forms are learnable, and, while the spark of ingeniousness is necessary to find the right topic, patience and hard work are more essential in producing a laudable specimen (vv. 386–390). For every Roman reader, and for much of the Horatian tradition, it was easily discernible that these compositional rules, and the virtues they required, were to serve as moral guidelines for the incipient poet as well.[15] Included in such work is an understanding of poetic genres as forms generated in a semimythical, yet certainly innerworldly past. Each of the genres has a first inventor, a *protos heuretes* who inaugurated the rules to be employed and the situations in which the poems find their place (Archilochos for the lyric, v. 79; Homer for the epic, v. 74; Thespis for tragedy, v. 276). So un-Platonic are these figures that Horace cracks his well-known joke that at times even Homer seems to have fallen asleep while composing hexameters. Learning genres also means learning to orient poetic productions toward a specific audience, one that might be distracted or haughty, vulgar or just ignorant, but nonetheless a legitimate force in the compositional process. To underscore this craftsmanlike approach to poetry, Horace takes great delight in ridiculing "romantic" and enthusiastic behavior of poets who, struck by divine inspiration, forget to shave and bathe (vv. 295–301). Lest anybody think this is all frivolous and superficial, Horace interrupts his discourse with a half line that shockingly insists on the link between poetry and finitude: "To death we owe ourselves and everything we have" (debemur morti nos nostraque, v. 63).

These low-key assumptions fit perfectly into the distanced, skeptical, and slightly melancholic culture of the European Enlightenment, and Horace remained the undisputed preceptor of style and taste long beyond the 1750s. It is remarkable in our context that aside from questions regarding the origin of genres and species, eighteenth-century preformationism shared key Horatian convictions. These include the idea that forms are essentially fixed, that growth (of an individual, of a poem) is to be understood as the gradual unfolding of a preexisting form rather than as continuous organic innovation, and that the elegance of an explanation has to count as an important criterion in the absence of indisputable truth. Perhaps the most lyrical defender of preformationism at the time, Charles Bonnet, was, for example, quite adamant about the reasons for his theoretical commitments: "I could not bring myself to abandon a theory so beautiful as that of the preexisting germs in order to embrace purely mechanistic explanations."[16] Equally obvious is the delight both Horatian poetics (not so much in the inaugural poem but in the many subsequent reworkings) and preformationist natural history take in classifying, categorizing, registering, and archiving their material. Nonetheless, the difference in their respective views on the origin of the forms that are instanced in the respective products is insuperable. Preformationists cannot get around admitting their transcendent origin directly from the hand of the creator, whereas any eighteenth-century poetaster, while acknowledging the antiquity of traditional genres and their requirements, would certainly shrink from ascribing them to the heavens of ideas, or to divine creation. It is in this question that the crossover between the two discourses occurs. Biology (to use an anachronistic title) will adopt the naturalness and immanence of forms, although with one significant difference from theorists of poetry that is derived from Aristotle's *de generatione animalium* rather than from his *Poetics*. Theorists of biological inheritance will explain the regularity of species forms by affirming the difference between the sexes. The constitutive, rather than the gradual, difference between the sexes requires a spiritual and uniform force to achieve their congress. Opposites have to seek each other out, desire and love one another.[17] Early Romantic poetics, on the other hand, will adopt the transcendence of form, but also with one important exception, later systematized by idealist philosophy. The exception lies in assumption that ideas (forms), without losing any of their universality, are not something that are espied by visionary philosophers, but are embodied in the mind and the products of the poet and artist.[18]

19.2 Gottsched's Rules

Gottsched's *Versuch einer Kritischen Dichtkunst* is an ambitious text, convinced as it is that it can summarize the past and predict the future of poetics, giving rules to future poets as well as to present readers.[19] Its division into two parts, general and special, its paragraphs and subparagraphs, its incessant quotations and references to other authorities all show evidence of the classificatory impulse of mid-eighteenth-century culture. And yet its overriding premise—poetry is, or at least ought to be, the imitation of nature—already points to the concerns of later theorists to unify and "organize" poetry around a single principle independent from, or at least not wholly beholden to, tradition and convention. Testimony to the inner diremption of Gottsched's approach is the double derivation of poetic genres, which occurs once in the general and once in the special sections of the treatise. The general part seeks to derive poetic genres from three basic varieties of imitation, and it does so in an ascending order of difficulty and involvement. First is simple description, then impersonation, and finally the invention of stories ("fables"). To the first correspond those forms of poetry that differ from historical narration only in their rhymed or metered form. Their general impulse is descriptive or ecphrastic, and their ultimate justification lies in the physicotheological desire to praise God through the lively description ("sehr lebhafte Schilderey," p. 142) of even the minutest of his creatures.[20] The second mode of imitation includes all manner of interior representation of affect and action. This is a superior and more involved mode because it requires intimate knowledge of the human heart, and the ability to distinguish between authentic and inauthentic affects.[21] The end point of this mode is dramatic poetry where knowledge of other minds and hearts has to be translated into the representation of consequent action. The third mode of imitation results in the invention of a fully self-contained story, a "fable" as Gottsched calls it. Only at this stage does poetry (*Dichtung*) come into its own. Only in the composition of a fable does the poet construct a counterworld that no longer directly depends on observation of the world.

Gottsched's modes of imitation do not correspond to traditional literary genres, but rather to states of intensity of the imagination. Following the leads of Plato, Aristotle, and Horace, literary scholarship had agreed that there were three main genres under which all works of poetry could be classified: the lyric, the epic, and the dramatic. Historical, psychological, even theological explanations of this tripartite structure abounded, and various attempts were made to establish a stable hierarchy, either according to the poem's purpose, or its provenance, or entirely based on formal characteristics.[22] Gottsched's hierarchy deliberately cut across the genres; he pointed to the famous ecphrases and the dialogues in Homer's epics to show that one genre can contain the others without resulting in the monstrous confusion of which Horace had warned. His explications of the three modes of imitation rather read like an inverted list of the first three "tasks of the orator" (*officia oratoris*). Descriptive poetry is chiefly concerned with the choice of words and the question of aptitude of expression. It is, therefore, an exercise in *elocutio*. Interior representation, because it is mostly the dramatization of a solitary or contentious deliberation, requires experience in shaping an argument so that it may be convincing, and at the same time expressive of a protagonist's character. It is, in short, concerned with the task of ordering, or *dispositio*. Finally, imitation that results in the composition of a complete fictional "fable" is modeled after the rhetorical task of *inventio*, the search for a suitable topic on which to exercise one's rhetorical prowess.

The purpose of the last task becomes clearer when Gottsched specifies that it is not any sort of fable that the poet should invent, but "the narration of a possible, yet not really occurred event under which a useful moral truth is hidden."[23] The process of *inventio* in traditional rhetoric is governed by an underlying purpose. The orator has to ask himself what topic would best suit his purpose and then search for that topic in the storehouse of tradition to which he gains access through his learning and his intimate acquaintance with exemplary works. In Gottsched's theory, the poet's purpose is the moral truth he wants to broadcast by hiding it in the fable. It is for this reason that the central chapter on the three modes of poetic imitation (*Von den drey Gattungen der poetischen Nachahmung und insonderheit von der Fabel*) is preceded by a chapter underscoring the literary training and aesthetic judgment or "taste" of the poet (*Vom guten Geschmacke eines Poeten*). For it is this judgment and taste that allows him to chose the right moral truth and to scan the possible events in which it could be presented. Yet even the finest taste and surest judgment would be in danger of remaining superficial were they not supported by the probity of the poet's character. For it is entirely possible that a poet with perfect technical abilities would chose the wrong topic because of low motifs, such as flattery, dishonesty, or unadulterated wickedness, on the score of which Gottsched had a few candidates in mind. Hence, a chapter on the character of the poet (*Vom Charactere eines Poeten*) opens the systematic part of the *Kritische Dichtkunst*.[24]

The moral rigorousness at the heart of Gottsched's poetics is supposed to contain the endless proliferation of imitation, the lateral sprawl of "literature," which in the eighteenth century—called the "ink-dripping century" by Schiller—threatened to overgrow the field of poetry. For if there is a moral core in poetry and if that core must needs be an expression of the poet's character, mere imitators who only imitate the outside characteristics of genres can easily be spotted and eliminated from consideration. Much like Horace—a translation in rhymed verse of the *ars poetica* serves as the introduction to the *Kritische Dichtkunst*—Gottsched insists on the moral responsibility of the author. In contrast to Horace, Gottsched locates this morality in the propositional core of the work rather than in the poet's character alone. By ordering the genres according to their proximity to this moral core, Gottsched sought to overcome their historical and formal contingency and give them an immanent cause. Such an attempt, however, is ultimately based on a rhetorical presupposition, the presupposition that the procedures for understanding a composed piece of language are simply the reverse of the procedures for formulating it. Dressing a moral precept in a fable, giving it a proper disposition, and decking it out in fine locutions makes sense only if the reader can undress it again without missing anything of its original impact.[25] Just as is the case with orators, Gottsched's poets need to be persuasive if they want to be good, and persuasiveness as he defines it must not be distorted by the irreducible twists of geniality or originality. This is the decisive difference between a poetics of the type Gottsched wanted to preserve, and Romantic hermeneutics, which, with a great deal of help from biblical scholarship, began to rise on the horizon. The latter allows for an authorial subject that, in the end, is accountable to nothing but his own work.

Because Gottsched hoped to introduce peace, once and for all, to the contested terrain of poetry through the obligation to be moral, he provoked, of course, the exact opposite. The *Kritische Dichtkunst* became the target of concerted attacks, which finally resulted in the overthrow of any system that relied on rhetorical rules and moral principles. In these campaigns—first launched from Zürich, but soon engulfing all learned German-speaking elites—generic stability and rhetorical encoding come under fire from the miraculous (*das Wunderbare*), a category that aimed to undermine conventions of both *inventio* and *elocutio*. Using Milton as one of their chief witnesses, "the Swiss" clamored for a much larger margin of freedom in the use of metaphors as well as in the invention of fables. Angels, they contended, not the monsters dreaded by Horace, were the product of a liberated imagination.[26] "The Swiss" argued that morality and history, even the natural world, are restrictive for poetry. The dogma of slavish poetic imitation of these subjects would ultimately be counterproductive if not reinvented and vitalized by an original mind who freely mused upon them a second time.

> Beautiful, Mother Nature, is the wealth of your invention spread across the meadows, more beautiful a happy face which thinks the great thought of your creations a second time[27]

sang Friedrich Gottlieb Klopstock. With the publication of his influential ode *Der Zürchersee* (1750) he would become an important conduit of poetic liberation to the *Sturm und Drang* poets and to the young Goethe. Following Klopstock, the genres especially should no longer be understood as molds into which a poet has to pour words from a limited store of locutions, but rather as occasions to shape a particular authorial mood, to give expression to a particular set of concerns. At the basis of every poem, whether short or long, dramatic or epic, should be the truth not of a proposition but of an individual.

The shift of genre theory away from the mixture of rational laws and historical contingencies toward a basis in authorial sensibility could not be but transitory, given the fragility of the idea of a "generic feeling." The figure to embody this catachresis was the poetic genius—a figure torn between being the passive vessel of transcendent inspiration and the relentless rebel against the sediments of tradition and convention.[28] The 1770s and early 1780s, the years in which the *Genie* rushes onstage in German-speaking countries, are a period of tremendous generic instability and innovation. Such venerable lyrical genres as the ode are charged with new, Promethean energies. Tragedy moves from the court and the drawing room (and from bloodless imitations of ancient models) to the representation of very contemporary emotional dilemmas. Novels begin to take the individuality of their protagonists seriously rather than sending them through set pieces of scenery and conflict. Yet this Storm and Stress–straining against the limits of genres proved, on the whole, to be disastrous, because it propelled authors ever further into uncharted and only subjectively recognizable territory. One indication of this loss of direction comes from the fact that no period in German literature is as easily parodied. Indeed characteristic works of the time like Friedrich Maximilian Klinger's epoch-naming *Sturm und Drang* read today, as they did only a few short years after their publication, like parodies. Only Goethe, with his interest in the natural sciences, his journey to Italy and hence introduction to the classical forms of antiquity, and his infinite power of renewal, was able to make the transition to a new acceptance of forms. Of course, as a practitioner he had no need to systematically argue the reasons behind his stupendous contributions to all major genres, but his insuperable example, paired with that of Schiller, forced theoreticians and admirers to reconsider the status of literary genres.[29]

19.3 The Ideals of Genre

There were many intermediary steps on the way to a systematic reconsideration of genres.[30] Wilhelm von Humboldt, a close friend of Goethe and Schiller and himself an accomplished anatomist and natural philosopher, tried to align the genres with definable moods or temperaments (*Stimmungen*) of the poet.[31] The epic, in Humboldt's scheme, requires a greater admixture of contemplation in the poetic process, whereas tragedy calls for more involvement of the will and other active elements of the soul. Although much more philosophically circumspect than the Swiss theorists' attempt to align genres with sensual feeling, Humboldt's lengthy disquisition still suffers from the fundamental tension in the concept of "generic feeling." He hesitates, as he does in his later philosophy of language, between the power of tradition and the individual's force to resist it.[32]

To overcome this tension, what needs to be shown is that genres are woven so deeply into the essence of human expression that they cannot but manifest themselves in the production of poetry. The scission between a largely contingent history of given forms and the inner necessity of moral conviction or individual feeling has to be eliminated. At the very least, the duality of these two sources of inspiration must be conceived in terms of individual experience, as in the struggle felt by the poet during the composition. The genres, such must be the argument, are necessary forms in which human poetic activity is expressed through history. They have their reason, or ground, in the way the human mind interacts with the world—not, as in previous conceptions, as the world's counterpart or antagonist, but as its part and expression. The reason for the overcoming of antagonism is the discovery that the world itself is mind, which is the basic message of German Idealism. Natural philosophy, aesthetics, religious practice, and ethical concerns—not to mention frustration over the absence of a credible nation-state—contributed to an all-inclusive, all-explaining system based on, or resulting in, unity.[33]

Among the Idealist system builders, Schelling was the first, and the most active, supporter of natural philosophy. Inaugurating full-blown idealist *Naturphilosophie*, Schelling claimed that without an investigation into the meaning of nature in every one of its manifestations, no philosophical system could ever hope to be complete. Significantly, Schelling was also the first of the idealists to conceive a fully articulated philosophy of art, which he viewed as complementary to his philosophy of nature. This relationship between nature and art is by no means strategic or, as Schelling does not tire of repeating, "external." Nature and art are two instances in which the same principle, the same essence, ultimately God, expresses itself in the world. In Schelling's system, history is another expression of this principle.[34]

This approach by Schelling places art and nature on the same level in a process of active creation and therefore sounds the death knell for all theories of art-as-imitation. In order for art to copy nature it is necessary to conceive of nature as holding still long enough to allow for it to be copied, and this in turn fails to realize that artist and nature are of the same dynamic essence.[35] This leads Schelling to decry the excesses of *Sturm und Drang* and elaborate philosophically on the notion of genius, adumbrated by Kant in the *Critique of Judgment* but not developed in all its consequences. Only a philosophy that begins with, and proceeds from, the Absolute can explain how it is that the genius is autonomous—that is, how a genius can give himself laws that are nonetheless not arbitrary.[36] Autonomy, understood literally as the free self-imposition of laws, is the pivotal concept that links works of art and products of nature in all idealist philosophies. Rather than accepting external rules or preformed shapes, art and nature generate their forms from the inside out—they carry the principles of their form within themselves, subjectively as artistic genius or objectively as drive or force. In poetry, these principles are the first, most general instances of the laws of genre, but at this stage of the deduction what is most important to Schelling is that they represent the coincidence, or the "synthesis" of freedom and necessity—a coincidence that, as we will see, is represented in, and by, the genre of tragedy.

To bolster such claims of genuine autonomy in art—incomprehensible as they are for a "merely mechanical understanding" (*Ueber Wissenschaft*, p. 574)—Schelling frequently appeals to the explanatory power and higher prestige of natural philosophy. For example, he argues that the drive to build things, literally the *Bildungstrieb* first introduced by Blumenbach, is an early avatar of the sexual drive, witnessed in the case of what were for Schelling sexless bees and spiders (*Philosophie der Kunst*, p. 402). Schelling's is a philosophy of productivity and generation, but he is quite insistent that there needs to be, as he expresses it (p. 402), an identity between producer and product that prevents the system from spiraling into empty infinity of the *Sturm und Drang* rebellion. A poet gives expression to certain aspects of the universe as much as he himself is an expression of the universe. This dialectic allows Schelling to argue that form, and indeed any kind of limitation of the boundless universe, is not limitation or negation, but measure and affirmation: "At first, you may think that the outlines of a body are a constraint that it suffers; but if you were to look at the generative force you would understand it as a measure, which it imposes on itself, and in which it manifests itself as a truly sensible force."[37]

Through the self-imposition of laws and limitations the dictates of imitation collapse, and art and nature, which is an unconscious artist, reemerge as active and indispensable participants in the drama of the universe. The link between the universal and the particular, between the most general forces of nature and the various species, between language and its instantiation in genres and individual works, needs to be necessary and "organic." This is in contrast to the rhetorical tradition, in which such relations were the result of a "mechanistic" application of linguistic adornment to given topics and forms. Schelling potentially has to be able to account for every genre, and for every artistic product that falls under it. Idealist philosophy of art, like its counterpart, *Naturphilosophie*, has to be able to give the "inner" reasons for, say, a tragedy's existence; it has to be able to explain the choices of composition in *Oedipus Rex*. And Schelling complies with this demand.

Gottsched and the normative critics had understood tragedy as a moral conflict, in which individual shortcomings, however innocent, run afoul of the demands of the world. Gottsched's infamous demand that every "fable," every product of fiction, contain within itself an unambiguous and decipherable moral proposition applied particularly to tragedy. Both in its traditional form with aristocratic or mythical protagonists and in its *bourgeois* incarnation of the 1770s and 1780s, the main purpose of tragedy was to reach out beyond the proscenium. The genre might act as a kind of mirror held up to the aristocracy and to the sovereign, or as a medium to generate compassion and thus galvanize the spectators into a moral community. In both cases, Schelling observes, the goal of tragedy lies outside itself—it serves as a means or medium for ethical, political, or psychological purposes that are heteronomous with respect to it. All of his philosophical instincts went against such a conception of art's subservience, even to the needs of ethics.

All linguistic arts, for Schelling, belong to the "ideal" series of artistic manifestations as opposed to the "real" arts, like sculpture, painting, and music. This is the same division on a more specific level as that between art and nature as the twin manifestations of the absolute in the world. The superiority of poetry over any other art form lies in its heightened degree of creativity—it creates something seemingly out of nothing. That human beings should be capable of this sort of creation, of the same kind of fiat that brought the world into being, shows that the spark of divinity, and thus our substantial relation to God is still active and alive.[38] It is from this perspective that Schelling gives a new reasoning for the traditional division of literary genres.

As we have seen, the division of poetry into lyric, epic, and drama was initially devised and argued for with formal or historical, sometimes with psychological criteria, in mind. In all of these divisions and classifications, the cause for the respective genre predated, as it were, the individual work. Schelling replaces this poetic preformationism with a deduction that purports to show how each genre represents a necessary, organic step in the unfolding of ideas in the world. In lyric poetry, from the Greeks onward, Schelling diagnoses a resistance against the constraints of the world. This kind of poetry (Schelling has Tyrtaios, Archilochos, Pindar, and Sappho in mind) is the medium of rebellion and invective, of desire, furor, and praise. The epic, on the other hand, represents pure necessity, the unchallenged and irresistible flow of fate in which the individual is utterly passive. Discussing these genres, Schelling goes into a good deal of philological detail, showing that the opposition between unbridled freedom and utter necessity also manifests itself in formal aspects such as speaker situation, length, and, above all, meter. It is this way of accounting for seemingly contingent detail in terms of the organic realization of ideas that distinguishes the philosophical aesthetics that Schelling proposes from the normative poetics of earlier times. And these internal reasons, finally, point to the fact that the true synthesis of the subject and world, the altercation between the lyrical individual and epic history, occurs in tragedy.

Since his precocious philosophical beginnings, tragedy has held a special place in Schelling's thought, as it did too in the thought and poetry of his Tübingen roommates Hegel and Hölderlin. In his early *Philosophical Letters on Dogmatism and Criticism* of 1795,[39] Schelling had cited the tragic conflict as an example for the unremitting strife between freedom and necessity. This conflict he understood not as a philosophical problem among others, but as the central struggle of being in and of the world. By the time he gave the lectures on the Philosophy of Art, this ontological turn in the interpretation of the literary genres had become even more pronounced and systematized. Schelling proceeded from the observation that the existence of Greek tragedy would be utterly incomprehensible were it to represent nothing but the preordained demise of an innocent. Gottsched, and many before and after him, had tried to obviate this danger of incomprehensibility by assuming at the heart of tragedy a character flaw, for which the hero is, however exorbitantly, punished. The hero's demise, in short, was a visible demonstration of a moral proposition of the form "thou shalt" or "thou shalt not," and tragedy thus an imitation, even if on the very highest level of abstraction.

In his lectures on *Philosophy of Art,* Schelling sharpened his earlier criticism of imitation in art in general, and in tragic poetry in particular, stating that it would be foolish to call freedom, what is rightly ordained to perish. Freedom would be illusion, rebellion, folly, but not a harbinger of an ingredient essential to the differentiation and development of the universe.[40] In fact, the entire process of unfolding the Absolute in the world, in nature as well as in art, had to be understood as the strife between freedom and necessity, in which both poles have to be preserved in their own right—otherwise it would be a mere spectacle for the eyes of bemused gods. Thus, the only way to represent unalienable freedom in the protagonists is to show them as subject to a cruel fate, and have them accept the punishment, however unjust it might be. It is Schelling's totalizing perspective, his "organic" deduction of tragedy from the overall exigencies of a universe conceived as a continuous process of identity and difference, that leads him to this extraordinary definition of freedom. Only where subjective freedom is renounced, only where it gives itself over to unjust punishment does it preserve itself in its pure form. Tragic freedom clearly is not a freedom to choose in the sense that the freedom of Gottsched's protagonists was understood, but a freedom to be. Furthermore, tragedy is not just one medium in which to represent this freedom as for Gottsched tragedy was one way to present a moral proposition, but the manifestation of this freedom itself.

Similar deductions (in a remarkable anticipation of present theories, Schelling likes to call them "constructions") could be, and are, performed for the other genres, for other art forms, and for other individual works. What the normative poetics from Horace to Gottsched could only explain in a contingent or historical manner, namely the inner relation between human existence and the forms of its artistic expression, Schelling's philosophy exposes as an "organic" and necessary link. In its ultimate consequence this means that artistic expressions of the conflict between fate and freedom—the three literary genres—are themselves not objects of free artistic choice. And since the artist can no longer freely choose how to express freedom, he slips into a position that is very much like that of the tragic hero. He demonstrates his freedom by accepting and submitting to the law of the genre. At this point, the inevitable question for Schelling, and for idealist philosophy in general, is whether or not there can be any true art that is not tragic. Both the internal structure of the systems, and the dispositions of their main exponents—Hölderlin, Hegel, Fr. Schlegel, and Solger to name but a few—suggest that the answer is no. The only way to attenuate the tragic bent of systematic idealism is to stretch it onto a historical frame within which tragic art is finally overtaken and comprehended by tragic philosophy. This, moreover, was the path chosen by Hegel.

19.4 Cultural Exchange

As we have seen, the attacks on normative poetics and its notion of genres as given, technically available, and disposable forms to convey an equally given content, or topic, began in the 1750s and culminated, around 1800, in the replacement of poetics by a philosophy, and later a philosophical history, of art. So thorough was this replacement that the idea of generic artifice in literature would forever vanish, to reappear only in such avant-garde rebellions as the surrealists' *écriture automatique*. The whole pathos of Romantic victim fantasies—the fear of being shackled by forms and conventions, the aversion to real rather than imagined traditions, the desire for originality and immediacy—came to affirm the destruction of poetics, of the idea that literature and poetry could be produced and understood by laying out the rules for their composition. Ultimately, it was the anteriority of forms before content, of genre before individual work, of history before present that the Romantic self could no longer tolerate.

If we align these developments with debates during the transition from preformationist to epigenetic ways of conceiving of generation and generic, obvious similarities suggest themselves. For example, in the often-used argument that preformationism cannot explain family resemblance we can detect the concern with valuing individuality over the regularity of the species. Caspar Friedrich Wolff's arguments against Albrecht von Haller's experiments with chicken eggs ultimately turn on the question of whether there is an essential relation between chick and egg, between form and content.[41] Blumenbach's use of the regenerative powers of polyps as an example for the immanence of generative force in organic processes relates to assertions that the formative powers of literary authors are both unconscious and regular.[42] The thrust in theories of poetics and of organic reproduction pushes in the same direction, and, as the chapters in this book demonstrate, there is a well-researched intellectual and social context that makes this development even more plausible.

And yet, one should keep in mind the chiastic structure of this development. As mentioned, systems of animal preformationism, like that of Charles Bonnet, essentially conceive of genera and species as immutable. In the end, only the fact that they are created guarantees that they are stable and can serve as the basis for classification and teleological explanation. This Christian Platonism stands in contrast to the explicit Aristotelianism of its contemporary poetics, where genres, as we have seen, are simply the accretion of historical experience, even if their origin lies in a mythical past, and their purpose can be explained with reference to the audience's expectation. Conversely, by the end of this development, epigenetic biology had returned to an essentially Aristotelian concept of species, where natural forms are generated by the interplay of fundamentally distinct formative forces. On the other hand, as we have seen in Schelling, poetics had returned to a concept of literary genres as transcendent, immutable, necessary forms (see note 4). This changeover, to be sure, is not perfectly symmetrical, because Romantic biology charged Aristotelian theories of epigenesis with a high degree of sacrality where nature appears as a divine artisan, while the newly enthroned poetic ideals leave a good margin for technical manipulation, for individual endowment by the poet. The technical conception of mechanisms of nature, although fundamentally achieved by making her forms accessible, is as imperfect as the naturalization of poetry, however much desired by practitioners and recipients.

If we take a step back and ask how this episode in literary history fits into the emergence of heredity as an operative concept in the eighteenth century, we can see that literary genres undergo the same process of "nominalization"[43] that can be observed in other areas of knowledge. In the handbooks of poetics, which, like Gottsched's *Versuch einer Kritischen Dichtkunst*, seek to regulate both the production and the reception of literature up until the middle of the eighteenth century, the transmission of essential generic qualities occurred through adjectival attachment. A topic is retrieved from the storehouse of possible propositions and problems and then fitted for dramatic, lyrical, or epic dress, with good judgment (*iudicium*) ensuring that particular topics are matched with particular genres so that any unseemly clash of attributes (Horace's first lines) is prevented. Each literary work is, properly speaking, an example of its genre, exhibiting the genre's traits without substantially sharing or exhausting it. If judiciousness guarantees the individual work's propriety for the respective genre, it is the imperative of imitation that ensures the survival of the genre as such. This includes imitation of recognizable topics, as well as imitation of unsurpassable models. The concept of imitation—always striving for, but, by definition, never achieving perfection—implies that producing an example of a genre can and must be learned. It helps to be inspired and talented, but perseverance, learning, and probity are just as important.

This latter conviction is the first to be overthrown by the new aesthetics that begin to take hold beginning in the late 1750s and are integrated into the idealist systems around 1800. The poetic genius is now seen as actuating the generic qualities either through inspiration or through free acts of generation. This *Sturm und Drang* form of heredity is trimmed into a more predictable shape when philosophers like Schelling conjoin the generative force of the author with an essentially productive conception of genre and species. Every work of art, every successful piece of literature, instantiates the genre completely and exhaustively; there is nothing left, so to speak, that could serve as an object of imitation. The ultimate reason for the full realization of the genre in every work of art lies in the grand vision of all idealist philosophers and aestheticians. They envisaged that the world itself can only be understood as a generative being that manifests itself in everything it contains. This vision obviates the question of how generic traits are transmitted, for work and genre, individual and species are one and the same, both manifestations of a nature that freely represents itself in generic forms. Only because he advocates imitation does Horace fear that art can mix genres and shapes, that it can produce monsters and failures; but, as any idealist interpreter would be quick to point out, he does so in a perfectly harmonious poem.

It is difficult to reconstruct an epoch's cultural economies in order to understand which group commands the greatest cultural authority, which lines of argument and disciplinary constructions appear as most convincing, which venues allow the greatest access to public opinion. This is because in hindsight we tend to favor the factors that are most easily accessible or quantifiable. The difficulty is multiplied in an epoch such as the turn toward the nineteenth century where new cultural and scientific authorities, new discourses, and new media were emerging in rapid succession. The subject matter and the investigations in this book, however, give support to the thesis advanced at the beginning of this chapter. The poetological and aesthetic debates of the last half of the eighteenth and the beginning of the nineteenth century crucially supported the speculative advances in the life sciences that would later coalesce in a theory of heredity as we know it today. This was not only a matter of mutual conceptual borrowings—notions such as "organism," "drive," "naturalness," or "harmony" do double duty in the life sciences and aesthetics—but of a more substantial, analogical support, which ensured that evidence in aesthetics would count in the life sciences and vice versa. Immanuel Kant had made that suggestion in the *Critique of Judgment* of 1790, without, however, fully integrating the aesthetic and scientific vocabularies. That was the impetus behind Romantic *Naturphilosophie*, of which Schelling is a prime exponent. His explanation of the coincidence of generic stability and individual variation in works of art borrows from, and influences in return, similar deliberations in the life sciences. We know that Goethe, Schiller, and the Humboldt brothers (to invoke only some of the most prominent names)—although much less speculatively than Schelling—essentially shared this view of an interpenetration of the natural and the aesthetic. Only when the infusion of aesthetic capital into the life sciences was no longer needed, did the process of differentiation and often violent rejection begin.

Two further aspects of this cultural exchange and its eventual collapse would be of interest in a larger study. First, how was this predominantly German phenomenon translated into the discussions in England and France? The comparative study of the aesthetics and poetics of heredity could provide an interesting means of assessing the emergence of scientific protocols in the European context. Second, one wonders whether it is specifically the question of heredity that attracts interventions from the humanities at various junctures. The debates in the feuilletons of recent years certainly suggest this attraction.

Notes

1. Weimar 2003.

2. See McLaughlin, chapter 12, this volume.

3. See Müller-Wille, chapter 8, this volume.

4. See Löw 1980, 229–232.

5. Blumenbach 1781.

6. This is a complicated process, which was essentially completed by the time of the annual meeting of the Gesellschaft deutscher Naturforscher und Ärzte 1828 in Berlin; see Humboldt and Lichtenstein 1829 as well as Müller-Sievers 2005. Many actors began as *Naturphilosophen*, but later abjured more or less violently their juvenile works. For Johannes Müller, one of the central figures in this process, see Gregory 1992. For a history of the institutional context see Lenoir 1992, 18–52.

7. I will be quoting from the 4th ed. of 1751.

8. I will be quoting from Schelling 1985.

9. For the use of epigenesis and preformation as quasi-aesthetic categories see Müller-Sievers 1997.

10. Q. Horati Flacci *ars poetica,* vv. 1–13, in Horatius Flaccius 1959, my translation. For a translation, commentary, and a history of reception see Hardison and Golden 1995. Unless otherwise noted, all translations are my own.

11. Aristotle 1943, xxxviii–xxxix.

12. Plato 1977, Ion 533c–535a.

13. See Hardison and Golden 1995, 33–40; Steidle 1967.

14. "Those rules of old discovered, not devised / are nature still, but nature methodised" (Alexander Pope, (1711), vv. 88–89).

15. See Oliensis 1998, 198–223.

16. Bonnet [1762] 1985, 22.

17. For the importance of desire and love in the epigenetic thought of Caspar Friedrich Wolff, see Müller-Sievers 1994, 44. Goethe in particular uncovered the codependence of spiritual and

biological generation in his novel *Die Wahlverwandtschaften*; see Müller-Sievers 1997, 137–169. An early testimony to the constitutive function of desire in epigenetic thought is Maupertuis' observation in his *Système de la Nature*: "Il faut avoir recours à quelque principe d'intelligence, à quelque chose semblable à ce que nous appellon désir, aversion, memoire" (quoted in Terral, chapter 11, this volume).

18. I am encouraged by the fact that in his extensive reconstruction of the concept of genre in the eighteenth and nineteenth centuries, Gottfried Willems (1981) also opposes the classicist catalog of (in-)imitable models to a "biologistic" concept of genre emerging at the beginning of the nineteenth century.

19. For a detailed discussion of Gottsched's conception of rhetorical critique see Ruediger Campe's (1990, 9–54) indispensable treatment.

20. Gottsched [1751] 1962, 142.

21. Gottsched [1751] 1962, 145: "Man muß hier die innersten Schlupfwinkel des Herzens ausstudiret, und durch eine genaue Beobachtung der Natur, den Unterscheid des gekünstelten, von dem ungezwungenen angemerket haben."

22. See Combe 1992.

23. Gottsched [1751] 1962, 150: "Die Erzählung einer möglichen, aber nicht wirklich vorgefallenen Begebenheit, darunter eine nützliche moralische Wahrheit verborgen liegt."

24. The very first chapter after the preface is titled "On the Origin and Growth of Poetry" (*Vom Ursprunge und Wachsthume der Poesie*), and gives a very cursory historical overview of the various genres and epochs.

25. For the reversibility of the rhetorical code in Gottsched, and the imminent split between poetics as guidelines for the production of texts (Gottsched) and hermeneutics as guidelines for reading texts (Breitinger), see Weimar 2003, 58–66. For the neurophysiological background discussions surrounding the rise of hermeneutics, see Koschorke 1999, 383–389.

26. For the development of this debate, see Crüger [1884] 1965, I–CI.

27. "Schön ist, Mutter Natur, deiner Erfindung Pracht / Auf die Fluren verstreut, schöner ein froh Gesicht, / das den großen Gedanken / Deiner Schöpfung noch einmal denkt" (Friedrich Gottlieb Klopstock, *Der Zürchersee* (1750) in Klopstock 1962).

28. Still indispensable: Schmidt 1985.

29. Goethe began to reflect intensively on the history of poetry, and of art in general, after his return from Italy. The most directly relevant writings in the present context are first his summary of a debate with Friedrich Schiller, published as *Über epische und dramatische Dichtung*, in Goethe 1977, vol. 14, 367–370, and second his *Noten und Abhandlungen zum besseren Verständnis des West-Östlichen Divans*, in Goethe 1977, vol. 3, especially 480–482 ("Naturformen der Dichtung").

30. There is a very thorough reconstruction of this transition (without special reference to natural philosophy) in Szondi 1974.

31. See Wellbery 2003, especially 711–712.

32. Throughout his career, Humboldt was preoccupied with this dialectic between power (*Macht*) and violence (*Gewalt*). It has its antecedent in Aristotle's distinction between material and formal cause, but found its latest expression in Immanuel Kant's analytic of the sublime: "Power is a faculty that can overcome big obstacles. It is called violence if it overcomes the resistance of that which has power" ("Macht ist ein Vermögen, welches großen Hindernissen überlegen ist. Ebendieselbe heißt Gewalt, wenn sie auch dem Widerstand dessen, was selbst Macht besitzt, überlegen ist") (Kant 1968, 260).

33. The best new reconstruction of the philosophical path from Kantian criticism to the idealist system is Pinkhardt 2002.

34. Friedrich Wilhelm Joseph von Schelling, *Ueber die Wissenschaft der Kunst, in Bezug auf das akademische Studium*, in Schelling 1985, 569–577.

35. Friedrich Wilhelm Joseph von Schelling, *Ueber das Verhältniß der bildenden Künste zu der Natur*, in Schelling 1985, 583–584.

36. *Ueber die Wissenschaft*, in Schelling 1985, 574.

37. *Ueber das Verhältniß*, in Schelling 1985, 593: "Gemeinhin denkst du freilich die Gestalt eines Körpers als eine Einschränkung, welche er leidet; sähest du aber die schaffende Kraft an, so würde sie Dir einleuchten als ein Maß, das diese sich selbst auferlegt, und in dem sie als wahrhaft sinnige Kraft erscheint."

38. In an earlier section, Schelling had said that the real world "is no longer the living word, the speaking of God, but only the spoken—coagulated—word" (*Philosophie der Kunst*, in Schelling 1985, 312).

39. The easiest way to get at the relevant parts of this text is through Peter Szondi's (1978, 157–161) commentary.

40. There is no true strife unless the possibility of victory exists on both sides ("Kein wahrhafter Streit ist, wo nicht die Möglichkeit obzusiegen auf beiden Seiten ist") (*Philosophie der Kunst*, in Schelling 1985, 517).

41. This debate can be followed in Roe 1981. Indispensable for a wider appreciation of the forces in this debate is Bernardi 1986.

42. Blumenbach 1781, 31–32.

43. The argument of a renewed Aristotelianism in late eighteenth-century natural philosophy is also made by Löw 1980, 229–232.

44. This is the process described by López-Beltrán, chapter 5, this volume.

References

Aristotle. 1943. *Generation of Animals*. Loeb Classical Library. Cambridge, MA: Harvard University Press.

Bernardi, Walter. 1986. *Le Metafisiche dell' Embrione: Scienze della Vita e Philosofia da Malpighi a Spallanzani*. Florence: Olschki.

Blumenbach, Johann Friedrich. 1781. *Über den Bildungstrieb und das Zeugungsgeschäfte*. Göttingen: Dietrich.

Bonnet, Charles. [1762] 1985. *Considerations sur les corps organisés*. Paris: Fayard.

Campe, Ruediger. 1990. *Affekt und Ausdruck: Zur Umwandlung der literarischen Rede im 17. und 18. Jahrhundert*. Tübingen: Niemeyer.

Combe, Dominique. 1992. *Les genres littéraires*. Paris: Hachette.

Crüger, Johannes, ed. [1884] 1965. *Joh. Christoph Gottsched und die Schweizer Joh. J. Bodmer und Joh. J. Breitinger*. Darmstadt: Wissenschaftliche Buchgesellschaft.

Goethe, Johann Wolfgang. 1977. *Sämtliche Werke*. Zürich: Artemis.

Gottsched, Johann Christoph. [1751] 1962. *Versuch einer Kritischen Dichtkunst*. 4th ed. Reprint. Darmstadt: Wissenschaftliche Buchgesellschaft.

Gregory, Frederick. 1992. Hat Müller die Naturphilosophie wirklich aufgegeben? In Michael Hagner and Bettina Wahrig-Schmidt, eds, *Johannes Müller und die Philosophie*, 143–154. Berlin: Akademie-Verlag.

Hardison, O. B., and Leon Golden. 1995. *Horace for Students of Literature*. Gainesville: University Press of Florida.

Horatius Flaccius, Quintus. 1959. *Q. Horati Flacci Opera*. Ed. E. Wickham. Oxford: Oxford University Press.

Humboldt, Alexander von, and Heinrich Lichtenstein. 1829. *Amtlicher Bericht über die Versammlung deutscher Naturforscher und Ärzte zu Berlin*. Berlin: Trautwein.

Kant, Immanuel. 1968. Kritik der Urtheilskraft. In *Kants Werke*. Akademie-Textausgabe, vol. V. Berlin: de Gruyter.

Klopstock, Friedrich Gottlieb. 1962. *Ausgewählte Werke*. Munich: Hanser.

Koschorke, Albrecht. 1999. *Körperströme und Schriftverkehr*. Munich: Fink.

Lenoir, Timothy. 1992. *Politik im Tempel der Wissenschaft*. Frankfurt am Main: Campus.

Löw, Reinhard. 1980. *Philosophie des Lebendigen*. Frankfurt am Main: Suhrkamp.

Müller-Sievers, Helmut. 1974. *Epigenesis: Naturphilosophie im Sprachdenken Wilhelm von Humboldts*. Paderborn: Schoeningh.

———. 1997. *Self-Generation: Biology, Philosophy, and Literature around 1800*. Stanford, CA: Stanford University Press.

———. 2005. Ablesen: Zur Genealogie des wissenschaftlichen Blickes. In Bernhard J. Dotzler and Sigrid Weigel, eds, *"Fülle der combination": Literaturforschung und Wissenschaftsgeschichte*, 305–318. Munich: Fink.

Oliensis, Ellen. 1998. *Horace and the Rhetoric of Authority*. Cambridge: Cambridge University Press.

Pinkhardt, Terry. 2002. *German Philosophy, 1760–1860: The Legacy of Idealism*. Cambridge: Cambridge University Press.

Pope, Alexander. 1928. *An essay on criticism*. San Francisco: Clark.

Plato. 1977. *Werke in acht Bänden, griechisch-deutsch*. Vol. 1. Darmstadt: Wissenschaftliche Buchgesellschaft.

Roe, Shirley A. 1981. *Matter, Life, and Generation: Eighteenth-Century Embryology and the Haller-Wolff Debate*. Cambridge: Cambridge University Press.

Schelling, F. W. J. 1985. *Ausgewählte Schriften*. Vol. II. Frankfurt am Main: Suhrkamp.

Schmidt, Jochen. 1985. *Die Geschichte des Genie-Gedankens in der deutschen Literatur, Philosophie und Politik 1750–1945*. Vol. I. Darmstdt: Wissenschaftliche Buchgesellschaft.

Steidle, Wolf. 1967. *Studien zur Ars Poetica des Horaz*. Hildesheim: Olms.

Szondi, Peter. 1974. *Poetik und Geschichtsphilosophie II*. Studienausgabe der Vorlesungen, vol. 3. Frankfurt am Main: Suhrkamp.

———. 1978. Versuch über das Tragische. In *Schriften I*, 149–260. Frankfurt am Main: Suhrkamp.

Weimar, Klaus. 2003. *Geschichte der deutschen Literaturwissenschaft*. Paderborn: Fink.

Wellbery, David. 2003. Stimmung. In *Ästhetische Grundbegriffe*, vol. 5, 703–733. Stuttgart: Metzler.

Willems, Gottfried. 1981. *Das Konzept der literarischen Gattung*. Tübingen: Niemeyer.

Contributors

Laure Cartron is lecturer at Fudan University, Shanghai, China. She is working on a thesis on the spreading of the concept of pathological heredity in regard to the changes in French society in the first half of the nineteenth century at the Institut d'histoire et de philosophie des sciences et des techniques, University of Paris I (Panthéon-Sorbonne).

Silvia De Renzi is lecturer in history of medicine at the Open University, Milton Keynes (UK). She works on medicine and natural history in seventeenth-century Rome and contributed several chapters to P. Elmer, ed., *The Healing Arts: Health, Disease and Society in Europe (1500–1800)* (Manchester: Manchester University Press, 2003). Among her recent articles are "Witnesses of the Body: Medico-Legal Cases in Seventeenth-Century Rome," *Studies in History and Philosophy of Science* 33A (2002): 219–242. She is the author of *Instruments in Print: Books from the Whipple Collection* (Cambridge: Whipple Museum for the History of Science, 2000).

François Duchesneau is Professor of History and Philosophy of Science at the Université de Montréal. His main scholarly contributions are *L'Empirisme de Locke* (The Hague: M. Nijhoff (Springer), 1973); *La Physiologie des Lumières: Empirisme, modèles et théories* (The Hague: M. Nijhoff (Springer), 1982); *Genèse de la théorie cellulaire* (Montréal: Bellarmin; Paris: Vrin, 1987); *Leibniz et la méthode de la science* (Paris: PUF, 1993); *La Dynamique de Leibniz* (Paris: Vrin, 1994); *Philosophie de la biologie* (Paris: PUF, 1997); *Les Modèles du vivant de Descartes à Leibniz* (Paris: Vrin, 1998).

Carlos López-Beltrán is Professor of Philosophy at the National Autonomous University of Mexico, Mexico City. He is author of *El Sesgo Hereditario: Ámbitos Históricos del Concepto de Herencia Biológica* (Mexico City: Universdad Nacional Antonónoma de México, Estudios sobre la Ciencia, 2004) and of numerous articles on the early history of the concept of heredity, including "In the Cradle of Heredity: French Physicians and 'l'hérédité naturelle' in the Early Nineteenth Century," *Journal of the History of Biology* 37 (2004): 39–72.

Renato G. Mazzolini is Professor of the History of Science at the University of Trento, Italy. He is the author of *The Iris in Eighteenth-century Physiology* (Bern: Huber, 1980), and (with Shirley A. Roe) *Science against the Unbelievers* (Oxford: Voltaire Foundation, 1986), as well as editor of *Non-Verbal Communication in Science prior to 1900* (Florence: Olschki, 1993).

Peter McLaughlin is Professor of Philosophy at the University of Heidelberg, Germany. He works on philosophy of science and on the relations between science and philosophy in the early modern period. His major publications are *Kants Kritik der teleologischen Urteilskraft* (Bonn: Bouvier, 1989); *What Functions Explain* (New York: Cambridge University Press, 2001); and with P. Damerow, G. Freudenthal, J. Renn *Exploring the Limits of Preclassical Mechanics* (New York: Springer, 2004).

Helmut Müller-Sievers is Professor of German, Comparative Literature, and Classics at Northwestern University in Evanston, Illinois. He works on the intersection of literature, philosophy, and the history of science. He has published *Epigenesis: Naturphilosophie im Sprachdenken Wilhelm von Humboldts* (Paderborn: Schoeningh, 1994), *Self-Generation: Science, Philosophy, and Literature around 1800* (Stanford: Stanford University Press, 1997), and *Desorientierung: Anatomie und Dichtung bei Georg Büchner* (Göttiugen: Wallstein, 2003).

Staffan Müller-Wille is research fellow at the ESRC Research Centre for Genomics in Society, University of Exeter, UK. He received his PhD in philosophy from the University of Bielefeld, Germany. He is working on the cultural history of heredity with a focus on classical genetics and structuralist anthropology. He is author of a book on Carolus Linnaeus's taxonomic philosophy (*Botanik und weltweiter Handel*, Berlin: VwB 1999). More recent publications are "Early Mendelism and the Subversion of Taxonomy: Epistemological Obstacles as Institutions," *Studies in History and Philosophy of Biological and Biomedical Sciences* 36 (2005): 465–487, and "Gene," in *The Stanford Encyclopedia of Philosophy*, winter 2004 edition, http://plato.stanford.edu (with Hans-Jörg Rheinberger).

Ohad Parnes is research fellow at the Centre for Literary and Cultural Research, Berlin, Germany. His major teaching and research interests focus largely on the history of biology and of medicine in the modern era. Recent research projects include the history of the notion of specificity in modern biomedicine and the history of autoimmune disease. He is currently working on a book-length study on the history of the inheritance of acquired characters from the nineteenth century to the present.

Nicolas Pethes is Professor of European Literature and Media Studies at the FernUniversität Hagen, Germany, and senior researcher on the cultural history of human experimentation at the University of Bonn, Germany. He is coeditor of *Gedächtnis und Erinnerung: Ein interdisziplinäres Lexikon* (Reinbek: Rowohlt, 2001). His major publications are "Terminal Men: Biotechnological Experimentation and the Reshaping of 'the human' in Medical Thrillers," *New Literary History* 36 (2005): 161–185, and *Zöglinge der Natur: Der literarische Menschenversuch des 18. Jahrhunderts* (Göttingen: Wallstein, 2007).

Marc J. Ratcliff is researcher at the Institut d'Histoire de la Médecine et de la Santé of the University of Geneva. He has published extensively on the history and epistemology of natural history. His recent article "Abraham Trembley's Strategy of Generosity and the Scope of Celebrity in the Mid-Eighteenth Century," *Isis* 95 (2004): 555–575, won the 2005 Derek Price/Rod Webster Award from the History of Science Society.

Hans-Jörg Rheinberger is a Scientific Member of the Max Planck Society and Director at the Max Planck Institute for the History of Science in Berlin. He has published numerous articles on molecular biology and the history of science. Among other books, he has written *Experiment, Differenz, Schrift* (Marburg: Basilisken-Presse, 1992), *Toward a History of Epistemic Things: Synthesizing Proteins in the Test Tube* (Stanford: Stanford University Press, 1997), and *Epistemologie des Konkreten* (Frankfurt: Suhrkamp, 2006). With P. Beurton and R. Falk he edited *The Concept of the Gene in Development and Evolution* (Cambridge: Cambridge University Press, 2000), and with Jean-Paul Gaudillière *The Mapping Cultures of Twentieth Century Genetics* (2 vols., London: Routledge, 2004).

David Warren Sabean is the Henry J. Bruman Professor of German History at the University of California, Los Angeles. His research centers on kinship and family in Europe, and he is currently engaged in a project on incest discourse in Europe and America from 1600 to the present. His major publications are *Property, Production and Family in Neckarhausen, 1700–1870* (Cambridge: Cambridge University Press, 1990); *Kinship in Neckarhausen, 1700–1870* (Cambridge: Cambridge University Press, 1998); "Inzestdiskurse vom Barock bis zur Romantik," *L'Homme: Zeitschrift für feministische Geschichtswissenschaft* 13 (2002): 7–28.

Mary Terrall is Associate Professor of History at the University of California, Los Angeles. She is the author of *The Man Who Flattened the Earth: Maupertuis and the Sciences in the Enlightenment* (Chicago: University of Chicago Press, 2002), which won the Pfizer Prize from the History of Science Society and the Gottschalk Prize from the American Society for Eighteenth-Century Studies. Her current research concerns the practice of natural history in the circles around Réaumur in the middle decades of the eighteenth century.

Ulrike Vedder is research fellow at the Centre for Literary Research, Berlin, Germany, where she is working in the project *Erbe, Erbschaft, Vererbung: Überlieferungskonzepte zwischen Natur und Kultur im historischen Wandel*. With Sigrid Weigel, Stefan Willer, and Ohad Parnes, she has edited *Generation: Zur Genealogie des Konzepts—Konzepte von Genealogie* (Munich: Fink 2005), and is author of *Geschickte Liebe: Zur Mediengeschichte des Liebesdiskurses im Briefroman "Les Liaisons dangereuses" und in der Gegenwartsliteratur* (Cologne/Weimar: Böhlau 2002).

Paul White is a Research Associate at the University of Cambridge. He teaches in the Department of History and Philosophy of Science and is an editor of *The Correspondence of Charles Darwin* (Cambridge: Cambridge University Press). He is the author of *Thomas Huxley: Making the "Man of Science"* (Cambridge: Cambridge University Press, 2003), and articles on the relationship between science and literature in nineteenth-century culture.

Stefan Willer is research fellow at the Centre for Literary Research, Berlin, Germany, where he is working in the project *Erbe, Erbschaft, Vererbung: Überlieferungskonzepte zwischen Natur und Kultur im historischen Wandel*. He is author of *Botho Strauß zur Einführung* (Hamburg: Junius, 2000) and *Poetik der Etymologie: Texturen sprachlichen Wissens in der Romantik* (Berlin: Akademie Verlag, 2003), and has coedited *Generation: Zur Genealogie des Konzepts—Konzepte von Genealogie* (Munich: Fink, 2005).

Philip K. Wilson is a historian of medicine and Associate Professor in the Department of Humanities at Penn State's College of Medicine in Hershey, Pennsylvania. He edited the five-volume series *Childbirth: Changing Ideas and Practices in Britain and America, 1600 to the Present* (New York: Garland, 1996). Other publications include *Surgery, Skin & Syphilis: Daniel Turner's London (1667–1741)* (Amsterdam: Rodopi Press, 1999), as well as chapters in *The Genius of Erasmus Darwin* (Aldershot: Ashgate, 2005); *Medicine and the History of the Body* (Tokyo: Ishiyaku EuroAmerica, 1999); *The Healing Arts: The Story of Medicine* (London: Ivy Press and Marlowe & Co, 1997); and *Venereal Disease in Eighteenth-Century France and Great Britain* (Lexington: University Press of Kentucky, 1996).

Roger J. Wood is in his retirement from an academic career at the University of Manchester, where he now holds an Honorary Readership in Genetics. As an experimentalist and theoretician, he specializes in the genetics of insect pests and vectors of disease. He is also known for historical research on the concept of heredity in relation to animal breeding. Recent publications include *Genetic Prehistory in Selective Breeding: a prelude to Mendel* (Oxford: Oxford University Press, 2001); "Scientific Breeding in Central Europe during the Early Nineteenth Century: Background to Mendel's Later Work," *Journal of the History of Biology* 38 (2005): 239–272; and "Essence and Origin of Mendel's Discovery," *Comptes Rendus de l'Académie des Sciences Paris, Sciences de la vie* 323 (2000): 1037–1041 (all coauthored with with V. Orel).

Index

Académie de Médecine, 159
Académie des Sciences, 207
Acquired characters, 4, 24, 124, 145, 198, 278, 281, 326. *See also* Character
  faculties and, 408–412
Adanson, Michel (1727–1806), 209, 212–213, 217, 221
Adaptation, 3–4
  Kant on, 280–288
  materialism and, 278–279
"Addition à l'article qui a pour titre, Variétés dans l'espèce humaine" (Buffon), 350
Adultery
  Erasmus Darwin and, 144
  illegitimacy and, 63–66
  Ius Commune and, 63–64
  Ius Proprium and, 63–64
  legal issues and, 61, 63–66
Adventures of Humphrey Clinker, The (Smollett), 137
Adventures of Peregrine Pickle, The (Smollett), 136–137
Aesthetics. *See* Poetics
Affinity. *See* Kinship
African blacks
  castas and, 355–366
  social issues and, 350–355
Agnosticism, 260
Albinism, 266
Alcoholism, 133, 135–136, 384–389
Alienism
  as mental medicine, 162–164
  moral causes and, 162
  social class and, 162–163
  statistics and, 163–167
Alliance, 186
  Delille and, 47–49
  principle of (Foucault), 29, 186
  property and, 41, 45–47
Allioni, Ludovico (1728–1804), 213
Amoreux, Pierre-Joseph (1741–1824), 112–114, 116, 118, 120–122
Anatomical Disputation Concerning the Movement of the Heart and Blood in Living Creatures (Harvey), 179
Anatomical Exercises on the Generation of Animals (Harvey), 3, 179
Anderson, James (1739–1808), 236
Animals
  alternation of generations and, 319–323
  blood circulation in, 179–180
  breeders and, 20 (*see also* Breeding)
  character and, 376–381
  generation theories and, 265–270
  Harvey and, 179–180
  horses, 377
  livestock farmers and, 209–210, 214–215, 377–381
  microscopy and, 259–261, 265
  natural faculties and, 69–71
  Paracentrotus lividus, 304
  poultry, 381–382

Animals (cont.)
race concept and, 215
Salpa, 319
sheep, 229–241
Anlagen (dispositions), 17, 240, 306–312, 325
Annales d'hygiène publique et de médecine légale, 165
Annales médico-psychologiques, 165
Anthropology, 12, 15
castas and, 350–360
caste and, 360–366
genius and, 419–434
savage children and, 22–23, 399–413
Apoplexy, 122
Arbitratiis iudicum quaestionibus, De (Menochio), 65
Archilochos (c.680 BC–c.645/630 BC), 447, 456
Aristocracy. *See* Social class
Aristotle (384 BC–322 BC), 6, 42, 62, 110
genealogy and, 177–178, 180
poetics and, 444, 458–459
women's role in generation and, 181
Aristotle's Masterpiece, Domestic Medicine (Anonymous), 141
Armand Velpeau, Alfred Louis (1795–1867), 155
Arnim, Achim von (1781–1831), 90–93, 97
Ars poetica (Horace), 445–449
Association of Friends, Experts and Supporters of Sheep Breeding, The, 238–241
Assyria, 66
Astral influences, 3–4
Atavism, 119, 124, 315
Austria, 44, 46
Aztecs, 353

Bach, Carl Philipp Emanuel (1714–1788), 427–429
Bach, Johann Sebastian (1685–1750)
as cultural icon, 432–434
education of, 430–431
genealogy of, 426–430
genius of, 419–420, 426–434
as mentor, 431
Bacteriology, 146–147
Baer, Karl Ernst von (1792–1876), 155, 304, 315
Baillarger, Jules (1809–1890), 165
Bakewell, Robert (1725–1795)
character and, 378, 388–389
sheep breeding and, 232–239
Balzac, Honoré de (1799–1850), 88–89, 93–97, 158
Barsanti, Giulio, 207
Bartenstein, E., 239, 241
Bartolo da Sassoferrato (1313–1357), 65–66
Bastardbefruchtung im Pflanzenreich (Wichura), 331
Bates, Thomas (1775–1849), 377
Battista, Giovanni, 61, 68–69
Baumann, Christian (1739–1803), 238
Belgium, 50–51
Belvisio, Jacopo de (1270–1335), 65
Berengher, Francesco, 352
Bernard, Claude (1813–1878), 21–22, 295
Bernoulli, Daniel (1700–1782), 163
Besitzstand (acquired state), 325
Bestreben (tendency), 328–329
Bésuchet (b. 1754), 117
Beurton, Peter, 8
Bibliotheca botanica (Haller), 217
Bicêtre Hospital, 163–165
Bildungstrieb (vital force), 21
cell theory and, 294–295, 297, 301–302, 310–311
poetics and, 454
Biology, 15
alternation of generations and, 319–323
cell theory and, 293–312
disease and, 111–123 (*see also* Disease)
Enlightenment and, 315–336
generation concept and, 20–21
inherent design and, 328
Kant on, 280–288
organization concept and, 20–22

Biometrics, 389–390
Blastemic crystallization, 293
Blending, 14–15, 240
Blood, 62, 115, 118, 123, 142, 147, 168, 181, 204, 296, 376
  breeding and, 229–239
  circulation of, 69, 179–180, 269
  and kinship, 9, 11, 41–54, 87, 188
  as metaphor, 41, 45
  race and, 354–359
  "unity of flesh and blood" and, 12
Blood line, 142–143, 230–232, 237
Blumenbach, Johann Friedrich (1752–1840), 21, 121
  adaptation and, 279
  castas and, 351–352, 354
  cell theory and, 294–295, 308
  poetics and, 444–445, 458
  race and, 362–365
Blunt, Lady Anne (1837–1917), 389
Blunt Lytton, Judith (1873–1957), 389
Blunt, Wilfrid (1840–1922), 389
Bonnaterre, Pierre-Joseph (1747–1804), 403–404
Bonnet, Charles (1720–1793), 117, 264–265, 352, 448, 458
Bonpland, Aimé (1773–1858), 353
Bory de Saint-Vincent, Jean-Baptiste (1778–1840), 364
Botanical gardens, 20–21
Botanists. *See* Breeding
Boulton, Matthew (1728–1809), 146
Bourgeoisie. *See* Social class
Bowler, Peter, 8
Boyle, Robert (1627–1691), 383
Brahmins, 363
Braun, Alexander (1805–1877), 321–322, 328–329, 331
Breeding, 12, 15, 20–21, 145, 253
  adaptation and, 277–288
  albinism and, 266
  Bakewell and, 232–239
  botanists and, 205–222
  character and, 376–381
  Charles Darwin and, 188–189
  color inheritance and, 209–210
  Culley and, 231–232, 235–236
  Duchesne and, 205–207, 210, 212–221
  environmental influences and, 230–232, 235
  eugenic charts and, 387–389
  evolutionary theory and, 212
  fixist system and, 212
  generation theories and, 265–270
  grafting and, 210
  growers' practices and, 207–211
  hybridization and, 205–210, 214, 221–222, 268, 331, 334
  improvement of pure race and, 240
  inbreeding and, 232–234, 237, 239
  individual trait selection and, 232–234
  interracial crossing and, 349–350
  "like begets like" and, 17, 105, 143–144, 177–178, 229–230
  Linnaeus and, 211–212
  livestock, 377–381
  locality and, 230–232
  managing varieties and, 211–213
  Meiners and, 361–363
  Mendel and, 3, 8, 19, 193, 240–241, 321, 335–336
  morphological changes and, 207–211
  naturalists and, 194
  Napp and, 240–241
  natural selection and, 381
  nobility and, 376–381
  pedigrees and, 331, 334, 376–381, 389
  polydactyly and, 265–266
  population thinking and, 234–236
  poultry, 267–268, 381–382
  powers of discrimination and, 381–383
  prepotency and, 237
  progeny testing and, 232–234
  race concept and, 236–237
  Ray's axiom and, 215–216
  sheep and, 229–242

Breeding (cont.)
species and, 236–237 (*see also* Species)
Suffolk Stud Book and, 377
Broca, Pierre-Paul (1824–1880), 355
Broussais, François-Victor (1772–1838), 166
Brown, John (1735–1788), 138–139
Brücke, Ernst Wilhelm von (1819–1892), 305
Buchan, William (1729–1805), 140–146
Buchez, Philippe-Joseph-Benjamin (1796–1866), 169
Buds, 22
Buffon, Georges-Louis Leclerc Comte de (1707–1788), 14, 20, 157, 192–193, 221
adaptation and, 278, 282
castas and, 352, 354, 359–360
epigenesis and, 262–264
experiment and, 259, 265, 268–269, 399
generation theories and, 257, 259–260
internal molds and, 259–263
interracial crossing and, 350
law of succession and, 95
materialistic theories of, 279
race concept and, 215, 282, 364–365
Burdach, Karl Friedrich (1776–1847), 307

Cabanis, Pierre-Jean-Georges (1757–1808), 111, 158, 161–162
Cadogan, William (1711–1797), 139–140
Calculus, 162–163
Caldani, Leopoldo Marc'Antonio (1725–1813), 352
Calvinism, 387
Campanella, Tommaso (1568–1639), 69
Camper's facial angle, 364–365
Cancer, 298
Canon law, 63–64
Caractères physiologiques des races humaines considérés dans leurs rapports avec l'histoire, Des (Edwards), 363–364
Caritat, Jean-Antoine-Nicolas de (Marquis de Condorcet) (1743–1794), 318
Carlson, Elof Axel, 8
Carlyle, Thomas (1795–1881), 383
Carol, Ann, 157
Carpenter, William (1813–1885), 320
Cartron, Laure, 125
Carus, Victor (1823–1903), 326–328
Castas, 14–15, 23
caste and, 360–365
concept of, 355, 358
historical caste systems and, 360–361
prejudice and, 359–360
Causes morales et physiques des maladies mentales, Des (Voisin), 167
Celibacy, 64
Cell in Development and Inheritance, The (Wilson), 307
Cell theory, 22, 305–310
attractive forces and, 297
Bildungstrieb and, 294–295, 297, 301–302, 310–311
circle of variation and, 306
cytoblastema and, 293, 295–305
differential proliferation and, 296–297
dispositions and, 308–312
establishment of, 293–294
holistic approach and, 293
Lebensprincip theory and, 298, 301, 310–311
materialism and, 305
Müller and, 298–307, 310–312
multiplying by division and, 309
as reductionism, 294–298
Remak and, 293, 304–305, 307–308, 311
Schwann and, 293–305, 310
type replication and, 298–305
Virchow and, 293, 304–308, 311
Cellularpathologie in ihrer Begründung auf physiologische und pathologische Gewebelehre, Die (Virchow), 304
Celtic race, 362
Century of the Gene, The (Fox Keller), 8
Chamisso, Adalbert von (1781–1838), 319–320
Character
alcoholism and, 384–389
breeding and, 376–381

causal role and, 375
diseases of, 384–387
as essential quality, 375
eugenic self-fashioning and, 387–390
hereditary genius and, 383–384
innate capacity and, 376
pedigrees and, 376–381
powers of discrimination and, 381–383
self-made man and, 376–381
social class and, 376
Suffolk Stud Book and, 377
"Characteristics" (Carlyle), 383
Children, 144–145, 161
illegitimacy and, 63, 366
mental health and, 168–169
naturales, 63–64
savage, 22–23, 319–413
spurii, 63–64
Church
caste systems and, 363
France and, 163
Gregorian reforms and, 9
inheritance and, 9
interracial crossings and, 351
pilgrims and, 64
Circulation, 24, 40
of blood, 69, 179–180, 269
Harvey and, 179
Linnaeus and, 185
Clan
affinity and, 45–51
agnatic relationships and, 38
alliance and, 41, 45–51
blood ties and, 41–42, 45
common ancestor and, 44–45
hierarchalization and, 45
incest and, 51–54
patrilines and, 38–39
primogeniture and, 38
property inheritance and, 37–45
vertical links and, 49
Code civile, 11, 53–54
Cohn, Ferdinand (1828–1898), 297
Colling, Charles (1751–1836), 378, 380
Colling, Robert (1749–1820), 378, 380
Collini, Stefan, 375
Colonialism, 141, 179, 190, 194, 282, 400
epistemic space and, 14, 17, 22
race and, 355, 358, 361, 364
Comprehensive Course of Agriculture (Rozier), 211
Comte, Auguste (1798–1857), 323–324
Comte, François-Charles-Louis (1782–1837), 364
Concept of the Gene in Development and Evolution, The (Beurton, Falk, and Rheinberger), 8
Concubines, 64
Condillac, Étienne Bonnot (1715–1780), 405–408, 412
Congregation of Orders, 44
Conjectures on Original Composition (Young), 423–424, 426, 432
Consanguinity, 9–10
Conseil des Cinq Cents, 162
Considération sur la nature et le traitement des maladies héréditaires ou maladies de familles (Portal), 160
Constant varieties, 193
Consumption, 19, 122, 133
Contrat de mariage, Le (Balzac), 93–96
Copy, 421–424, 432–434
Corbichon, Jean de, 217
Courtet de l'Isle, Victor (1813–1867), 364
Cousin marriage, 47–51
Creoles, 353
Cretinism, 158
Criollos, 358–359
Critique of Judgment (Kant), 423, 425–426, 454, 460
Cuba, 353
Cullen, William (1710–1790), 134, 136, 138
Culley, George, 231–232, 235–236
Cuvier, Georges (1769–1832), 94, 364–365

d'Ardène, Jean-Paul de Rome (1689–1769), 209–210, 218

Darwin, Charles (1809–1882), 12, 22, 24, 133
adaptation and, 277
argument from design and, 279
breeding and, 188–189, 381–382
cell theory and, 308
community of descent and, 186–190
divergence of character and, 189–190
Galton's critique of, 384
genealogy and, 186–191, 195
generation and, 186–190
geological formations and, 190
gout and, 138
natural selection and, 381
social class and, 382–383
Darwin, Charles (1758–1778), 135
Darwin, Erasmus (1731–1802), 124
Buchan and, 140–146
disease and, 133–146
gout and, 133–140
heredity and, 133–146
illegitimate daughters of, 144
nonnaturals and, 134, 137
Darwin, George (1845–1912), 51
Darwin, Leonard (1850–1943), 147
Darwin, Robert Waring (1766–1848), 135
Darwin, William (c.1573–1644), 138
Daumer, Georg Friedrich (1800–1875), 413
Death, 89–90
Defoe, Daniel (c.1660–1731), 402
Degeneration
defining, 156–157
environmental changes and, 156–159
France and, 156–159
mental health and, 162–169
Delille, Gérard, 47–49
De Renzi, Silvia, 44, 117
Derouet, Bernard, 40–41, 47
Descartes, René (1596–1650), 4, 6, 260
Desmoulins, Louis Antoine (1794–1828), 364–365
"Determination of the Concept of a Race of Humans" (Kant), 280
De Vries, Hugo (1848–1935), 308
Diathesis, 18, 137, 156, 161
Dictionnaire des Sciences Médicales, 124
Diderot, Denis (1713–1784), 254–259
Different Human Races, On the (Kant), 280
Differentia specifica (specific differences), 185
Differenzierungsreihe (series of differentiation), 326–328
Dilthey, Wilhelm (1833–1911), 324–325
Discourse on the Manner of Studying and Expounding Natural History (Buffon), 20
Discours sur l'origine et les fondements de l'inégalité parmi les hommes (Rousseau), 399
Disease, 5, 278
alcoholism and, 133, 135–136, 384–389
atavism and, 119, 124
bacteriology and, 146–147
Brunonian medicine and, 138–139
Buchan and, 140–146
Cadogan and, 139–140
cancer, 298
causal issues and, 107–111
of character, 384–387
classification of, 107–111, 134–135
congenital, 18, 115
cretinism and, 158
degeneration fears and, 19, 156–159
diathesis and, 18, 137, 156, 161
environmental changes and, 156–159, 163
epilepsy, 122, 133, 136, 160
Erasmus Darwin and, 133–146
fetal transmission of, 114–115
four orders of, 122–123
French Revolutionary Era and, 111–123
Galenists and, 110
gout and, 19, 106, 122–123, 133–140, 142, 378
habit and, 141–142, 145–146
homochrony and, 118, 124
humoralists and, 114–119
industrial issues and, 146
latent causes and, 119–121, 124
Linnaeus and, 134

marriage and, 143
mental health and, 162–169
nervous diseases and, 134–135
nosography and, 12
phthisis, 19, 160
physiological influences and, 107–111
politics and, 19
predisposition toward, 122
resemblance and, 120–121
rickets, 160
scrofula, 160
secondary triggering causes and, 119–120
semen and, 108–109
social class and, 19, 133–147
solidists and, 16–17, 115–118, 121–123
statistical approach to, 139
temperament and, 68, 108, 160
transmission of, 107–123
worker health and, 146
Dishley Society, 234, 238
Dispute, La (Marivaux), 402
Dissertation . . . sur la question . . . comment se fait la transmission des maladies hereditaires? (Louis), 110
Divination, 69
Doctor Thorne (Trollope), 385–386
Domestic Medicine (Buchan), 141
Dower, 39–40, 61
Downward devolution, 45
Drugs, 163
Dual theory of seed, 67
Duchesne, Antoine-Nicolas (1747–1827), 210
botany and, 212–213, 216–218
constancy and, 214–216
experiments on new races and, 218–221
Linnaeus and, 205, 207
methodology of, 221–222
race concept and, 213–221
sexual reproduction and, 214
social class of, 218
species and, 214–216
Duchesneau, François, 293–314
Ducray-Duminil, François-Guillaume (1761–1819), 402–403, 408–410, 412
Duhamelle, Christoph, 44
Duméril, André Marie Constant (1774–1860), 364–365
Dunn, Leslie Clarence, 8

Eaton, John Matthews, 381–382
École de Médicine, 160
Economy of nature, 193–194
Edgeworth, Maria (1767–1849), 400
Edgeworth, Richard (1744–1817), 400
Education, 318, 383–384
Bach and, 430–431
emotion faculties and, 409
pedagogical transmission and, 443
poetics and, 445–461
savage children and, 401–413
Edwards, William-Frédéric (1777–1842), 363–364
Ego, 11
Egypt, 361–363
Eliot, George (1819–1880), 375
Emancipated son, 180–181
Endogamy, 45–51
England, 38, 45
Enlightenment and, 133–147
Erasmus Darwin and, 133–146
heredity and, 133–147
self-made man and, 376–381, 387
sheep breeding and, 232–238
Enlightenment, The
epistemology and, 315–336
education and, 318
England and, 133–147
First Memorandum on Public Instruction and, 318
generation theories of, 253–271
microscopy and, 259–261
reactions to epigenesis and, 262–265
social flux during, 323–326
transmission theories and, 253–271
Entailed estate. *See* Majorat

Epigenesis, 256, 267, 271, 281, 288, 311, 444, 459
generation and, 183
genius and, 422
reactions to, 262–265
Epilepsy, 122, 133, 136, 160
Epistemic space, 24–25
biological concepts and, 16–23, 326–328
distribution of early modern heredity and, 8–13
generation and, 3–8
as knowledge regime, 13–16
Esquirol, Jean-Etienne-Dominique (1772–1840), 156, 158, 164–166
Essai politique sur le royaume de la Nouvelle-Espagne (von Humboldt), 352–353
Essai sur les maladies héréditaires (Petit), 160
Essai sur l'origine des connaissances humaines (Condillac), 407
Essay Concerning Human Understanding (Locke), 177–178
Essegern, Ute, 39–40
Eugenic charts, 387–389
Eugenics Society, 387–388
Evolution, 133
breeding and, 212
Charles Darwin and, 186–190
constancy of species and, 177–178
Galton's critique of, 384
hereditary adaptation and, 277–288
Kant on, 280–288
"like begets like" and, 17, 105, 143–144, 177–178, 229–230, 255, 323–324
Linnaeus and, 182–186
natural selection and, 381
Experiment, 12, 21
breeding and, 23, 265–270 (*see also* Breeding)
disease transmission studies and, 107–126
Duchesne and, 205–207, 210, 212–221
human ancestors and, 22–23
microscopy and, 259–261, 265
philosophers and, 22, 258–259
savage children and, 399–413

Facultas formatrix, 69–71
Fairchild, Thomas (1667–1729), 211
Falconer, William (1744–1824), 140
Falk, Raphael, 8
Falret, Jean-Pierre (1794–1870), 164–165
Fanfan et Lolotte (Ducray-Duminil), 402, 406
fascinatio (spells), 71
Fascinatione libri tres, De (Vairo), 71
Fernel, Jean (c.1497–1558), 5, 105
Ferrus, Marie-André (1784–1861), 162
Fertility, 143
Fetus, 4, 114–115. *See also* Generation
Feuerbach, Anselm (1775–1833), 413
Filiatio. *See* Resemblance
Fink, J. H. (1730–1807), 231
First Momorandum on Public Instruction (de Caritat), 318
Fixist system, 212
Florists. *See* Breeding
Fodéré, François-Emmanuel (1764–1835), 158, 160–161, 164
Forbes, Edward (1815–1854), 320
Forkel, Johann Nikolaus (1749–1818), 419–423, 426–434
Forster, Georg (1754–1794), 280, 363
Foucault, Michel, 20, 186
Foundling Hospital of London, 139
Foville, Achille-Louis (1799–1878), 165
Fowler, Orson Squire (1809–1887), 105–106
Fox Keller, Evelyn, 8
France, 16, 38, 40–41, 141, 364
alienism and, 162–167
asylum populations of, 158
Carbonarie movement and, 169
civil code of, 11, 53–54
degeneration fears and, 156–159
Duchesne's strawberries and, 205–207, 210, 213–221
endogamous marriage and, 50–51
First Empire and, 156

historical hereditary disease concepts and, 111–123
intellectual hierarchy and, 159
majorat and, 86
medical science in, 155–169
mental health and, 158, 162–169
morcellement and, 87–88
physician regime of, 159–164
Franz Sternbald's Journey (Tieck), 428
French Revolution, 11, 22, 46
First Memorandum on Public Instruction and, 318
mobilization and, 157
physicians and, 161–164

Gachupines, 353
Galenists, 41–42, 62
disease transmission and, 110
four human races and, 285
health maintenance and, 134
innate heat and, 67–68
women's role in generation and, 181
Gall, Franz Joseph (1758–1828), 410–412
Galton, Francis (1822–1911), 390
acquired character and, 383–384
bacteriology and, 146
eugenic policy and, 389
genius and, 419–420
heredity and, 6–7, 12, 16, 23–24
Gardener's Dictionary (Miller), 214
Garden of Eden, 285–286
Gaskell, Elizabeth (1810–1865), 387
Gayon, Jean, 334
Geiringer, Karl (1899–1989), 420
Geisslern, Ferdinand (1751–1824), 238
Gemmules, 22
Gender, 3–4
castas and, 351
disease and, 163, 167
genius and, 425–428
male line and, 11, 12, 38–44, 87, 91, 94–95, 389
property inheritance and, 38–45
sheep breeding and, 232
Gene, The: A Critical History (Carlson), 8
Genealogy
alliance and, 186
Bach and, 426–430
constancy of species and, 177–178
convention for, 329–335
Darwin, Charles, and, 186–190
Duchesne and, 205–207, 210, 212–221
economy of nature and, 193–194
Harvey and, 179–183
Linnaeus and, 182–186, 190–193
majorat and, 86–98
as oldest logic, 177
philosophers and, 177–178, 180
resemblance and, 61–75, 120–121
species and, 182–186
General History of Music (Forkel), 430, 432
Genera plantarum (Linnaeus), 375
Generation, 3–8
alternation of generations and, 319–323
animal experiments and, 265–270
breeding and, 207–211 (*see also* Breeding)
castas and, 355–361
cause and, 182
cell theory and, 293–312
character and, 375–380
concepts for, 20–21, 316–320, 323–326
convention for, 329–335
Darwin, Charles, and, 186–190
differentiation series and, 326–328
disease transmission and, 107–126
divergence of character and, 189–190
Duchesne and, 205–207, 210, 212–221
economy of nature and, 193–194
emancipated son concept and, 180–181
Enlightenment theories of, 253–271
epigenesis and, 183, 256, 262–265
equivocal, 256–257
essence of organism and, 322
exchange concept and, 185–186
four human races and, 283–286

Generation (cont.)
genius and, 421–426
Harvey and, 179–183
hereditary adaptation and, 277–288
hybridization and, 331, 334
imagination and, 69
individuality and, 315–316, 319–323
interracial crossings and, 350–355
law of reproduction and, 323–324
law of succession and, 85–98
"like begets like" and, 17, 105, 143–144, 177–178, 229–230, 255, 323–324
Linnaeus and, 182–186, 190–193
livestock farmers and, 209
maternal imagination and, 62–63, 65, 69–74
mathematics and, 326–328
Mendel and, 335–336
microscopy and, 259–261
monsters and, 62
paternity and, 61–75
pedigree and, 326, 331, 334
poetics of, 422
progeny testing and, 232–234
pulsating point and, 179–181
resemblance and, 61–75, 120–121
savage children and, 399–413
sheep breeding and, 229–241
social scientists and, 323–326
species and, 182–186
stable/nonstable traits and, 315
temperament and, 63, 67–68
transmission problem and, 316–319
Generationswechsel, 319
Generis humani varietate nativa, De (Blumenbach), 351–352
Genetics, 13
differentiation series and, 326–328
heredity concept and, 315–316 (*see also* Heredity)
Mendel and, 3, 8, 19, 193, 241, 321, 335–336
Genetic Studies of Genius (Terman), 420
Genitura, De (Hippocrates), 178
Genius, 383–384
Bach and, 419–420, 426–434
concepts of, 421–426
Forkel and, 419–423, 426–434
Galton and, 419–420
gender and, 425–428
innateness and, 423
physiology of, 421–426
"Genius" (Hildebrand), 421
Gentry. *See* Social class
Geoffroy Saint-Hilaire, Étienne (1772–1844), 94–95
Geology, 188–190
Georget, Etienne (1795–1828), 165
German Dictionary (Grimm), 421
Germany, 9
endogamous marriage and, 46–51
majorat and, 86
property inheritance and, 38, 40, 45–49
Germs, 24–25. *See also* Generation
animal experiments and, 265–270
Anlagen (dispositions), 17, 240, 306–312, 325
Buchan and, 141–142
cell theory and, 293–312
epigenesis and, 262–265
four human races and, 283–286
generation theories and, 263
interracial crossing and, 350–355
Mendel and, 241
preformation and, 182–186
species and, 182–186
Geschlecht (gender), 425
Gmelin, Johann Georg (1709–1755), 212, 221
Gobineau, Arthur de (1816–1882), 363, 365
God, 71–72, 182, 381
as Creator, 279–280
as divine watchmaker, 183
Enlightenment and, 255–256
generation theories and, 256–257
poetics and, 453–455

Goethe, Johann Wolfgang von (1749–1832), 422, 428, 444, 452–453, 460
Golden mean, 446
Goody, Jack, 9
Gottsched, Johann Christoph (1700–1766)
 poetics and, 445, 447, 449–452, 455–459
 rules of, 449–452
Gout, 19, 378
 Buchan and, 142
 Erasmus Darwin and, 133–140
 medical science and, 106, 122–123
 noble disease and, 19, 144
Grant, William, 140
Greeks, 17, 444–449, 458
Gregorian reforms, 9
Growth of Biological Thought, The (Mayr), 8
Grundriß der Geschichte der Menscheit (Meiners), 361–362
Guettard, Jean-Etienne (1715–1786), 207
Gumilla, José (1686–1750), 351, 354, 359–360
Guyénot, Emile, 207, 214

Haller, Albrecht von (1708–1777), 205, 217, 262–263, 270
Handbuch der Physiologie des Menschen (Müller), 298–299, 306
Harvey, William (1578–1657)
 Enlightenment and, 255
 exchange concept of, 185
 generation and, 3, 6, 24, 179–182
 inheritance and, 179–183, 185, 190–191, 195
Hauser, Kaspar (c.1812–1833), 400–401, 413
Heber, Reginald (1783–1826), 363
Heeren, Arnold (1760–1842), 363
Hegel, Georg Wilhelm Friedrich (1770–1831), 52–53, 444, 456
Helvétius, Claude Adrien (1715–1771), 399
Herder, Johann Gottfried (1744–1803), 422–423
Hereditary Genius (Galton), 23, 383–384, 419
Hereditary predisposition, 161
Hereditary Talent (Galton), 419
Heredity
 acquired character and, 375–390
 adaptation and, 277–288
 astral influences and, 3–4
 atavism and, 315
 breeding and, 229–241 (*see also* Breeding)
 castas and, 14–15, 355–361
 cell theory and, 293–312
 concepts for, 3, 12–13, 16–23, 105–107, 234, 315–316
 constancy of species and, 16–17
 court evidence for, 68–73
 differentiation series and, 326–328
 disease and, 5, 12, 18 (*see also* Disease)
 distribution of early modern, 8–13
 early modern physiology and, 107–111
 engendering and, 3–4
 Enlightenment England and, 133–147
 as epistemic space, 16–23, 24–25, 326–328
 Erasmus Darwin and, 133–146
 fertility and, 143
 Garden of Eden and, 285–286
 generation and, 3–8, 20–21 (*see also* Generation)
 genius and, 419–434
 human ancestors and, 22–23
 improving offspring and, 144–145
 individuality and, 315–316, 319–323
 influence model and, 23
 inherent design and, 328
 intemperance and, 135–137
 Kant and, 280–288
 as knowledge regime, 13–16
 law of succession and, 85–98
 "like begets like" and, 17, 105, 143–144, 177–178, 229–230, 255, 323–324
 maternal imagination and, 62–63, 65, 69–74
 medical origins of, 105–126
 medieval view of, 9–12
 Mendel and, 3, 8, 19, 193, 241, 321, 335–336
 mental character and, 3–5, 162–169
 as metaphor, 3–7

Heredity (cont.)
modern genetics and, 315–316
natural history and, 190–195
nature/nurture debate and, 4
organization concept and, 20–22
original sin and, 280–281
poetics and, 443–461
resemblance and, 61–75, 121
savage children and, 399–413
social scientists and, 323–326
soft vs. hard, 13–14, 18, 134, 147
stable/nonstable traits and, 315
subspecific levels and, 17–18
as succession, 4–5
temperament and, 160
Hertwig, Oscar (1849–1922), 308
Hildebrand, Rudolf (1824–1894), 421
Hippocrates (c.460 BC–c.370 BC), 110, 178
Histoire de l'eugénisme en France (Carol), 157
Histoire naturelle (Buffon), 95, 263, 279
Histoire naturelle des fraisiers (Duchesne), 205, 207, 210, 213, 219
Hoffmann, E. T. A. (1776–1822), 88–90, 428
Hölderlin (1770–1843), 456
Homer, 447
Homochrony, 118, 124
Horace (65BC–8BC), 445–449, 451, 460
Hörisch, Jochen, 401
Horticulturalists. *See* Breeding
Hospital of Santo Spirito, 64
Howard, Charles, 135–136
Howard, Mary (1739–1770), 135–136
Humboldt, Alexander von (1769–1859), 352–353, 361
Humboldt, Wilhelm von (1767–1835), 354, 453, 460
Hume, David (1711–1776), 279
Humoralism, 114–119
Hunt, John, 237
Hunter, John (1728–1793), 124
Huxley, Thomas (1825–1895), 315, 382–383
Hybrids, 12, 193–194
breeding and, 205–210 (*see also* Breeding)
castas and, 355–361
interracial crossing and, 350–355
Hydra, 257–258
Hysteria, 160

Ideas on the Aesthetic of Tonal Art (Schubart), 430
Illegitimacy, 63–66, 366
Imagination
resemblance and, 62–63, 65, 69
specific plan and, 327–328
Impressions of Theophrastus Such, The (Eliot), 375
Inbreeding, 232–234, 237, 239
Incest, 51–53
India, 361–363
Indios, 353
Industrialization, 17
Individuality, 141
alternation of generations and, 319–322
character and, 375–390
stable/nonstable traits and, 315–316
Infidelity, 402
Inheritance, 4–5
acquired characteristics and, 375–390
agnatic relationships and, 38
alienation and, 24–25
alliances and, 41, 45–51
arranged, 140–141
bilateral, 268
blood ties and, 41–42, 45
Buchan and, 140–146
castas system and, 14–15
Church and, 9
circulation and, 24–25
Darwin, Charles and, 186–190
degeneration fears and, 156–159
of disease, 133–147 (*see also* Disease)
exchange and, 182, 185, 194
flower color and, 209–210
forms of government and, 317
formula of (*Erbschaftsformel*), 334
French civil code and, 11, 53–54

Galenists and, 41–42
genius and, 419–434
illegitimate children and, 63–66
knowledge regimes and, 13–16
law of succession and, 85–98
legal issues and, 140 (*see also* Property)
majorat and, 86–98
medieval view of, 9, 11, 37–45
paternity and, 61–75
patrilines and, 38–39, 46–49
poetics and, 443–461
primogeniture and, 11–12, 38, 87–88
resemblance and, 61–66
restricting of, 37–45
retrait lignager concept and, 42–43
semantics of, 11–12
theological model of, 41–42
women and, 38–39
Inheritance capacity, 241
Innate heat, 67–68
Inquiétude automate (Automatic restlessness), 258
Internal molds, 259–261
Ipse, 11
Italy, 44, 46, 48
Counterreformation and, 64
endogamous marriage and, 49–51
illegitimate children and, 63–66
Roman Law and, 9, 38, 61–75
Itard, Jean (1774–1838), 403–410, 412
Ius commune, 63–64
Ius proprium, 63–64

Jacob, François, 3, 7–8, 20
Jardin du Roi, 159
Jauffret, Louis-François (1770–1850), 401–403, 405
Jefferson, Thomas (1743–1826), 317
Jesus, 281
Johann Sebastian Bach's Life, Art, and Works of Art, On (Forkel), 421, 426–427
Johnson, Christopher, 52
Johnson, Samuel (1709–1784), 140
Jointure, 39–40
Journal économique, 207, 209–210, 218
Juan y Santacilia, Jorge (1713–1773), 351
Jussieu, Bernard de (1699–1777), 205, 210
Justinus, Johann Christoph, 239

Kant, Immanuel (1724–1804), 17–18, 21
concepts of, 280–281, 286–287
generation and, 318–319
genius and, 423, 425–426
heredity adaptation and, 282–286
marriage and, 51
original sin and, 280–281
poetics and, 454, 460
race and, 280–282
skin color and, 283–286
Vererbung and, 280–282
Kerr, Norman (1834–1899), 386
Kielmeyer, Carl Friedrich (1765–1844), 20, 193
King's Evil, 133
Kinship, 45–51
affines, 9, 12, 47, 49–50
agnatic relationships and, 38
alliances and, 41, 45–51
blood ties and, 41–42, 45
cadets, 45
cognates, 9
cousins, 47–51, 188
degree of, 9
French civil code and, 53–54
Galenists and, 41–42
Germanic system of, 9
horizontal links and, 45–46
incest and, 51–53
law of succession and, 85–98
patrilines and, 38–39, 46–49
property inheritance and, 37–45
rechaining and, 47, 49
resemblance and, 61–75
Roman system of, 9
Klopstock, Friedrich Gottlieb (1724–1803), 452

Knowledge regimes, 13–16
Koelreuter, Joseph Gottlieb (1733–1806), 209, 221
Kölliker, Albert (1817–1905), 304, 321
Kühn, Arthur, 50
Kurze Geschichte der Genetik (Stubbe), 8

Labour Party, 387
Lactation, 3
Ladevère, 114
Lamarck, Jean-Baptiste de (1744–1829), 145–146
Laplace, Pierre-Simon de (1749–1827), 162
Law of reproduction, 323–324
Law of succession
 Arnim and, 90–93
 Balzac and, 93–96
 Hoffmann and, 89–90
 literature on, 85–98
 nature and, 85–89
Lawrence, William (1783–1867), 354–355
Leben der Gattung (life of the species), 20
Lebensprincip theory, 298, 301, 310–311
Leçons sur les phénomènes de la vie communs aux animaux et aux végétaux (Bernard), 21–22
Lectures on Physiology, Zoology, and the Natural History of Man (Lawrence), 354
Leeuwenhoek, Antoni van (1632–1723), 255
Legal issues, 13–15, 22, 140
 affinity and, 45–51
 castas and, 14–15
 dower rights and, 39–40, 61
 expert witnesses and, 61
 illegitimate children and, 63–66
 incest and, 51–53
 inheritance restricting and, 37–45
 Ius Commune and, 63–64
 Ius Proprium and, 63–64
 jointure and, 39
 law of succession and, 85–98
 majorat and, 86–98
 maternal imagination and, 62–63, 65, 69, 74
 medical evidence and, 61–63, 67–75
 medieval view and, 37–45
 natural evidence and, 68–73, 85–89
 paternity and, 61–75
 patrilines and, 38–39, 46–47
 portion and, 39–40
 primogeniture and, 11–12, 38
 property inheritance and, 37–45 (*see also* Property)
 resemblance and, 61–66
 retrait lignager concept and, 42–43
 Roman Law and, 38, 61–75
 Saxony and, 39–40
 Strict settlement and, 140–141
Leges generationis (laws of generation), 185
Lenoir, Timothy, 298
Lessing, Gotthold Ephraim (1729–1781), 422
Lévi-Strauss, Claude, 178
Lewes, George (1817–1878), 375, 389
Lexicon (Zedler), 281
Lichtenberg, Georg Christoph (1742–1799), 362
Life and Opinions of Tristram Shandy, Gentleman, The (Sterne), 4
Life and Works of Goethe, The (Lewes), 375, 389
Lignac, Lelarge de (1710–1762), 264
Lignage, 11
"Like begets like," 17, 105, 323–324
 concept of, 143–144
 Diderot and, 255
 inheritance and, 177–178
 sheep breeding and, 229–230
Linnaeus, Carolus (1707–1778), 20, 23, 134, 190, 195, 282
 breeding and, 205, 207, 215, 217, 221
 character and, 375
 classification and, 134, 185, 211–212
 creation theory of, 184, 191
 Duchesne and, 205, 207
 economy of nature and, 193
 exchange concept of, 185–186, 192–193

four races of, 285
reproduction and, 183–184
savage children and, 399–400
species and, 182–186, 191–192
Littré, Emile (1801–1881), 306–307
Livestock, 377–381
Locke, John (1632–1704), 177–178, 406
Logique du vivant, La (Jacob), 3, 8
Lohff, Brigitte, 298
Lombroso, Cesare (1836–1909), 384
Long, Edward (1734–1813), 355
Longue durée development, 7–8
Lonitzer, Adam, 217
López-Beltrán, Carlos, 147, 253
degeneration and, 155, 160
heredity and, 12
resemblance and, 62–63
Louis, Antoine (1723–1792), 110–111
Louis, Pierre-Charles-Alexandre (1787–1872), 165–166
Louis XIV, King of France (1638–1715), 218
Louis XV (1710–1774), King of France, 205
Louis XVI (1754–1793), King of France, 168
Lucas, Prosper (1805–1885), 12, 155, 188–189, 307
Lunar Society, 136, 143, 146
Lyell, Charles (1797–1875), 381
Lymphatic temperament, 160
Lynx-girl, 269–270

McLaughlin, Peter, 277–291, 443
Maimon, Salomon (1753–1800), 431
Majorat, 5
Arnim on, 90–93
Balzac on, 93–96
Hoffman on, 89–90
property issues and, 86–98
Majorat, Das (Hoffmann), 88–90
Majoratsherren, Die (Arnim), 88, 90–93
Maladies mentales considérées sous les rapports médical, hygiénique et médico-légal, Des (Esquirol), 158, 165
Malatesta, Sigismondo (1417–1468), 66
Malesherbes, Chrétien Guillaume Lamoignon de (1721–1794), 212
Malpighi Marcello (1628–1694), 255
Mania, 122, 160
Manuel de botanique (Duchesne), 212
Marchant, Jean (1650–1738), 212, 221
Marriage, 9, 11–12, 349. *See also* Kinship
bloodline purity and, 142–143
of cousins, 47–48
cretinism and, 158
disease and, 143, 168–169
dower rights and, 39–40, 61
endogamous, 45–51
Galenists and, 41–42
illegitimate children and, 63–66
incestuous, 51–53
jointure and, 39–40
legislation and, 155
paternity and, 61–75
patrilines and, 38–39, 46–49
philosophers on, 51–52
physicians and, 155–156
portion and, 39–40
property inheritance and, 38–45
Saxony and, 39–40
sentiment and, 51–54
social class and, 142–144
Marriages of Plants (Linnaeus), 183
Marshall, William (1745–1818), 236
Marx, Karl (1818–1883), 324
Mascardo, Giuseppe, 66
Materialistic theories, 278–279, 305
Maternal imagination
German physicians and, 74
resemblance and, 62–63, 65, 69–74
Mathematics, 162–163
alternation of generations and, 322–323
Carus and, 326–328
Maupertuis, Pierre Louis Moreau de (1698–1759), 263, 271, 399
animal experiments and, 265–268
experiment and, 259, 262
generation theories and, 257–258

Mayr, Ernst, 8, 178–179, 193, 335–336
Mazzolini, Renato G., 349–373
Médecin de Campagne, Le (Balzac), 158
Medical Mandarins, The (Weisz), 159
Medical science, 254
alienism and, 162–167
atavism and, 119
bacteriology and, 146–147
blood circulation and, 179
Brunonian, 138–139
Buchan and, 140–146
Cabanis and, 158, 161–162
Cadogan and, 139–140
causal issues and, 107–111, 113
character and, 383–387
classification and, 134–135
constancy of species and, 177–178
cretinism and, 158
Darwin, Erasmus and, 133–146
degeneration fears and, 156–159
disease and, 107–111 (*see also* Disease)
early modern transmission physiologies and, 107–111
Enlightenment England and, 133–147
environmental changes and, 163
Esquirol and, 164–167
Fodéré and, 158, 160–161
French historical perspective on, 111–123, 155–169
genealogy and, 177–178
humoralists and, 114–119
intellectual hierarchy and, 159
marriage and, 155–156
maternal imagination and, 62–63, 65, 69–74
mathematics and, 162
mental health and, 162–169
natural evidence and, 68–73
Petit and, 160–161
Pinel and, 163, 164
Portal and, 159–160
Pussin and, 163
resemblance and, 61–63, 67–75
self-healing and, 163
solidists and, 16–17, 115–118, 121–123
statistics and, 139, 156, 164–167
Vandermonde and, 269
Zacchia and, 61–63, 67–75
Medicina (Fernel), 5
Medieval times, 9–10
dower rights and, 39–40
jointure and, 39–40
portion and, 39–40
property inheritance and, 37–45
retrait lignager concept and, 42–43
Saxony and, 39–40
Meiners, Christoph (1747–1810), 361–363
Mendel, Gregor (1822–1884), 3, 8, 19, 193, 321, 335–336
sheep breeding and, 240–241
Mendelian Revolution, The (Bowler), 8
Menochio, Giacomo (1532–1607), 65–66
Mental health, 19, 133, 136
alienism and, 162–167
controlling transmission of, 167–169
French asylum populations and, 158
mania and, 122, 160
marriage and, 168–169
scrofula and, 160
Merlino, Clemente, 66
Mestizos, 350, 353–354, 359
Mexico, 353
Michelet, Jules (1798–1874), 364
Microscopy
alternation of generations and, 319–323
experiment and, 259–261, 265
Mikroskopische Untersuchungen über die Übereinstimmung in der Struktur und dem Wachsthum der Thiere und Pflanzen (Schwann), 294, 298–299
Mill, John Stuart (1806–1873), 324
Miller, Philip (1691–1771), 211, 214
Millot, Jacques André (1728–1811), 155
Milton, John (1608–1674), 451
Mirabeau, Gabriel-Honoré Riquetti, Comte de (1749–1791), 157

Mobilization, 17, 157
Mohl, Hugo von (1805–1872), 297
Mongoloids, 362
Monsters, 62. *See also* Savage children
Montaigne, Michel de (1533–1592), 108–109
Monti, Giuseppe (1682–1760), 212–213
Moral physicians, 163
Moravia, 238–241
Morcellement, 87–88
Morel, Bénédicte-Augustin (1809–1873), 156
Morgan, Charles, 377
Morgenstern, Karl (1770–1852), 428
Morton, Samuel G. (1799–1851), 355
Moss, Lenny, 8
Mulatos, 349–355, 359
Müller, Johannes (1801–1858), 293, 307, 310–312
  cell life and, 298–299
  individual organismic potential and, 306
  Lebensprincip theory and, 298, 301
  Schwann's cell theory and, 298–305
Museum d'Histoire Naturelle, 159
Musicology, 419

Nägeli, Carl (1817–1891), 304, 306, 308, 321, 334
Napoleon (1769–1821), 156, 161
Napp, Cyrill (1792–1867), 240–241
Natural history, 12, 15, 19, 22–23
  alternation of generations and, 319–323
  animal experiments and, 265–270
  breeding and, 207–211 (*see also* Breeding)
  Duchesne and, 205–207, 213–221
  economy of nature and, 193–194
  epigenesis and, 262–265
  generation theories and, 253–271 (*see also* Generation)
  heredity and, 190–195 (*see also* Heredity)
  interracial crossing and, 350–355
  Kant and, 280–288
  Linnaeus and, 185–186, 190–193
  poetics and, 445–461
  savage children and, 399–413
Natural selection, 189–190, 381
Nature/nurture dichotomy, 4, 134, 399
Naturphilosophie (Schelling), 453–457, 460
Naturzweck (natural purpose), 21
Naudin, Charles (1815–1899), 329
Needham, John Turberville (1713–1781), 259, 261, 265
Negro, 357, 360
Newton, Isaac (1643–1727), 138, 260
New Werther, The (Pearson), 389
Nitzsch, Carl Ludwig (1751–1831), 51–53
Noble savage, 157
Nonnaturals, 134, 137
Norway, 50–51
Nosology, 12, 134–135
Nothis, Spurrisque filiis liber, De (Paleotti), 64

Observations on Livestock (Culley), 232
Octavons, 350–355, 359
Olby, Robert, 8
Omnis cellula e cellula principle, 293, 304–305
"On Generation" (Erasmus Darwin), 133
Organization concept, 20–22
Original Composition (Bach), 431
Original sin, 280–281
Origin of Species (Charles Darwin), 186, 277, 381–382
Origins of Mendelism (Olby), 8
Orinoco ilustrado, El (Gumilla), 351
Oxford English Dictionary, 277

Pagès, Jean-François (1760–1803), 112–113, 118
Paine, Thomas (1737–1809), 317
Paleotti, Gabriele (1522–1597), 64–65
Paley, William (1743–1805), 279
Panckoucke, Charles Joseph (1736–1798), 260–261
Paracelsians, 108
Pariahs, 363
Paris Academy of Sciences, 268, 270
Parisano, Emilio (1567–1643), 69
Paris Council of Hygiene and Salubrity, 158

Paris Faculty of Medicine, 163
Parnes, Ohad S., 168, 306, 315–346
Parturition, 3
Paternity
adultery and, 61
illegitimate children and, 63–66
improving offspring and, 144–145
majorat and, 86–98
maternal imagination and, 62–63, 65, 69
medical evidence and, 61–63, 67–75
natural evidence and, 68–73
resemblance and, 61–63, 61–75
temperament and, 63, 67–68
Patriarchy, 179–182
Patrilines, 38–41, 46–49, 86–87, 93, 180, 182, 427
Pauw, Cornélius de (1739–1799), 350–352, 354
Pearson, Karl (1857–1936), 389–390
Peculiar Musical Life of Musician Joseph Berglinger (Wackenroder), 428
Pedigrees, 331, 334
character and, 376–381
social class and, 389
women and, 389
Petit, Antoine, 124–125, 160–161
Phenomenology of mind, The (Hegel), 52–53
Philosophers, 17–18, 21–22, 142, 253–254. *See also specific individuals*
argument from design and, 279
Enlightenment England and, 133–147. *See also* Enlightenment, The
epigenesis and, 262–265
experimentation and, 22
four human races and, 283–286
genealogy and, 177–178, 180
generation theories and, 253–271
genius and, 421–426
materialism and, 278–279, 305
positivism and, 323–324
reductionism and, 294–298
savage children and, 399–413
tâtonnement and, 258–259
Philosophica botanica (Linnaeus), 191, 205, 213
Philosophical Letters on Dogmatism and Criticism (Schelling), 456
Philosophie der Kunst (Schelling), 445
Philosophie Zoologique (Lamarck), 145
Philosophy of Art (Schelling), 456–457
Physicians. *See* Medical science
Physiochemical processes, 21–22
Physiologie als Erfahrungswissenschaft, Die (Burdach), 307
Pickering, Charles (1805–1878), 363
Pigeons, 382
Pilgrims, 64
Pindar (c.518 BC–445 BC), 456
Pinel, Philippe (1745–1826), 111, 156, 163–165, 404–405, 408
Pinel, Scipion (1795–1829), 165
Plan for the Conduct of Female Education in Boarding Schools, A (Erasmus Darwin), 144
Plants
alternation of generations and, 319–323
breeders and, 20, 23 (*see also* Breeding)
growers' practices and, 207–211
Linnaeus and, 182–186
microscopy and, 259–261
strawberries, 205–207, 213–221
Plato (429–347), 444–445, 458
Pliny (23–79), 65–66, 354
Poetics, 422
aesthetics and, 444–461
Aristotle and, 444, 458–459
beauty and, 446–447
Bildungstrieb and, 454
cultural exchange and, 457–461
epigenetic art concept and, 445, 458–459
genre theory and, 443, 451–457
golden mean and, 446
Gottsched and, 449–452, 455–457, 459
Greeks and, 444–449, 458–459
Horace and, 445–449
imitation of nature and, 449–452
life sciences and, 443–444

mechanical understanding and, 454
nominalization and, 459
orator tasks and, 450
pedagogical transmission and, 443
Plato and, 444–445, 447, 458
Romanticism and, 445–449, 458–460
Schelling and, 453–460
social class and, 451–452
Sturm und Drang approach and, 452, 454, 459–460
the Swiss and, 451–452
Verebung concept and, 443
Politica Indiana (Solórzano Pereira), 358–359
Politics, 317
disease and, 19
epistemic space and, 9, 11–12
socialism and, 387–388
Polydactyly, 266
Polyps, 257–258
Pomata, Gianna, 19
Population thinking, 234–236
Portal, Antoine (1742–1832), 159–160
Porter, Ted, 389
Portion, 39–40
Portuguese, 355–361
Positivism, 323–324
Pouchet, Félix-Archimède (1800–1872), 155
Practical Education (Edgeworth), 400
Preformation, 12, 22, 183, 256, 264, 278, 280, 287, 349, 352, 445, 448, 456, 458
Prepotency, 237
Primitive Physic (Wesley), 141
Primogeniture, 11–12, 38, 87–88
Principles of Surgery (Hunter), 124
Progeny testing, 232–234
Prolegomena (Müller), 298
Property, 9, 23, 140
affinity and, 45–51
alliances and, 41, 45–51
cadet, 45
dower rights and, 39–40, 61
exclusion of women and, 38–39
French civil code and, 11, 53–54
illegitimate children and, 63–66
inheritance restriction and, 37–45
jointure and, 39–40
law of succession and, 85–98
majorat and, 86–98
morcellement and, 87–88
patrilines and, 38–39, 46–47
portion and, 39–40
primogeniture and, 11–12, 38, 87–88
retrait lignager concept and, 42–43
Saxony and, 39–40
strict settlement and, 39, 140–141
Prostitutes, 64
Prussia, 87, 231
Psychiatrists, 107
Puchuella, 350–351
Pujol, Alexis (1739–1804), 112–123
Pulsating point, 179–181
Purkynje, Jan Evangelista (1787–1869), 322
Pussin, Jean-Baptiste (1745–1811), 163

Quaestiones Medico-Legales (Zacchia), 61, 63, 67–69
Quarterons, 350–355, 359
Quinterons, 354

Race, 8–9, 213–221
Camper's facial angle and, 364–365
castas and, 355–361
caste and, 360–366
categorizations for, 350–351, 353
concept of, 364–365
improvement of, 240, 388
interracial crossing and, 350–355
Kant on, 280–286
Schlag (strain) and, 183
skin color and, 283–286, 349–355
Spielart (sport) and, 183
Races of Man, The (Pickering), 363
Ray, John (1627–1704), 211
Réaumur, René-Antoine Ferchault de (1683–1757), 263–271
Rechaining, 47, 49

Recherches philosophiques sur les Américains (de Pauw), 350
Reductionism, 294–298
Reflections on the Philosophy of the History of Mankind (Herder), 422
Regelpoetik, 446
Regeneration, 157
Reichert, Carl Bogislaus (1811–1883), 321, 326
Reif, Heinz, 44
Relación histórica del viage a la América meridional (Juan y Santacilia and de Ulloa), 351
Religion, 113, 366, 384
 agnosticism and, 260
 alcoholism and, 387
 alienism and, 163
 celibacy and, 64
 endogamous marriage and, 50–51
 four human races and, 285–286
 Garden of Eden and, 285–286
 inheritance and, 9, 41–42
 original sin and, 280–281
 pilgrims and, 64
 poetics and, 458
 witchcraft and, 71
Religion within the Bounds of Mere Reason (Kant), 280
Remak, Robert (1815–1865), 293, 304–308, 311
Reproduction, 3–5, 6, 8, 15, 20–24, 86, 96, 140, 155, 178, 183, 190–195, 253, 257, 328, 389, 444, 458
 Buffon and, 192, 259–260
 of cells, 293, 301–311, 321
 constancy of species and, 210, 212, 214, 216
 disease and, 108
 of genus, 425
 "law of" (Comte), 323
 Lyell and, 381
 of power, 186
 self-, 421
 of species, 185, 193, 321
Resemblance, 120–121
 adultery and, 61
 differing views of, 61–63
 disease transmission and, 120–121
 dual theory of seed and, 67
 as evidence, 63–66
 illegitimate children and, 63–66
 imagination and, 65, 69
 innate heat and, 67–68
 legal issues and, 61–66
 maternal imagination and, 62–63, 65, 69–74
 medical evidence and, 61–63, 67–75
 natural evidence and, 68–73
 paternity and, 61–63
 temperament and, 63, 67–68
Restoration, 161
Rey, Guillaume (1687–1756), 117
Richard, Louis Claude (1754–1821), 210
Richter, Jean Paul Friedrich (1763–1825), 422, 424–425
Rights of Man, The (Paine), 317
Risley, Herbert Hope (1851–1911), 361
Robinson Crusoe (Defoe), 402
Roman law, 9, 38
 illegitimate children and, 63–66
 Ius commune and, 63–64
 Ius proprium and, 63–64
 resemblance and, 61–75
 Sacra Rota and, 61, 66, 68–69
Romanticism
 Enlightenment and, 133, 429–430
 poetics and, 445–449, 458–460
Rougemont, Joseph-Claude (1756–1818), 112–114
Rousseau, Jean-Jacques (1712–1778), 142, 157, 399, 411–412
Royal Society of London, 270
Rozier, François Abbé (1734–1793), 211, 219–220

Sabean, David Warren, 98, 141
Sacra Rota, 61, 66, 68–69

Salpa, 319
Salpêtrière, La, 164–165, 167
Sappho (630/612 BC–c.570), 456
Sars, Michael, 320
Sassoferrato, Bartolo da, 65
Sauvages, François Boissier de, 134
Savage children, 22–23
 education and, 401–413
 Gall and, 410–412
 idiocy and, 404, 406, 410–412
 Itard and, 403–410, 412
 literature on, 401–405
 nobility of, 404–405
 pretended savage and, 410–412
 as spectacles, 400–401
 Spurzheim and, 410–412
 wolf children and, 399–401
Saxony, 39–40
Schelling, Friedrich Joseph Wilhelm (1775–1854), 444–445, 453–460
Schiller, Friedrich (1759–1805), 422, 444, 453, 460
Schlag (strain), 183
Schlegel, Karl Wilhelm Friedrich von (1772–1829), 444
Schlegel, August Wilhelm von (1767–1845), 444
Schleiermacher, Friedrich (1768–1834), 319
School for Aesthetics (Paul), 424
Schubart, Christian Friedrich Daniel (1739–1791), 430
Schwann, Theodor (1810–1882), 293–305, 310, 321
Scrofula, 133, 160
Sebright, John Saunders (1767–1846), 236–238
Secord, Anne, 387
Secord, James, 381
Segalen, Martine, 47
Self-made man, 376–381, 387
Semen, 180. *See also* Generation
 cell theory and, 293–312
 epigenesis and, 262–265
 pangenetic route and, 108–109
Semiramis, Queen of Assyria, 66
sensibilité sourde (muffled sensibility), 258
Serres, Olivier de (1539–1619), 211
Sex. *See also* Gender
 reproduction and, 22, 24, 62, 69, 181, 183, 210, 214, 230–232, 237, 241, 253–269, 278, 309, 388, 402, 425, 448
 sexual cells and, 308–309
 sexual selection and, 133, 354
Sexuality,
 desire and, 51, 53, 454
 dispositif of (Foucault), 186, 190
 dispositive, 29
 sexual act, 160, 181, 301
Short History of Genetics, A (Dunn), 8
Shorthorn Transactions, 380
Sicard, Roche-Ambroise (1742–1822), 404–405
Sinclair, John (1754–1835), 231, 236, 238
Skin color, 22–23, 266, 349
 castas and, 355–361
 Enlightenment and, 283–286
 interracial crossing and, 350–355
 Malpighian layer and, 352
Slav race, 362
Smiles, Samuel (1812–1904), 377, 385, 387
Smollett, Tobias (1721–1771), 136–137
Social class, 25, 52
 alcoholism and, 136, 384–389
 alienism and, 162–163
 alliances and, 45–51
 bloodline purity and, 142–143
 Camper's facial angle and, 364–365
 castas and, 355–361
 caste and, 360–366
 character and, 376
 Darwin Charles and, 382–383
 degeneration fears and, 156–159
 disease and, 19, 133–147
 Duchesne and, 218
 education and, 318
 Enlightenment and, 323–326
 eugenic charts and, 387–389

Social class (cont.)
- four human races and, 283–286
- incest and, 51–54
- industrialization and, 146
- innate capacity and, 376
- interracial crossing and, 350–355
- livestock breeding and, 377–381
- majorat and, 86–98
- marriage and, 142–144
- medical hierarchy and, 159
- morcellement and, 87–88
- patricians and, 44
- pedigrees and, 377–381, 389
- poetics and, 451–452
- poultry breeding and, 381–382
- property and, 37–45, 140
- resemblance and, 65–66
- royal genealogies and, 317
- savage children and, 399–413
- self-made man and, 376–381, 387
- Strict settlement and, 140–141
- work and, 383–384

Socialism, 387–388

Société des Observateurs de l'homme Jauffret and, 401–403, 405
- savage children and, 401–410

Société Royale de Médecine, 18, 111–112, 122–123, 163

Sociology
- dead govern the living and, 323–324
- Enlightenment and, 323–326
- law of reproduction and, 323–324

Solidism, 115–123

Solórzano Pereira, Juan de (1575–1655), 358–359

Spallanzani, Lazzaro (1729–1799), 209

Spaniards, 44, 355–361

Species
- Charles Darwin and, 186–190
- constancy of, 177–178
- divergence of character and, 189–190
- Duchesne and, 214–216
- economy of nature and, 193–194
- four human races and, 283–286
- Kant and, 280–288
- "like begets like" and, 17, 105, 143–144, 177–178, 229–230, 255, 323–324
- Linnaeus and, 185–186, 191–193, 211–212
- natural selection and, 381
- sheep breeding and, 229–241
- stable/nonstable traits and, 315–316

Spencer, Herbert (1820–1903), 308

Sperma, 41–42

Spielart (sport), 183

Spitta, Phillip (1801–1859), 420

Sponsalia planatarum (Marriages of Plants) (Linnaeus), 183

Spring, Eileen, 38–39, 42, 45

Spurzheim, Johann Christoph (1776–1832), 410–412

Stafford, Barbara Maria, 400–401

Statistics, 139, 156
- alienism and, 163–167
- biometrics and, 389–390

Steenstrup, Johannes Jupetus (1813–1897), 320

Stephenson, George (1781–1848), 377

Sterne, Laurence (1713–1768), 4

Stirp, 6–7

Stone, Lawrence, 140–141, 144

Strasburger, Eduard (1844–1912), 308

Strict settlement, 39, 140–141

Stubbe, Hans, 8

Stuttering, 135

Succession, 4–5

Suffolk Stud Book, 377

Sun, 3, 182

Sweden, 50–51

Synopsis Nosologiae Methodicae (Cullen), 134–135

Syphilitic virus, 160

Systema naturae (Linnaeus), 23, 207, 400

Tartaric-Caucasians, 362

Taxonomy, 11–12
- disease and, 134–135
- four human races and, 283–286

knowledge regimes and, 13–16
skin color and, 283–286
Teichmeyer, Hermann Friedrich (1685–1746), 74
Temperament, 18, 160
disease and, 68
lymphatic, 160
resemblance and, 63, 67–68
transmission of, 108
Temple of Nature; or, The Origin of Society (Erasmus Darwin), 133, 144
Terman, Lewis M. (1877–1956), 420
Terrall, Mary, 253–275
Theophrastus (c.370 BC–285 BC), 375
Theoria generationis (Wolff), 422
Theory of Heredity, A (Galton), 6, 16, 24
Theory of sensory education, 405–406, 408
Thespis (6th century BC), 447
Thierry, Amédée (1797–1873), 363
Tieck, Ludwig (1773–1853), 428
Tiers-Etat, 157
Tours, Jacques Moreau de (1804–1884), 165
Traité de goître et du crétinisme (Fodéré), 158
Traité de législation (Comte), 364
Traité de médecine légale et d'hygiène publique ou de police de santé (Fodéré), 160–161
Traité des dégénérescences (Morel), 156
Traité des tulipes (d'Ardène), 209
Traité philosophique et physiologique de l'hérédité naturelle dans les états de santé et de maladie du système nerveux (Lucas), 155, 189, 307
Trélat, Ulysse (1828–1890), 165
Trembley, Abraham (1710–1784), 257
Trollope, Anthony (1815–1882), 385–386
Trotter, Thomas (1760–1832), 138
Truncus, 11
Tuberculosis, 19, 122, 133
Tyndall, John (1820–1893), 383
Tyrtaios (7th century BC), 456

Über den Bildungstrieb und das Zeugungsgeschäfte (Blumenbach), 444–445
Ulloa, Antonio de (1716–1795), 351
Unger, Franz (1800–1870), 321
Uniqueness, 3
United States, 141
Untersuchungen über die Entwickelung der Wirbelthiere (Müller), 304
Urbanization, 17

Vairo, Leonardo, 71
Valentini, Michael Bernard (1657–1729), 74
Valerius Maximus (1st century BC/1st century AC), 65–66
van der Lugt, Maaike, 106
Vandermonde, Charles Augustin (1727–1762), 268–269
Variation of Animals and Plants under Domestication (Darwin), 24, 381
Vedder, Ulrike, 39, 45
Vererbung concept, 17, 443
Versuch einer Kritischen Dichtkunst (Gottsched), 445, 449–452
Versuche über Pflanzenhybriden (Mendel), 335
Verzichtserklärung (bride portion), 40
Vicq d'Azyr, Felix (1748–1794), 111–112, 118
Victor, ou l'enfant de la forêt (Ducray-Duminil), 403
Victorian ideology, 23
Victor (Savage of Aveyron), 400
Bonnaterre and, 403–404
classification of, 403–404
Itard and, 403–410
language training and, 408–409
Virchow, Rudolf (1821–1902), 293, 304–308, 311
Virey, Julien-Joseph (1775–1846), 157, 364–365
Vittorio, Marco Antonio, 61, 68–69, 72–73
Vogel, Rudolph August (1724–1774), 134
Voisin, Félix (1794–1872), 165, 167

Wackenroder, Wilhelm Heinrich (1773–1798), 428–431
Waller, John C., 147

Watt, James (1736–1819), 146
Wechsel der Generationen (generational alternation), 319–323
Weckherlin, August von (1794–1868), 229
Wedgwood, Josiah (1730–1795), 11, 146
Weismann, August (1834–1914), 308, 387–388
Weisz, George, 159
Wesen (essence), 322
Wesley, John (1703–1791), 141
What Genes Can't Do (Moss), 8
Wichura, Max (1817–1866), 331, 334
Wild Peter of Hanover, 399–400
Wilhelm Meister (Goethe), 428
Wilson, Edmund B. (1856–1939), 307
Wilson, Philip K., 53, 124, 160
Wiseman, Nicholas (1802–1865), 363
Witchcraft, 69, 71
Wolf children, 399–401
Wolff, Caspar Friedrich (1733–1794), 183, 422, 458
Women
  character and, 388–389
  dower rights and, 39–40, 61
  education and, 144
  Erasmus Darwin and, 144
  fetal disease transmission and, 114–115
  genius and, 425–428
  Harvey and, 181
  illegitimate children and, 63–66
  jointure and, 39
  maternal imagination and, 62–63, 65, 69–74
  pedigrees and, 389
  temperament and, 63, 67–68
Wood, Roger J., 376

Youatt, William (1776–1847), 376
Young, Arthur (1741–1820), 235
Young, Edward (1683–1765), 423–424, 426, 431–432

Zacchia, Paolo (1584–1659), 61
  innate heat and, 67–68
  medical evidence and, 61–63, 67–75
  temperament and, 63, 67–68
Zambos, 353–354
Zedler, Johann Heinrich (1706–1763), 281
Zeugung (procreation), 316
Zonabend, Françoise, 47
Zooids, 319
Zoological Society of London, 376
Zoology
  alternation of generations and, 319–323
  character and, 376–381
Zoonomia: or, The Laws of Organic Life (Darwin), 133–134, 141, 145
Zürchersee, Der (Klopstock), 452